# Nitrogen Cycle

Ecology, Biotechnological Applications and Environmental Impacts

T0203799

*Editors*

Jesus Gonzalez-Lopez and Alejandro Gonzalez-Martinez
Institute of Water Research
Department of Microbiology
University of Granada
Spain

CRC Press
Taylor & Francis Group
Boca Raton London New York

CRC Press is an imprint of the
Taylor & Francis Group, an **informa** business

A SCIENCE PUBLISHERS BOOK

Cover credit: The figure has been reproduced with permission of the authors of Chapter 12 (Castellano-Hinojosa et al.).

First edition published 2021
by CRC Press
6000 Broken Sound Parkway NW, Suite 300, Boca Raton, FL 33487-2742

and by CRC Press
2 Park Square, Milton Park, Abingdon, Oxon, OX14 4RN

*Library of Congress Cataloging-in-Publication Data*

Names: Gonzalez-López, Jesús, editor. | Gonzalez-Martinez, Alejandro, 1986- editor.
Title: Nitrogen cycle : ecology, biotechnological applications and environmental impacts / editors, Jesus Gonzalez-Lopez and Alejandro Gonzalez-Martinez, Institute of Water Research, Department of Microbiology, University of Granada, Spain.
Description: First edition. | Boca Raton : CRC PRESS, 2021. | Includes bibliographical references and index.
Identifiers: LCCN 2020053245 | ISBN 9780367260378 (hardcover)
Subjects: LCSH: Nitrogen cycle.
Classification: LCC QH344 .N526 2021 | DDC 577/.145--dc23
LC record available at https://lccn.loc.gov/2020053245

ISBN: 978-0-367-26037-8 (hbk)
ISBN: 978-0-367-71436-9 (pbk)
ISBN: 978-0-429-29118-0 (ebk)

Typeset in Times
by TVH Scan

# Preface

There is no doubt that there are numerous books that describe in a very different way the Nitrogen Cycle, its ecological importance and its biological aspects. Even some of them are very accurate in terms of the information they contain. The question is then why is a new book necessary right now?

In recent years, the importance of the anthropogenic processes that have clearly altered the N-cycle has been the subject of a huge scientific study and social concern. This has determined a huge amount of information that sometimes makes it difficult to acquire information for PhD students, scientists and professionals from many different areas. That is why when we write this book we try to discuss and solve a series of current problems that may be useful to all those professionals interested in this area of research and study. We also try to emphasize the importance that biological knowledge of the Nitrogen Cycle has a role in solving problems in the field of climate change.

It is in this context that the present book aims to provide the importance of new biotechnological processes applied to engineer technologies, which can represent new strategies for the resolution of specific problems related to the imbalance originated in the N-cycle. Aspects such as new conceptions in agriculture, wastewater treatment and greenhouse gas emissions are raised in this book under a multidisciplinary vision. A team of authors from different countries, such as the United States, the United Kingdom, Finland, Italy, Mexico, Ireland and Spain, with wide experience in the area has contributed up-to-date review in 12 chapters devoted to different biological and applied aspects of the N-cycle.

It has therefore been our intention to write this book that highlights those important scientific principles, discusses their environmental importance and integrates different biotechnological processes that can be used to solve environmental problems generated by the anthropogenic alteration of the N-cycle. If we have succeeded in these objectives, this book can be useful not only for academic lectures or for researchers specializing in this area but also for graduate and post-graduate students of different fields, such as biology, agriculture, biotechnology and engineering, who need this information for the study of the subsidiary and multidisciplinary subjects.

Finally, we would like to sincerely thank all the authors, who contributed the chapters compiled in this book, for contributing their outstanding expertise in the field as well as for the many comments and suggestions they provided in the preparation of this volume. The enormous effort and devotion of all the authors have made the final version of the book possible.

**Alejandro Gonzalez-Martinez**
**and Jesus Gonzalez-Lopez**

# Contents

# 1

# The Nitrogen Cycle: An Overview

## Massimiliano Fenice

## 1.1 INTRODUCTION

Earth represents an extremely complex physicochemical system that, under the thermodynamic point of view, is 'open' concerning energy and 'closed' while considering the global balance of matter. Although the total quantities of chemical elements of the planet are practically constant (negligible variations could be due to nuclear reactions and/or by inputs from meteorites), their location, combination, and physical/oxidation state change continuously due to a series of cyclic reactions. Hence, all chemical elements are removed from natural reservoirs (generally located in the lithosphere) and transferred to other compartments (hydrosphere and atmosphere) where they are available for all living organisms (biosphere), which actively participate to most of these reactions and some of them also contribute to regenerate the reservoirs.

Thus, the equilibrium state of life on Earth is strongly dependent on the recycling of essential chemical elements (principally carbon, nitrogen, oxygen, sulphur, iron, and manganese) occurring through a series of physical transformations, chemical and biochemical reactions and combinations of these processes. Since all living organisms influence the flow of chemical elements on Earth, the global turnover reactions of the elements on Earth are linked to the so-called 'bio-geochemical cycles'.

The various human activities (i.e., deforestation and wood/coil combustion) always affected the cycles (those of carbon and nitrogen in particular). However, the most important unbalance of the cycles (mainly carbon's, due to the $CO_2$ release) caused by anthropic activities started during the Industrial Revolution. Particularly relevant was the massive use of electricity, chemistry, and fossil fuels, which started from the second half of the nineteenth century, accompanied by subsequent further extensive deforestation due to intensive farming and livestock production. For these highly negative footprints on the environment, caused by humans in the last two hundred years and producing environmental degradation at a very high rate, a new 'geological era' has been defined by some scientists: 'The Anthropocene' (Crutzen 2006, Bellarby et al. 2008, Mondelaers et al. 2009, de Vries and de Boer 2010, Fowler et al. 2013, Lewis and Maslin 2015, Crutzen 2016).

In any case, all nutrient cycles are somehow linked and the various transformations of the elements correlated to the metabolism of living organisms, make life possible. However, due to the high amount of carbon and nitrogen present in all organisms, the cycles of these elements are more strictly associated and this association has a great environmental significance. For example, the fixed nitrogen is the principal limiting factor to primary production (carbon fixation) and it has been demonstrated that small variations in the ratio of nitrogen fixation to denitrification can significantly change atmospheric carbon dioxide concentrations (Falkowski 1997). Nitrogen is a key element that controls species composition, diversity, and dynamics; it regulates the functioning of most terrestrial, freshwater, and marine ecosystems (Vitousek et al. 1997).

Department of Ecology and Biology (DEB), University of Tuscia, Viterbo Italy.
E-mail: fenice@unitus.it

However, due to their metabolic versatility, microorganisms (prokaryotes in particular) undertake the most important and active role in these transformations which move the elements through the biosphere between the various compartments. The role of microorganisms within the bio-geochemical cycles is even more important since a number of metabolic processes, very important at a global level, can be almost exclusively carried out by Bacteria and/or Archaea (i.e., nitrogen fixation, methanogenesis, methanotrophy, and sulphate reduction).

## 1.2   THE NITROGEN AND ITS CYCLE

Nitrogen is the fifth most common element in the solar system and our planet; it exists both in organic and inorganic forms and also in a number of oxidation states (from +5 to –3) (Table 1.1).

It is the most abundant element in the atmosphere where it is present principally in its diatomic form ($N_2$, ca. 78% of air composition, pairs to approximately $4 \times 10^{15}$ tons) with traces of $NH_3$ and various oxides, such as $N_2O$, NO, and $NO_2$, resulting from the combustion and some biological processes (i.e. denitrification). The atmosphere plays a central role in the redistribution of nitrogen compounds (N-compounds) in terrestrial and marine ecosystems. In some cases, it passively behaves as a carrier; while in other circumstances, it participates in a pattern of complex chemical reactions, which transform the various N-compounds before they return to Earth (soil and waters) with different deposition velocities and various ecological and environmental effects (Galbally and Roy 1983, Nielsen et al. 1996, Hertel et al. 2011, Fowler et al. 2013).

**TABLE 1.1**  Different oxidation states (N° of oxidation) of nitrogen and corresponding compounds

| Oxidation State | Formula | Common Name |
|---|---|---|
| –3 reduced | $NH_3$, $NH_4^+$, $N^{3-}$ | Ammonia, ammonium, nitride |
| –2 | $NH_2$-$NH_2$ | Hydrazine |
| –1 | $NH_2$-OH | Hydroxylamine |
| 0 | $N_2$ | Dinitrogen |
| +1 | $N_2O$ | Nitrous oxide |
| +2 | NO | Nitric oxide |
| +3 | $N_2O_3$, $HNO_2$, $NO_2^-$ | Nitrous anhydride, nitrous acid, nitrite |
| +4 | $NO_2$, $N_2O_4$ | Nitrogen dioxide, dinitrogen tetroxide |
| +5 oxidized | $N_2O_5$, $HNO_3$, $NO_3^-$ | Nitric anhydride, nitric acid, nitrate |

It has been also estimated that a lot of nitrogen ($> 2 \times 10^{17}$ tons) is present in the Earth most inner layers, but only a limited amount is transferred to the surface or to the atmosphere. This transfer is mainly due to volcanic activities that contribute to the global cycle. By contrast, in the Earth's crust, nitrogen concentration is rather low and this element contributes scarcely to the compositions of minerals. The evaporitic nitrate deposits represent not only the most important nitrogen-containing minerals but also various minor ammonium minerals. In addition to this, a rather limited amount of nitrogen (<0.1-2%) is contained in solid and/or liquid fossil fuels, while natural gas could contain even higher amounts of this element (in general <5%). Soils represent the most active reservoirs of nitrogen in the lithosphere, where its concentration is extremely variable in function of various pedological, geographical, and physicochemical parameters (Johnson and Goldblatt 2015, Zedgenizov and Litasov 2017).

The nitrogen cycle has important gaseous components and it is essentially driven by the biosphere. Nitrogen is one of the six essential elements known as 'CHNOPS' and the fourth most abundant element in cell biomass and, after carbon, is the most important nutrient for life. All cells require this element in rather large quantities for the synthesis of numerous bio-molecules used for structural, metabolic, hereditary, and other functions. Nitrogen is principally found in proteins and nucleic acids, and also in a wide array of other organic molecules, including vitamins, chlorophylls (the key pigments needed by photosynthetic organisms) and some polysaccharides, such as chitin and chitosan, which are among the most abundant renewable resources on Earth and can be used as a non-conventional nitrogen source by microorganisms (Blank and Hinman 2016, Fenice 2016, Stein and Klotz 2016, Takai 2019).

Most of the reactions within the cycle largely depend upon the much diversified microbial metabolism, and some of them are of exclusive competence of peculiar microbial groups. In this context, however, the knowledge of the microorganisms and/or the enzymatic processes, transforming nitrogen into compounds having different oxidation states within the cycle is still incomplete (Stein and Klotz 2016).

On the other hand, on a global scale, the nitrogen cycle is heavily modified by a number of human activities (i.e., increased use of fossil fuels, increased demand for fertilisers in agriculture and nitrogen compounds in the industry). There has been increasing anthropogenic production of reactive nitrogen compounds; a lot of this nitrogen is released to the atmosphere, hydrosphere, and soil affecting directly and/or indirectly air and water quality and generating a series of environmental and health problems. For example, the anthropic perturbation of the cycle has caused extensive eutrophication of freshwaters and coastal areas and increased the release of nitrous oxide, a potent greenhouse gas (Vitousek et al. 1997, Galloway et al. 2008, Singh and Singh 2008, Aneja et al. 2009, Schlesinger 2009, Canfield et al. 2010, Fowler et al. 2013).

Over time, the metabolic competences of various microorganisms could restore the nitrogen cycle balance, but the heavy damages caused by humans will persist for decades or even centuries if active actions and sustainable environmental strategies are not planned in time. It has been broadly recognised that the doom of humanity would be sturdily related to its ability to control the nitrogen cycle (Canfield et al. 2010, Stein and Klotz 2016).

The nitrogen cycle was traditionally subdivided into three main processes: nitrogen fixation, nitrification, and denitrification. As a consequence, the microorganisms involved in nitrogen transformation were defined as nitrogen fixers, nitrifiers, and denitrifiers assigning them to specific microbial guilds performing a single process. This simplistic vision of the cycle was called into question by several microbial-ecologists according to strong evidence of various additional and more differentiated reactions brought about by other specialised microbial groups. Hence, the current understanding of the cycle consists of a much articulate flow of nitrogen transformations mediated by various microbial groups (Stein and Klotz 2016).

The key transformations of nitrogen and the involvement of microorganisms in these reactions (the majority of them within redox systems), occurring both in oxic and anoxic environments are summarised in Figure 1.1. The various steps of the cycle (nitrogen fixation, ammonification, nitrification, denitrification, and nitrate assimilative reduction) will be addressed in detail in the following paragraphs.

Briefly, gaseous $N_2$ is reduced to ammonia by free or symbiotic N-fixing microorganisms (belonging only to the domains of Archaea and Bacteria); ammonia is assimilated in the $NH_2$ groups of proteins, both in oxic and anoxic environments, or it can be transformed into nitrate by the nitrification process. This could require different groups of microorganisms performing separately the oxidation from ammonia to nitrite and from nitrite to nitrate or by microorganisms that could perform both reactions (COMAMMOX or Complete Ammonia Oxidiser). Moreover, nitrate can be assimilated, by microorganisms and plants, in the proteins ($NH_2$ groups) or subject to various reduction processes. In addition to this, organic $NH_2$ groups can be converted into ammonia, in oxic or anoxic environments, by the ammonification process. Nitrate can be reduced to ammonia by the process of 'dissimilative reduction of nitrate to ammonia' (DNRA) or reduced to nitrite. Nitrite is subject to further reduction leading to various gaseous nitrogen compounds ($NO_2$, $N_2O$, and $NH_3$), including its molecular diatomic $N_2$ form, in the so-called denitrification process. In the ANAMMOX reaction (Anaerobic Ammonium Oxidation), nitrite and ammonium are converted directly into $N_2$ and water.

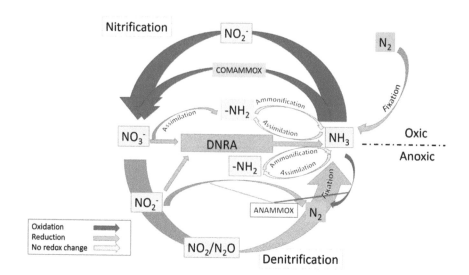

**FIGURE 1.1** Overview of the global reactions within the nitrogen cycle in oxic and anoxic conditions.

## 1.3 NITROGEN FIXATION

As said, the main nitrogen reservoir of the planet is represented by its gaseous diatomic form $N_2$ (ca. 78% V/V in the atmosphere; ca. 1.8% V/V in the water at 1 bar and 15°C), which, due to its triple covalent bond (945,3 kJ/mol), is the most stable and less reactive form of this element. For example, its chemical reduction to ammonia ($N_2 + 3H_2 \rightleftarrows 2NH_3$) could be obtained in the laboratory only at high temperatures (>400°C) and pressure (>300 bars) using a catalyser (Fe activated by alkaline-metals oxides). The industrial process of this reaction, having a rather low yield (< 20%) was developed in the early twentieth century (Haber-Bosch process 1909).

The interchange between atmospheric $N_2$ and 'reactive nitrogen' (those nitrogen compounds that support or are products of cellular metabolism and growth) is almost entirely controlled by a wide array of microbial activities. A certain amount of nitrogen is removed from the atmosphere by both chemical (natural and not natural processes) and biological fixation (fixation = combination with other elements). Biological nitrogen fixation (ca. $150 \times 10^6$ t/year) represents ca. 80% of the global process, while the chemical fixation is definitely of a much lower entity. The high energy required for the chemical nitrogen fixation in the environment can be supplied by lightning (electric energy) or cosmic radiation when nitrogen oxides or ammonia are produced (Stein and Klotz 2016). The abiotic nitrogen fixation has certain environmental importance; for example, the relationship between some crop yield and the annual abundance of lightning has been noted, however these phenomena account (theoretically) for less than 10% of the entire process (Takai 2019).

In spite of its great availability on Earth, only a limited number of specialised prokaryotes (various genera of Bacteria and Archaea), called diazotrophs, can use $N_2$ as a nitrogen source for their nutrition and are key players both in the global cycle of the element and on various scales in biological communities since life formation started about four billion years ago. All these microorganisms can assimilate nitrogen by a reductive process: gaseous nitrogen is first reduced into $NH_3$ and subsequently assimilated in proteins ($-NH_2$ groups) and other organic molecules. In any case, genera of prokaryotes presenting various species of nitrogen-fixing organisms are spread in all environments from Antarctica to Arctic regions, including many different temperate and tropical areas like in aquatic and rock and soil ecosystems (Pajares and Bohannan 2016, Park et al. 2016, Ampomah et al. 2017, Baker et al. 2017, Yousuf et al. 2017, Thajudeen et al. 2017, Timperio et al. 2017, Barghini et al. 2018, Amarelle et al. 2019, Gorrasi et al. 2019, Takai 2019).

The biological nitrogen reduction to ammonia (see the global reaction below) requires a lot of metabolic energy and progressively proceeds through different steps: nitrogen > diimide > hydrazine > ammonia.

$$(N_2 + 8H^+ + 8e^- + 16/24 \text{ ATP} \rightarrow 2NH_3 + H_2 + 16/24 \text{ ADP} + 16/24 \text{ P}_i)$$

Most organisms involved in the process obtain this energy by oxidising organic molecules. Photosynthetic microorganisms can use the sugars produced by photosynthesis, while non-photosynthetic free-living microorganisms must obtain these molecules from other organisms. Associative and symbiotic microorganisms obtain these nutrients from the rhizospheres of the host plant.

The key enzymes catalysing the reaction is the nitrogenase complex consisting of two proteins (nitrogenase and nitrogenase-reductase) that are present with rather minor differences in all nitrogen-fixing organisms (see below).

## 1.3.1  Nitrogen-Fixing Microorganisms

Nitrogen-fixing microorganism has certain ecological advantages somehow allowing them to break free from the dependence of fixed nitrogen forms. This is particularly true when there is a short supply of nitrogen sources and growth of the non-fixing organism is strongly limited by nitrogen availability. Thus, in many environments, the nitrogen-fixing organisms can flourish and predominate just for the scarce availability of fixed nitrogen.

As mentioned above, nitrogen-fixing microorganisms could be subdivided into different groups in the function of their ecological role as free or symbiotic organisms (Table 1.2):

### • *Free-Living Fixators (Aerobes and Anaerobes Archaea and Bacteria)*

They do not associate with other organisms and can fix a rather low amount of nitrogen (up to ca. 50 kg/ha per year*). A number of heterotrophic soil bacteria can fix nitrogen with no direct interactions with plants or other organisms. They can obtain energy by the oxidation of organic compounds (Wagner 2012) produced by other organisms or by decomposition of organic matter. Among free-living nitrogen-fixing microorganisms, we have photosynthetic bacteria (i.e., cyanobacteria, purple and green sulphur bacteria). Moreover, some N-fixing chemiolithotrophic organisms obtain energy from the oxidation of inorganic compounds. Since the nitrogenase can be inhibited by oxygen, the behaviour of free-living organisms could be microaerophilic or anaerobic during the nitrogen fixation process. In any case, the free-living nitrogen-fixing bacteria contribute to the nitrogen balance on soils providing in certain cropping systems up to 50% of the total nitrogen needs (Vadakattu and Paterson 2006, Herridge et al. 2008, De Bruijn 2015).

### • *Free-Living Associative (Or Semi-Symbiotic) Fixators*

They had been found both in temperate and tropical areas and, are associated with low specificity to the roots of various plants (including some cereal crops which are important at the agronomical level) but do not produce nodules. Their capacity of fixation is higher than that of the free-living nitrogen-fixing bacteria (up to ca. 200 kg/ha per year*) and is determined by various factors including soil temperature, oxygen, and nitrogen concentration and specific efficiency of the nitrogenase system. The most limiting factor is the availability (quality and quantity) of organic nutrients, which are in general supplied by the associated plant; in general, they prefer carboxylic acids, malate in particular (Vlassak and Reynders 1979).

### • *Symbiotic Fixators*

They are strictly associated with another organism (i.e., plants, lichenized fungi). Those associated with plants produce root nodules. The host provides nutrients (i.e., sugars from the photosynthesis), which are used by the bacterium as an energy source for the fixation and receive fixed nitrogen for its growth. Their symbiosis is, in general, highly species-specific and the fixation capacity is quite high (up to ca. 400 Kg/ha per year).

**TABLE 1.2**  Different groups of nitrogen-fixing organisms and some prototypical genera, containing species active in nitrogen fixation, belonging to each group

| Free-Living and Free-Living Associative* (1 = Anaerobe, 2 = Archaea) | | |
|---|---|---|
| **Phototrophs** | **Chemoorganotrophs** | **Chemoauthotrophs** |
| *Anabaena, Gleocapsa, Nostoc, Trichodesmium, Cylindrospermum, Rhodobacter[1], Rhodospirillum[1], Rhodomicrobium[1], Rhodospesudomonas[1], Chlorobium[1], Chromatium[1]* | *Azotobacter*, Azospirillium*, Bacillus, Beijerinckia, Clostridium[1], Enterobacter*, Klebsiella*, Pseudomonas, Nocardia, Desulfovibrio[1]* | *Alcaligenes, Methanosarcina[1,2], Methanococcus[1,2], Methylococcus[1], Methylosinus[1], Thiobacillus* |
| **Symbiotic** | | |
| **With Leguminous Plants** | | **With other Plants/Organisms** |
| *Allorhizobium, Azorhizobium, Rhizobium, Bradyrhizobium, Sinorhizobium, Burkholderia, Mesorhizobium* | | *Anabaena, Frankia, Richelia* |

The most important and studied (i.e., for its implication for humans) symbiotic associations between plants and nitrogen-fixing bacteria are those regarding species belonging to the genus *Rhizobium* (and similar genera such as *Azorhizobium, Bradyrhizobium*) with different legumes representing very important crops worldwide (i.e., alfalfa, beans, chickpeas, clover, peanuts, peas, and soybeans). These symbiotic bacteria are also used as bio-fertilisers in order to increase the efficiency of the naturally-occurring nitrogen fixation and improve the yield of various crops (Van Rhyn and Vanderleyden 1995, Long 1996, Vance 2001, Somasegaran and Hoben 2012, Matkarimov et al. 2019, Sulewska et al. 2019).

Rhizobia are rod-shaped, mobile, and Gram-negative bacteria (dimensions ca. 0.5-09 × 1.2-3.0 µm). They produce the development of nodules on plant roots or sometimes on stems. The bacterial endosymbionts populate nodules and acts as nitrogen-fixers. The early stages of this process include gene expression in the bacterium and cell growth, division and differentiation in the host plant. Nodulation has been characterised as occurring in a series of stages, each of which may be influenced by one or more genes in each symbiotic partner. These events are mediated by an exchange of signals between the plant and the bacteria and many changes in the biochemistry and structure of both organisms occur during the nodule development. *Rhizobia* actively suppress the immune response of the host plant, permitting the establishment of the infection and symbiosis; the management of plant defences by the bacterium is important for the maintenance of the symbiotic relationships. The plant produces a signal molecule (i.e., a flavonoid) that induces gene expression in the bacterium, which subsequently produces a signal triggering early nodule development on the plant. The nodule morphology is determined by the plant and not by the bacterium. The plant maintains various mechanisms to control the nutrient supplied to the symbionts and the number of nodules to prevent the bacteria to become too oppressive. One of the main nodule forms is the indeterminate type (sometimes called cylindrical or meristematic). This nodule typology is formed for example on the roots of alfalfa, clover, and pea. A second main type is a determinate or a spherical nodule, which is developed in *Phaseolus and Glycine* (soya) (Long 1996, Somasegaran and Hoben 2012, Gourion et al. 2015, Cao et al. 2017, Appelbaum 1990).

Nodulation occurs through different steps needing a time course of about one month (Figure 1.2), and the mechanism described below regards the symbiosis of Rhizobia with terrestrial leguminous while it is different in those bacteria infecting aquatic leguminous and the stem-nodulating *rhizobia* that infect various leguminous in tropical areas.

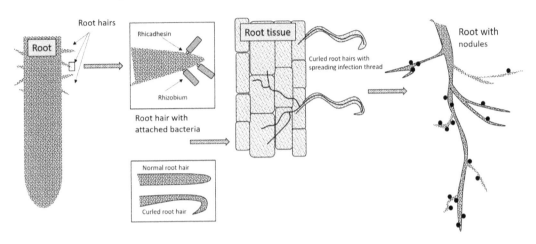

**FIGURE 1.2** Nodulation of leguminous roots by the infection with Rhizobia. See text for details (pictures not in scale).

The different steps of the *rhizobial* infection leading to nodulation are resumed below:

(i) Recognition (mutual) and attachment of the bacterial cell to the host root hairs mediated by the bacterial adhesion protein 'rhicadhesin' and other factors (i.e., lectines), including specific plant receptors. Recognition is somehow species-specific and various compounds secreted by the plant roots affect the composition of the rhizosphere microbial composition.

(ii) Secretion of Nod factors (i.e., oligosaccharides) by the bacterium causing curling of root hairs.

(iii) Root invasion in which bacteria penetrate root hairs within an 'infection thread' produced by the root cell under the induction of the rhizobium.

(iv) Growth of the infection threat inside the root; the bacterium invaded also nearby cells, which are induced to divide.

(v) Bacterial cells modify into a *bacteroid* state (cells with a swollen, misshapen, and branched aspect) within the root cells (Figure 1.2). Their division induces the formation of nodules, which are tumor-like tissues where modified plant cells are filled with these modified bacterial cells. Within the nodules bacteroids are enclosed by portions of the plant membrane producing a peculiar structure called symbiosome (Figure 1.3); at this stage, nitrogen fixation can begin. In order to remove oxygen, which inhibits N-fixation enzymes (nitrogenases) legumes produce a specific oxygen-binding protein (leghaemoglobin or legoglobin) in response to root colonisation by *Rhizobium* in the nodulation process. Leghaemoglobin has the same high affinity for oxygen and function as another haemoprotein, the 'cyanoglobin', found in cyanobacteria (Appleby 1984, Thorsteinsson et al. 1999).

The genes involved in nodulation are called *nod* genes; they encode for proteins producing polysaccharides (Nod factors), which bind to plant cell surface receptors (NFR) eliciting nodule organogenesis; for example, the mentioned induction of root hair curling and plant cell division. These genes (i.e., *nodABC*) have been identified in various Rhizobia.

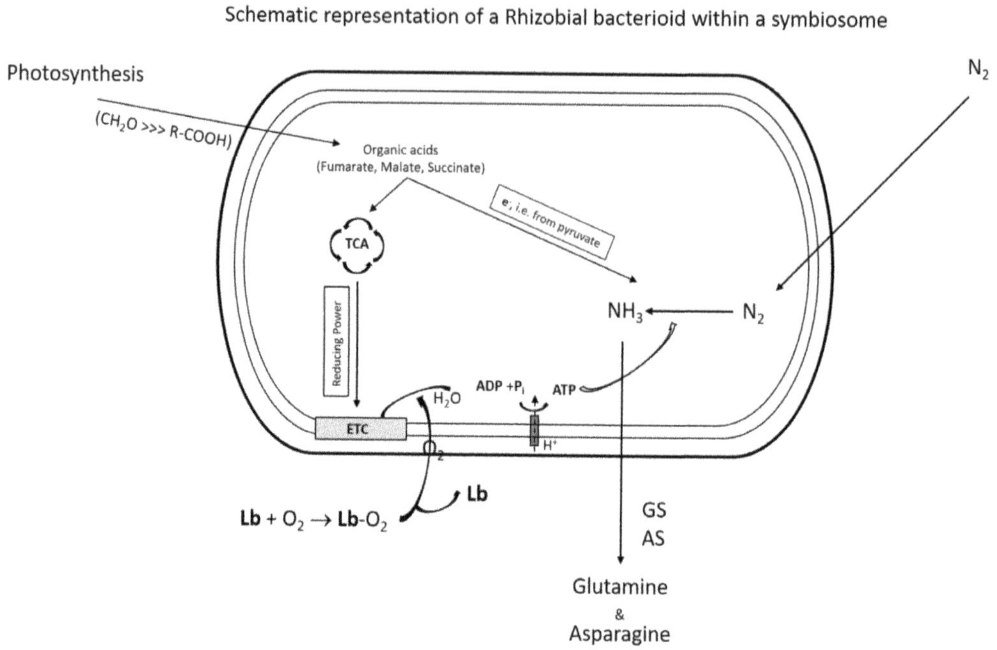

Schematic representation of a Rhizobial bacterioid within a symbiosome

**FIGURE 1.3** Representation of bacteroid metabolism within the nodule (symbiosome). AS = asparagine synthetase, GS = glutamine synthetase, LB = leghaemoglobin, ETC = electron transport chain. For details of the nitrogenase complex see Figure 1.4.

Other important symbiotic relationships occur between nitrogen-fixing bacteria and non-leguminous plants. For example, the cyanobacterium *Anabaena azollae* establishes a symbiosis with species of the aquatic fern *Azolla*. The bacterium colonises cavities at the base of the fern leaves and fixes significant amounts of nitrogen in specialised cells (heterocysts). *Azolla-Anabaena* symbiosis has been used for many centuries as a natural fertiliser in Southeast Asia rice paddies, which are typically covered by the fern; fixation yield could reach 600 kg/ha* (*Very different data from different authors) per years during the growing period (Postgate 1982, Fattah 2005).

Another symbiotic relationship deserving to be mentioned is that of soil actinomycetes species of the genus *Frankia* and various actinorhizal plants. These belong to various classes of angiosperms (i.e., Betulaceae, Casuarinaceae, Coriariaceae, Elaeagnaceae, Myricaceae, Rhamnaceae, and Rosaceae) that are commonly found in various environments including alpine and glacial areas, arctic tundra, coastal dunes, forests, riparian, and xeric areas (Benson and Silvester 1993, Pawlowski and Sirrenberg 2003).

## 1.3.2 The Role of Nitrogenases in Nitrogen Fixation

The nitrogenase enzymes are required for all nitrogen-fixing microorganisms being present in both Archaea and Bacteria. This class of enzymes, which has been highly conserved through evolution, seems to be the only one able to catalyse the process (Zehr et al. 2003). The nitrogenase system is a complex constituted by two proteins:

- The major component is an iron/molybdenum-containing protein ($MoFe_{nitr}$, nitrogenase)-, a heterotetrameric nitrogenase that reduces nitrogen to ammonia (Figure 1.4).
- An iron-containing homodimeric protein ($Fe_{nitr}$, nitrogenase reductase, or Fe-only nitrogenase) with a high reducing power, which is responsible to catch the electrons necessary for the reduction from a reducing agent (i.e., ferredoxin or flavodoxin) for transferring them to the $MoFe_{nitr}$. Electron transfer requires energy that is supplied by ATP: hydrolysis of 2 or 3 ATP

molecules is necessary for each electron that is transferred. Alternative nitrogenases contain vanadium in place of molybdenum (V-nitrogenase) but also non-molybdenum, non-vanadium nitrogenases are known, with still scarce knowledge about their distribution and function.

The various structural components of these enzymes are expressed by various operons within the various '*nif*' regulons (i.e., *nif*, Mo-nitrogenases; *anf*, alternative $N_2$-fixation, or Fe-only nitrogenases and *vnf*, V-nitrogenase). Nitrogen fixation is firmly regulated; the process is controlled by the concentration of available nitrogen sources (i.e., ammonium) and oxygen. In diazotrophic bacteria, the metabolic switch from the $N_2$ fixation-repressed conditions to the $N_2$-fixing must be carefully controlled mainly because of the high energy requested by the process (Gussin et al. 1986, Leigh 2002, Rubio and Ludden 2002, Siemann et al. 2002, Zehr et al. 2003, Poza-Carrión et al. 2014, Zhang et al. 2016, McRose et al. 2017).

## The Nitrogenase complex

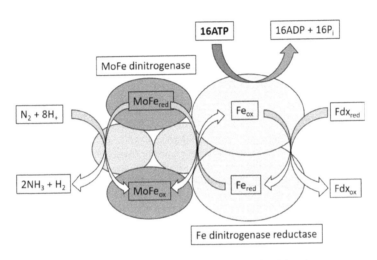

**FIGURE 1.4** Overview of the nitrogen fixation process and role of the nitrogenase complex.

The process of nitrogen fixation is repressed both by fixed forms of nitrogen ($NH_3$, $NO_3^-$) and oxygen. Since oxygen rapidly and irreversibly inactivate the majority of these microbial enzymes (in particular the nitrogenase reductases), the various aerobic nitrogen-fixing microorganisms need specific mechanisms to protect them from oxygen inactivation. Some bacteria (i.e., *Azobacter* spp. and *Azomonas* spp.) use both an enhanced respiration rate and/or a specific conformational protein that reversibly binds the nitrogenase causing a reversible inactivation.

Another strategy is the production of thick layers of protective slimy or gummy compounds (i.e., those found in *Azobacter* spp., *Azomonas* spp. *Beijerinkia* spp. and *Gluconacetobacter* spp.). Some microorganisms solve the problem of localising the reaction in specially modified cells, such as the apical vesicles of *Frankia* spp. or the heterocysts of cyanobacteria, where anoxic conditions are established and N-fixation is compartmentalised. Specific haemoproteins with a high affinity for oxygen has been discussed above.

Heterocysts, present only in filamentous cyanobacteria, deserve special attention for their high specialisation. They are derived from a vegetative cell modified by specific gene expression driven by limitation of fixed nitrogen and initiated by 2-oxoglutarate (2-Ogl) accumulation. In Fixed-N starving conditions, the increased concentration of 2-Ogl activates a global transcriptional regulator (NtcA), which activates the *hetR* gene expressing the principal transcriptional regulator of heterocyst formation (HetR). This protein activates a cascade of genes for the heterocyst differentiation and production of the N-fixation biochemical machinery.

Heterocysts have additional multilayer cell-walls to avoid oxygen permeation, to remove oxygen trace in the heterocyst and a cytochrome oxidase is expressed also. Photosynthesis within the heterocyst occurring only trough photosystem I, produce the energy necessary for the process (ATP). Energy is also obtained by respiration (Figure 1.5).

Connections (plasmodesms) between vegetative cells and heterocysts are established to permit the exchange of gases (i.e., $N_2$) and metabolites, including a specific protein (PatS) that migrates from the heterocyst to nearby vegetative cells in order to inhibit their transformation in heterocysts (Figure 1.5). A PatS gradient is established along with subsequent vegetative cells and a new heterocyst formation is only possible when its concentration becomes low enough (typically every eight vegetative cells) (Wolk et al. 1994, Herrero et al. 2016).

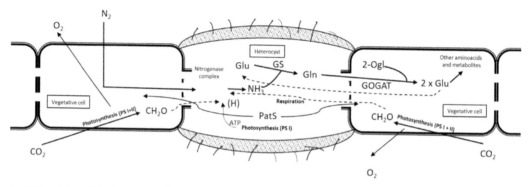

**FIGURE 1.5** Simplified reactions of nitrogen fixation and assimilation within the heterocyst and relations with adjacent vegetative cells. Vegetative photosynthetic cells transfer fixed carbon to the heterocyst and nitrogen diffuses from vegetative cells to the heterocyst where it is fixed by the nitrogenase complex, assimilated and subsequently shared with vegetative cells. The inhibition of heterocyst formation by adjacent vegetative cells is driven by the protein PatS (blue arrows) produced by the heterocyst. Glu = Glutamate, Gln = Glutamine, 2-Ogl = 2-Oxoglutarate, GS = Glutamine synthetase, GOGAT = glutamate synthetase.

## 1.4 ASSIMILATION OF AMMONIA/AMMONIUM AND AMMONIFICATION

As said, nitrogen fixed by nitrogen-fixing microorganisms is transformed in ammonia (or ammonium at neutral pH). This inorganic compound can be directly used by the microorganisms and/or by their symbiotic partners through a process called ammonia assimilation (incorporation into organic compounds). This is a very important step in the nitrogen cycle, and a large variety of living forms somehow depend upon nitrogen fixation and its subsequent assimilation.

During assimilation, ammonia is incorporated as $NH_2$- group in the glutamate that is transformed in glutamine by the enzyme glutamine synthetase; the reaction requires ATP hydrolysis. Glutamine can be subsequently transformed into other amino acids. Glutamine synthetase (GS) is a ubiquitous enzyme being present both in prokaryotes and eukaryotes. Cyclically glutamine is then re-converted in glutamate (Figure 1.6) by glutamate synthetase (GOGAT) and from glutamate in other amino acids and then to nucleic acids and other organic compounds containing nitrogen. The reaction catalysed by GS is considered the principal route for the assimilation of inorganic nitrogen, but alternative pathways that are known, such as the reductive amination of α-ketoglutarate, are catalysed by glutamate dehydrogenase, GDH, found in *Saccharomyces cerevisiae* (Hirel and Lea 2001, Magasanik 2003).

It is worth noting that, however, not all the ammonia available for assimilation is produced through $N_2$ fixation. Practically all living organisms can transform organic nitrogen into ammonia/ammonium (ammonification) through the breakdown (decomposition) of a wide array of nitrogenous compounds, such as proteins, amino acids, nucleic acids, and nucleotides, amino-sugars and their polysaccharides. Various enzymes are involved in the hydrolysis of these compounds (i.e., proteases, nucleases, decarboxylases, deaminases, and chitinolytic enzymes) and a large variety of nitrogen-containing products are released together with ammonia during the decomposition process.

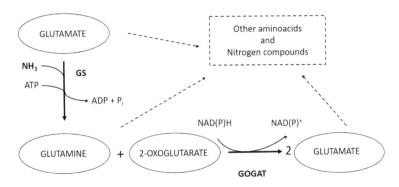

**FIGURE 1.6** Assimilation of ammonia and role of glutamine synthetase (GS) and glutamate synthetase (GOGAT).

In different animals, the catabolism of organic nitrogen compounds produces different nitrogen waste products such as ammonia (i.e., from aquatic invertebrates and teleosts), urea (i.e., from elasmobranches, terrestrial vertebrates and amphibian) or uric acid (i.e., from birds, reptiles and insects). The last two compounds can be used as nitrogen sources by various aerobic and anaerobic microorganisms (bacteria, fungi, and algae) having specific enzymes, such as ureases and/or uricases (urate oxidase) (Singer 2003).

$NH_3$ can also be generated by the respiratory reduction of nitrate (and nitrite); the process, called dissimilative nitrate reduction to ammonia (DNRA), plays an important energetic role in habitats showing enrichment in electron donors and/or limitation of electron acceptors. It dominates in some anoxic environments, such as some marine sediments and coastal environments, the human gut, and so on, but it is present also in soils and activated sludge wastewater plants. Nitrate-reducing bacteria use this pathway principally in nitrate limiting conditions; the process is also known to be active in some fungi such as *Fusarium oxisporum* and *Aspergillus nidulans* (Morozkina et al. 2007, Rütting et al. 2011, Giblin et al. 2013, Takai 2019).

In soil, the majority of the ammonium released by aerobic decomposition is rapidly assimilated (converted to amino acids and other N compounds) by plants and microorganisms. Nevertheless, some ammonia is lost in alkaline conditions due to its volatility.

In any case, ammonia represents only ca. 15-20% of the total nitrogen released to the atmosphere that is mainly constituted by $N_2$ or $N_2O$ produced during denitrification. However, these nitrogen volatile compounds are subject to various reactions, the products of which are brought back to the soil by rain.

## 1.5    AMMONIUM OXIDATION IN AEROBIC (NITRIFICATION) AND ANAEROBIC (ANAMMOX) CONDITIONS

Nitrification, the oxidation of reduced nitrogen, principally ammonia/ammonium, to nitrate (see reaction below), plays an important role in the global nitrogen cycle; important intermediates to nitrification include hydroxylamine, nitric oxide, and nitrite.

$$NH_4^+ + 2O_2 \rightarrow NO_3^- + 2H^+ + H_2O$$

This process occurs at neutral or slightly alkaline pH in oxic terrestrial and aquatic ecosystems including well-drained agricultural, grassland, forest and alpine soils, sediments, and other aquatic environments. Nitrification is carried out only by microorganisms (both Bacteria and Archaea), which can rapidly transform the ammonia obtained from the mineralisation of organic matter; they are ubiquitously distributed in nature and exist virtually everywhere in the presence of ammonia and oxygen (Stein and Nicol 2001). The increased content of ammonia in the environment, such as due to the addition of manure, can increase the rate of nitrification and consequent acidification of the environment since the oxidation produces the transformation of a cation into an anion ($NH_4^+$ >>> $NO_3^-$).

In soil, although acidification improves mineral solubility, nitrification is considered as a negative event for sol fertility since nitrates are highly soluble and can be easily leached if not promptly adsorbed by plant roots. By contrast, ammonium can be retained by clay particles and humic substances due to their negative charge.

The oxidation of ammonia to nitrate (via nitrite) has been considered for decades to be a two-step process carried out by different functional groups of chemolithoautotrophic microorganisms which can oxidase ammonia and nitrite separately, although complete nitrification is energetically feasible or even convenient. Conversely, some recently discovered bacteria (i.e., members of the genus *Nitrospira*), can perform the complete oxidation of ammonium to nitrate ('comammox' bacteria); these bacteria have a broad environmental distribution (Daims et al. 2015, Van Kessel et al. 2015, Hu and He 2017, Stein and Nicol 2001).

Nevertheless, very often the reaction is split into two semi-reactions carried out by different microbial groups. Various species of ammonium-oxidising Archaea (AOA; i.e., *Nitrosopumilus* spp.) and Bacteria (AOB; i.e., *Nitrosomonas* spp., *Nitrosospira* spp.) can oxidise ammonium only to nitrite; however, archaeal communities are always very abundant and sometimes predominate in soil and oceans; very likely they have a predominant role to control ammonium oxidation rates in the ecosystem (Leininger et al. 2006, Agogué et al. 2008, Schleper and Nicol 2010).

By contrast, to date, only Bacteria (i.e., *Nitrobacter* spp., *Nitrococcus* spp., *Nitrospina* spp., *Nitrospira* spp., *Nitrotoga* spp., *Thiocapsa* spp.) are known for their ability to obtain nitrate from nitrite (nitrite-oxidising bacteria, NOB). Yet, the two reactions are somehow strictly coupled and it is difficult to observe the accumulation of nitrite in the environment (Costa et al. 2006, Abdulrasheed et al. 2018).

The key enzymes in the nitrification process are the ammonium monoxygenase (AMO), the hydroxylamine oxidoreductase (HAO), and the nitrite oxidoreductase (NOR or NXR); a putative enzyme for nitric oxide oxidation to nitrite is still to be characterised (Stein and Nicol 2001).

In ammonia-oxidising bacteria, AMO, the first membrane-associated enzyme (integral), oxidises the cytoplasmic ammonia to hydroxylamine that is subsequently oxidised to nitrite in the periplasm by the HAO. Elements of the electron transport chain (ETC), such as some cytochromes and ubiquinone, are involved. The electrons donated by ammonia enter the ETC establishing a proton motive force (PMF), which could drive ATP production by ATPase. A reverse electron flow can be used to produce NADH.

Nitrite is oxidised to nitrate by nitrite-oxidising bacteria through a short ETC (cytochromes) by the nitrite oxidoreductase. Oxygen is the terminal electron acceptor and also this reaction produces proton motive force. A reverse electron flow can be used here as well.

The above-mentioned nitrifying microorganisms are generally aerobes and their metabolism is predominantly autotrophic: the pathways used by Bacteria to fix $CO_2$ are the Calvin-Benson and the reductive TCA cycles, while the Archaea use the 3-hydroxypropionate/4-hydroxybutyrate pathway. However, they can also have alternative mechanisms to conserve energy (chemoorganotrophy on glucose or other organic substrates). Some ammonia-oxidising bacteria could be either chemolithotrophs or mixotrophs, while some autotrophic ammonia-oxidising Archaea use a variation of the hydroxypropionate pathway. The specialisation of these prokaryotes permits also the colonisation of unique niches, but in general their growth is somewhat slow since the energy produced by the mentioned processes is rather low. By contrast, their ecological role is quite important in all aquatic and terrestrial environments. The conversion of ammonia to nitrate is important since this is a key nutrient for plants.

However, in some acidic forest soils, their efficiency is rather low and their function is thought to be taken by heterotrophic microorganisms, fungi in particular (heterotrophic nitrification). A wide range of fungi (i.e., species of the genera *Acremonium, Aspergillus, Verticillium, Penicillium, Phoma* and *Trichoderma*), bacteria (i.e., *Absidia, Arthrobacter, Bacillus, Pseudomonas*) and Archaea (*Nitrosopumilus, Nitrososphaera, Nitrosotalea*) are known as heterotrophic nitrifiers, but the physiological and ecological role of this process, the phylogenetic diversity of the involved microorganisms and their interactions are yet to be completely understood (Lang and Jagnow 1986, Hayatsu et al. 2008, Zhu et al. 2015, Isobe et al. 2018, Li et al. 2018, Stein and Nicol 2001).

Ammonia can be oxidised also in anoxic conditions within the 'ANAMMOX' (Anaerobic Ammonium Oxidation) process by different genera of Bacteria (*Anammoxoglobus, Brocadia, Jettenia, Kuenenia, Scalindua*), which were discovered in the early 1990s in wastewater plant sludges. Even with some metabolic diversities seen among the various bacterial groups, ammonia is oxidised anaerobically by nitrite (or less preferably nitrate) producing gaseous nitrogen as a final product (overall reaction: $NH_4^+ + NO_2^- \rightarrow N_2 + 2H_2O$). In this catabolic reaction, nitrite acts as an electron acceptor, but it could also be an electron donor in the coupled anabolic reaction with carbon dioxide, which autotrophically produces organic matter (biomass) ($2NO_2^- + CO_2 + 2H_2O \rightarrow NO_3^- + CH_2O$).

The anammox cell is quite peculiar (Figure 1.7); their coccoid cell wall has no peptidoglycan and a large intracytoplasmic compartment (anammoxosome) confined by a membrane, showing very uncommon lipid composition (ladderane lipids), is present. This pseudo-organelle is the site of anammox process, in which nitrite is used as the electron acceptor in the conversion of ammonium to $N_2$ gas and water. The compartmentation is necessary also to isolate the toxic intermediate hydrazine ($N_2H_4$) from the cell. In addition to this, the anammoxosome membrane contains an electron transfer chain producing PMF that can generate ATP trough an ATPase.

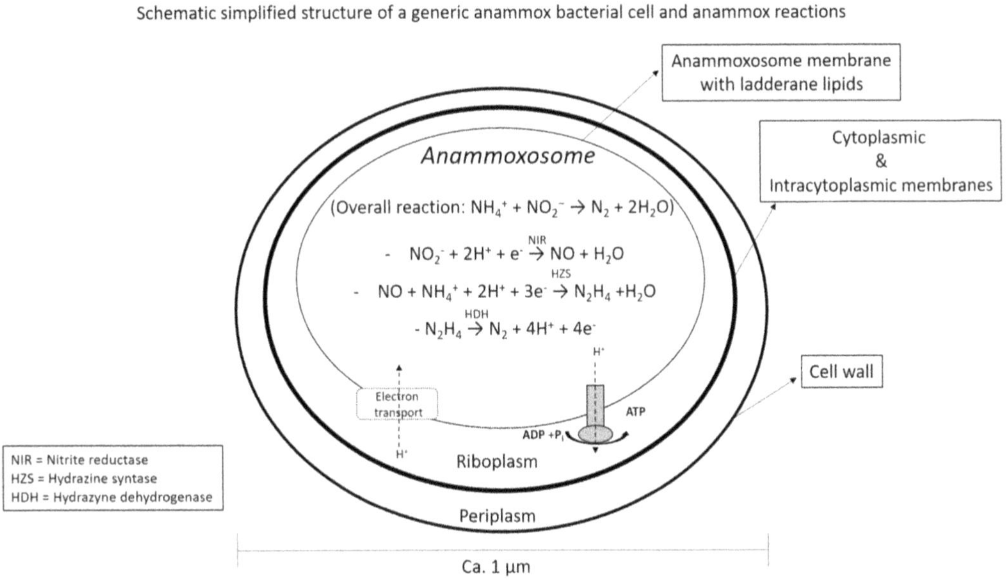

Schematic simplified structure of a generic anammox bacterial cell and anammox reactions

**FIGURE 1.7**  Simplified structure of an anammox bacteria.

The principal enzymes involved in the anammox catabolic reaction are the nitrite reductase (NIR), reducing nitrite to nitric oxide; the hydrazine synthase (HZS or hydrazine hydrolase [HH]), which combines ammonium with nitrogen oxide producing hydrazine; and the hydrazine dehydrogenase (HDH or hydrazine oxidoreductase HAO), which produce diatomic nitrogen by hydrazine oxidation (Figure 1.7). The hydrazine hydrolase (HH), can combine ammonium and hydroxylamine to produce hydrazine. The whole system is linked with an ETC generating a proton motive force.

Anammox, occurring in every anoxic situation where both ammonium and nitrite are present, is considered a principal process both in natural environments, such as marine anoxic sediments, and artificial environments, such as sewage plants (anammox bioreactors had been designed already years ago), where depletion of ammonia, nitrite, and amines is a very important goal for de-pollution purposes (van Niftrik et al. 2004, Dalsgaard et al. 2005, Kuenen 2008, Karlsson et al. 2009, de Almeida et al. 2015, Gonzalez Martinez et al. 2015).

## 1.6 DISSIMILATIVE (DENITRIFICATION) AND ASSIMILATIVE REDUCTION OF NITRATE

Denitrification somehow represents the opposite process of nitrification since it consumes nitrate. Excluding the limited amount of nitrate produced by physical phenomena in the atmosphere (oxidation of nitric oxide), which is transferred to the lithosphere and hydrosphere by atmospheric precipitations and is contained in ornithogenic soils due to bird guano deposition; the majority of nitrate in nature is produced by the above mentioned ammonium-oxidising microorganisms (Schreiber and Burger 2001).

Nitrate can be a microbial alternative electron acceptor in the anaerobic respiration. During denitrification, it can be transformed into different reduced nitrogen forms: nitrite, nitric and nitrous oxides, and diatomic nitrogen ($NO_3^- \rightarrow NO_2^- \rightarrow NO \rightarrow N_2O \rightarrow N_2$). The last three compounds are gaseous and can diffuse to the atmosphere from soils and waters, which are depleted of bio-available (fixed) nitrogen. Thus, denitrification is considered as a very negative process in agriculture since it removes the nitrates, which are often added to soils for crop fertilisation. For this reason, coupled with correct agricultural procedures (i.e., crop rotation), soils must be kept well aerated and drained to avoid the establishment of anoxic conditions leading to denitrification and consequent economic loss together with environmental concerns.

Nitrogen oxides are strong pollutants with marked negative environmental effects at the global level (climate change). Nitrous oxide exerts a strong greenhouse effect causing much higher environmental warming than carbon dioxide. It has been estimated that more than 66% of this gas emission from soils is originated by microbial processes. Furthermore, its photochemical oxidation in the atmosphere produces nitric oxide that is submitted to various transformation reacting with oxygen and ozone (it is considered the principal oxygen depleting substance released in the twenty-first century); nitrite and nitrate can be produced which return to Earth as acid rains ($HNO_2/HNO_3$) with possible effects to the soil community function and structure and consequent soil fertility and plant diversity.

Hence, the denitrification produces a series of environmental negative effects contributing to ozone diminution, acid rains production and the increasing of global warming (Parkin 1987, Philippot et al. 2007, Castellano-Hinojosa et al. 2018a, Putz et al. 2018).

On the other hand, denitrification is helpful in wastewater treatment technology since nitrate depletion can reduce the eutrophication problems caused by the uncontrolled release of effluents with high fixed-nitrogen content. Also, in drinking water, the high concentration of nitrates (and nitrites) could cause problems (i.e., methemoglobinemia in infants). Thus, denitrifying microorganisms are widely present in most water/wastewater treatment plants.

Among the various denitrifying-microorganisms, we can ascribe both bacteria (various phyla) and fungi (in particular various classes of Ascomycota), even though within these two microbial groups, the process is rather different at the cellular level (Figure 1.8); denitrification in Archaea has been reported both in extreme and non-extreme environments, but it is still being investigated in details. Denitrification has been observed also in some protists (i.e., foraminifers) (Ravishankara et al. 2009, Stein and Klotz 2016, Regan et al. 2017, Castellano-Hinojosa et al. 2018b, Kaurin et al. 2018, Delgado et al. 2019).

Different bacteria, mostly belonging to the phylum *Proteobacteria*, act within the denitrification process. Most of them are facultative anaerobe (i.e., *Aeromonas, Alcaligenes, Bacillus, Escherichia, Nocardia, Staphylococcus and Vibrio*) and can only partially oxidise nitrate to nitrite, which is secreted. The oxidation is carried out in a short ETC by the nitrate-reductase, a membrane enzymatic complex. Other genera (i.e., *Halomonas, Pseudomonas, Paracoccus*) can perform the complete nitrate reduction to $N_2$ ($NO_3^- \rightarrow NO_2^- \rightarrow NO \rightarrow N_2O \rightarrow N_2$) through a more complete ETC, including the nitrite-reductase, the nitric oxide-reductase, and the nitrous oxide-reductase complexes. Only these bacteria can be defined as true denitrifying and the process, which was formerly considered only possible in anaerobic conditions, can also occur under aerobic conditions. Moreover, a co-respiration process of oxygen and nitrate has been recently proposed as an adaptation for the degradation of toxic nitrogen forms and seems to be present in a wide pattern of natural or artificial environments, such as soil, fresh and marine waters, and wastewater treatment plants. The majority of denitrifying bacteria show a chemoorganotrophic metabolism and are facultative anaerobes (Takaya et al. 2003, Barnard 2005, Knowles 1982, Yang et al. 2015, Marchant et al. 2017).

Denitrification is also widely present among fungi where a large number of genera present this metabolic ability (i.e., *Aspergillus, Chaetomium, Clonostachys, Fusarium, Penicillium, Trichoderma*, and many others) and contribute to the release of the greenhouse gas $N_2O$. Most of these fungi are present both as terrestrial and marine species in a wide array of different habitats (Tempesta et al. 2003, Cathrine and Raghukumar 2009, Mouton et al. 2012, Richards et al. 2012 Pasqualetti et al. 2019a, b).

The process is split between mitochondria and cytoplasm. Nitrate is reduced to nitrite by nitrate reductase (NAR) and subsequently to nitric oxide by NIR in the mitochondrion; the latter being reduced to nitrous oxide by NOR in the cytoplasm; $N_2O$ is then excreted. The system functions during anaerobic respiration. Phylogenetic analyses of the genes involved in the process showed that the fungal denitrifying system has a common ancestor with the bacterial equivalent indicating its possible proto-mitochondrial origin. Some fungi have both assimilative and dissimilative NARs. (Maeda et al. 2015, Novinscak et al. 2016, Shoun and Fushinobu 2016).

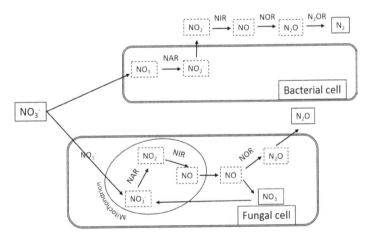

**FIGURE 1.8** Denitrification in bacteria and fungi. NAR, nitrate reductase; NIR, nitrite reductase; NOR, nitrous oxide reductase; $N_2OR$, nitric oxide reductase.

Nitrate biological reduction is also used to assimilate nitrogen; its incorporation in numerous bacteria fungi and algae as organic nitrogen is called assimilative nitrate reduction. Nitrate is more accessible to plants than ammonium due to its negative charge that does not permit to be retained by soil (opposite than ammonium). During the assimilation process, quite expensive under the energetic point of view, a specific permease permits nitrate transport inside the cell where it is reduced to nitrite and subsequently to ammonia by two specific enzymes the assimilative nitrate and nitrite reductases ($NO_3^- \rightarrow NO_2^- \rightarrow NH_3$), which are not inhibited by oxygen. Ammonium is then assimilated as discussed above.

Since, as said, the process is energy-consuming and requires a lot of reducing power; in order to avoid metabolic energy dispersion, its regulation is quite efficient and driven by the availability of ammoniacal nitrogen.

Besides the mentioned anaerobic respiration process, nitrate reduction could also occur in anaerobic conditions within a fermentative metabolism (re-oxidation of NADH) both in bacteria (i.e., *Ammonifex, Clostridium*) and fungi (i.e., *Aspergillus, Fusarium*). In the 'ammonia fermentation' the production of ammonium is coupled to the catabolic oxidation of an organic electron donor (i.e., ethanol to acetate) and consequent substrate-level phosphorylation. In other words, the ammonium fermentation couples dissimilatory and assimilatory mechanisms. In *F. oxysporum* two pathways of dissimilatory nitrate reduction exist and are alternatively expressed in response to environmental $O_2$ concentration. The fungus prefers aerobic respiration when oxygen is sufficient and it was discovered as the first facultative anaerobe eukaryotic known to express one of three distinct metabolic energy mechanisms strictly depending on $O_2$ concentration. Ammonia fermentation occurs in many other soil fungi, suggesting that facultative anaerobes are widely distributed among fungi that have been generally considered aerobic organisms (Huber et al. 1996, Zhou et al. 2002, Takasaki et al. 2004).

## Textbooks and Research Papers and Reviews

- Barbieri, P., Bestetti, G., Galli., E. and Zannoni, D. 2008. Microbiologia ambientale ed elementi di ecologia microbica. Casa Editrice Ambrosiana (Mi).
- Deho, G. and Galli, E. 2018. Biologia dei microrganismi. Casa Editrice Ambrosiana (Mi).
- Madigan, M.T., Bender, K.S., Buckley, D.H., Sattley and Stahl, D.A. 2019. Brock Biology of Micoorganisms, 15th edition, Global Edition. Pearson (NY).
- Willey, J., Sherwood, L. and Woolverton, J. 2016. Prescott's Microbiology, 10th Edition. Mc Graw-Hill Higher Education (NY).

## 1.7 REFERENCES

Abdulrasheed, M., Ibrahim, H.I., Maigari, F.U., Umar, A.F. and Ibrahim, S. 2018. Effect of soil pH on composition and abundance of nitrite-oxidising bacteria. Journal of Biochemistry, Microbiology and Biotechnology 6(1): 27-34.

Agogué, H., Brink, M., Dinasquet, J. and Herndl, G.J. 2008. Major gradients in putatively nitrifying and non-nitrifying Archaea in the deep North Atlantic. Nature 456(7223): 788.

Amarelle, V., Carrasco, V. and Fabiano, E. 2019. The hidden life of antarctic rocks. pp. 221-237. *In*: Castro-Sowinski, S. (ed.). The Ecological Role of Micro-organisms in the Antarctic Environment. Springer, Cham.

Ampomah, O.Y., Mousavi, S.A., Lindström, K. and Huss-Danell, K. 2017. Diverse Mesorhizobium bacteria nodulate native Astragalus and Oxytropis in arctic and subarctic areas in Eurasia. Systematic and Applied Microbiology 40(1): 51-58.

Aneja, V.P., Schlesinger, W.H. and Erisman, J.W. 2009. Effects of agriculture upon the air quality and climate: research, policy, and regulations. Environmental Science and Technology 43: 4234-4240.

Appelbaum, E. 1990. The Rhizobium/Bradyrhizobium-legume symbiosis. pp. 131-158. *In*: Gresshof, P.M. (ed.). Molecular Biology of Symbiotic Nitrogen Fixation. CRC Press, Boca Raton, Florida (USA).

Appleby, C.A. 1984. Leghemoglobin and rhizobium respiration. Annual Review of Plant Physiology 33: 443-478.

Baker, J., Riester, C., Skinner, B., Newell, A., Swingley, W., Madigan, M., et al. 2017. Genome sequence of *rhodoferax antarcticus* ANT. BRT; a psychrophilic purple nonsulfur bacterium from an antarctic microbial mat. Microorganisms 5(1): 8.

Barghini, P., Pasqualetti, M., Gorrasi, S. and Fenice, M. 2018. Bacteria from the "Saline di Tarquinia" marine salterns reveal very atypical growth profiles with regards to salinity and temperature. Mediterranean Marine Science 19(3): 513-525.

Barnard, R., Leadley, P.W. and Hungate, B.A. 2005. Global change, nitrification, and denitrification: a review. Global Biogeochemical Cycles 19(1) doi: https://doi.org/10.1029/2004GB002282.

Bellarby, J., Foereid, B. and Hastings, A. 2008. Cool farming: climate impacts of agriculture and mitigation potential. pp. 12-43. Green Peace International, Amsterdam, NL.

Benson, D.R. and Silvester, W.B. 1993. Biology of Frankia strains, actinomycete symbionts of actinorhizal plants. Microbiological Reviews 57: 293-319.

Blank, C.E. and Hinman, N.W. 2016. Cyanobacterial and algal growth on chitin as a source of nitrogen; ecological, evolutionary, and biotechnological implications. Algal Research 15: 152-163.

Canfield, D.E., Glazer, A.N. and Falkowski, P.G. 2010. The evolution and future of Earth's nitrogen cycle. Science 330(6001): 192-196.

Cao, Y., Halane, M.K., Gassmann, W. and Stacey, G. 2017. The role of plant innate immunity in the legume-rhizobium symbiosis. Annual Review of Plant Biology 68: 535-561.

Castellano-Hinojosa, A., González-López, J. and Bedmar, E.J. 2018a. Distinct effect of nitrogen fertilisation and soil depth on nitrous oxide emissions and nitrifiers and denitrifiers abundance. Biology and Fertility of Soils 54(7): 829-840.

Castellano-Hinojosa, A., Maza-Márquez, P., Melero-Rubio, Y., González-López, J. and Rodelas, B. 2018b. Linking nitrous oxide emissions to population dynamics of nitrifying and denitrifying prokaryotes in four full-scale wastewater treatment plants. Chemosphere 200: 57-66.

Cathrine, S.J. and Raghukumar, C. 2009. Anaerobic denitrification in fungi from the coastal marine sediments off Goa, India. Mycological Research 113(1): 100-109.

Costa, E., Pérez, J. and Kreft, J.U. 2006. Why is metabolic labour divided in nitrification? Trends in Microbiology 14(5): 213-219.

Crutzen, P.J. 2006. The "anthropocene". pp. 13-18. *In*: Ehler, E. and Krafft, T. (eds.). Earth System Science in the Anthropocene. Springer, Berlin, Heidelberg.

Crutzen, P.J. 2016. Geology of mankind. pp. 211-215. *In*: Paul, J. Crutzen (ed.). A Pioneer on Atmospheric Chemistry and Climate Change in the Anthropocene. Springer, Cham.

Daims, H., Lebedeva, E.V., Pjevac, P., Han, P., Herbold, C., Albertsen, M., et al. 2015. Complete nitrification by Nitrospira bacteria. Nature 528(7583): 504.

Dalsgaard, T., Thamdrup, B. and Canfield, D.E. 2005. Anaerobic ammonium oxidation (anammox) in the marine environment. Research in Microbiology 156(4): 457-464.

de Almeida, N.M., Neumann, S., Mesman, R.J., Ferousi, C., Keltjens, J.T., Jetten, M.S., et al. 2015. Immunogold localization of key metabolic enzymes in the anammoxosome and on the tubule-like structures of Kuenenia stuttgartiensis. Journal of Bacteriology 197(14): 2432-2441.

De Bruijn, F.J. 2015. Biological nitrogen fixation. pp. 215-224. *In*: Lugtenberg, B. (ed). Principles of Plant-Microbe Interactions. Springer, Cham.

de Vries, M. and de Boer, I.J. 2010. Comparing environmental impacts for livestock products: a review of life cycle assessments. Livestock Science 128(1-3): 1-11.

Delgado, M.J., Hidalgo-Garcia, A., Torres, M.J., Salas, A., Bedmar, E.J. and Girard, L. 2019. Rhizobium etli produces nitrous oxide by coupling the assimilatory and denitrification pathways. Frontiers in Microbiology 10: 980.

Elliott, G.N., Chen, W.M., Chou, J.H., Wang, H.C., Sheu, S.Y., Perin, L., et al. 2007. Burkholderia phymatum is a highly effective nitrogen-fixing symbiont of Mimosa spp. and fixes nitrogen ex planta. New Phytologist 173(1): 168-180.

Falkowski, P.G. 1997. Evolution of the nitrogen cycle and its influence on the biological sequestration of $CO_2$ in the ocean. Nature 387(6630): 272.

Fattah, Q.A. 2005. Plant Resources for Human Development. Third International Botanical Conference. Bangladesh Botanical Society, Dhaka, Bangladesh.

Fenice, M. 2016. The psychrotolerant Antarctic fungus Lecanicillium muscarium CCFEE 5003: a powerful producer of cold-tolerant chitinolytic enzymes. Molecules 21(4): 447.

Fowler, D., Coyle, M., Skiba, U., Sutton, M.A., Cape, J.N., Reis, S., et al. 2013. The global nitrogen cycle in the twenty-first century. Philosophical Transactions of the Royal Society B: Biological Sciences 368(1621): 20130164.

Galbally, I.E. and Roy, C.R. 1983. The fate of nitrogen compounds in the atmosphere. pp. 265-284. *In*: Freney, J.R. and Simpson, J.R. (eds.). Gaseous Loss of Nitrogen from Plant-Soil Systems. Springer, Dordrecht.

Galloway, J.N., Townsend, A.R., Erisman, J.W., Bekunda, M., Cai, Z., Freney, J.R., et al. 2008. Transformation of the nitrogen cycle: recent trends, questions, and potential solutions. Science 320(5878): 889-892.

Giblin, A.E., Tobias, C.R., Song, B., Weston, N., Banta, G.T. and Rivera-Monroy, V.H. 2013. The importance of dissimilatory nitrate reduction to ammonium (DNRA) in the nitrogen cycle of coastal ecosystems. Oceanography 26(3): 124-131.

Gonzalez-Martinez, A., Osorio, F., Morillo, J.A., Rodriguez-Sanchez, A., Gonzalez-Lopez, J., Abbas, B.A., et al. 2015. Comparison of bacterial diversity in full scale anammox bioreactors operated under different conditions. Biotechnology Progress 31(6): 1464-1472.

Gorrasi, S., Pesciaroli, C., Barghini, P., Pasqualetti, M. and Fenice, M. 2019. Structure and diversity of the bacterial community of an Arctic estuarine system (Kandalaksha Bay) subject to intense tidal currents. Journal of Marine Systems 19: 77-85.

Gourion, B., Berrabah, F., Ratet, P. and Stacey, G. 2015. Rhizobium–legume symbioses: the crucial role of plant immunity. Trends in Plant Science 20(3): 186-194.

Gussin, G.N., Ronson, C.W. and Ausubel, F.M. 1986. Regulation of nitrogen fixation genes. Annual Review of Genetics 20(1): 567-591.

Hayatsu, M., Tago, K. and Saito, M. 2008. Various players in the nitrogen cycle: diversity and functions of the microorganisms involved in nitrification and denitrification. Soil Science and Plant Nutrition 54(1): 33-45.

He, H., Zhen, Y., Mi, T., Fu, L. and Yu, Z. 2018. Ammonia-oxidizing Archaea and Bacteria differentially contribute to ammonia oxidation in sediments from adjacent waters of Rushan Bay, China. Frontiers in Microbiology 9: 116.

Herrero, A., Stavans, J. and Flores, E. 2016. The multicellular nature of filamentous heterocyst-forming cyanobacteria. FEMS Microbiology Reviews 40(6): 831-854.

Herridge, D.F., Peoples, M.B. and Boddey, R.M. 2008. Global inputs of biological nitrogen fixation in agricultural systems. Plant and Soil 311: 1-18.

Hertel, O., Reis, S., Skjoth, C.A., Bleeker, A., Harrison, R., Cape, J.N., et al. 2011. Nitrogen processes in the atmosphere. pp. 177-207. *In*: Sutton, Mark A., Howard, Clare M., Erisman, Jan Willem, Billen, Gilles, Bleeker, Albert, Grennfelt, Peringe, van Grinsven, Hans, Grizzetti, Bruna. (eds.). The European Nitrogen Assessment: Sources, Effects and Policy Perspectives. Cambridge, Cambridge University Press.

Hirel, B. and Lea, P.J. 2001. Ammonia assimilation. pp. 79-99. *In*: Lea, P.J. and Morot-Gaudry, J.-F. (eds.). Plant Nitrogen. Springer, Berlin, Heidelberg.

Hu, H.W. and He, J.Z. 2017. Comammox—a newly discovered nitrification process in the terrestrial nitrogen cycle. Journal of Soils and Sediments 17(12): 2709-2717.

Huber, R., Rossnagel, P., Woese, C.R., Rachel, R., Langworthy, T.A. and Stetter, K.O. 1996. Formation of ammonium from nitrate during chemolithoautotrophic growth of the extremely thermophilic bacterium ammonifex degensii gen. nov. sp. nov. Systematic and Applied Microbiology 19(1): 40-49.

Isobe, K., Ikutani, J., Fang, Y., Yoh, M., Mo, J., Suwa, Y., et al. 2018. Highly abundant acidophilic ammonia-oxidizing archaea causes high rates of nitrification and nitrate leaching in nitrogen-saturated forest soils. Soil Biology and Biochemistry 122: 220-227.

James, E.K. 2000. Nitrogen fixation in endophytic and associative symbiosis. Field Crops Research 65(2-3): 197-209.

Ji, B., Yang, K., Zhu, L., Jiang, Y., Wang, H., Zhou, J. and Zhang, H. 2015. Aerobic denitrification: a review of important advances of the last 30 years. Biotechnology and Bioprocess Engineering 20(4): 643-651.

Johnson, B. and Goldblatt, C. 2015. The nitrogen budget of Earth. Earth-Science Reviews 148: 150-173.

Karlsson, R., Karlsson, A., Bäckman, O., Johansson, B.R. and Hulth, S. 2009. Identification of key proteins involved in the anammox reaction. FEMS Microbiology Letters 297(1): 87-94.

Kaurin, A., Mihelič, R., Kastelec, D., Grčman, H., Bru, D., Philippot, L., et al. 2018. Resilience of bacteria, archaea, fungi and N-cycling microbial guilds under plough and conservation tillage, to agricultural drought. Soil Biology and Biochemistry 120: 233-245.

Knowles, R. 1982. Denitrification. Microbiological Reviews 46(1): 43.

Kuenen, J.G. 2008. Anammox bacteria: from discovery to application. Nature Reviews Microbiology 6(4): 320.

Lang, E. and Jagnow, G. 1986. Fungi of a forest soil nitrifying at low pH values. FEMS Microbiology Ecology 2(5): 257-265.

Leigh, G.J. (ed.). 2002. Nitrogen Fixation at the Millennium. Elsevier, Amsterdam, NL.

Leininger, S., Urich, T., Schloter, M., Schwark, L., Qi, J., Nicol, G.W., et al. 2006. Archaea predominate among ammonia-oxidizing prokaryotes in soils. Nature 442(7104): 806.

Lewis, S.L. and Maslin, M.A. 2015. Defining the anthropocene. Nature 519(7542): 171.

Li, Y., Chapman, S.J., Nicol, G.W. and Yao, H. 2018. Nitrification and nitrifiers in acidic soils. Soil Biology and Biochemistry 116: 290-301.

Long, S.R. 1996. Rhizobium symbiosis: nod factors in perspective. The Plant Cell 8(10): 1885.

Maeda, K., Spor, A., Edel-Hermann, V., Heraud, C., Breuil, M.C., Bizouard, F., et al. 2015. $N_2O$ production, a widespread trait in fungi. Scientific Reports 5: 9697.

Magasanik, B. 2003. Ammonia assimilation by Saccharomyces cerevisiae. Eukaryotic Cell 2(5): 827-829.

Marchant, H.K., Ahmerkamp, S., Lavik, G., Tegetmeyer, H.E., Graf, J., Klatt, J.M., et al. 2017. Denitrifying community in coastal sediments performs aerobic and anaerobic respiration simultaneously. The ISME Journal 11(8): 1799.

Matkarimov, F., Jabborova, D. and Baboev, S. 2019. Enhancement of Plant Growth, Nodulation and Yield of Mungbean (Vigna radiate L.) by Microbial Preparations. International Journal of Current Microbiological Applied Sciences 8(8): 2382-2388.

McRose, D.L., Zhang, X., Kraepiel, A.M. and Morel, F.M. 2017. Diversity and activity of alternative nitrogenases in sequenced genomes and coastal environments. Frontiers in Microbiology 8: 267.

Mondelaers, K., Aertsens, J. and Van Huylenbroeck, G. 2009. A meta-analysis of the differences in environmental impacts between organic and conventional farming. British Food Journal 111(10): 1098-1119.

Morozkina, E.V. and Kurakov, A.V. 2007. Dissimilatory nitrate reduction in fungi under conditions of hypoxia and anoxia: a review. Applied Biochemistry and Microbiology 43(5): 544-549.

Mouton, M., Postma, F., Wilsenach, J. and Botha, A. 2012. Diversity and characterization of culturable fungi from marine sediment collected from St. Helena Bay, South Africa. Microbial Ecology 64(2): 311-319.

Nielsen T., Pilegaard, K., Egeløv, A.H., Granby, K., Hummelshøj, P., Jensen, N.O., et al. 1996. Atmospheric nitrogen compounds: occurrence, composition and deposition. Science of the Total Environment 189: 459-465.

Novinscak, A., Goyer, C., Zebarth, B.J., Burton, D.L., Chantigny, M.H. and Filion, M. 2016. Characterization of the prevalence and diversity of soil fungal denitrifiers using a novel P450nor gene detection assay. Applied and Environmental Microbiology 82(15): 4560-4569.

Pajares, S. and Bohannan, B.J. 2016. Ecology of nitrogen fixing, nitrifying, and denitrifying microorganisms in tropical forest soils. Frontiers in Microbiology 7: 1045.

Park, C.H., Kim, K.M., Kim, O.S., Jeong, G. and Hong, S.G. 2016. Bacterial communities in Antarctic lichens. Antarctic Science 28(6): 455-461.

Parkin, T.B. 1987. Soil microsites as a source of denitrification variability. Soil Science Society of America Journal 51(5): 1194-1199.

Pasqualetti, M., Barghini, P., Giovannini, V. and Fenice, M. 2019a. High production of chitinolytic activity in halophilic conditions by a new marine strain of clonostachys rosea. Molecules 24(10): 1880.

Pasqualetti, M., Giovannini, V., Barghini, P., Gorrasi, S. and Fenice, M. 2019b. Diversity and ecology of culturable marine fungi associated with Posidonia oceanica leaves and their epiphytic algae Dictyota dichotoma and Sphaerococcus coronopifolius. Fungal Ecology 44: 100906.

Pawlowski, K. and Sirrenberg, A. 2003. Symbiosis between Frankia and actinorhizal plants: root nodules of non-legumes. Indian Journal of Experimental Biology 41: 1165-1183.

Philippot, L., Hallin, S. and Schloter, M. 2007. Ecology of denitrifying prokaryotes in agricultural soil. Advances in Agronomy 96: 249-305.

Postgate, J.R. 1982. The Fundamentals of Nitrogen Fixation. New York, NY: Cambridge University Press.

Poza-Carrión, C., Jiménez-Vicente, E., Navarro-Rodríguez, M., Echavarri-Erasun, C. and Rubio, L.M. 2014. Kinetics of nif gene expression in a nitrogen-fixing bacterium. Journal of Bacteriology 196(3): 595-603.

Putz, M., Schleusner, P., Rütting, T. and Hallin, S. 2018. Relative abundance of denitrifying and DNRA bacteria and their activity determine nitrogen retention or loss in agricultural soil. Soil Biology and Biochemistry 123: 97-104.

Ravishankara, A.R., Daniel, J.S. and Portmann, R.W. 2009. Nitrous oxide ($N_2O$): the dominant ozone-depleting substance emitted in the 21st century. Science 326(5949): 123-125.

Regan, K., Stempfhuber, B., Schloter, M., Rasche, F., Prati, D., Philippot, L., et al. 2017. Spatial and temporal dynamics of nitrogen fixing, nitrifying and denitrifying microbes in an unfertilized grassland soil. Soil Biology and Biochemistry 109: 214-226.

Richards, T.A., Jones, M.D., Leonard, G. and Bass, D. 2012. Marine fungi: their ecology and molecular diversity. Annual Review of Marine Science 4: 495-522.

Rubio, L.M. and Ludden, P.W. 2002. The gene products of the nif regulon. pp. 101-136. *In*: Leigh, G.J. (ed.). Nitrogen fixation at the millennium, 1st ed. Elsevier Science, Amsterdam, NL.

Rütting, T., Boeckx, P., Müller, C. and Klemedtsson, L. 2011. Assessment of the importance of dissimilatory nitrate reduction to ammonium for the terrestrial nitrogen cycle. Biogeosciences 8(7): 1779-1791.

Schleper, C. and Nicol, G.W. 2010. Ammonia-oxidising archaea–physiology, ecology and evolution. Advances in Microbial Physiology 57: 1-41.

Schlesinger, W.H. 2009. On the fate of anthropogenic nitrogen. Proceedings of the National Academy of Sciences 106(1): 203-208.

Schreiber, E.A. and Burger, J. (eds.). 2001. Biology of Marine Birds. CRC Press, New York, USA.

Shoun, H. and Fushinobu, S. 2016. Denitrification in fungi. pp. 331-348. *In*: Moura, I., Moura, J.J.G., Pauleta, S.R. and B. Maia, L.B. (eds.). Metalloenzymes in Denitrification. Applications and Environmental Impacts. RDC Metallobiology Series 9 (ISBN: 978-1-78262-376-2) Vol. 39. The Royal Society of Chemistry (UK).

Siemann, S., Schneider, K., Dröttboom, M. and Müller, A. 2002. The Fe-only nitrogenase and the Mo nitrogenase from Rhodobacter capsulatus: a comparative study on the redox properties of the metal clusters present in the dinitrogenase components. European Journal of Biochemistry 269(6): 1650-1661.

Singer, M.A. 2003. Do mammals, birds, reptiles and fish have similar nitrogen conserving systems? Comparative Biochemistry and Physiology Part B: Biochemistry and Molecular Biology 134(4): 543-558.

Singh, B. and Singh, Y. 2008. Reactive nitrogen in Indian agriculture: inputs, use efficiency and leakages. Current Science 94(11) 1382-1393.

Somasegaran, P. and Hoben, H.J. 2012. Handbook for Rhizobia: Methods in Legume-Rhizobium Technology. Springer Science and Business Media, New York, USA.

Stein, L.Y. and Nicol, G.W. 2001. Nitrification. In eLS. Edited by. http://www.els.net. John Wiley and Sons Ltd, Chichester, UK.

Stein, L.Y. and Klotz, M.G. 2016. The nitrogen cycle. Current Biology 26(3): R94-R98.

Stewart, W.D.P. (ed.). 1975. Nitrogen Fixation by Free-Living Micro-Organisms (Vol. 6). CUP Archive.

Sulewska, H., Ratajczak, K., Niewiadomska, A. and Panasiewicz, K. 2019. The use of microorganisms as bio-fertilizers in the cultivation of white lupine. Open Chemistry 17(1): 813-822.

Sutton, M.A., Howard, C.M., Erisman, J.W., Billen, G., Bleeker, A., Grennfelt, P., et al. 2011. The European Nitrogen Assessment: Sources, Effects and Policy Perspectives. Cambridge University Press.

Takai, K. 2019. The nitrogen cycle: a large, fast, and mystifying cycle. Microbes and Environments 34(3): 223-225.

Takasaki, K., Shoun, H., Yamaguchi, M., Takeo, K., Nakamura, A., Hoshino, T., et al. 2004. Fungal ammonia fermentation, a novel metabolic mechanism that couples the dissimilatory and assimilatory pathways of both nitrate and ethanol role of acetyl CoA synthetase in anaerobic ATP synthesis. Journal of Biological Chemistry 279(13): 12414-12420.

Takaya, N., Catalan-Sakairi, M.A.B., Sakaguchi, Y., Kato, I., Zhou, Z. and Shoun, H. 2003. Aerobic denitrifying bacteria that produce low levels of nitrous oxide. Applied and Environmental Microbiology 69(6): 3152-3157.

Tempesta, S., Pasqualetti, M., Fonck, M. and Mulas, B. 2003. Succession of microfungi in Phillyrea angusti-folia litter in a Mediterranean maquis in Sardinia. Plant Biosystems 137(2): 149-154.

Thajudeen, J., Yousuf, J., Veetil, V.P., Varghese, S., Singh, A. and Abdulla, M.H. 2017. Nitrogen fixing bacterial diversity in a tropical estuarine sediment. World Journal of Microbiology and Biotechnology 33(2): 41.

Thorsteinsson, M.V., Bevan, D.R., Potts, M., Dou, Y., Eich, R.F., Hargrove, M.S., et al. 1999. A cyanobacterial hemoglobin with unusual ligand binding kinetics and stability properties. Biochemistry 38(7): 2117-2126.

Timperio, A.M., Gorrasi, S., Zolla, L. and Fenice, M. 2017. Evaluation of MALDI-TOF mass spectrometry and MALDI BioTyper in comparison to 16S rDNA sequencing for the identification of bacteria isolated from Arctic sea water. PloS One 12(7): e0181860.

Vadakattu, G. and Paterson, J. 2006. Free-living bacteria lift soil nitrogen supply. Farming Ahead 169: 40.

Vance, C. 2001. Symbiotic nitrogen fixation and phosphorus acquisition. Plant nutrition in a world of declining renewable resouces. Plant Physiology 127: 391-397.

Van Kessel, M.A., Speth, D.R., Albertsen, M., Nielsen, P.H., den Camp, H.J.O., Kartal, B., et al. 2015. Complete nitrification by a single microorganism. Nature 528(7583): 555.

van Niftrik, L.A., Fuerst, J.A., Damsté, J.S.S., Kuenen, J.G., M.S. Jetten and Strous, M. 2004. The anammoxosome: an intracytoplasmic compartment in anammox bacteria. FEMS Microbiology Letters 233(1): 7-13.

Van Rhyn, P. and Vanderleyden, J. 1995. The Rhizobium-plant symbiosis. Microbiological Reviews 59: 124-142.

Vitousek, P.M., Aber, J.D., Howarth, R.W., Likens, G.E., Matson, P.A., Schindler, D.W., et al. 1997. Human alteration of the global nitrogen cycle: sources and consequences. Ecological Applications 7(3): 737-750.

Vlassak, K. and Reynders, R. 1979. Agronomic aspects of biological dinitrogen fixation by Azoxpirillum spp. pp. 93-102. *In*: Vose, P.B. and Ruschel, A.P. (eds.). Associative $N_2$ Fixation. Volume I. CRC Press, Boca Raton, FL.

Wagner, S.C. 2012. Biological nitrogen fixation. Nature Education Knowledge 3(10): 15.

Walker, R., Agapakis, C.M., Watkin, E. and Hirsch, A.M. 2015. Symbiotic nitrogen fixation in legumes: perspectives on the diversity and evolution of nodulation by *Rhizobium* and *Burkholderia* species. Biological Nitrogen Fixation 2: 913-923.

Wolk, C.P., Ernst, A. and Elhai, J. 1994. Heterocyst metabolism and development. pp. 769-823. *In*: Bryant, D.A. (ed.). The Molecular Biology of Cyanobacteria. Springer, Dordrecht.

Yousuf, J., Thajudeen, J., Rahiman, M., Krishnankutty, S., P. Alikunj, A. and A. Abdulla, M.H. 2017. Nitrogen fixing potential of various heterotrophic Bacillus strains from a tropical estuary and adjacent coastal regions. Journal of Basic Microbiology 57(11): 922-932.

Zedgenizov, D.A. and Litasov, K.D. 2017. Looking for "missing" nitrogen in the deep Earth. American Mineralogist 102(9): 1769-1770.

Zehr, J.P., Jenkins, B.D., Short, S.M. and Steward, G.F. 2003. Nitrogenase gene diversity and microbial community structure: a cross-system comparison. Environmental Microbiology 5(7): 539-554.

Zhang, X., McRose, D.L., Darnajoux, R., Bellenger, J.P., Morel, F.M. and Kraepiel, A.M. 2016. Alternative nitrogenase activity in the environment and nitrogen cycle implications. Biogeochemistry 127(2-3): 189-198.

Zhou, Z., Takaya, N., Nakamura, A., Yamaguchi, M., Takeo, K. and Shoun, H. 2002. Ammonia fermentation, a novel anoxic metabolism of nitrate by fungi. Journal of Biological Chemistry 277(3): 1892-1896.

Zhu, T., Meng, T., Zhang, J., Zhong, W., Müller, C. and Cai, Z. 2015. Fungi-dominant heterotrophic nitrification in a subtropical forest soil of China. Journal of Soils and Sediments 15(3): 705-709.

# The Nitrogen Fixation Dream: The Challenges and the Future

Allen G. Good[1,*] and Ray Dixon[2]

## 2.1 INTRODUCTION

The use of chemical fertilizers has been a prime factor in the consistent increase in crop yields over the last 100 years, and the ability to synthesize ammonia (to manufacture nitrogenous fertilizers) by the Häber-Bosch process has arguably been more beneficial to humanity than any other discovery of the twentieth century (Erisman et al. 2008, Smil 2004). Together with advances in plant breeding, the increased use of chemical fertilizers has provided the foundation of the 'Green Revolution' (Smil 2004, Borlaug 1972).

The global value of N fertilizers has increased rapidly since 1960 with current costs of over $120M Mt annually, having increased by 38% in the last 15 years (FAO 2017). However, there are significant challenges associated with artificial fertilizers, environmentally and in terms of their contribution to the increase in greenhouse gases, at a time when mankind is becoming aware of the huge consequences of global warming (Good and Beatty 2011). In particular, nitrous oxide has a global warming potential almost 300 times greater than carbon dioxide. It has been estimated that fertilizer production contributed approximately 6% of global GHG emissions as $CO_2$ and $N_2O$, but $N_2O$ emissions from soils, caused in part by fertilizer applications, contributed approximately 32% of global emissions (Oertel et al. 2016).

Interestingly, the N problem is unique for several reasons. First, although the problem is globally significant, there are distinct differences between the developed world (where N is freely available) and the developing world (e.g., Africa, where N fertilizers are too expensive for farmers). Second, there is no lack of available N; the problem is that most of the N is in the gaseous form of $N_2$ and as such it is not biologically active (i.e., it needs to be fixed by converting $N_2$ into $NH_3$). Third, the problem can be clearly defined in a way that should allow scientific solutions to be developed. Finally, on this day when climate change has become both a major concern and a political focus, one would hope that any potential solution would be properly evaluated. However, this requires a commitment, money and time, all of this especially when society seems unable to coordinate any kind of realistic approach to addressing greenhouse gas emissions (IPCC 2019).

Given the concern over the contribution of greenhouse gases, we have a unique opportunity to change the way agriculture is conducted and to provide safe and environmentally sensible solutions to the problem of nitrogen limitation in crop plants. Moreover, addressing the N issue has the potential to provide substantial benefits in terms of food security. First, reducing input costs

---

[1] Department of Biological Sciences, University of Alberta, Edmonton, Alberta, T6G 2E9 Canada.
[2] Department of Molecular Microbiology, John Innes Centre, The John Innes Centre, Norwich Research Park, Norwich, NR4 7UH, UK.
* Corresponding author: alleng@ualberta.ca

should help to reduce the burden on poor people in the world. Second, having local food that can be produced sustainably is one of the keys to global food security. The combination of increasing fertilizer prices, their poor efficiency of use, concerns over their impacts on the environment and all these points to the need to develop alternative approaches to improve the availability of the plant nutrients. Clearly, as we intensify agricultural production (to satisfy the demands of the burgeoning world population for the consumption of cereals, oilseeds and meat), replacing these soil nutrients is going to be essential.

In order to reduce mankind's dependence on N fertilizers, several approaches have been considered. Firstly, from a policy perspective, it is important to identify and encourage the adoption of policies that will reduce the excess application of N based fertilizers. Secondly, from a crops' perspective, it would be valuable to improve the efficiency with which plants capture and utilize N. Thirdly, we could engineer nitrogen fixation in cereals (Beatty and Good 2011). This review will focus solely on this approach.

One hundred years ago Burrill and Hanson asked the question, "Is symbiosis possible between legume bacteria and non-legume plants?" (Burrill and Hansen 1917). They realized that "the benefit could ordinary farm crops be enabled to utilise atmospheric nitrogen would be inestimable" but unfortunately their attempts to encourage legume bacteria to inoculate non-legume plants were unsuccessful. It was not until the advent of genetic engineering that the idea of endowing crop plants with the ability to fix their N from atmospheric $N_2$, thereby synchronizing plant N demand with plant N supply, could be given serious consideration (Dixon et al. 1997, Merrick and Dixon 1984). Certainly, in recent years, this idea has resurfaced as a more viable research goal due to a number of reasons, including technological advances and our understanding of how biological nitrogen fixation works at the genetic and biochemical levels (Beatty and Good 2011).

Several approaches have been suggested to do this, one way is to improve endophytic interactions between diazotrophic rhizosphere bacteria and cereals. Since free-living diazotrophs are usually unwilling to give up fixed N, one idea is to synthesize novel diazotrophic bacteria: plant associations that would benefit both partners (Bueno and Dixon 2019, Mus et al. 2016, Geddes et al. 2015). These would result in synthetic symbioses in which N provided by the bacteria would be exchanged for carbon by the plant. The additional signal exchange between the plant and microbe could be built in to engender more efficient colonization by the diazotroph and controlled nutrient exchange between the partners. A second more sophisticated approach would be to engineer cereals with the capacity to form root nodules, thus providing an ideal environment for symbiosis with nitrogen-fixing bacteria (Rogers and Oldroyd 2014, Oldroyd and Dixon 2014, Charpentier and Oldroyd 2010). The third approach involves transforming the plant with the genes required to express functional nitrogenase enzyme replete with its cofactors, therefore allowing the plant to directly fix $N_2$ into ammonia (Good 2018, Buren and Rubio 2018, Oldroyd and Dixon 2014, Curatti and Rubio 2014).

## 2.2  THE APPROACHES

### 2.2.1  The Enhancement of Endophytic Diazotrophic Bacteria and Cereal Crop Associations

There are many different living arrangements between diazotrophic bacteria and plants, different types of benefits to the plant and varying levels of $N_2$ fixation. Diazotrophic bacteria can be free-living, rhizosphere- and root surface-associated or endophytic (colonizing and living within plants). Depending on the plant species or cultivar, they provide the plants with low to modest amounts of fixed N (Yanni and Dazzo 2010, Shrestha and Ladha 1996, Hurek et al. 2002). In non-leguminous plants, such as sugarcane, nitrogen fixation contributes substantial N to benefit the growth of the crop (Boddey et al. 2003). There are also some cases where specific cereal varieties can potentially provide an appropriate environment to support significant nitrogen fixation by diazotrophs. For example, a landrace of maize native to Mexico has been recently reported to obtain up to 80% of plant N from a diazotrophic community resident in a sugar-rich mucilage associated with its aerial roots (Van Deynze et al. 2018). However, in most cases, it is not clear to what extent the diazotrophic bacteria-cereal associations actively transport fixed N to the plant partner. Plants benefit from these associations, but this may be due to the production of plant growth-promoting regulators and/or improving the plants' efficiency of N uptake (Yanni and Dazzo 2010, Reinhold-Hurek and Hurek 2011). Nevertheless, a recent re-estimation of global N budgets for cereals indicates that 48% of crop N is derived from sources other than fertilizer or soil N with non-symbiotic nitrogen fixation contributing to 24% of total N to the crop (Ladha et al. 2016).

There have been many reports in the literature on natural endophytic bacterial-cereal associations and their benefits to plant growth promotion under controlled growth experiments (Yanni and Dazzo 2010). There are also now many commercial biofertilizers available to farmers; however, many of these promote plant growth via plant hormone induction or increase the efficiency of the plants to take up soil N but do not necessarily provide fixed N to the plant. In addition to this, bio-fertilizers are not uniform in their quality and these products would benefit from more research into their efficacy and use. Nitrogen fixation by plant-associated diazotrophs is controlled by complex regulatory circuits to ensure that nitrogenase activity is tightly coupled to N assimilation so that the ammonia produced by the enzyme is utilized for cell proliferation rather than being altruistically released to the environment (Bueno and Dixon 2019). There is, therefore, a potential for enhancement of the association by engineering ammonium excreting strains, but this is likely to incur a fitness penalty as N is lost from the cell and such strains may not be competitive in the rhizosphere. There are many examples where genetic manipulation has resulted in ammonia excretion by diazotrophic bacteria, either by specifically targeting nitrogen fixation (*nif*), gene regulation/ammonia assimilation or a combination of both examples (Machado et al. 1991, Ortiz-Marquez et al. 2014, Barney et al. 2015, Brewin et al. 1999, Ortiz-Marquez et al. 2012, Setten et al. 2013, Bali et al. 1992). In some cases, ammonium excreting strains have been shown to provide significant plant growth promotion when inoculated on model plants grown under sterile conditions (Setten et al. 2013, Pankievicz et al. 2015, Santos et al. 2016) and in non-sterile soil under greenhouse conditions (Fox et al. 2016, Van Dommelen et al. 2009, Christiansen-Weniger and Van Veen 1991). However, there are a few examples where ammonium excreting strains have been shown to robustly benefit crop growth in the field. Recent experiments in India with ammonia excreting strain of *Azotobacter chroococcum* exhibited an increase in wheat grain yield of ~ 20% when 40 kg N ha$^{-1}$ was also applied (Bageshwar et al. 2017). Field trials in Brazil with ammonium excreting mutant of *Azospirrilum brasilense*, which in model experiments provided substantial N to the plant (Pankievicz et al. 2015), gave increases in grain yield of maize, varying between 4.7-29%, equivalent to an increase of 460 to 1,769 kg ha$^{-1}$, when provided with base fertilization of 40 kg N ha$^{-1}$ (Pedrosa et al. 2019). Given the growth penalty associated with ammonium excretion, these yield increases are unexpected but would provide substantial benefit if they can be reproduced in regions of the world with chronically N poor soils and low availability of N fertilizers, such as smallholder farms in sub-Saharan Africa. Notably; however, such yield increases are likely to be reliably obtained only if a base level of N fertilizer is provided in addition to inoculation with the mutant strain.

Further enhancement of the association might include engineering of strains in which assimilation of fixed N would be retained in the free-living state but specifically repressed *in planta*, thus retaining the competitiveness of inoculant strains in soils and enabling controlled release of ammonia to plants. This would be paralleled by approaches to identify and engineer plant processes that would facilitate hosting of endophytes, provision of carbon substrates to support nitrogen fixation and the active transfer of fixed N from the bacterial symbiont. Plant engineering will also require signaling mechanisms to attract and control nitrogen fixation by root-associated diazotrophs. Proof of principle for the latter has now been demonstrated through the development of a transkingdom signaling system in which transgenic plants synthesize rhizopine and exude this into the rhizosphere to control gene expression in bacteria via a rhizopine biosensor (Geddes et al. 2019). Given rapid progress towards the establishment of synthetic symbioses, we anticipate this approach will come on stream before the other systems discussed below. Even if these artificial symbioses provide only 25 kg N ha$^{-1}$ crop$^{-1}$, as mentioned above, this could make a major contribution to agriculture in developing countries where N fertilizers are unaffordable.

## 2.2.2    Developing the Legume-Rhizobium Symbiosis in Cereals

Legumes have evolved one of the most complex and productive nitrogen-fixing symbioses in the living world. This results from rhizobia colonizing the plant cell and developing a symbiosis within plant nodules; a structure similar to an organelle. Given the advances in understanding how these symbioses are established with legumes, scientists can now begin to consider how to engineer this process into cereals.

The direct approach to developing a rhizobial symbiosis in cereals is dependent upon the detailed fundamental knowledge of the legume nodule symbiosis. While we currently do not understand the N-fixing symbiosis in its entirety, we do understand the key genetic steps needed for symbiosis to work in plants and have determined that cereals already have some of the genes required to form symbioses (Rogers and Oldroyd 2014, Oldroyd and Dixon 2014, Charpentier and Oldroyd 2010). Three major steps required to engineer efficient N-fixing cereals and phylogenomic analysis of the number of genes required are detailed below. A fourth step required to ensure that cereals provide the appropriate environment for nitrogen fixation in engineered nodules is difficult to evaluate at present until significant progress in the first three steps has been achieved.

### 2.2.2.1    Engineering Nodulation Signaling Pathways in Cereals

There is molecular crosstalk between rhizobia and legumes where the plant secretes flavonoids that activate the bacteria to secrete Nod factors. The plant recognizes these Nod factors (lipochitooligosaccharides or LCOs) and the process of bacterial infection and nodule formation begin. Recently, the discovery was made that Myc factors, which are part of molecular crosstalk between plants and arbuscular mycorrhizal (AM) endosymbiotic fungi, are very similar in structure to Nod factors (Maillet et al. 2011). It appears that *Rhizobium* (and even *Frankia*) species adopted or mimicked the Myc factors and recognition pathway from AM fungi to form their symbioses with plants at some point during their evolution (Gherbi et al. 2008). Therefore, many common features between AM fungi and rhizobial symbioses exist. Nod and Myc factors are recognized by LysM receptors, which form complexes with LysM receptor kinases within membrane microdomain platforms (Kelly et al. 2017). A heteromer of LysM receptor-like kinases has been recently demonstrated to mediate the perception of AM fungi in rice. Engineering chimeric receptors in which the ectodomains of the AM rice receptors are replaced by their legume Nod factor counterparts results in Nod factor perception in rice, an important step towards engineering cereals that can recognize rhizobia (He et al. 2019).

A symbiosis signaling pathway, common to both the AM fungi and the rhizobial symbioses, acts downstream of the receptor complexes to activate calcium oscillations in the nucleus (Oldroyd 2013). Several of the rice signaling components in this pathway can complement mutations in the equivocal genes in legumes for both mycorrhization and nodulation (Banba et al. 2008). The extensive conservation of this common pathway suggests that major re-engineering of components may not be required to achieve nodulation signaling in cereals. However, the pathway diverges downstream of calcium signaling to drive activation of transcription factors specific to the expression of nodulation or mycorrhization genes. These transcription factor cascades are complex and in the legume symbiosis, they play roles in both nodule organogenesis (Soyano and Hayashi 2014) and bacterial infection (Diédhiou and Diouf 2018, Liu et al. 2019). Therefore stacking of nodule-specific transcription factors will probably be required to engineer nodulation in cereals.

### 2.2.2.2   Nodule Organogenesis

It has been known for some time that there are steps in common between the development of lateral roots and nodules. These communalities have been investigated with a high temporal and spatial resolution recently and demonstrated to converge on the accumulation of the plant hormones auxin and cytokinin and the transcription factor LBD16, which was found to be required for both nodule and lateral root initiation (Soyano et al. 2019, Schiessl et al. 2019). In legumes, the promotion of auxin biosynthesis by LBD16 was activated by the master nodule transcription factor NIN, which itself is activated via a second plant hormone cytokinin. Hence, NIN activates the formation of nodules through cytokinin induction of local auxin biosynthesis to promote root organogenesis. This finding emphasizes the conservation of components in the two developmental pathways and suggests that future engineering of nodulation in cereals could focus on cytokinin activation of root development via NIN rather than the need to engineer the entire nodulation developmental program.

### 2.2.2.3   Bacterial Infection

While Nod factor recognition is important for infection, other components are also clearly necessary. Moreover, the requirements for intracellular accommodation of bacteria have not yet been defined. Most *Rhizobia* enter legume root cells via an infection thread (IT), which is a highly evolved plant structure formed by tubular invagination of root cells. However, the mutant legumes that cannot form ITs can still be colonized by *Rhizobia* via intercellular infection or crack entry (Madsen et al. 2010). These two other means of infection allow for N fixation, therefore these more primitive and less genetically complex infection methods might be used to engineer a cereal symbiosis in parallel with further studies to elucidate the mechanism of infection thread development.

In addition to the Nod factor pathway for bacterial perception, rhizobial exopolysaccharides (EPS) are also important determinants of the symbiotic interaction. Mutant bacterial strains found deficient in the later stages of EPS biosynthesis are defective in IT development, and they induce the formation of only small uninfected nodule-like structures (Kelly et al. 2013). A plant EPS receptor, EPR3, was identified, which has a striking similarity to the Nod factor-dependent LysM receptors (Kawaharada et al. 2015). However, expression of the *epr3* gene is itself dependent on activation by the Nod factor signaling pathway. This, therefore, sets up a second molecular dialogue between the partners after Nod Factor recognition that is important for bacterial invasion.

Engineering IT development in cereals is likely to be a difficult challenge since a few components required for IT initiation and progression only have so far been characterized, although downstream transcription factors and several genes directly involved in infection have been identified (Murray 2011), including those subject to regulation by the master transcription factor NIN (Liu et al. 2019). In addition to this, two proteins, VPY and LIN, known to be required for IT formation, have been shown to co-localize with an exocyst subunit during rhizobial infection, to form a protein complex required for polar growth of the IT (Liu et al. 2019).

### 2.2.2.4   How Many Genes are Required?

Comparative phylogenetics and phylogenomics provide powerful tools to trace the evolutionary path of nitrogen-fixing symbioses. These symbioses have arisen at least ten times in the evolution of land plants, affording an opportunity to identify key genomic innovations that drive evolution and the emergence of root nodule symbiosis (Delaux et al. 2015). Comparative genomics of the non-legume plant *Parasponia*, which can form nitrogen-fixing nodules in symbiosis with rhizobium, has provided evidence for multiple independent loss of nodulation genes amongst the nitrogen-fixing plant clade in accordance with the hypothesis that nodulation probably evolved only once in an ancestor of this clade. Comparisons with *Medicago truncatula* identified 26 identified putative orthologs of legume symbiosis genes including key players such as NIN, *RHIZOBIUM-DIRECTED POLAR GROWTH* (*RPG*) and LysM-type Nod receptors (van Velzen et al. 2018). A parallel study in which ten plant species covering the nitrogen-fixing clade were sequenced, reached similar conclusions and validated multiple independent loss of nodule symbiosis in four orders of this clade (Griesmann et al. 2018). This study also identified parallel loss of NIN and RPG in non-nodulating species. These findings have important implications for engineering the nitrogen fixation symbiosis in cereals, which may require the introduction of far fewer genes than originally anticipated, particularly if the host already associates with AM fungi and therefore encodes the building blocks and the common signaling pathway for the mycorrhizal symbiosis. More extensive phylogenomics analysis will help to more clearly define the number of genes required for engineering the nodule symbiosis.

### 2.2.3    Introducing Nitrogenase Into a Plant Organelle

It has been suggested some time ago that the absence of plant organelles that perform nitrogen fixation is puzzling, given the need for fixed nitrogen and the costs of acquiring it (Merrick and Dixon 1984). Since nitrogenase requires high levels of energy and reductants to function, there has been a long-standing interest in introducing nitrogenase into either mitochondria or chloroplasts in view of the roles of these organelles in energy conversion. This, in essence, would create novel organelles that can fix N within plant cells. One of the benefits of this type of approach is that it would provide germline transmission. The technology will be in the seed. Moreover, this approach should be generic, if it works in one plant, it should work in other plant/platforms too.

There are two key steps to introducing nitrogenase into plants. The first is to engineer the minimal genetic requirements for the assembly of active nitrogenase enzymes in plant organelles. The second and more challenging problem is to create an appropriate environment that will enable nitrogenase to function. Since nitrogenase is an extremely oxygen-sensitive enzyme, this is particularly difficult in the case of chloroplasts, which evolve oxygen during photosynthesis. Nevertheless, cyanobacteria have evolved various mechanisms to reconcile the incompatibility between oxygen-sensitive nitrogen fixation and oxygenic photosynthesis, including expression of nitrogenase in differentiated cells called heterocysts that lack photosystem 2 or temporal separation of nitrogenase from photosynthesis during the dark and the light periods respectively in non-heterocystous cyanobacteria (Fay 1992, Gallon 1992). These strategies could potentially be mimicked in transgenic cereals to engineer diazotrophic plant plastids. It is also important to note that an oxygen-sensitive nitrogenase-like enzyme, termed light-independent protochlorophyllide oxidoreductase (DPOR), which performs the last step in chlorophyll biosynthesis in the dark, has been retained throughout the evolution of photosynthetic eukaryotes from algae to gymnosperms but has been lost in flowering plants (Cvetkovska et al. 2019, Vedalankar and Tripathy 2019, Reinbothe et al. 2010). DPOR is structurally and functionally similar to nitrogenase and has likely been evolved from a common ancestor (Boyd and Peters 2013). Since DPOR is encoded within the chloroplast genomes of algae and non-flowering plants, it seems probable that this highly oxygen-sensitive enzyme is protected from oxygen damage during the dark period. It will, therefore, be of interest to determine if DPOR is active when engineered into angiosperms as this will give clues as to whether oxygen-protective mechanisms, to support dark dependent enzymes, have been retained in flowering plants or not.

In contrast to the oxygen-protection issue in the chloroplast, although mitochondria carry out oxidative respiration, the mitochondrial matrix is predicted to be oxygen deplete and there are a number of oxygen-sensitive enzymes in mitochondria including those required to synthesize iron-sulfur clusters. On the other hand, mitochondria cannot easily be manipulated by DNA transformation, so the normal route to introduce foreign proteins, involves the insertion of genes into the nuclear genome and the use of signal peptides to target their products to mitochondria. Chloroplast engineering is not restricted to this route, since the plastid transformation is a well-established methodology, although it is not yet routine in cereals.

Nitrogenase is a complex and extremely oxygen-sensitive enzyme, which contains three distinct metalloclusters and requires multiple genes essential for assembly and activity. The best-studied nitrogenase (containing molybdenum in its active site co-factor) consists of two component-proteins: Fe protein (NifH), a dimer that contains a [4Fe-4S] cluster bridged between the two subunits and the MoFe protein (NifDK), an $\alpha_2\beta_2$-heterotetramer that contains two metallocenters per $\alpha\beta$-subunit pair: a P-cluster located at each $\alpha/\beta$-subunit interface and FeMoco, one of the most complex co-factors in biology containing Mo, Fe, S, C and homocitrate, which is located within each $\alpha$-subunit (8). The minimal genetic requirements that can sustain nitrogenase activity depend on the physiology of the host organism, with, in general, more genes being necessary for *aerobic diazotrophs*. For example, the minimal gene set for full activity of *Klebsiella oxytoca* Mo nitrogenase in *E. coli* includes genes for the nitrogenase component proteins (the Fe protein and MoFe protein), encoded by *nifH*, *nifD* and *nifK*, a chaperone for correct folding of the Fe protein (*nifM*), electron transfer proteins (*nifF, J*), accessory factors and cofactor assembly proteins (*nifE, N, X, U, S, V, B, Q, W, T, Z, Y*) (Wang et al. 2013). The transfer of a minimal gene cluster containing 9 *nif* genes from *Paenibacillus* strain WLY78 to *E. coli* did result in some nitrogenase activity, but the introduction of additional genes was required to achieve 50% activity in this host (Li et al. 2016, Wang et al. 2013). To understand the requirements for nitrogenase activity in photosynthetic cyanobacteria, large gene clusters have been transferred from nitrogen-fixing non-heterocystous cyanobacteria to the non-diazotrophic model strain *Synechocystis sp.* PCC 6803 (Tsujimoto et al. 2018, Liu et al. 2018). This identified minimal gene clusters containing 25 essential *nif* genes and resulted in nitrogenase activity in the new host, which increased under reducing conditions. Interestingly, the oxygen tolerance of nitrogenase in *Synechocystis* increased when additional genes for an uptake hydrogenase were also introduced, thus providing an important pointer towards engineering nitrogen fixation in chloroplasts (Liu et al. 2018).

The development of synthetic biology tools has expedited the engineering of *nif* gene clusters, enabling refactoring of genes and inducible control of expression (Wang et al. 2013, Temme et al. 2012). Combinatorial design, assembly and optimization of parts have significantly facilitated the move from the expression of genes in bacterial polycistronic operons to the monocistronic expression format in which each gene is controlled by individual promoters and terminators, suitable for engineering in eukaryotes (Buren et al. 2017b, Smanski et al. 2014). However, even with vast combinatorial libraries, it is difficult to achieve a balanced expression of protein components in eukaryotes, particularly when targeting the proteins encoded by transgenes in the nucleus into organelles. This is particularly the case when considering the requirements for assembly and activity of nitrogenase, where stoichiometric expression of proteins in complexes and expression ratios of other components are extremely important. To simplify the expression in eukaryotes, a polyprotein strategy derived from a plant RNA virus has been utilized to assemble *nif* coding sequences into giant genes with each coding sequence flanked by TEV protease cleavage sites. This enables an operon-like expression of coding sequences from giant genes, allowing stoichiometric expression of individual proteins after cleavage with TEV protease. Using this methodology, the 18 gene *K. oxytoca nif* cluster was re-engineered into five giant genes, which maintained balanced levels of Nif components after cleavage with the protease and enabled diazotrophic growth (Yang et al. 2018).

A major breakthrough in engineering nitrogenase component proteins in eukaryotes was achieved after *nifH* and *nifM* from *Azotobacter vinelandii* were introduced into the nuclear genome of the yeast *Saccharomyces cerevisiae* and their protein products were targeted to the mitochondria. NifH/NifM co-expression was able to produce active nitrogenase Fe protein in aerobically grown yeast, even without the Nif-specific Fe-S cluster assembly proteins (NifS and NifU) (Lopez-Torrejon et al. 2016). Since the Fe protein is the most oxygen-sensitive component of nitrogenase, this demonstrated that mitochondria can provide a suitable environment to protect the enzyme from oxygen damage in aerobically grown yeast and that the native mitochondrial Fe-S cluster assembly machinery can successfully assemble and insert the Fe-S clusters required for Fe protein function. A parallel approach was taken in plants, whereby the *nifH* and *nifM* genes were introduced by plastid transformation into the tobacco chloroplast genome and the transplastomic plants were able to produce low but detectable levels of Fe protein activity when incubated under low oxygen conditions (10% $O_2$) (Ivleva et al. 2016). The low levels of activity observed under these conditions presumably reflect oxygen inactivation of the Fe protein in the chloroplast.

In contrast to the assembly of nitrogenase Fe protein, pathways for the maturation of the MoFe protein and the biosynthesis of its metalloclusters are complex, reflecting many challenges to the successful engineering of nitrogenase in eukaryotic organelles. Although, a number of proteins required for MoFe protein assembly and activity have been targeted to yeast mitochondria (Buren et al. 2017b) or transiently expressed in tobacco mitochondria (Allen et al. 2017), issues of solubility and stability have been observed. There is a particular issue with NifD, the α-subunit of the MoFe protein, which has been shown to be unstable in mitochondria, although the NifDK tetramer has been detected (Buren et al. 2017a). Another issue has arisen with NifB, an S-adenyl methionine dependent enzyme required for the synthesis of NifB-co, a key intermediate in the biosynthesis of FeMoco and the cofactors of alternative nitrogenases. *A. vinelandii* NifB was found to be insoluble in both yeast and tobacco mitochondria, even when co-expressed with NifU, NifS and FdxN which are involved in cluster assembly and the activity of NifB. However, thermophilic NifB from the methanogen *Methanocaldococcus infernus* was found to be soluble in the mitochondria of both yeast and plants and when purified from yeast was active in an *in vitro* FeMoco biosynthesis assay after reconstitution with Fe and S (Buren et al. 2017a). Furthermore, archaeal NifB proteins have been demonstrated to produce active NifB-co without further reconstitution when co-expressed in yeast with NifU, NifS and FdxN (Buren et al. 2019). These findings highlight the need to screen Nif components from diverse phylogenetic backgrounds in order to achieve stable and soluble expression of proteins required for nitrogen fixation in plant organelles.

We now know that yeast and plants can make the most oxygen labile component of nitrogenase, the Fe protein in different organelles and that this component is functional when combined with bacterially produced MoFe protein *in vitro*. We can also use *in vitro* reconstitution assays to systematically access the activities of Nif components expressed in organelles and resolve bottlenecks due to loss of functionality. However, we still do not know how many genes are required to support nitrogenase activity in eukaryotes. For example, do organelles accumulate the necessary metals and co-factors to support nitrogen fixation? It may be necessary to introduce transporters and storage proteins in order to traffic molybdenum into FeMoco, or make use of the alternative iron-only nitrogenase that requires fewer genes than the molybdenum system when reconstituted in *E. coli* (Yang et al. 2014). It is logical to expect that more genes will be required for nitrogen fixation in chloroplasts in order to provide the necessary oxygen protection. On the other hand, it may be feasible to utilize plant counterparts of Nif components in order to support nitrogenase activity. For example, although prokaryotic diazotrophs utilize specific electron transfer components as electron donors to nitrogenase, these can be replaced by plant-derived electron transport chains from both chloroplasts and mitochondria (Yang et al. 2017). In our view, although there will be bottlenecks along the way, the expression of active nitrogenase in eukaryotes is highly feasible. However, will this lead to crops in the field that fix substantial amounts of nitrogen without a yield penalty? Given that nitrogenase uses eight moles of ATP for every mole of ammonia generated, this is a significant carbon/energy cost. However, a similar argument can be made for legumes, which suffer no yield penalty when using biological nitrogen fixation, compared with growth on nitrate (Vance and Heichel 1991). It has also been inferred that cereals, such as rice grains have the capacity to provide the necessary carbon to support nitrogen fixation (Rosenblueth et al. 2018).

## 2.3  FUTURE CHALLENGES

Each of these approaches has scientific, management and funding challenges. There are a number of scientific bottlenecks that continue to exist, which have been discussed above. Once transgenic plants have been developed that have the ability to fix nitrogen in the laboratory, there will be significant challenges to moving the technology to the field. This may well end up representing the greatest challenge.

Regardless of the approach used, to solve these challenges, researchers need the level of funding for these projects to be increased and novel models of funding considered. Over $100B a year is spent on fertilizers, an investment of 1% would provide $1B of funding annually, which would allow the different approaches to this problem to be seriously addressed. In terms of $CO_2$ equivalents, the IPCC* recognizes that 1% of the N fertilizer is lost in the form of NOx. This would translate into a $CO_2$ equivalent of 300 Mmt at a current value of $4.5B annually. Finally, such an ambitious project requires a focused team with the management that understands these challenges. This approach will preclude using the typical short term grant funding and will require the kind of approach that allowed NASA to put a man on the moon in less than a decade.

## 2.4 REFERENCES

Allen, R.S., Tilbrook, K., Warden, A.C., Campbell, P.C., Rolland, V., Singh, S.P., et al. 2017. Expression of 16 nitrogenase proteins within the plant mitochondrial matrix. Frontiers in Plant Science 8: 287.

Bageshwar, U.K., Srivastava, M., Pardha-Saradhi, P., Paul, S., Gotahndapani, S., Jaat, R.S., et al. 2017. An environmentally friendly engineered Azotobacter strain that replaces a substantial amount of urea fertilizer while sustaining the same wheat yield. Applied and Environmental Microbiology 83: e00590-17.

Bali, A., Blanco, G., Hill, S. and Kennedy, C. 1992. Excretion of ammonium by a nifL mutant of *Azotobacter vinelandii* fixing nitrogen. Applied and Environmental Microbiology 58: 1711-1718.

Banba, M., Gutjahr, C., Miyao, A., Hirochika, H., Paszkowski, U., Kouchi, H., et al. 2008. Divergence of evolutionary ways among common sym genes: CASTOR and CCaMK show functional conservation between two symbiosis systems and constitute the root of a common signaling pathway. Plant and Cell Physiology 49: 1659-1671.

Barney, B.M., Eberhart, L.J., Ohlert, J.M., Knutson, C.M. and Plunkett, M.H. 2015. Gene deletions resulting in increased nitrogen release by *Azotobacter vinelandii*: application of a novel nitrogen biosensor. Applied and Environmental Microbiology 81: 4316-4328.

Beatty, P.H. and Good, A.G. 2011. Future prospects for cereals that fix nitrogen. Science 333: 416-417.

Boddey, R.M., Urquiaga, S., Alves, B.J.R. and Reis, V. 2003. Endophytic nitrogen fixation in sugarcane: present knowledge and future applications. Plant and Soil 252: 139-149.

Borlaug, N.E. 1972. The green revolution, peace and humanity. pp. 1951-1970. *In*: Haberman, F.W. (ed.). Nobel Lectures, Peace. Amsterdam: Elsevier.

Boyd, E.S. and Peters, J.W. 2013. New insights into the evolutionary history of biological nitrogen fixation. Frontiers in Microbiology 4: 201.

Brewin, B., Woodley, P. and Drummond, M. 1999. The basis of ammonium release in *nifL* mutants of *Azotobacter vinelandii*. Journal of Bacteriology 181: 7356-7362.

Bueno Batista, M. and Dixon, R. 2019. Manipulating nitrogen regulation in diazotrophic bacteria for agronomic benefit. Biochemical Society Transactions 47: 603-614.

Buren, S., Jiang, X., López-Torrejón, G., Echavarri-Erasun, C. and Rubio, L.M. 2017a. Purification and *in vitro* activity of mitochondria targeted nitrogenase cofactor maturase NifB. Frontiers in Plant Science 8: 1567.

Buren, S., Young, E.M., Sweeny, E.A., Lopez-Torrejon, G., Veldhuizen, M., Voigt, C.A., et al. 2017b. Formation of nitrogenase NifDK tetramers in the mitochondria of *Saccharomyces cerevisiae*. ACS Synthetic Biology 6: 1043-1055.

Buren, S. and Rubio, L.M. 2018. State of the art in eukaryotic nitrogenase engineering. FEMS Microbiology Letters 365: fnx274.

Buren, S., Pratt, K., Jiang, X., Guo, Y., Jimenez-Vicente, E., Echavarri-Erasun, C., et al. 2019. Biosynthesis of the nitrogenase active-site cofactor precursor NifB-co in *Saccharomyces cerevisiae*. Proceedings of the National Academy of Sciences USA 116: 25078-25086.

Burrill, T.J. and Hansen, R. 1917. Is symbiosis possible between legume bacteria and non-legume plants? Bulletin (University of Illinois (Urbana-Champaign campus) Agricultural Experiment Station); no 202.

Charpentier, M. and Oldroyd, G. 2010. How close are we to nitrogen-fixing cereals? Current Opinion in Plant Biology 13: 1-9.

Christiansen-Weniger, C. and Van Veen, J.A. 1991. $NH_4^+$-excreting Azospirillum brasilense mutants enhance the nitrogen supply of a wheat host. Applied and Environmental Microbiology 57: 3006-3012.

Curatti, L. and Rubio, L.M. 2014. Challenges to develop nitrogen-fixing cereals by direct nif-gene transfer. Plant Science 225: 130-137.

Cvetkovska, M., Orgnero, S., Hüner, N.P.A. and Smith, D.R. 2019. The enigmatic loss of light-independent chlorophyll biosynthesis from an Antarctic green alga in a light-limited environment. New Phytologist 222: 651-656.

Delaux, P.-M., Radhakrishnan, G. and Oldroyd, G. 2015. Tracing the evolutionary path to nitrogen-fixing crops. Current Opinion in Plant Biology 26: 95-99.

Diédhiou, I. and Diouf, D. 2018. Transcription factors network in root endosymbiosis establishment and development. World Journal of Microbiology and Biotechnology 34: 37.

Dixon, R., Cheng, Q., Shen, G.-F., Day, A. and Dowson-Day, M. 1997. *Nif* gene expression in chloroplasts: prospects and problems. Plant and Soil 194: 193-203.

Erisman, J.W., Sutton, M.A., Galloway, J., Klimont, Z. and Winiwarter, W. 2008. How a century of ammonia synthesis changed the world. Nature Geoscience 1: 636.

FAO. 2017. World fertilizer trends and outlook to 2020. Food and Agricultural Organisation of the United Nations (Rome). http://www.fao.org/3/a-i6895e.pdf.

Fay, P. 1992. Oxygen relations of nitrogen fixation in cyanobacteria. Microbiological Reviews 56: 340-373.

Fox, A.R., Soto, G., Valverde, C., Russo, D., Lagares, A., Zorreguieta, Á., et al. 2016. Major cereal crops benefit from biological nitrogen fixation when inoculated with the nitrogen-fixing bacterium *Pseudomonas protegens* Pf-5 X940. Environmental Microbiology 18: 3522-3534.

Gallon, J. 1992. Tansley review no. 44. Reconciling the incompatible: nitrogen fixation and oxygen. New Phytologist 122: 571-609.

Geddes, B.A., Ryu, M.-H., Mus, F., Garcia Costas, A., Peters, J.W., Voigt, C.A., et al. 2015. Use of plant colonizing bacteria as chassis for transfer of $N_2$-fixation to cereals. Current Opinion in Biotechnology 32: 216-222.

Geddes, B.A., Paramasivan, P., Joffrin, A., Thompson, A.L., Christensen, K., Jorrin, B., et al. 2019. Engineering transkingdom signalling in plants to control gene expression in rhizosphere bacteria. Nature Communications 10: 3430.

Gherbi, H., Markmann, K., Svistoonoff, S., Estevan, J., Autran, D., Giczey, G., et al. 2008. SymRK defines a common genetic basis for plant root endosymbioses with arbuscular mycorrhiza fungi, rhizobia, and Frankia bacteria. Proceedings of the National Academy of Sciences 105: 4928-4932.

Good, A.G. and Beatty, P.H. 2011. Fertilizing nature: a tragedy of excess in the commons. PLoS Biology 9: e1001124.

Good, A.G. 2018. Toward nitrogen-fixing plants. Science 359: 869-870.

Griesmann, M., Chang, Y., Liu, X., Song, Y., Haberer, G., Crook, M.B., et al. 2018. Phylogenomics reveals multiple losses of nitrogen-fixing root nodule symbiosis. Science 361: eaat1743.

He, J., Zhang, C., Dai, H., Liu, H., Zhang, X., Yang, J., et al. 2019. A LysM receptor heteromer mediates perception of arbuscular mycorrhizal symbiotic signal in rice. Molecular Plant 12: 1561-1576.

Hurek, T., Handley, L., Reinhold-Hurek, B. and Piché, Y. 2002. *Azoarcus* grass endophytes contribute fixed nitrogen to the plant in an unculturable state. Molecular Plant Microbe Interactions 15: 233-242.

IPCC. 2019. IPCC Intergovernmental Panel on Climate Change. https://www.ipcc.ch/.

Ivleva, N.B., Groat, J., Staub, J.M. and Stephens, M. 2016. Expression of active subunit of nitrogenase via integration into plant organelle genome. PLoS One 11: e0160951.

Kawaharada, Y., Kelly, S., Wibroe Nielsen, M., Hjuler, C.T., Gysel, K., Muszyński, A. et al. 2015. Receptor-mediated exopolysaccharide perception controls bacterial infection. Nature 523: 308.

Kelly, S., Radutoiu, S. and Stougaard, J. 2017. Legume LysM receptors mediate symbiotic and pathogenic signalling. Current Opinion in Plant Biology 39: 152-158.

Kelly, S.J., Muszyński, A., Kawaharada, Y., Hubber, A.M., Sullivan, J.T., Sandal, N., et al. 2013. Conditional requirement for exopolysaccharide in the mesorhizobium–lotus symbiosis. Molecular Plant Microbe Interactions 26: 319-329.

Ladha, J.K., Tirol-Padre, A., Reddy, C.K., Cassman, K.G., Verma, S., Powlson, D.S., et al. 2016. Global nitrogen budgets in cereals: a 50-year assessment for maize, rice, and wheat production systems. Scientific Reports 6: 19355.

Li, X.-X., Liu, Q., Liu, X.-M., Shi, H.-W. and Chen, S.-F. 2016. Using synthetic biology to increase nitrogenase activity. Microbial Cell Factories 15: 1-11.

Liu, C., Breakspear, A., Stacey, N., Findlay, K., Nakashima, J., Ramakrishnan, K., et al. 2019. A protein complex required for polar growth of rhizobial infection threads. Nature Communications 10: 2848.

Liu, C.-W., Breakspear, A., Guan, D., Cerri, M.R., Jackson, K., Jiang, S., et al. 2019. NIN acts as a network hub controlling a growth module required for rhizobial infection. Plant Physiology 179: 1704-1722.

Liu, D., Liberton, M., Yu, J., Pakrasi, H.B. and Bhattacharyya-Pakrasi, M. 2018. Engineering nitrogen fixation activity in an oxygenic phototroph. MBio 9: e01029-18.

Lopez-Torrejon, G., Jimenez-Vicente, E., Buesa, J.M., Hernandez, J.A., Verma, H.K. and Rubio, L.M. 2016. Expression of a functional oxygen-labile nitrogenase component in the mitochondrial matrix of aerobically grown yeast. Nature Communications 7: 11426.

Machado, H.B., Funayama, S., Rigo, L.U. and Pedrosa, F.Q. 1991. Excretion of ammonium by azospirillum-brasilense mutants resistant to ethylenediamine. Canadian Journal of Microbiology 37: 549-553.

Madsen, L.H., Tirichine, L., Jurkiewicz, A., Sullivan, J.T., Heckmann, A.B., Bek, A.S., et al. 2010. The molecular network governing nodule organogenesis and infection in the model legume Lotus japonicus. Nature Communications 1: 10.

Maillet, F., Poinsot, V., André, O., Puech-Pagès, V., Haouy, A., Gueunier, M., et al. 2011. Fungal lipochitooligosaccharide symbiotic signals in arbuscular mycorrhiza. Nature 469: 58-63.

Merrick, M. and Dixon, R. 1984. Why don't plants fix nitrogen? Trends in Biotechnology 2: 162-166.

Murray, J.D. 2011. Invasion by invitation: rhizobial infection in legumes. Molecular Plant Microbe Interactions 24: 631-639.

Mus, F., Crook, M.B., Garcia, K., Garcia Costas, A., Geddes, B.A., Kouri, E.D., et al. 2016. Symbiotic nitrogen fixation and the challenges to its extension to nonlegumes. Applied and Environmental Microbiology 82: 3698-3710.

Oertel, C., Matschullat, J., Zurba, K., Zimmermann, F. and Erasmi, S. 2016. Greenhouse gas emissions from soils—a review. Chemie der Erde-Geochemistry 76: 327-352.

Oldroyd, G.E.D. 2013. Speak, friend, and enter: signalling systems that promote beneficial symbiotic associations in plants. Nature Reviews Microbiology 11: 252-263.

Oldroyd, G.E.D. and Dixon, R. 2014. Biotechnological solutions to the nitrogen problem. Current Opinion in Biotechnology 26: 19-24.

Ortiz-Marquez, J.C.F., Do Nascimento, M., Dublan, M.d.l.A. and Curatti, L. 2012. Association with an ammonium-excreting bacterium allows diazotrophic culture of oil-rich eukaryotic microalgae. Applied and Environmental Microbiology 78: 2345-2352.

Ortiz-Marquez, J.C.F., Do Nascimento, M. and Curatti, L. 2014. Metabolic engineering of ammonium release for nitrogen-fixing multispecies microbial cell-factories. Metabolic Engineering 23: 154-164.

Pankievicz, V.C., do Amaral, F.P., Santos, K.F., Agtuca, B., Xu, Y., Schueller, M.J., et al. 2015. Robust biological nitrogen fixation in a model grass-bacterial association. The Plant Journal 81: 907-919.

Pedrosa, F.O., Oliveira, A.L.M., Guimarães, V.F., Etto, R.M., Souza, E.M., Furmam, F.G., et al. 2019. The ammonium excreting Azospirillum brasilense strain HM053: a new alternative inoculant for maize. Plant and Soil 451: 45-56.

Reinbothe, C., Bakkouri, M.E., Buhr, F., Muraki, N., Nomata, J., Kurisu, G., et al. 2010. Chlorophyll biosynthesis: spotlight on protochlorophyllide reduction. Trends in Plant Science 15: 614-624.

Reinhold-Hurek, B. and Hurek, T. 2011. Living inside plants: bacterial endophytes. Current Opinion in Plant Biology 14: 435-443.

Rogers, C. and Oldroyd, G.E.D. 2014. Synthetic biology approaches to engineering the nitrogen symbiosis in cereals. Journal of Experimental Botany 65: 1939-1946.

Rosenblueth, M., Ormeño-Orrillo, E., López-López, A., Rogel, M.A., Reyes-Hernández, B.J., Martínez-Romero, J.C., et al. 2018. Nitrogen fixation in cereals. Frontiers in Microbiology 9: 1794.

Santos, K.F.D.N., Moure, V.R., Hauer, V., Santos, A.R.S., Donatti, L., Galvão, C.W., et al. 2016. Wheat colonization by an Azospirillum brasilense ammonium-excreting strain reveals upregulation of nitrogenase and superior plant growth promotion. Plant and Soil 415: 245-255.

Schiessl, K., Lilley, J.L.S., Lee, T., Tamvakis, I., Kohlen, W., Bailey, P.C., et al. 2019. NODULE INCEPTION recruits the lateral root developmental program for symbiotic nodule organogenesis in medicago truncatula. Current Biology 29: 3657-3668.e5.

Setten, L., Soto, G., Mozzicafreddo, M., Fox, A.R., Lisi, C., Cuccioloni, M., et al. 2013. Engineering *Pseudomonas protegens* Pf-5 for nitrogen fixation and its application to improve plant growth under nitrogen-deficient conditions. PLoS One 8: e63666.

Shrestha, R.K. and Ladha, J.K. 1996. Genotypic variation in promotion of rice dinitrogen fixation as determined by nitrogen-15 dilution. Soil Science Society of America Journal 60: 1815-1821.

Smanski, M.J., Bhatia, S., Zhao, D., Park, Y., Woodruff, L.B.A., Giannoukos, G., et al. 2014. Functional optimization of gene clusters by combinatorial design and assembly. Nature Biotechnology 32: 1241-1249.

Smil, V. 2004. Enriching the Earth: Fritz Haber, Carl Bosch, and the Transformation of World Food Production. Cambridge, MA: MIT Press.

Soyano, T. and Hayashi, M. 2014. Transcriptional networks leading to symbiotic nodule organogenesis. Current Opinion in Plant Biology 20: 146-154.

Soyano, T., Shimoda, Y., Kawaguchi, M. and Hayashi, M. 2019. A shared gene drives lateral root development and root nodule symbiosis pathways in Lotus. Science 366: 1021-1023.

Temme, K., Zhao, D. and Voigt, C.A. 2012. Refactoring the nitrogen fixation gene cluster from Klebsiella oxytoca. Proceedings of the National Academy of Sciences 109: 7085-7090.

Tsujimoto, R., Kotani, H., Yokomizo, K., Yamakawa, H., Nonaka, A. and Fujita, Y. 2018. Functional expression of an oxygen-labile nitrogenase in an oxygenic photosynthetic organism. Scientific Reports 8: 7380.

Van Deynze, A., Zamora, P., Delaux, P.-M., Heitmann, C., Jayaraman, D., Rajasekar, S., et al. 2018. Nitrogen fixation in a landrace of maize is supported by a mucilage-associated diazotrophic microbiota. PLOS Biology 16: e2006352.

Van Dommelen, A., Croonenborghs, A., Spaepen, S. and Vanderleyden, J. 2009. Wheat growth promotion through inoculation with an ammonium-excreting mutant of Azospirillum brasilense. Biology and Fertility of Soils 45: 549-553.

van Velzen, R., Holmer, R., Bu, F., Rutten, L., van Zeijl, A., Liu, W., et al. 2018. Comparative genomics of the nonlegume Parasponia reveals insights into evolution of nitrogen-fixing rhizobium symbioses. Proceedings of the National Academy of Sciences 115: E4700-E4709.

Vance, C.P. and Heichel, G.H. 1991. Carbon in $N_2$ fixation: limitation or exquisite adaptation. Annual Review of Plant Physiology and Plant Molecular Biology 42: 373-390.

Vedalankar, P. and Tripathy, B.C. 2019. Evolution of light-independent protochlorophyllide oxidoreductase. Protoplasma 256: 293-312.

Wang, L., Zhang, L., Liu, Z., Zhao, D., Liu, X., Zhang, B., et al. 2013. A minimal nitrogen fixation gene cluster from Paenibacillus sp. WLY78 enables expression of active nitrogenase in Escherichia coli. PloS Genetics 9: e1003865.

Wang, X., Yang, J.-G., Chen, L., Wang, J.-L., Cheng, Q., Dixon, R., et al. 2013. Using synthetic biology to distinguish and overcome regulatory and functional barriers related to nitrogen fixation. PLoS One 8: e68677.

Yang, J., Xie, X., Wang, X., Dixon, R. and Wang, Y.P. 2014. Reconstruction and minimal gene requirements for the alternative iron-only nitrogenase in Escherichia coli. Proceedings of the National Academy of Sciences 111: E3718-E3725.

Yang, J., Xie, X., Yang, M., Dixon, R. and Wang, Y.P. 2017. Modular electron-transport chains from eukaryotic organelles function to support nitrogenase activity. Proceedings of the National Academy of Sciences 114: E2460-E2465.

Yang, J., Xie, X., Xiang, N., Tian, Z.X., Dixon, R. and Wang, Y.P. 2018. Polyprotein strategy for stoichiometric assembly of nitrogen fixation components for synthetic biology. Proceedings of the National Academy of Sciences 115: E8509-E8517.

Yanni, Y.G. and Dazzo, F.B. 2010. Enhancement of rice production using endophytic strains of Rhizobium leguminosarum bv. trifolii in extensive field inoculation trials within the Egypt Nile delta. Plant and Soil 336: 129-142.

# Nitrogen Cycle in Agriculture: Biotic and Abiotic Factors Regulating Nitrogen Losses

María López-Aizpún[1], Antonio Castellano-Hinojosa[2],
Jesús González-López[3], Eulogio J. Bedmar[4], Nadine Loick[5],
Harry Barrat[5], Yan Ma[6], Dave Chadwick[6] and Laura M. Cardenas[1,*]

## 3.1 INTRODUCTION

Nitrogen (N) is a key element for life since it is a basic component of proteins and nucleic acids. Molecular nitrogen ($N_2$) accounts for approximately 78% of the Earth's atmosphere. However, this N chemical form is not usable by most organisms due to the significant amount of energy required to break the triple bond between the two N atoms. Only certain specialized microorganisms (rhizobial and actinorhizal symbioses, free-living and symbiotic cyanobacterial fixers and free-living and symbiotic heterotrophic bacteria) have developed the ability to convert atmospheric $N_2$ into ammonium ($NH_4^+$) and subsequently other reactive nitrogen species (Nr)[*] via biological nitrogen fixation (BNF) (Vitousek et al. 2013). For millennia, little Nr was accumulated in environmental reservoirs as Nr formation was balanced by deep sedimentation and the conversion of Nr back to $N_2$ by denitrification, anammox and other processes (Galloway et al. 2013). Consequently, N has often been the limiting factor to increase production for both terrestrial and aquatic ecosystems (Vitousek et al. 2010). In an N-limited world, practices such as crop rotation, use of legumes for BNF or application of livestock manure and guano have been traditionally used to increase crop production and thus meet the growing demand for food and ensure the population's sustenance (Galloway et al. 2013).

---

[*] Nr includes all biologically active, photochemical reactive, and radiatively active N compounds in the atmosphere and biosphere of Earth: inorganic reduced forms of N (e.g., $NH_3$ and $NH_4^+$), inorganic oxidized forms (e.g., $NO_x$, $HNO_3$, $N_2O$, and $NO_3^-$), and organic compounds (e.g., urea, amines, proteins).

---

[1] Rothamsted Research, North Wyke, Okehampton, Devon, EX20 2SB, UK.
[2] University of Florida IFAS Southwest Florida Research and Education Center, 2685 State Rd 29N, Immokalee, FL.
[3] Department of Microbiology, Faculty of Pharmacy, University of Granada. Campus Cartuja, 18071-Granada, Spain.
[4] Department of Soil Microbiology and Symbiotic Systems, Estación Experimental del Zaidín, CSIC. P.O. Box 419. 18080-Granada, Spain.
[5] Rothamsted Research, North Wyke, Okehampton, Devon, EX20 2SB, UK.
[6] School of Natural Sciences, Bangor University, Bangor, Gwynedd, LL57 2UW, UK.
[*] Corresponding author: laura.cardenas@rothamsted.ac.uk

However, in the last two centuries, human actions have dramatically altered the natural functioning of the N-cycle as a consequence of changes in the energy and food production patterns during the eighteenth century Industrial Revolution and the 19th and 20th centuries Agricultural Revolutions. In this sense, the change from wood and other bio-fuels to fossil fuels during the Industrial Revolution resulted in a remarkable increase in the creation of unintended Nr, specifically the formation of N oxides ($NO_x$) (Hertel et al. 2012). On the other hand, the growing demand for food motivated the emergence of the industrial processes of synthesis of N-fertilizers (Haber-Bosch and Whöhler processes) (Erisman et al. 2008). It is estimated that more than a quarter of the world population over the past century has been fed by synthetic N-fertilizers (Ramankutty et al. 2018). Therefore, the benefits that N-fertilizers represent for the sustenance of the population at the global scale are undeniable. However, the nitrogen-use efficiency (NUE) in agriculture (the amount of nitrogen retrieved in food produced per unit of nitrogen applied) is extremely low (EU Nitrogen Expert Panel, 2015). Volatilization, leaching, soil erosion and denitrification processes result in the loss of most of the applied N, leading to field recoveries that rarely exceed 50% (Smil 2011). Figure 3.1 shows an outline of the N-cycle in agriculture.

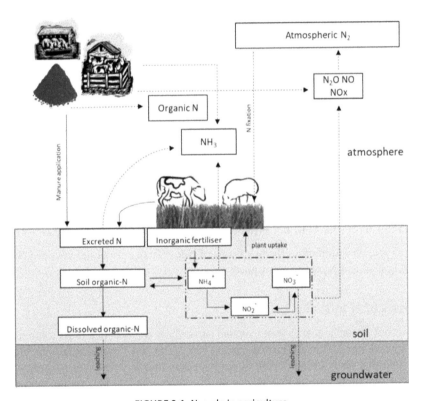

**FIGURE 3.1** N-cycle in agriculture.

Until the emergence of the industrial processes of synthesis of N-fertilizers, BNF and denitrification [the sequential reduction of nitrate ($NO_3^-$) to $N_2$] processes had similar yields, about 110 tons of $NH_4^+$ produced from $N_2$ compared to 108 tons of $NO_3^-$ eliminated as nitrous oxide ($N_2O$) or $N_2$ by denitrification (Gruber and Galloway 2008). However, excessive and repeated applications of N-fertilizers have increased the amount of Nr to about 120-160 Tg N $yr^{-1}$ (Fowler et al. 2015). Consequently, denitrification cannot eliminate the excess Nr produced, resulting in N loss to the environment. This N lost has led to several environmental problems that are not only related to $NH_3$ and $N_2O$ emissions to the atmosphere and $NO_3^-$ and organic N compounds losses to water but also to the waste of energy (and hence the associated greenhouse gas emission) that was invested in N production. Amongst the effects are increases in the emission of greenhouse gases, nitrate contamination of waters and sediments, acidic deposition, eutrophication and loss of biodiversity.

The EU Nitrogen Expert Panel published a report (EU Nitrogen Expert Panel 2015) recommending boundaries for the use of the indicator NUE. This indicator considers inputs and outputs as well as the changes in soil stock and the three main targets: maximum NUE (90%), maximum N surplus (80 kg N ha$^{-1}$) and minimum N offtake (80 kg N ha$^{-1}$). The desired NUE values are between 50 and 90%. Below these values, the risk of pollution increases, above them, soil stocks are mined. Cardenas et al. (2019) applied this approach on 5 field grassland sites and found that increasing N rates increased NUE; however, N surplus was higher than the 80 kg N ha$^{-1}$ recommended by the EU Expert Panel, resulting in increased N$_2$O emissions.

In agricultural soils, the application of inorganic N-fertilizers leads to the interaction of multiple factors and processes which are mainly associated with changes in soil physicochemical properties, emission of greenhouse gases and microbial ecology. After the application of an N-fertilizer, the microbial processes of nitrification and denitrification are responsible for the main reactions driving the conversion of NH$_4^+$ and NO$_3^-$ to N$_2$O respectively. As suggested by recent meta-analyzes (Zhou et al. 2017; Ouyang et al. 2018), combating the negative impacts of increasing N$_2$O fluxes poses considerable challenges and will be ineffective without incorporating microbial-regulated N$_2$O processes into mitigation strategies.

The world population is expected to increase to 9.7 billion by 2050 (WPP 2019), thereby putting extreme pressure on the land for agriculture use that is forecast to increase N losses to the environment with catastrophic consequences for the future of the human race (Shibata et al. 2015). This generation is facing the great challenge of providing the understanding of processes to optimize N use and develop strategies and new technologies to avoid losses and remove those that are avoidable.

This chapter aims to provide an overview of the nitrogen cycle in agriculture and the N losses derived from the use of N-fertilizers as well as give a comprehensive description of the factors that drive these losses.

## 3.2 LOSSES OF N TO THE ENVIRONMENT FOLLOWING N APPLICATION

The following section summarizes the key forms of N loss from agriculture systems: NO$_3^-$ to water and NH$_3$, NO$_x$, N$_2$O and N$_2$ to the atmosphere.

### 3.2.1 Losses of N to Water

Large amounts of N can be lost to watercourses particularly as NO$_3^-$ due to its solubility (ca. 30%) and its negative charge (IPCC 2006) but nitrite (NO$_2^-$) and NH$_4^+$ can also be lost. Oenema et al. (2009) reported that Nr released during agriculture activities as NO$_3^-$ was roughly equal to that emitted as NH$_3$, highlighting the importance of this Nr income into the environment. Factors that contribute to NO$_3^-$ losses from agricultural land include over-fertilization of crops, prolonged periods of bare soil during the winter drainage period, an excessive number of grazing livestock per unit area of land and consequent overloading of fields with manure and inefficient use of manures as well as lack of synchronicity between crop demand and N supply (Jarvis and Menzi 2004, Schröder et al. 2010, Cardenas et al. 2011).

Environmental concerns in aquatic ecosystems due to N losses are mainly related to eutrophication (Voss et al. 2013) and water acidification (Camargo and Alonso 2006). Gaseous losses from leached N as N$_2$O also occur (Tian et al. 2019). National and international legislation aiming at improving water quality restrict application rates and timing of manures in order to limit NO$_3^-$ concentrations in surface and groundwater. The European Commission (EC) Nitrates Directive requires areas of land that drain into waters polluted by nitrates to be designated as Nitrate Vulnerable Zones (NVZs).

## 3.2.2    Losses of N to the Atmosphere

After being applied to the soil, N can be lost to the atmosphere in the form of $NH_3$, $NO_x$, $N_2O$, nitrous acid (HONO) $N_2$ and organic N with some of these forms having detrimental impacts on the air quality due to the formation of pollutants such as particulate matter (PM), ozone ($O_3$) and impacting climate change through the emissions of greenhouse gases ($N_2O$).

### 3.2.2.1    Ammonia

Ammonia is the main unintended gaseous compound released during agricultural activities (Jarvis et al. 2011). $NH_3$ emissions are attributed to two main sources; those from urea-based fertilizer to produce crops and $NH_3$ emissions from managed animal manure and urine and dung deposited by grazing livestock. In 2015, the EU-28 agricultural sector emitted a total of 3,751 kilotons (kt) of $NH_3$ and was responsible for 94% of total $NH_3$ emissions across the region. Livestock excreta (from livestock housing, manure storage, urine and dung deposition in grazed pastures or after manure spreading into land) were responsible for 52% of agricultural ammonia emissions, whereas urea-based fertilizer applications accounted for the rest (Eurostat 2017). Inventories from China (Zhou et al. 2015) and North America (Bittman and Mikkelsen 2009) report that livestock is the dominant contributing source of $NH_3$ emission, proving this is a global-scale problem. Ammonium containing fertilizers such as ammonium nitrate (AN) and urea, manure, livestock slurry, urine and dung deposited during grazing, release $NH_3$ within a short time after application. Misselbrook et al. (2014) estimated emissions of $NH_3$-N of 1.9 and 24.7% for AN and urea, respectively, while emissions from slurry application represent were 22.1% of the applied N.

### 3.2.2.2    Nitrous Oxide

Along with $NH_3$, agricultural activities are a source of $N_2O$, a powerful greenhouse gas accounting for approximately 6% of the current global warming (WMO 2019). It has a long atmospheric residence time (minimum lifetime of about 20 years) and a global warming potential (GWP) 296 times higher than carbon dioxide ($CO_2$) (Erisman et al. 2011). Furthermore, the long residence time of $N_2O$ in the atmosphere provides the opportunity for its transport to the stratosphere where, after photolysis producing NO, removes $O_3$.

According to data compiled by Snyder et al. (2009) and Fowler et al. (2015), atmospheric concentrations of $N_2O$ have risen from ~270 ppb during the pre-industrial era to 332 ppb in 2019 (https://www.n2olevels.org). Davidson and Kanter (2014) quantified the global natural $N_2O$ emissions from 10 to 12 Tg N $yr^{-1}$, whereas about 5.3 Tg N $yr^{-1}$ were considered as anthropogenic. Although several sources might contribute to this budget, agriculture is the largest one, representing 60% of such emissions (Syakila and Kroeze 2011).

In agriculture, $N_2O$ is mainly emitted from fertilized soils and animal excreta. N-fertilizer application, whether organic or synthetic, results in $N_2O$ emissions due to the transformation of N compounds added to the soil (Aguilera et al. 2013). Thus, $N_2O$ emissions from agricultural soils are mainly a function of N applied as fertilizer. $N_2O$ is produced in soils via several processes: (i) nitrification that has been reported as autotrophic ($NH_4^+$ oxidation) and heterotrophic (organic N oxidation) (Zhang et al. 2018), (ii) denitrification due to the incomplete reduction of $NO_3^-$ under $O_2$ limiting conditions (Attard et al. 2011), (iii) nitrifier denitrification (Zhu et al. 2013) and (iv) chemodenitrification as a non-biological process (Stanton et al. 2018) (Figure 3.2). Emissions through nitrification and denitrification account for up to 70% of the annual $N_2O$ emitted worldwide (Butterbach-Bahl et al. 2013). The production of $N_2O$ and the processes driving this production are controlled by several abiotic and biotic factors (see comprehensive reviews by (Baggs and Philippot 2010, Syakila and Kroeze 2011, Butterbach-Bahl et al. 2013, Hu et al. 2015). These factors are described in detail in the next section. Besides $N_2O$ production, recent evidence of $N_2O$ consumption in grasslands via denitrification have been reported in the literature (Hu et al. 2015a).

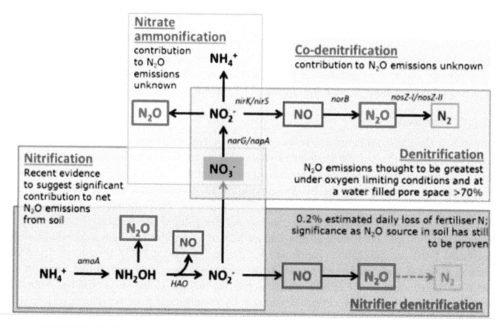

**FIGURE 3.2** N$_2$O formation via microbial terrestrial pathways. Modified from Baggs (2008).

### 3.2.2.3  Nitric Oxide

In addition to N$_2$O, agricultural soils are a dominant source of nitric oxide (NO) emissions (IPCC 2007, Ravishankara et al. 2009). While NO is not a GHG as such, it catalyzes the formation of ground-level ozone affecting human health and vegetation (Crutzen 1981) and takes part in the formation of acid rain and the eutrophication of semi-natural ecosystems. Like N$_2$O, NO is produced in soils predominantly by nitrification and denitrification (Butterbach-Bahl et al. 2013, Pilegaard 2013). With typical amounts of NO produced of between 1-4% of the oxidized NH$_4^+$ in well-aerated soils (Hutchinson and Brams 1992), most studies indicate that emitted NO is mainly produced from hydroxylamine (NH$_2$OH) during nitrification by ammonium oxidizers (Pilegaard 2013). During denitrification, NO is a direct intermediate and Remde et al. (1989) showed that NO production rates are one to two orders of magnitude larger under anaerobic than under aerobic conditions (Figure 3.2). However, NO is a very reactive gas as well as being toxic to most organisms (Richardson et al. 2009).

### 3.2.2.4  Nitrogen Gas

In contrast to N$_2$O and NO, a considerable fraction of the applied N is likely to be lost as N$_2$. Emissions of the inert gas N$_2$ are important as it represents an inefficient use of N sources. Measurements of N$_2$ emissions are challenging due to the large atmospheric background but represent a financial loss to the farmer. A number of techniques have been developed to measure emissions of N$_2$, allowing indirect determination via inhibiting N$_2$O reduction to N$_2$ using acetylene; direct determination using the He/O$_2$ atmosphere technique or use of $^{15}$N isotopes. Results from laboratory studies using the He/O$_2$ technique show losses of up to 70% of the N applied when conditions are optimal for denitrification (Cardenas et al. 2017), so they could be an important N loss that should be considered when assessing the efficiency of N use. Ratios of N$_2$O/N$_2$ can be used to estimate N$_2$ emissions, for example, values obtained in optimal denitrifying conditions after application of nitrate and glucose are reported by Cardenas et al. (2003) and Bol et al. (2003) range between 1.8 and 5.0. Values from the slurry application are generally < 1 (Cardenas et al. 2007). This implies that there is more potential for further reduction of $N_2O$ to $N_2$ with the application of organic fertilizers compared to mineral fertilizer.

### 3.2.2.5   Organic Nitrogen

Agriculture activities are also a source of organic N (i.e., free amino acids and proteinaceous compounds such as urea and amines (Huang et al. 2009). These compounds can be released both in the application of organic fertilizers mainly in the form of urea (Zhang et al. 2012) and in animal husbandry in the form of amines (Ge et al. 2011). Although these compounds are known to have a detrimental impact on air quality and have implications in the N biogeochemistry (Cape et al. 2001) little attention has been paid to these compounds and the organic fraction is less known and less understood in the atmospheric N-cycle (Fowler et al. 2015).

### 3.2.2.6   N Loss Accounting

The IPCC recommends default values to estimate direct and indirect losses of N from agricultural soils as $NH_3$ + NO and $N_2O$ (IPCC 2019). Emissions of $NH_3$ + NO are accounted for under the category volatilization in the inventory, as they contribute to $N_2O$ emissions indirectly. Direct emissions of $N_2O$ are considered for losses from the soil due to mineral and organic fertilizer application, incorporation of crop residues and cultivation of organic soils. The default values provided by the IPCC to estimate emissions are, however, general and not necessarily applicable to all climates, soil types and management practices. Country specific values need to be developed to improve the accuracy of estimates of losses. This has been applied in the UK with new $N_2O$ emission factors at high spatial disaggregation (Chadwick et al. 2018). Losses arise from synthetic fertilizer application to arable crops and grassland, grazing livestock due to their urine and dung deposited to the soil and their manure that is managed and later applied to the soil. Emissions also include crop residues incorporated to soil and cultivated histosols. The losses of NO, $N_2O$ and $NH_3$ in the UK for 2016 are shown in Table 3.1, representing 22% of the total synthetic fertilizer applied (Brown et al. 2018).

**TABLE 3.1** Losses of N to the atmosphere and the water in the UK for 2016 after fertilizer application

|  | N Amount | Units |
| --- | --- | --- |
| **Total fertilizer N input** | 1,090,343,900.89 | kg N yr$^{-1}$ |
| **Total N$_2$O** | 30.42 | kt N$_2$O-N |
| **Total NH$_3$** | 200.88 | kt NH$_3$-N |
| **Total NO$_x$** | 3.44 | kt NO$_x$-N |
| **Total N leached** | 3.07 | kt NO$_3$-N |
| **Total N losses** | 237.80 | kt N |
|  | 237,804,222.85 | kg N |

It is important to note that emissions of different gases are differently influenced by both biotic and abiotic factors. Furthermore, factors interact making it difficult to estimate emissions purely based on individual controlling factors. Figure 3.3 shows a modified scheme from Firestone and Davidson (1989) attempting to illustrate these interactions. In the following section, both abiotic and biotic main factors influencing N emissions are described in detail.

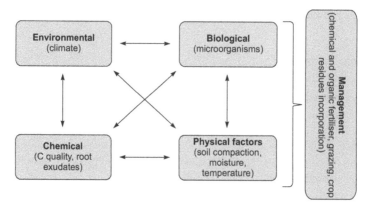

**FIGURE 3.3** Interactions between factors regulating N emissions. Modified from Firestone and Davidson (1989).

## 3.3 BIOTIC AND ABIOTIC FACTORS INFLUENCING N LOSSES

### 3.3.1 Abiotic Factors

#### 3.3.1.1 Soil pH

Soil pH is one of the most important factors controlling rates and product accumulation from nitrification, nitrate ammonification and denitrification. Moreover, it has an indirect effect on organic carbon and mineralized N with both becoming reduced under acidic conditions (pH < 7). Although the critical threshold for nitrification is pH 5.0, it has been shown to occur at pH 4.5 due to acid-adapted nitrifiers (Bouwman 1990), which makes it possible for the production of $N_2O$ in acid soils (Martikainen and Boer 1993). Denitrification reductases are inhibited at a pH lower than 7, especially nitrous oxide reductase, which converts $N_2O$ to $N_2$ (Knowles 1982). This reversible inhibitory effect seems to be originated by the formation of protonated species of the toxic nitrous acid ($HNO_2$) (Glass and Silverstein 1999). Nevertheless, denitrification can still occur in acid soils at pH values as low as 3.5 (Saggar et al. 2013). A number of studies have shown that optimum pH for the maximum conversion to $N_2$ ranges from 7.0 to 8.0, demonstrating that lower pH values retard the activity of denitrification enzymes, particularly nitrous oxide reductase. Also, it has been reported that the lower the pH, the more likely $N_2O$ will be produced without further reduction to $N_2$ (Sahrawat and Keeney 1986). For example, Blackmer and Bremner (1978) recorded an inhibition of $N_2O$ reduction of 60% at pH 6.8 ($N-NO_3^-$ 10 µg $g^{-1}$) and 88% at pH 5.7. Although it is true that a lower pH will more likely result in a higher $N_2O{:}N_2$ ratio, in a more recent study Šimek and Cooper (2002a) pointed out that it is misleading to suggest an optimum pH range without specifying the denitrifier community. The optimum denitrifying enzyme activity (DEA) might be acidic depending on the organisms responsible and their associated natural soil conditions (Šimek et al. 2002b). For example, Scholefield et al. (1997) incubated soil cores at varying pH levels 5.1 (natural conditions) to 9.4 and denitrification decreased with pH > 5.1. Furthermore, it is important to note that pH is both a distal and proximal control. Over the long term, it will determine the community structure and therefore the absolute denitrification rate, but the ratio of denitrification products ($N_2O/N_2$) is regulated by immediate (proximal) pH conditions (Čuhel and Šimek 2011).

#### 3.3.1.2 Soil Temperature

Temperature controls biochemical reaction rates, which typically increase with temperature. This is also the case for denitrification in temperate and hot climates. Dorland and Beauchamp (1991) conducted laboratory experiments measuring denitrification rates from −2 to 25°C and found that the rate decreased with temperature. Similar to pH variations, soil denitrifying microorganisms seem to be adapted to the naturally occurring soil temperatures (Powlson et al. 1988). Temperature also directly affects dissolved oxygen concentrations (DO) (higher temperature results in less DO). Veraart et al. (2011) measured denitrification's dependence on temperature during microcosm and field experiments in freshwater ecosystems, concluding that the rate of increase was exponential with 1°C temperature rise leading to 24% to 28% rise in the denitrification rate. This has future implications as climate change is predicted to warm temperate regions and increase the frequency of flooding, which could result in much higher denitrification rates and a greater release of $N_2O$ (Stocker 2014).

### 3.3.1.3   Soil Moisture and Oxygen Availability

Soil moisture impacts $N_2O$ emissions because it affects temperature and oxygen concentrations. Oxygen concentration determines whether the predominant N pathway in soils is $O_2$ limited or aerobic (i.e., denitrification and nitrification, respectively) through regulating the reactions of oxidation and reduction (Bollmann and Conrad 1998). The main factors determining $O_2$ concentration are the soil water content and the $O_2$ consumption by plant roots and microorganisms (respiration). Soil moisture can directly or indirectly influence denitrification by i) providing a suitable environment for microbial growth and activity, ii) preventing the supply of $O_2$ to microsites by filling soil pores, iii) releasing available C and N substrates during wetting and drying cycles and iv) through a provision of a diffusion medium through which substrates and products are moved to and away from soil microorganisms. It has been shown that after rainfall, irrigation and flooding, the denitrification rate increases due to the decrease of $O_2$ diffusion into the soil (Ruser et al. 2006, Sanchez-Rodriguez et al. 2019). A closer relationship was found between the water-filled pore space (WFPS) and $N_2O$ emission since this value also takes total pore space into account (Danielson and Sutherland 1986). $N_2O$ production from nitrification typically occurs within the range of 30-60% WFPS and denitrification dominates in wet soils with WFPS > 70% (Braker and Conrad 2011, Hu et al. 2015b). However, it has been shown that in arable systems, this threshold value can also be higher (Ruser et al. 2006). The aerobic denitrification is mediated by facultative anaerobes which under low oxygen concentrations, typically above 60% WFPS, respire using $NO_3^-$ and other reduced N species as alternative electron acceptors and carbon (C) as an electron donor to synthesize adenosine triphosphate (ATP). Naturally, this process is limited by $NO_3^-$ concentrations and available carbon, as the communities is heterotrophic. Moreover, high $NO_3^-$ concentrations inhibit the conversion of $N_2O$ to $N_2$, as $NO_3^-$ is a preferred electron acceptor over $N_2O$, resulting in the production of nitrate reductase over that of nitrous oxide reductase (Baggs et al. 2003). Even though $O_2$ is the most suitable electron acceptor for respiration, aerobic denitrification is possible, and some denitrifiers such as *Pseudomonas* sp. have proven to produce $N_2$ under hypoxic conditions (Zhu et al. 2012). Continuous aerobic conditions seem to suppress the production of nitrous oxide reductase (NosZ), thus the process produces a higher ratio of $N_2O$: $N_2$ (Takaya et al. 2003). Aerobic denitrification is rarely considered as a key source of $N_2O$, despite the possibility of high concentrations of denitrifiers' resistance to hypoxic conditions existing within natural environments (Patureau et al. 2000, Baggs 2011).

### 3.3.1.4   Soil Carbon Content

Carbon affects N cycling, particularly denitrification as it acts as an electron donor for this reductive process. Increasing soil organic C enhances nitrification and denitrification reactions (Saggar et al. 2013) because it can stimulate microbial growth and activity and also provide the organic C needed by soil denitrifiers (Cameron et al. 2013). It is usually accepted that increasing C supply decreases the $N_2O$: $N_2$ ratio (Cameron et al. 2013, Saggar et al. 2013). Whether C is applied with manure, urea and crop residues or released in the root exudates, C and its availability is a major control factor. The quality of the carbon compounds determines its availability, the more available, the larger the potential for further reduction of $N_2O$ to $N_2$ (Morley and Baggs 2010 and references therein).

Sources of C such as glucose and ethanol stimulate denitrification whilst succinate and methanol are less stimulatory. Morley and Baggs (2010) found a close relationship between the C source and the availability of oxygen in the soil; in the case of slurries, $O_2$ reduction stimulated denitrification while carbohydrates promoted dissimilatory nitrate reduction to ammonium (DNRA). The amount of NO emissions depends on the area/soil volume that received $NO_3^-$ and C fertilizer, while the scale of $N_2O$ and $N_2$ emissions depends on the amount of the applied $NO_3$ and the available C (Loick et al. 2017). Though the amounts of emitted NO were low (< 1% of applied N) and the contribution of this gas to the total gaseous emissions of N being negligible, with mitigation strategies reducing emissions of $N_2O$, NO will become of more interest in the future and different factors influencing its emission will need to be considered and incorporated into mitigation strategies.

### 3.3.1.5  C:N Ratio of Applied Substrates

The ratio C:N of applied substrates has also been used as an indicator of the process rate. N transformations in soils include two important biological processes: immobilization (or assimilation), this is the uptake of N by microorganisms and its conversion into organic N, and mineralization (or ammonification), this is the conversion of organic N to $NH_4^+$. The balance between mineralization and immobilization depends on the soil C:N ratio and the residues added. Soil and residues with a C:N ratio lower than 30:1 present dominance of mineralization over the immobilization, and the available N can be absorbed by plants or used in microbial processes. The presence of a high C:N ratio in the soil may increase the immobilization of the N-fertilizer applied (Baggs et al. 2000), thus decreasing the denitrification reactions and $N_2O$ emissions. When a low C:N ratio is present in the soil, N immobilization is reduced and more N will be available for nitrification and denitrification processes, thus increasing $N_2O$ emissions (Baggs et al. 2000). In this sense, Baggs et al. (2000) found a strong influence of the C:N ratio of applied crop residues to the UK soil. In the case of grass/clover swards and Italian ryegrass (with ratios 13:1 and 25:1, respectively) emissions of $N_2O$ were similar. When incorporating lettuce residues (ratio 7.5:1), emissions were larger probably due to the promotion of mineralization; with larger ratios (38:1) as in the case of winter wheat straw, emissions decreased probably due to immobilization of N. Other ratios such as the already mentioned $N_2O:N_2$ (or $N_2:N_2O$) have been used to indicate the source process of emissions. Reduction to $N_2$ is the only sink of $N_2O$ in the troposphere.

### 3.3.1.6  Other Soil Physical Characteristics

A soil network of pores and their connectivity ultimately determines water and gas movement, mesofauna and microfauna movement, the rooting structure of plants and the cycling of important nutrients like N and C. Bulk density, which is the amount of dry matter (g) in a $cm^3$ is typically used as a shorthand to describe pore structure, with high bulk densities greater than 1.6 g $cm^{-3}$ resulting in less space for water and air to move and even restrictive root growth.

If the pore spaces are small and spaced far from each other because of compaction or a soil texture with a small particle size like clay, then anaerobic conditions are more likely, encouraging denitrification. Soils that more aerated, maybe because of soil texture with a larger particle size like sand, are more aerobic encouraging nitrification. For example, Harrison-Kirk et al. (2013) compared $N_2O$ emissions between a silt-loam and a clay loam after rewetting the soil and they found that a smaller pore structure produced greater emissions. Another good example of how soil porosity affects $N_2O$ emissions is the adoption of no-till agriculture. In the first years of no-tillage, the soil's bulk density can increase and therefore the emissions can be greater due to smaller and less connected pores. However, as the soil develops from organic matter build-up, root growth and mesofauna movement, the connectivity increases and the resulting $N_2O$ emissions decrease (Ball 2013, Zhao et al. 2016).

## 3.3.2   Biotic Factors

Although N emissions are influenced by abiotic factors as described above, soil N fluxes are the result of biological processes and, therefore, primarily driven by microbial pathways which are controlled at the molecular level (Singh et al. 2010).

### 3.3.2.1   Bacterial and Archaeal Nitrifiers

Chemolithoautotrophic microorganisms are responsible for the oxidation of $NH_4^+$ to $NO_3^-$ with $NO_2^-$ as an intermediate under aerobic conditions. $NH_4^+$ is converted via $NH_2OH$ to $NO_2^-$ by nitrification. $NO_2^-$ is then further oxidized to $NO_3^-$. $N_2O$ can be formed by two biochemical pathways (Figure 3.4): Firstly, as a bio-product during the $NH_4^+$ oxidation, hydroxylamine is spontaneously decomposed to $N_2O$. This process is regarded as the main source of $N_2O$ from nitrification. Secondly, it can be formed by the so-called nitrifier denitrification (Kool et al. 2011, Wrage-Mönnig et al. 2018), where $N_2O$ is an intermediate of the reduction of $NO_2^-$ to $N_2$.

The most studied groups of nitrifiers are the ammonium oxidizing bacteria (AOB) and the nitrite oxidizing bacteria (NOB). The AOB are classified into three genera based on their bacterial 16S ribosomal RNA gene sequences (Erguder et al. 2009): *Nitrosomonas*, *Nitrosospina* and *Nitrosococcus*. Fewer studies have been done on the classification of NOB, which are classified into four genera: *Nitrobacter*, *Nitrospina*, *Nitrococcus* and *Nitrospira* (Daims et al. 2016).

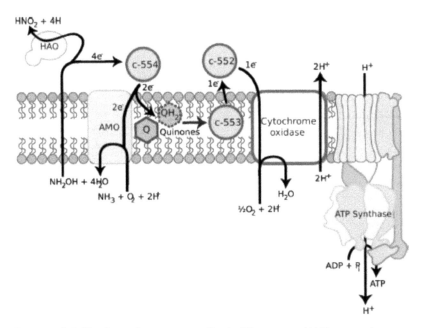

**FIGURE 3.4** Enzymes of nitrification and energy generation in *Nitrosomonas* (AMO: ammonium monooxygenase, HAO hydroxylamine oxidoreductase, c: cytochromes (Kowalchuk and Stephen 2001).

In addition to chemolithotrophic bacteria, some ammonia-oxidizing archaea (AOA) can nitrify (Prosser and Nicol 2012). The mechanism by which AOA produces $N_2O$ remains unclear (Hink et al. 2017, Hink et al. 2018). Some authors suggest that $N_2O$ may be produced abiotically by oxidation of compounds such as $NH_2OH$, NO or $NO_2^-$ (Harper et al. 2015). Classified initially by *rrs* gene phylogeny as Crenarchaeota, comparative genomics and phylogeny of concatenated genes placed these microorganisms into the new archaeal phylum *Thaumarchaeota* (Hatzenpichler et al. 2008, Prosser and Nicol 2012). Although members of two different domains of life, AOB and AOA exploit homologs ammonia monooxygenases (*amoA* gene) that are members of the copper (Cu)-containing membrane-bound monooxygenase (CuMMOs) enzyme family (Hatzenpichler 2012).

Generally, AOA seem to dominate ammonia oxidation in soil under low N availability (< 15 μg $NH_4^+$-Ng dry soil$^{-1}$), whereas AOB become more competitive at higher N loads (Jia and Conrad 2009, Prosser and Nicol 2012). The nature of the $NH_4^+$ source might be of relevance for niche and physiological differentiation of archaeal and bacterial ammonia oxidizers (Hink et al. 2017, Hink et al. 2018). AOA activity was detected when $NH_4^+$ was supplied as mineralized organic N derived from composted manure or soil organic matter, whereas AOB-dominated activity was measured with $NH_4^+$ originating from inorganic fertilizer (Schleper and Nicol 2010). In addition to that, (meta-) genome analyses (Pester et al. 2012) and environmental studies (Herndl et al. 2005, Ingalls et al. 2006) indicate that AOA might be able to switch from autotrophic ammonia oxidation to a mixotrophic, and possibly even heterotrophic lifestyle, a capacity that may contribute to their numerical dominance in soils.

### 3.3.2.2 *Bacterial Denitrifiers*

Denitrification is a facultative respiratory pathway where $NO_3^-$ or $NO_2^-$ is stepwise reduced to $N_2$ via NO and $N_2O$ under oxygen-limiting conditions (Zumft 1997). Each step is coupled to the electron transport chain and electrons from reductants can be passed on to different N oxides, allowing for the generation of a proton gradient across the membrane (Figure 3.5).

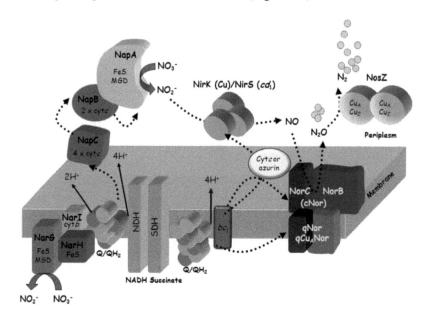

**FIGURE 3.5** Denitrification pathway and enzymes involved in $N_2O$ formation. The membrane-bound (NarGHI), and periplasmic, (NapABC) nitrate reductases as well as nitrite reductases (Cu-type or cd1-type), nitric oxide reductases (cNor, qNor, and qCuANor), and nitrous oxide reductase (NosZ) are shown (Bueno et al. 2012).

The complete denitrification pathway is catalyzed by a series of different enzymes, which may be functionally redundant. The first step, the conversion of $NO_3^-$ to $NO_2^-$ is catalyzed by either a membrane-associated nitrate reductase, encoded by the *narG* gene, or its soluble periplasmic homolog encoded by the *napA* gene. The Nar enzyme is present in members of the phyla Proteobacteria, Firmicutes, Actinobacteria and the Archaea domain, while Nap is present only in Proteobacteria (Bru et al. 2007, Simpson et al. 2010, Sparacino-Watkins et al. 2014). Both types of enzymes have been found in the genome of Fungi (Shoun et al. 2012). Nar is an integral membrane enzyme composed of three subunits called NarGHI encoded by the genes of the *narGHJI* operon. The Nap enzyme is composed of three subunits of which NapA and NapB are located in the periplasm and a third, called NapC, is an integral membrane protein.

The conversion of $NO_2^-$ to NO via nitrite reductases is considered the defining step in denitrification (Zumft 1997, Shapleigh 2006). This step is carried out by one of two different NO-forming $NO_2^-$ reductases, encoded by the *nirK* or *nirS* genes. Despite having similar functional role and localization in the cell, each protein has completely different structural features. The NirK protein is a member of the multi-Cu oxidase family, with Cu ions as ligands within the catalytic center. By contrast, the NirS protein contains two different heme ligands in the active centers of the enzyme. Previous studies have shown the functional redundancy of the *nir* genes in denitrifiers (Glockner et al. 1993). The *nirK* gene has been identified both in prokaryotes and eukaryotes (Long et al. 2015), while the *nirS* gene has been identified only in the Bacteria and Archaea domains (Mardanov et al. 2015). Although *nirS* and *nirK* are not related in evolutionary terms, the *nirK* gene prevails in Alphaproteobacteria, Firmicutes and Bacteroidetes; *nirS* is more common in Betaproteobacteria and there are no differences in its abundance in the case of Gammaproteobacteria (Heylen et al. 2006).

The reduction of NO to $N_2O$ is carried out by nitric oxide reductase (Nor), a membrane-associated protein. There are three types of enzymes Nor, one is dependent on a citrochrome *c* or pseudoazurin (cNor), another that uses quinol (qNor) and the third, which is called qCuANor, is a qNor enzyme that contains a different active Cu center (de Vries and Pouvreau 2007, Tosha and Shiro 2013). Although there is no clear prevalence between the *cnorB* and *qnorB* genes among the different phylogenetic groups, the Alphaproteobacteria only present the *cnorB* gene, while the rest of the bacterial classes have one or another type of gene (Jones et al. 2008). The ability to reduce NO to $N_2O$ is not unique to denitrification, as NO is highly toxic and a powerful intracellular signaling compound and thus microorganisms may possess Nor as a means of detoxification (Zumft 2005).

Finally, the conversion of $N_2O$ to $N_2$ is carried out by the *nosZ* gene product, thus closing the N-cycle. The operon of the genes is conserved in most of the microorganisms and usually includes the genes *nosRZDFYLX* (Wunsch et al. 2003, Bueno et al. 2012, Pauleta et al. 2013). The *nosZ* gene encodes the catalytic subunit of nitrous oxide reductase, an enzyme that contains two domains, CuA involved in the transfer of electrons and CuZ which contains Cu and S in the catalytic center (Pauleta et al. 2013). The remaining genes of the operon encode proteins required for the transcription and assembly of the active Cu centers. Recently, a second clade of nitrous oxide reductase has been identified, named *nosZII* (Sanford et al. 2012, Jones et al. 2013) which includes a large fraction of non-denitrifying $N_2O$ reducers, which could act as $N_2O$ sinks without major contribution to $N_2O$ formation (Hallin et al. 2018).

Physiologically, denitrification depends on the availability of $O_2$, N oxides and suitable reductants to drive electron transport. In most denitrifying species, expression of denitrification genes is tightly regulated by $O_2$ levels due to its status as the preferred electron acceptor in most cases. However, the exact level of anoxia required for denitrification gene expression can differ substantially among organisms (Ollivier et al. 2011) and denitrification activity can persist in the presence of $O_2$ in environments that have shifted from anaerobic to aerobic (Morley et al. 2008, Ji et al. 2015).

### 3.3.2.3  *Fungal Denitrifiers*

Although denitrification was thought over a long time to be restricted to prokaryotes, Shoun and Tanimoto (1991) showed that many fungi and yeasts (eukaryotes) also exhibit denitrification traits (Shoun and Fushinobu 2017). The essential enzymes for catalyzing denitrification by fungi are nitrite reductase (encoded by *nirK* gene) and nitric oxide reductase cytochrome P450nor (encoded by the *p450nor* gene) (Shoun and Fushinobu 2017).

Fungi can play an important role in $N_2O$ emission from various ecosystems, such as vegetable fields under intensive management, grasslands, forests, croplands and wetlands (Crenshaw et al. 2008, Seo and DeLaune 2010, Rütting et al. 2013, Chen et al. 2014, Ma et al. 2017). Denitrifying fungal communities have a higher threshold oxygen demand (Zumft 1997, Zhou et al. 2001) and dominate bacteria under sub-anoxic conditions and vice versa. Bacteria dominate fungi under strictly anaerobic conditions (Seo and DeLaune 2010, Chen et al. 2015). Also, fungi exhibit wider pH ranges for optimal growth, prevailing at acidic pH and bacteria at neutral and alkaline pH (Herold et al. 2012, Chen et al. 2015). However, the real role of fungi to produce $N_2O$ remains largely unknown (Maeda et al. 2015).

## 3.4 ROLE OF THE PLANT

Nitrogen use efficiency can be improved by achieving a better synchronization between the supply of N and plant demand as the more the N is uptaken by the plant, the less is lost. To assess this, it is important to understand the factors that drive N uptake by the plants. Plants have influence on N-cycling not only due to the uptake of N to be assimilated in plant tissue (shoots and roots) but also due to the release of nutrients from the plants in the rhizosphere and to the emission of N-gases from the plants. Besides, roots can induce changes in soil structure modifying the oxygen availability and consequently, denitrification rates.

### 3.4.1 N Uptake by the Plants and Biological $N_2$ Fixation

The uptake of N in plant tissue is mainly associated with $NH_4^+$. The preferred form in which N is taken up by the plant depends on the plant adaptation to soil conditions being $NH_4^+$ and amino acid the main N source for plants adapted to low pH and anaerobic conditions form and $NO_3^-$ in plants adapted to higher pH and more aerobic soils (Masclaux-Daubresse et al. 2010). Several studies have shown that organic N can be absorbed by leaves similar to $NO_3^-$ and $NH_4^+$ (Umana and Wanek 2010, Uscola et al. 2014). Chadwick et al. (2000) found N offtake values of 56 and 37% of the organic N added from layer manure and a pig slurry. The lowest values were from dairy cow slurry and beef FYM (< 2 and 6%, respectively). N uptake by plants is also influenced by species composition. In this sense, Abalos et al. (2014) found that shoot, root and total biomass, as well as shoot N uptake and C/N ratio and shoot/root ratio, were significantly affected (P < 0.001) by plant community composition and species identity in a mesocosm experiment.

Atmospheric $N_2$ can be fixed via BNF by the symbiotic association between legumes and *Rhizobium* species. It has been reported that over 200 kg ha$^{-1}$ year$^{-1}$ of plant-available N can be produced with white clover (Burchill et al. 2014). Estimates of BNF from white clover show that annual above ground BNF ranged from 14 kg N ha$^{-1}$ under 280 N fertilizer applications to 128 kg N ha$^{-1}$ under 0 N fertilizer application showing the detrimental impact of N fertilizer on BNF (Burchill et al. 2014).

### 3.4.2 Nutrient Release by the Plants

Plants can also be a source of nutrients due to the release of nutrients in the rhizosphere. C and N can be released from root exudates and for example in a study by Paynel et al. (2001) in mixtures of legumes and ryegrass, it was shown that $NH_4^+$ and amino acids were released from root exudates, with the former in much higher amounts. The main amino acids from the mixed root exudates were serine and glycine, with asparagine been the major amino acid in clover roots, while glutamine, glutamate and aspartate were the major amino acids in ryegrass roots. The role of these compounds has been mentioned as potentially influencing nutrient mobilization in the rhizosphere (affecting microflora population) and increasing C availability that will affect denitrification and consequently,

$N_2O$ emissions. Besides, the rhizo- and endosphere contain beneficial plant-associated bacteria that support plant health and growth (Compant et al. 2010). However, as well as beneficial organisms, root exudates and mucilage derived nutrients also attract deleterious bacteria. Additional characteristics such as bacterial flagella, quorum sensing and enzyme production, affect contact with microbes in the rhizosphere and regulate functions, including $N_2$ fixation (Compant et al. 2010).

### 3.4.3  Plant-Induced Changes in Soil Structure

The roots of the plants can modify soil structure as they modify structure creating channels for nutrients and water to move. Holtham et al. (2007) observed changes in the soil water retention curves caused by root-induced soil structuring under white clover compared to ryegrass. Changes in soil water retention derive in changes in oxygen availability affecting denitrification rates and $N_2O$ formation (Laudone et al. 2011).

### 3.4.4  Direct Emissions of N-Gases From the Plant

Plants are known to emit N gases in the form of NO, $N_2O$ and $NH_3$. The mechanisms for losses of N from the plants can be directly from the leaves through photo assimilation of $NO_2^-$ in the leaves or from the soil solution gases that can be transported through the plant via the transpiration stream and emitted from the leaf surface with transpiration. A study by Hakata et al. (2003) reported that there is a conversion of the $NO_3^-$ and nitrogen dioxide ($NO_2$) to $N_2O$ in plants. The conversion varied between species (assimilation of $N_2O$ had over 600-fold variation between 217 plant taxa). Emissions have been reported from wheat, canola, barley and beech as well as for grasses such as *Lolium perenne* (ryegrass), *Holcus lanatus* and *Paspalum dilatatum*. Baruah et al. (2012) reported emissions from wheat varieties ranging from 40 µg $N_2O–N$ $m^{-2}$ $h^{-1}$ to 295 µg $N_2O–N$ $m^{-2}$ $h^{-1}$ and suggested that movement of $N_2O$ along with the transpirational water flow may be an important mechanism for its transport and emission through wheat plants. Besides, emission of $N_2O$ from wheat leaves can also occur during the photoassimilation of $NO_2^-$ in the chloroplast (Smart and Bloom 2001). Rice plants are also known to act as an effective pathway for $N_2O$ transport through aerenchyma cells in submerged soils and during day time, transport of $N_2O$ from roots to shoots takes place with the transpiration stream and released through open stomata. These emissions are dependent on the growth stage of the plant (Baruah et al. 2010). Regarding grasses, $N_2O$ emissions have been reported for ryegrass with values that range between 0.8 and 13.3 mg $N_2O$-N $m^{-2}$ $day^{-1}$ (Chen et al. 1999) but it was not clear whether $N_2O$ was produced by the plants themselves or whether the ryegrass served as a conduit for $N_2O$ produced in the soil.

Plants can act as a source or as a sink of $NH_3$. The direction of the $NH_3$ flux between plant leaves and the atmosphere depends mainly on what is known as the stomatal $NH_3$ compensation point (χs) of leaves, which is the atmospheric $NH_3$ concentration where $NH_3$ emission and deposition are balanced and no net exchange occurs (Husted et al. 2002). $NH_3$ is emitted by the plant when the concentration of $NH_3$ in the atmosphere is lower than that of $NH_3$ in the substomatal cavities of leaves (Schjoerring et al. 2000). The amount of $NH_3$ emitted varies within plant species, being arable crops known to be sources of $NH_3$ whereas grassland act as $NH_3$ sinks (Herrman et al. 2009). In the case of wheat, oilseed rape and barley, Schjoerring and Mattsson (2001) observed that the crop foliage was a net source of $NH_3$ to the atmosphere with $NH_3$ emissions on a seasonal basis between 1 and 5 kg $NH_3–N$ $ha^{-1}$ which constituted between 1 and 4% of the applied N. By contrast, Horváth et al. (2005) found a weak $NH_3$ emission during the vegetation period compared to the high $NH_3$ deposition in the dormant season over a grassland (0.37 and 5.00 kg N $ha^{-1}$ $year^{-1}$, respectively).

Finally, NO emitted by plants constitute an important source of biological NO emission in the terrestrial ecosystem (Singh and Bhatla 2019). NO serves important functions in physiological processes such as germination, root development, flowering and senescence (Farnese et al. 2017). Besides, NO is known to be a stress indicator in plants (Wimalasekera et al. 2011).

## 3.5 LINKAGE BETWEEN N-FERTILIZATION, GENE ABUNDANCE AND MICROBIAL DIVERSITY

Recent meta-analyzes studies have revised our understanding of the effect of N-fertilization on microbial gene abundance and/or bacterial community composition in agricultural ecosystems (Geisseler and Scow 2014, Carey et al. 2016, Zhou et al. 2017, Ouyang et al. 2018, Wang et al. 2018). A meta-analysis conducted by Zhou et al. (2017) showed that N addition inhibits the microbial growth but when the N addition rate is lower than 100 kg N ha$^{-1}$ year$^{-1}$, a positive response of microbial growth was found. Wang et al. (2018) reported that N application decreases soil microbial diversity whose changes depend on the ecosystem analyzed, the initial soil properties, the duration of the treatment, the N addition rates and the changes in soil organic C. Ouyang et al. (2018) concluded that N fertilization significantly increased the abundance of nitrifiers (*amoA* AOA, *amoA* AOB genes) and denitrifiers (*nirK*, *nirS* and *nosZ* genes) and that the fertilizer form and duration, crop rotation and soil pH were main factors regulating the response of the N-cycling genes.

To date, there is no consensus on which gene of denitrification could be considered as a molecular marker of this process (Philippot et al. 2011, Levy-Booth et al. 2014, Hu et al. 2015a, Hallin et al. 2018). Recently, Castellano-Hinojosa et al. (2018, 2019, 2020), found that the N$_2$O emission was mainly related with the abundance of the *norB* gene in an agricultural soil fertilized with urea, ammonium sulfate or potassium nitrate. Results suggested the *norB* as a key gene for studies on N$_2$O production. Calculation of the ratio between genes involved in N$_2$O production (*amoA* AOB + *amoA* AOA + *nirK* + *nirS* + *norB*) and reduction (*nosZ*I + *nosZ*II) explained the changes in the fluxes and cumulative N$_2$O emission during the experimental period.

Previous studies have shown that N fertilization alters or not the structure and composition of the bacterial community (Ramirez et al. 2010, Zhou et al. 2017, Wang et al. 2018). Castellano-Hinojosa et al. (2018, 2019 and 2020) reported that application of N changes the structure of the bacterial community. Values of the Shannon diversity index were similar in the N-fertilized soils with a significant decrease in the number of Operational Taxonomic Units (OTUs) with a relative abundance lower than 1% in the N-treated soils compared to the unfertilized soils. Calculation of the Simpson diversity index showed that the bacterial community became less diverse or dominated by a small group of OTUs in agreement with previously published meta-analyzes (Zhou et al. 2017, Ouyang et al. 2018). Taken together, the results on the structure and composition of the bacterial community indicate that N-fertilization decreases soil biodiversity and that its response depends on the type of the N-fertilizer.

## 3.6 MITIGATION STRATEGIES

The main challenge in reducing N losses from the agriculture sector activities is to optimize the amount of added fertilizer or manure N that is used by the crop. For that purpose, several mitigation strategies are proposed to improve NUE in cropland and grassland systems. In the case of synthetic fertilizer, optimizing the application rates, splitting the annual N rate required, improving the application method, applying fertilizer at the right time, are the most common. In the case of N applied to the soil via animal excreta, the use of nitrification inhibitors as well as changes in animal diets to reduce the excreta-N are implemented. In this section, several mitigation strategies are described.

### 3.6.1 Use of Inhibitors

Mitigation strategies to reduce NO$_3^-$ leaching include the application of nitrification inhibitors (NIs). These compounds act by deactivating the enzyme responsible for the first step of nitrification, the oxidation of NH$_4^+$ to NO$_2^-$ (Subbarao et al. 2006). The most common compounds are manufactured dicyanamide (DCD) and 3,4-dimethylpyrazole phosphate (DMPP) (Liu et al. 2013). The result of this is the retention of N as $NH_4^+$, a less mobile form of soluble N, that can be retained in the soil, particularly those with high clay content. Nevertheless, Lam et al (2017) found that the application of NIs resulted in 3 to 65% increases in $NH_3$ emissions. Other potential effects of using inhibitors are for example the reduction in $N_2O$ emissions as was the case in Linzmeier et al. (2001) and Misselbrook et al. (2014) for synthetic fertilizer when using DMPP and DCD, respectively. Similarly, Merino et al. (2005) and Misselbrook et al. (2014) reported a reduction in $N_2O$ emissions after organic manure application when using DMPP and DCD, respectively. Other inhibitors are also reported, such as urease inhibitors (UIs). The compound N-(n-butyl) thiophosphoric triamide (nBTPT or Agrotein®) slows down the hydrolysis of urea reducing $NH_3$ volatilization (Chen et al. 2008). Chambers and Dampney (2009) showed from 39 experiments including winter cereal and grassland sites, that nBTPT added at increasing rates (250, 500, 1,000 mg nBTPT/kg urea) caused a 25-100% reduction in $NH_3$ emissions from granular urea.

Natural inhibitors are found as a defene mechanism of plants when N availability is low. Tropical grasses of the genus *Brachiaria* have shown effective biological nitrification inhibitory (BNI) activity where N is maintained as $NH_4^+$ (Subbarao et al. 2007). The activity is linked to root exudates releasing biological nitrification inhibitors (BNI) compounds, a tightly regulated physiological process, which is stimulated solely by external exposure of roots to $NH_4^+$ with extensive genetic variability found in selected crops and pasture grasses. Brachialactone, a recently discovered BNI compound, contributed 60-90% of the inhibitory activity released from the roots of the tropical forage grass *Brachiaria humidicola* (Subbarao et al. 2009). The BNI compounds in the shoot tissue of this tropical grass were identified as linoleic acid and linolenic acid, which resulted in low $NO_3^-$ accumulation and maintenance of soil inorganic N in the $NH_4^+$ form (Subbarao et al. 2008). Methyl 3-(4-hydroxyphenyl) propionate (Zakir et al. 2008) is the first compound identified from the root exudates of sorghum, followed by sakuranetin and sorgoleone (Subbarao et al. 2013). A new BNI compound, 1,9-decanediol, was identified from rice root exudates (Sun et al. 2016), which was effective in inhibiting the soil AOA and AOB (Lu et al. 2019). Recent evidence verified the BNI ability in wheat landraces (*Triticum aestivum*) (O'Sullivan et al. 2016). On the other hand, some plants such as *Plantago lanceolata* have the capacity to reduce $N_2O$ emissions from urine when consumed by cattle not only by reducing the N output in excreta (O'Connell et al. 2016) but also by the release of plant secondary metabolites (PSMs) with an inhibiting effect on biological nitrification (Gardiner et al. 2018). Indeed, further studies found that these PSMs, which are known to have beneficial properties on animal health (Esteban-Ballesteros et al. 2019), can also be excreted in urine from animals fed with plants containing these compounds (Cheng et al. 2017).

Inhibitors can be applied in a variety of ways, including separate applications from fertilizer application or coated/blended with fertilizer granule. In the case of grazed grassland, experimental studies have dosed livestock drinking water with DCD (Welten et al. 2014), included DCD in rumen boluses (Ledgard et al. 2008) or added DCD to animal feed (Luo et al. 2015) for specifically targeting the inhibitor to urine patches. Nitrification inhibitors, such as DCD, can also be added to slurry stores prior to slurry application (Minet et al. 2016). Mixing inhibitors have been recommended, such as combining a nitrification and urease inhibitor was reported to increase the efficiency in reducing $N_2O$ emissions compare to NI alone when urine was applied to the soil (Zaman and Blennerhassett 2010).

### 3.6.2   Modes of Application of N

The slurry application method is an important management factor that affects $NH_3$ volatilization, with band spreading reducing $NH_3$ emissions by 25 and 70% compared with surface broadcasting of digestate and slurry (Nicholson et al. 2017). Table 3.2 shows the effect of mitigation strategies on $N_2O$, $NH_3$ and leaching from several published studies. Generally, the use of inhibitors is effective but other measures do not show consistent results.

**TABLE 3.2** Examples of studies showing the effect of mitigation strategies on $N_2O$, $NH_3$ and leaching

| ID | N Application | Baseline Method | Mitigation Option | Effect on $NH_3$ | Effect on $N_2O$ | Effect on Leaching |
|---|---|---|---|---|---|---|
| Nicholson, Year is missing | Slurry | Broadcasting | Bandspread | –71% (s) | +64% (ns) | |
| Nicholson, Year is missing | Digestate | Broadcasting | Bandspread | –90% (ns) | –91% (ns) | |
| Chambers, Year is missing | | Autumn application | Spring application | | | |
| Misselbrook, 2014 | Fertilizer | Ammonium Nitrate | DCD* | ns | –66 to –85% | ns |
| Misselbrook, 2014 | Fertilizer | Urea | DCD | ns | –31 to –69% (one site increased by 10% | ns |
| Misselbrook, 2014 | Excreta | Urine | DCD | ns | –11 to –95% | –7% in one site, ns the others |
| Misselbrook, 2014 | Excreta | Slurry | DCD | +30% in one site, ns the others | +55 to +88% | ns |
| Misselbrook, 2014 | Urine | Slurry autumn application | Slurry spring or summer application | ns | One site increases, the other decreases | +73% |
| Misselbrook | Excreta | Slurry autumn application | Slurry spring application | ns | One site increases, the other decreases | ns |

* Dicyandiamide

## 3.7   WAY FORWARD

The current need for fertilizer in the agriculture sector to meet global food needs is undeniable. However, equally significant is the fact that due to the low NUE in agriculture, most of the applied N ends up being lost to the environment causing several detrimental impacts. Therefore, we suggest that in order to optimize the N applied via fertilizer, further exploration is needed to enhance the efficiency fertilizers as well as to optimize manure processing. Besides, sensing technologies could be used to enhance synchronicity between plant demand and N supply and novel plant varieties could be tested. Models are useful resources that if optimized can be used to make predictions and estimate rates of reactions. We suggest that as much as it is possible, experiments should cover a range of soils and explore the controlling factors but that it is combined with available models that can be validated and applied for different scenarios, i.e., changing climate.

## 3.8   REFERENCES

Abalos, D., De Deyn, G.B., Kuyper, T.W. and van Groenigen, J.W. 2014. Plant species identity surpasses species richness as a key driver of $N_2O$ emissions from grassland. Global Change Biology 20: 265-275.

Aguilera, E., Lassaletta, L., Sanz-Cobena, A., Garnier, J. and Vallejo, A. 2013. The potential of organic fertilizers and water management to reduce $N_2O$ emissions in Mediterranean climate cropping systems: a review. Agriculture Ecosystem and Environment 164: 32-52.

Attard, E., Recous, S., Chabbi, A., De Berranger, C., Guillaumaud, N., Labreuche, J., et al. 2011. Soil environmental conditions rather than denitrifier abundance and diversity drive potential denitrification after changes in land uses. Global Change Biology 17: 1975-1989.

Baggs, E.M., Rees, R.M., Smith, K.A. and Vinten, A.J.A. 2000. Nitrous oxide emission from soils after incorporating crop residues. Soil Use Management 16: 82-87.

Baggs, E.M., Richter, M., Cadisch, G. and Hartwig, U.A. 2003. Denitrification in grass swards is increased under elevated atmospheric $CO_2$. Soil Biology and Biochemistry 35: 729-732.

Baggs, E.M. 2008. A review of stable isotope techniques for $N_2O$ source partitioning in soils: recent progress, remaining challenges and future considerations. Rapid Communications in Mass Spectrometry 22: 1664-1672.

Baggs, E.M. 2011. Soil microbial sources of nitrous oxide: recent advances in knowledge, emerging challenges and future direction. Current Opinion in Environmental Sustainability 3: 321-327.

Baggs, E. and Philippot, L. 2010. Microbial terrestrial pathways to $N_2O$. pp. 4-35. *In*: Smith, K. (ed.). Nitrous Oxide and Climate Change. Earthscan, London, UK.

Ball, B.C. 2013. Soil structure and greenhouse gas emissions: a synthesis of 20 years of experimentation. European Journal of Soil Science 64: 357-373.

Baruah, K.K., Gogoi, B. and Gogoi, P. 2010. Plant physiological and soil characteristics associated with methane and nitrous oxide emission from rice paddy. Physiology and Molecular Biology of Plants 16: 79-91.

Baruah, K.K., Gogoi, B., Borah, L., Gogoi, M. and Boruah, R. 2012. Plant morphophysiological and anatomical factors associated with nitrous oxide flux from wheat (Triticum aestivum). Journal of Plant Research 125: 507-516.

Bittman, S. and Mikkelsen, R. 2009. Ammonia emissions from agricultural operations: livestock. Better Crops 93: 28-31.

Blackmer, A.M. and Bremner, J.M. 1978. Inhibitory effect of nitrate on reduction of $N_2O$ to $N_2$ by soil microorganisms. Soil Biology and Biochemistry 10: 187-191.

Bol, R., Toyoda, S., Yamulki, S., Hawkins, J., Cardenas, L. and Yoshida, N. 2003. Dual isotope and isotopomer ratios of $N_2O$ emitted from a temperate grassland soil after fertiliser application. RCM 17: 2550-2556.

Bollmann, A. and Conrad, R. 1998. Influence of $O_2$ availability on NO and $N_2O$ release by nitrification and denitrification in soils. Global Change Biology 4: 387-396.

Bouwman, A.F. 1990. Exchange of greenhouse gases between terrestrial ecosystems and the atmosphere. pp. 61-127. *In*: Bouwman, A.F. (ed.) Soils and the Greenhouse Effect. John Wiley and Sons, New York.

Braker, G. and Conrad, R. 2011. Diversity, structure, and size of $N_2O$ producingmicrobial communities in soils – what matters for their functioning? Advances in Applied Microbiology 75: 33-70.

Brown, P., Broomfield, M., Cardenas, L., Choudrie, S., Kilroy, E., Jones, L., et al. 2018. UK Greenhouse Gas Inventory, 1990 to 2016, Annual Report for Submission under the Framework Convention on Climate Change. Ricardo Energy and Environment The Gemini Building, Fermi Avenue Harwell Didcot Oxfordshire, OX11 0QR, UK.

Bru, D., Sarr, A. and Philippot, L. 2007. Relative abundances of proteobacterial membrane-bound and periplasmic nitrate reductases in selected environments. Applied and Environmental Microbiology 73: 5971-5974.

Bueno, E., Mesa, S., Bedmar, E.J., Richardson, D.J. and Delgado, M.J. 2012. Bacterial adaptation of respiration from oxic to microoxic and anoxic conditions: redox control. Antioxidants and Redox Signaling 16: 819-852.

Burchill, W., James, E.K., Li, D., Lanigan, G.J., Williams, M., Iannetta, P.P.M., et al. 2014. Comparisons of biological nitrogen fixation in association with white clover (Trifolium repens L.) under four fertiliser nitrogen inputs as measured using two $^{15}N$ techniques. Plant and Soil 385: 287-302.

Butterbach-Bahl K., Baggs E.M., Dannenmann M., Kiese R. and Zechmeister-Boltenstern S. 2013. Nitrous oxide emissions from soils: how well do we understand the processes and their controls? Philosophical Transactions of the Royal Society B: Biological Sciences 368: 20130122.

Camargo, J.A. and Alonso, Á. 2006. Ecological and toxicological effects of inorganic nitrogen pollution in aquatic ecosystems: a global assessment. Environment International 32: 831-849.

Cameron, K., Di, H. and Moir, J. 2013. Nitrogen losses from the soil/plant system: a review. Annals of Applied Biology 162: 145-173.

Cape, J.N., Kirika, A., Rowland, A.P., Wilson, D.R., Jickells, T.D. and Cornell, S. 2001. Organic nitrogen in precipitation: real problem or sampling artefact? The Scientific World Journal 2: 230-237.

Cardenas, L., Hawkins, J., Chadwick, D. and Scholefield, D. 2003. Biogenic gas emissions from soils measured using a new automated laboratory incubation system. Soil Biology and Biochemistry 35: 867-870.

Cardenas, L., Chadwick, D., Scholefield, D., Fychan, A., Marley, C., Jones, R., et al. 2007. The effect of diet manipulation on nitrous oxide and methane emissions from manure application to grassland soils. Atmospheric Environment 41: 7096-7107.

Cardenas, L., Bol, R., Lewicka-Szczebak, D., Gregory, A., Matthews, G., Whalley, W., et al. 2017. Effect of soil saturation on denitrification in a grassland soil. Biogeosciences 14: 4691-4710.

Cardenas, L.M., Cuttle, S.P., Crabtree, B., Hopkins, A., Shepherd, A., Scholefield, D., et al. 2011. Cost effectiveness of nitrate leaching mitigation measures for grassland livestock systems at locations in England and Wales. Science of the Total Environment 409: 1104-1115.

Cardenas, L.M., Bhogal, A., Chadwick, D.R., McGeough, K., Misselbrook, T., Rees, R.M., et al. 2019. Nitrogen use efficiency and nitrous oxide emissions from five UK fertilised grasslands. Science of the Total Environment 661: 696-710.

Carey, C.J., Dove, N.C., Beman, J.M., Hart, S.C. and Aronson, E.L. 2016. Meta-analysis reveals ammonia-oxidizing bacteria respond more strongly to nitrogen addition than ammonia-oxidizing archaea. Soil Biology and Biochemistry 99: 158-166.

Castellano-Hinojosa, A., González-López, J. and Bedmar, E.J. 2018. Distinct effect of nitrogen fertilisation and soil depth on nitrous oxide emissions and nitrifiers and denitrifiers abundance. Biology and Fertility of Soils 54: 829-840.

Castellano-Hinojosa, A., González-López, J. and Bedmar, E.J. 2019. Effect of nitrogen fertilisation on nitrous oxide emission and the abundance of microbial nitrifiers and denitrifiers in the bulk and rhizosphere soil of *Solanum lycopersicum* and *Phaseolus vulgaris*. Plant and Soil 451: 107-120. https://doi.org/10.1007/s11104-019-04188-6.

Castellano-Hinojosa, A., Correa-Galeote, D., González-López, J. and Bedmar, E.J. 2020. Effect of nitrogen fertilisers on nitrous oxide emission, nitrifier and denitrifier abundance and bacterial diversity in closed ecological systems. Applied Soil Ecology 145: 103380.

Chadwick, D.R., Pain, B.F. and Brookman, S.K.E. 2000. Nitrous oxide and methane emissions following application of animal manures to grassland. Journal of Environmental Quality 29: 277-287.

Chadwick, D.R., Cardenas, L.M., Dhanoa, M.S., Donovan, N., Misselbrook, T., Williams, J.R., et al. 2018. The contribution of cattle urine and dung to nitrous oxide emissions: quantification of country specific emission factors and implications for national inventories. Science of the Total Environment 635: 607-617.

Chambers, B.J. and Dampney, P.M.R. 2009. Nitrogen efficiency and ammonia emissions from urea-based and ammonium nitrate fertilisers. Proceedings of the International Fertiliser Society 657: 1-20.

Chen, D., Suter, H., Islam, A., Edis, R., Freney, J., Walker, C. 2008. Prospects of improving efficiency of fertiliser nitrogen in australian agriculture: a review of enhanced efficiency fertilisers. Soil Research 46: 289-301.

Chen, H., Mothapo, N.V. and Shi. W. 2014. The significant contribution of fungi to soil $N_2O$ production across diverse ecosystems. Applied Soil Ecology 73: 70-77.

Chen, X., Boeckx, P., Shen, S. and Van Cleemput, O. 1999. Emission of $N_2O$ from rye grass (Lolium perenne L.). Biology and Fertility of Soils 28: 393-396.

Chen, Z., Wang, C., Gschwendtner, S., Willibald, G., Unteregelsbacher, S., Lu, H., et al. 2015. Relationships between denitrification gene expression, dissimilatory nitrate reduction to ammonium and nitrous oxide and dinitrogen production in montane grassland soils. Soil Biology and Biochemistry 87: 67-77.

Cheng. L., McCormick, J., Hussein, A.N., Logan, C., Pacheco, D., Hodge, M.C., et al. 2017. Live weight gain, urinary nitrogen excretion and urination behaviour of dairy heifers grazing pasture, chicory and plantain. The Journal of Agricultural Science 155: 669-678.

Compant, S., Clément, C. and Sessitsch, A. 2010. Plant growth-promoting bacteria in the rhizo- and endosphere of plants: their role, colonization, mechanisms involved and prospects for utilization. Soil Biology and Biochemistry 42: 669-678.

Crenshaw, C.L., Lauber, C., Sinsabaugh, R.L. and Stavely, L.K. 2008. Fungal control of nitrous oxide production in semiarid grassland. Biogeochemistry 87: 17-27.

Crutzen, P.J. 1981. Atmospheric chemical processes of the oxides of nitrogen, including nitrous oxide. pp. 17-44. *In*: Delwiche, C.C. (ed.). Denitrification, Nitrification, and Atmospheric Nitrous Oxide. John Wiley and Sons Inc., New York, N.Y.

Čuhel, J. and Šimek, M. 2011. Proximal and distal control by pH of denitrification rate in a pasture soil. Agriculture, Ecosystems and Environment 141: 230-233.

Daims, H., Lucker, S. and Wagner, M. 2016. A new perspective on microbes formerly known as nitrite-oxidizing bacteria. Trends in Microbiology 24: 699-712.

Danielson, R.E. and Sutherland, P.L. 1986. Porosity. pp. 443-461. *In*: Klute, A. (ed.). Methods of Soil Analysis. Part I. Physical and Mineralogical Methods. Agronomy Monograph No. 9, American Society of Agronomy, Soil Science Society of America, Madison, WI.

Davidson, E.A. and Kanter, D. 2014. Inventories and scenarios of nitrous oxide emissions. Environmental Research Letters 9: 12.

de Vries, S. and Pouvreau, L.A.M. 2007. Nitric oxide reductase: structural variations and catalytic mechanism. pp. 57-66. *In*: Bothe, H., Ferguson, S.J. and Newton, W.E. (eds.). Biology of the Nitrogen Cycle. Elsevier, Amsterdam.

Dorland, S. and Beauchamp, E.G. 1991. Denitrification and ammonification at low soil temperatures. CJSS 71: 293-303.

Erguder, T.H., Boon, N., Wittebolle, L., Marzorati, M. and Verstraete, W. 2009. Environmental factors shaping the ecological niches of ammonia-oxidizing archaea. FEMS Microbiology Reviews 33: 855-869.

Erisman, J.W., Sutton, M.A., Galloway, J., Klimont, Z. and Winiwarter, W. 2008. How a century of ammonia synthesis changed the world. Nature 1: 636-639.

Erisman, J.W., Galloway, J., Seitzinger, S., Bleeker, A. and Butterbach-Bahl, K. 2011. Reactive nitrogen in the environment and its effect on climate change. Current Opinion in Environmental Sustainability 3: 281-290.

Esteban-Ballesteros, M., Sanchis, J., Gutiérrez-Corbo, C., Balaña-Fouce, R., Rojo-Vázquez, F.A., González-Lanza, C., et al. 2019. *In vitro* anthelmintic activity and safety of different plant species against the ovine gastrointestinal nematode Teladorsagia circumcincta. Research in Veterinary Science 123: 153-158.

EU Nitrogen Expert Panel 2015. Nitrogen Use Efficiency (NUE) - An Indicator for the Utilization of Nitrogen in Agriculture and Food Systems. Wageningen University, Alterra, PO Box 47, NL-6700 Wageningen, Netherlands.

Eurostat 2017. Agriculture–ammonia emission statistics. https://ec.europa.eu/eurostat/statistics-explained/index.php/Agri-environmental_indicator_-_ammonia_emissions.

Farnese, F.S., Oliveira, J.A., Paiva, E., Menezes-Silva, P.E., da Silva, A.A., Campos, F.V., et al. 2017. The involvement of nitric oxide in integration of plant physiological and ultrastructural adjustments in response to arsenic. Frontiers in Plant Science 8: 516.

Firestone, M. and Davidson, E. 1989. Microbiological basis of NO and $N_2O$ production and consumption in soil. pp. 47. *In*: Andreae, M.O. and Schimel, D.S. (eds.). Exchange of Trace Gases between Terrestrial Ecosystems and the Atmosphere. New York, NY.

Fowler, D., Steadman, C.E., Stevenson, D., Coyle, M., Rees, R.M., Skiba, U.M., et al. 2015. Effects of global change during the 21st century on the nitrogen cycle. Atmospheric Chemistry and Physics 15: 13849-13893.

Galloway, J.N., Leach, A.M., Bleeker, A. and Erisman, J.W. 2013. A chronology of human understanding of the nitrogen cycle. Philosophical Transactions of the Royal Society B 368: 20130120.

Gardiner, C.A., Clough, T.J., Cameron, K.C., Di, H.J., Grant, R., Edwards, G.R., et al. 2018. Potential inhibition of urine patch nitrous oxide emissions by Plantago lanceolata and its metabolite aucubin. New Zealand Journal of Agricultural Research 61: 495-503.

Ge, X., Wexler, A.S. and Clegg, S.L. 2011. Atmospheric amines – part I. A review. Atmospheric Environment 45: 524-546.

Geisseler, D. and Scow, K.M. 2014. Long-term effects of mineral fertilizers on soil microorganisms – a review. Soil Biology and Biochemistry 75: 54-63.

Glass, C. and Silverstein, J. 1999. Denitrification on high-nitrate high-salinity wastewater. Water Research 33: 223-229.

Glockner, A.B., Jungst, A. and Zumft, W.G. 1993. Copper-containing nitrite reductase from Pseudomonas aureofaciens is functional in a mutationally cytochrome cd1-free background (NirS-) of Pseudomonas stutzeri. Archives of Microbiology 160: 18-26.

Gruber, N. and Galloway, J.N. 2008. An earth-system perspective of the global nitrogen cycle. Nature 451: 293-296.

Hakata, M., Takahashi, M., Zumft, W., Sakamoto, A. and Morikawa, H. 2003. Conversion of the nitrate nitrogen and nitrogen dioxide to nitrous oxides in plants. Acta Scientific Biotechnology 23: 249-257.

Hallin, S., Philippot, L., Löffler, F.E., Sanford, R.A. and Jones, C.M. 2018. Genomics and ecology of novel N$_2$O-reducing microorganisms. Trends in Microbiology 26: 43-55.

Harper, W.F., Takeuchi, Y., Riya, S., Hosomi, M. and Terada, A. 2015. Novel abiotic reactions increase nitrous oxide production during partial nitrification: modeling and experiments. Chemical Engineering Journal 281: 1017-1023.

Harrison-Kirk, T., Beare, M.H., Meenken, E.D. and Condron, L.M. 2013. Soil organic matter and texture affect responses to dry/wet cycles: effects on carbon dioxide and nitrous oxide emissions. Soil Biology and Biochemistry 57: 43-55.

Hatzenpichler, R., Lebedeva, E.V., Spieck, E., Stoecker, K., Richter, A., Daims, H., et al. 2008. A moderately thermophilic ammonia-oxidizing crenarchaeote from a hot spring. Proceedings of the National Academy of Sciences 105: 2134-2139.

Hatzenpichler, R. 2012. Diversity, physiology, and niche differentiation of ammonia-oxidizing archaea. Applied and Environmental Microbiology 78: 7501-7510.

Herndl, G.J., Reinthaler, T., Teira, E., van Aken, H., Veth, C., Pernthaler, A., et al. 2005. Contribution of Archaea to total prokaryotic production in the deep Atlantic Ocean. Applied and Environmental Microbiology 71: 2303-2309.

Herold, M.B., Baggs, E.M. and Daniell, T.J. 2012. Fungal and bacterial denitrification are differently affected by long-term pH amendment and cultivation of arable soil. Soil Biology and Biochemistry 54: 25-35.

Herrmann, B., Mattsson, M., Jones, S.K., Cellier, P., Milford, C., Sutton, M.A., et al. 2009. Vertical structure and diurnal variability of ammonia exchange potential within an intensively managed grass canopy. Biogeosciences 6: 15-23.

Hertel, O., Skjøth, C.A., Reis, S., Bleeker, A., Harrison, R.M., Cape, J.N., et al. 2012. Governing processes for reactive nitrogen compounds in the European atmosphere. Biogeosciences 9: 4921-4954.

Heylen, K., Gevers, D., Vanparys, B., Wittebolle, L., Geets, J., Boon, N., et al. 2006. The incidence of nirS and nirK and their genetic heterogeneity in cultivated denitrifiers. Environmental Microbiology 8: 2012-2021.

Hink, L., Lycus, P., Gubry-Rangin, C., Frostegard, A., Nicol, G.W., Prosser, J.I., et al. 2017. Kinetics of NH$_3$-oxidation, NO-turnover, N$_2$O-production and electron flow during oxygen depletion in model bacterial and archaeal ammonia oxidisers. Environmental Microbiology 19: 4882-4896.

Hink, L., Gubry-Rangin, C., Nicol, G.W. and Prosser, J.I. 2018. The consequences of niche and physiological differentiation of archaeal and bacterial ammonia oxidisers for nitrous oxide emissions. The ISME Journal 12: 1084-1093.

Holtham, D.A.L., Matthews, G.P. and Scholefield, D.S. 2007. Measurement and simulation of void structure and hydraulic changes caused by root-induced soil structuring under white clover compared to ryegrass. Geoderma 142: 142-151.

Horváth, L., Asztalos, M., Führer, E. and Mészáros, R. 2005. Measurement of ammonia exchange over grassland in the Hungarian Great Plain. Agricultural and Forest Meteorology 130: 282-298.

Hu, H.W., Chen, D. and He, J.Z. 2015a. Microbial regulation of terrestrial nitrous oxide formation: understanding the biological pathways for prediction of emission rates. FEMS Microbiology Reviews 39: 729-749.

Hu, J., Inglett, K.S., Clark, M.W., Inglett, P.W. and Ramesh, R.K. 2015b. Nitrous oxide production and consumption by denitrification in a grassland: effects of grazing and hydrology. Science of the Total Environment 532: 702-710.

Huang, G., Hou, J. and Zhou, X. 2009. A measurement method for atmospheric ammonia and primary amines based on aqueous sampling, OPA derivatization and HPLC analysis. Environmental Science and Technology 43: 5851-5856.

Husted, S., Mattsson, M., Möllers, C., Wallbraun, M. and Schjoerring, J.K. 2002. Photorespiratory NH$_4^+$ production in leaves of wild-type and glutamine synthetase 2 antisense oilseed rape. Plant Physiology 130: 989-998.

Hutchinson, G.L. and Brams, E.A. 1992. NO versus N$_2$O emissions from an NH$_4^+$-amended Bermuda grass pasture. Journal of Geophysical Research: Atmospheres 97: 9889-9896.

Ingalls, A.E., Shah, S.R., Hansman, R.L., Aluwihare, L.I., Santos, G.M., Druffel, E.R.M., et al. 2006. Quantifying archaeal community autotrophy in the mesopelagic ocean using natural radiocarbon. Proceedings of the National Academy of Sciences 103: 6442-6447.

IPCC 2006. $N_2O$ emissions from managed soils, and $CO_2$ emissions from lime and urea application. Guidelines for National Greenhouse Gas Inventories, prepared by the National Greenhouse Gas Inventories Programme IGES, Hayama, Japan.

IPCC 2007. Climate Change 2007 - Mitigation of Climate Change: Working Group III contribution to the Fourth Assessment Report of the IPCC.

IPCC 2019. $N_2O$ emissions from managed soils, and $CO_2$ emissions from lime and urea application. Refinement to the 2006 IPCC Guidelines for National Greenhouse Gas Inventories prepared by the National Greenhouse Gas Inventories Programme IGES, Hayama, Japan.

Jarvis, S. and Menzi, H. 2004. Optimising best practice for N management in livestock systems: meeting environmental and production targets. Grass and Forage Science 9: 361-372.

Jarvis, S., Hutchings, N., Brentrup, F., Olesen, J. and Van der Hoek, K. 2011. Nitrogen flows in farming systems across Europe. pp. 211-228. *In*: Sutton, M.A., Howard, C.M., Erisman, J.W., Billen, G., Bleeker, A., Grennfelt, P., van Grinsven, H. and Grizzetti, B. (eds.). The European Nitrogen Assessment: Sources, Effects and Policy Perspectives. Cambridge, UK: Cambridge University Press.

Ji, B., Yang, K., Zhu, L., Jiang, Y., Wang, H., Zhou, J., et al. 2015. Aerobic denitrification: a review of important advances of the last 30 years. Biotechnology and Bioprocess Engineering 20: 643-651.

Jia, Z. and Conrad, R. 2009. Bacteria rather than Archaea dominate microbial ammonia oxidation in an agricultural soil. Environmental Microbiology 11: 1658-1671.

Jones, C.M., Stres, B., Rosenquist, M. and Hallin, S. 2008. Phylogenetic analysis of nitrite, nitric oxide, and nitrous oxide respiratory enzymes reveal a complex evolutionary history for denitrification. Molecular Biology and Evolution 25: 1955-1966.

Jones, C.M., Graf, D.R., Bru, D., Philippot, L. and Hallin, S. 2013. The unaccounted yet abundant nitrous oxide-reducing microbial community: a potential nitrous oxide sink. ISME Journal: Multidisciplinary Journal of Microbial Ecology 7: 417-426.

Knowles, R. 1982. Denitrification. Microbiology Reviews 46: 43-70.

Kool, D.M., Dolfing, J., Wrage, N. and Van Groenigen, J.W. 2011. Nitrifier denitrification as a distinct and significant source of nitrous oxide from soil. Soil Biology and Biochemistry 43: 174-178.

Lam, S.K., Suter, H., Mosier, A.R. and Chen, D. 2017. Using nitrification inhibitors to mitigate agricultural $N_2O$ emission: a double-edged sword? Global Change Biology 23: 485-489.

Laudone, G.M., Matthews, G.P., Bird, N.R.A, Whalley, W.R., Cardenas, L.M. and Gregory, A.S. 2011. A model to predict the effects of soil structure on denitrification and $N_2O$ emission. Journal of Hydrology 409: 283-290.

Ledgard, S.F., Menneer, J.C., Dexter, M.M., Kear, M.J., Lindsey, S., Peters, J.S., et al. 2008. A novel concept to reduce nitrogen losses from grazed pastures by administering soil nitrogen process inhibitors to ruminant animals: a study with sheep. Agriculture Ecosystem and Environment 125: 148-158.

Levy-Booth, D.J., Prescott, C.E. and Grayston, S.J. 2014. Microbial functional genes involved in nitrogen fixation, nitrification and denitrification in forest ecosystems. Soil Biology and Biochemistry 75: 11-25.

Linzmeier, W., Gutser, R. and Schmidhalter, U. 2001. Nitrous oxide emission from soil and from a nitrogen-15-labelled fertilizer with the new nitrification inhibitor 3,4-dimethylpyrazole phosphate (DMPP). Biology and Fertility of Soils 34: 103-108.

Liu, C., Wang, K. and Zheng, X. 2013. Effects of nitrification inhibitors (DCD and DMPP) on nitrous oxide emission, crop yield and nitrogen uptake in a wheat-maize cropping system. Biogeosciences Discussions 10: 711-737.

Loick, N., Dixon, E., Abalos, D., Vallejo, A., Matthews, P., McGeough, K., et al. 2017. "Hot spots" of N and C impact nitric oxide, nitrous oxide and nitrogen gas emissions from a UK grassland soil. Geoderma 305: 336-345.

Long, A., Song, B., Fridey, K. and Silva, A. 2015. Detection and diversity of copper containing nitrite reductase genes (nirK) in prokaryotic and fungal communities of agricultural soils. FEMS Microbiology Ecology 91: 1-9.

Lu, Y., Zhang, X., Jiang, J., Kronzucker, H.J., Shen, W. and Shi, W. 2019. Effects of the biological nitrification inhibitor 1,9-decanediol on nitrification and ammonia oxidizers in three agricultural soils. Soil Biology and Biochemistry 129: 48-59.

Luo, J., Ledgard, S., Wise, B., Welten, B., Lindsey, S., Judge, A., et al. 2015. Effect of dicyandiamide (DCD) delivery method, application rate, and season on pasture urine patch nitrous oxide emissions. Biology and Fertility of Soils 51: 453-464.

Ma, S., Shan, J. and Yan, X. 2017. $N_2O$ emissions dominated by fungi in an intensively managed vegetable field converted from wheat–rice rotation. Applied Soil Ecology 116: 23-29.

Maeda, K., Spor, A., Edel-Hermann, V., Heraud, C., Breuil, M.C., Bizouard, F., et al. 2015. $N_2O$ production, a widespread trait in fungi. Scientific Reports 5: 9697.

Mardanov, A.V., Slododkina, G.B., Slobodkin, A.I., Beletsky, A.V., Gavrilov, S.N., Kublanov, I.V., et al. 2015. The Geoglobus acetivorans genome: Fe(III) reduction, acetate utilization, autotrophic growth, and degradation of aromatic compounds in a hyperthermophilic archaeon. Applied and Environmental Microbiology 81: 1003-1012.

Martikainen, P.J. and de Boer, W. 1993. Nitrous oxide production and nitrification in acidic soil from a dutch coniferous forest. Soil Biology and Biochemistry 25: 343-347.

Masclaux-Daubresse, C., Daniel-Vedele, F., Dechorgnat, J., Chardon, F., Gaufichon, L. and Suzuki, A. 2010. Nitrogen uptake, assimilation and remobilization in plants: challenges for sustainable and productive agriculture. Annals of Botany 105: 1141-1157.

Merino, P., Menéndez, S., Pinto, M., González-Murua, C. and Estavillo, J.M. 2005. 3,4-dimethylpyrazole phosphate reduces nitrous oxide emissions from grassland after slurry application. Soil Use Management 21: 53-57.

Minet, E.P., Jahangir, M.M.R., Krol, D.J., Rochford, N., Fenton, O., Rooney, D., et al. 2016. Amendment of cattle slurry with the nitrification inhibitor dicyandiamide during storage: a new effective and practical $N_2O$ mitigation measure for landspreading. Agriculture Ecosystem and Environment 215: 68-75.

Misselbrook, T.H., Cardenas, L.M., Camp, V., Thorman, R.E., Williams, J.R., Rollett, A.J., et al. 2014. An assessment of nitrification inhibitors to reduce nitrous oxide emissions from UK agriculture. Environment Research Letters 9: 115006.

Morley, N., Baggs, E.M., Dorsch, P. and Bakken, L. 2008. Production of NO, $N_2O$ and $N_2$ by extracted soil bacteria, regulation by $NO_2^{(-)}$ and $O_2$ concentrations. FEMS Microbiology Ecology 65: 102-112.

Morley, N. and Baggs, L. 2010. Carbon and oxygen controls on $N_2O$ and $N_2$ production during nitrate reduction. Soil Biology and Biochemistry 42: 1864-1871.

Nicholson, F., Bhogal, A., Cardenas, L., Chadwick, D., Misselbrook, T., Rollet, A., et al. 2017. Nitrogen losses to the environment following food-based digestate and compost applications to agricultural land. Environmental Pollution 228: 504-516.

Oenema, O., Witzke, H.P., Klimont, Z., Lesschen, J.P. and Velthof, G.L. 2009. Integrated assessment of promising measures to decrease nitrogen losses from agriculture in EU-27. Agriculture Ecosystem and Environment 133: 280-288.

Ollivier, J., Towe, S., Bannert, A., Hai, B., Kastl, E.M., Meyer, A., et al. 2011. Nitrogen turnover in soil and global change. FEMS Microbiology Ecology 78: 3-16.

O'Connell, C.A., Judson, H.G. and Barrell, G.K. 2016. Sustained diuretic effect of plantain when ingested by sheep. Proceedings of the New Zealand Society of Animal Production 76: 14-17.

O'Sullivan, C.A., Fillery, I.R.P., Roper, M.M. and Richards, R.A. 2016. Identification of several wheat landraces with biological nitrification inhibition capacity. Plant and Soil 404: 61-74.

Ouyang, Y., Evans, S.E., Friesen, M.L. and Tiemann, I.K. 2018. Effect of nitrogen fertilization on the abundance of nitrogen cycling genes in agricultural soils: a meta-analysis of field studies. Soil Biology and Biochemistry 127: 71-78.

Patureau, D., Zumstein, E., Delgenes, J.P. and Moletta, R. 2000. Aerobic denitrifiers isolated from diverse natural and managed ecosystems. Microbial Ecology 39: 145-152.

Pauleta, S.R., Dell'Acqua, S. and Moura, I. 2013. Nitrous oxide reductase. Coordination Chemistry Reviews 257: 332-349.

Paynel, F., Murray, J.P. and Bernard Cliquet, J. 2001. Root exudates: a pathway for short-term N transfer from clover and ryegrass. Plant and Soil 229: 235-243.

Pester, M., Rattei, T., Flechl, S., Grongroft, A., Richter, A., Overmann, J., et al. 2012. amoA-based consensus phylogeny of ammonia-oxidizing archaea and deep sequencing of amoA genes from soils of four different geographic regions. Environmental Microbiology 14: 525-539.

Philippot, L., Andert, J., Jones, C.M., Bru, D. and Hallin, S. 2011. Importance of denitrifiers lacking the genes encoding the nitrous oxide reductase for $N_2O$ emissions from soil. Global Change Biology 17: 1497-1504.

Pilegaard, K. 2013. Processes regulating nitric oxide emissions from soils. Philosophical Transactions of the Royal Society B: Biological Sciences 368: 20130126.

Powlson, D., Saffigna, P. and Kragt-Cottaar, M. 1988. Denitrification at sub-optimal temperatures in soils from different climatic zones. Soil Biology and Biochemistry 20: 719-723.

Prosser, J.I. and Nicol, G.W. 2012. Archaeal and bacterial ammonia-oxidisers in soil: the quest for niche specialisation and differentiation. Trends in Microbiology 20: 523-531.

Ramankutty, N., Mehrabi, Z., Waha, K., Jarvis, L., Kremen, C., Herrero, M., et al. 2018. Trends in global agricultural land use: implications for environmental health and food security. Annual Review of Plant Biology 69: 789-815.

Ramirez, K.S., Lauber, C.L., Knight, R., Bradford, M.A. and Fierer, N. 2010. Consistent effects of nitrogen fertilization on soil bacterial communities in contrasting systems. Ecology 91: 3463-3470.

Ravishankara, A.R., Daniel, J.S. and Portmann, R.W. 2009. Nitrous oxide ($N_2O$): the dominant ozone-depleting substance emitted in the 21st century. Science 326: 123-125.

Remde, A., Slemr, F. and Conrad, R. 1989. Microbial production and uptake of nitric oxide in soil. FEMS Microbiology Letter 62: 221-230.

Richardson, D., Felgate, H., Watmough, N., Thomson, A. and Baggs, E. 2009. Mitigating release of the potent greenhouse gas $N_2O$ from the nitrogen cycle–could enzymic regulation hold the key? Trends in Biotechnology 27: 388-397.

Ruser, R., Flessa, H., Russow, R., Schmidt, G., Buegger, F. and Munch, J.C. 2006. Emission of $N_2O$, $N_2$ and $CO_2$ from soil fertilized with nitrate: effect of compaction, soil moisture and rewetting. Soil Biology and Biochemistry 38: 263-274.

Rütting, T., Huygens, D., Boeckx, P., Staelens, J. and Klemedtsson, L. 2013. Increased fungal dominance in $N_2O$ emission hotspots along a natural pH gradient in organic forest soil. Biology and Fertility of Soils 49: 715-721.

Saggar, S.N., Deslippe, J.J., Bolan, N.S., Luo, J., Giltrap, D.L., Kim, D.G., et al. 2013. Denitrification and $N_2O$: $N_2$ production in temperate grasslands: processes, measurements, modelling and mitigating negative impacts. Science of the Total Environment 465: 173-195.

Sahrawat, K.L. and Keeney, D.R. 1986. Nitrous Oxide Emission from Soils. pp. 103-148. In: Stewart, B.A. (ed.). Advances in Soil Science. Springer New York.

Sanchez-Rodriguez, A.R., Nie, C., Hill, P.W., Chadwick, D.R. and Jones, D.L. 2019. Extreme flood events at higher temperatures exacerbate the loss of soil functionality and trace gas emissions in grassland. Soil Biology and Biochemistry 130: 227-236.

Sanford, R.A., Wagner, D.D., Wu, Q., Chee-Sanford, J.C., Thomas, S.H., Cruz-García, C., et al. 2012. Unexpected nondenitrifier nitrous oxide reductase gene diversity and abundance in soils. Proceedings of the National Academy of Sciences 109: 19709-19714.

Schjoerring, J.K., Husted, S., Mäck, G., Nielsen, K.H., Finnemann, J. and Mattsson, M. 2000. Physiological regulation of plant-atmosphere ammonia exchange. Plant and Soil 221: 95-102.

Schjoerring, J.K. and Mattsson, M. 2001. Quantification of ammonia exchange between agricultural cropland and the atmosphere: measurements over two complete growth cycles of oilseed rape, wheat, barley and pea. Plant and Soil 228: 105-115.

Schleper, C. and Nicol, G.W. 2010. Ammonia-oxidising archae. Physiology, ecology and evolution. pp. 1-41. In: Poole, R.K. (ed.). Advances in Microbial Physiology. Academic Press, Oxford.

Scholefield, D., Hawkins, J.M.B. and Jackson, S.M. 1997. Use of a flowing helium atmosphere incubation technique to measure the effects of denitrification controls applied to intact cores of a clay soil. Soil Biology and Biochemistry 29: 1337-1344.

Schröder, J.J., Assinck, F.B.T., Uenk, D. and Velthof, G.L. 2010. Nitrate leaching from cut grassland as affected by the substitution of slurry with nitrogen mineral fertilizer on two soil types. Grass and Forage Science 65: 49-57.

Seo, D.C. and DeLaune, R.D. 2010. Fungal and bacterial mediated denitrification in wetlands: influence of sediment redox condition. Water Research 44: 2441-2450.

Shapleigh, J.P. 2006. The Prokaryotes. pp. 769-792. In: Falkow, S., Rosenberg, E., Schlieger, K.H. and Stackebrant, E. (eds.). A Handbook on the Biology of Bacteria. New York, Springer.

Shibata, H., Branquinho, C., McDowell, W.H., Mitchell, M.J., Monteith, D.T., Tang, J., et al. 2015. Consequence of altered nitrogen cycles in the coupled human and ecological system under changing climate: the need for long-term and site-based research. Ambio 44: 178-193.

Shoun, H. and Tanimoto, T. 1991. Denitrification by the fungus Fusarium oxysporum and involvement of cytochrome P-450 in the respiratory nitrite reduction. Journal of Biological Chemistry 266: 11078-11082.

Shoun, H., Fushinobu, S., Jiang, L., Kim, S.W. and Wakagi, T. 2012. Fungal denitrification and nitric oxide reductase cytochrome P450nor. Philosophical Transactions of the Royal Society B: Biological Sciences 367: 1186-1194.

Shoun, H. and Fushinobu, S. 2017. Metalloenzymes in Denitrification: Applications and Environmental Impacts. pp. 331-348. *In*: Moura, I.M., Pauleta, S.R. and Maia, L.B. (eds.). Royal Society of Chemistry. United Kingdom.

Šimek, M., Jíšová, L. and Hopkins, D.W. 2002a. What is the so-called optimum pH for denitrification in soil? Soil Biology and Biochemistry 34: 1227-1234.

Šimek, M. and Cooper, J.E. 2002b. The influence of soil pH on denitrification: progress towards the understanding of this interaction over the last 50 years. European Journal of Soil Science 53: 345-354.

Simpson, P.J., Richardson, D.J. and Codd, R. 2010. The periplasmic nitrate reductase in Shewanella: the resolution, distribution and functional implications of two NAP isoforms, NapEDABC and NapDAGHB. Microbiology 156(Pt 2): 302-312.

Singh, B.K., Bardgett, R.D., Smith, P. and Reay, D.S. 2010. Microorganisms and climate change: terrestrial feedbacks and mitigation options. Nature Reviews Microbiology 8: 779-790.

Singh, N. and Bhatla, S.C. 2019. Hemoglobin as a probe for estimation of nitric oxide emission from plant tissues. Plant Methods 15: 39.

Smart, D.R. and Bloom, A.J. 2001. Wheat leaves emit nitrous oxide during nitrate assimilation. Proceedings of the National Academy of Sciences 98: 7875-7878.

Smil, V. 2011. Nitrogen cycle and world food production. World Agriculture 2: 9-13.

Snyder, C.S., Bruulsema, T.W., Jensen, T.L. and Fixen, P.E. 2009. Review of greenhouse gas emissions from crop production systems and fertilizer management effects. Agriculture Ecosystem and Environment 133: 247-266.

Sparacino-Watkins, C., Stolz, J.F. and Basu, P. 2014. Nitrate and periplasmic nitrate reductases. Chemical Society Reviews 43: 676-706.

Stanton, C.L., Reinhard, C.T., Kasting, J.F., Ostrom, N.E., Haslun, J.A., Lyons, T.W., et al. 2018. Nitrous oxide from chemodenitrification: a possible missing link in the Proterozoic greenhouse and the evolution of aerobic respiration. Geobiology 16: 597-609.

Stocker, T. 2014. Climate change 2013: the physical science basis: Working Group I contribution to the Fifth assessment report of the Intergovernmental Panel on Climate Change, New York: Cambridge University Press.

Subbarao, G.V., Ito, O., Sahrawat, K.L., Berry, W.L., Nakahara, K., Ishikawa, T., et al. 2006. Scope and strategies for regulation of nitrification in agricultural systems—challenges and opportunities. Critical Reviews in Plant Sciences 25: 303-335.

Subbarao, G.V., Rondon, M., Ito, O., Ishikawa, T., Rao, I.M., Nakahara, K., et al. 2007. Biological nitrification inhibition (BNI) - Is it a widespread phenomenon? Plant and Soil 294: 5-18.

Subbarao, G.V, Nakahara, K. and Ishikawa, T. 2008. Free fatty acids from the pasture grass Brachiaria humidicola and one of their methyl esters as inhibitors of nitrification. Plant and Soil 313: 89-99.

Subbarao, G.V., Nakahara, K., Hurtado, M.P., Ono, H., Moreta, D.E., Salcedo, A.F., et al. 2009. Evidence for biological nitrification inhibition in Brachiaria pastures. Proceedings of the National Academy of Sciences 106: 17302-17307.

Subbarao, G.V, Nakahara, K., Ishikawa, T., Ono, H., Yoshida, M., Yoshihashi, T., et al. 2013. Biological nitrification inhibition (BNI) activity in sorghum and its characterization. Plant and Soil 366: 243-259.

Sun, L., Lu, Y., Yu, F., Kronzucker, H.J. and Shi, W. 2016. Biological nitrification inhibition by rice root exudates and its relationship with nitrogen-use efficiency. New Phytologist 212: 646-656.

Syakila, A. and Kroeze, C. 2011. The global nitrous oxide budget revisited. Greenhouse Gas Measurement and Management 1: 17-26.

Takaya, N., Catalan-Sakairi, M.A.B., Sakaguchi, Y., Kato, I., Zhou, Z. and Shoun, H. 2003. Aerobic denitrifying bacteria that produce low levels of nitrous oxide. Applied and Environmental Microbiology 69: 3152-3157.

Tian, L.L., Cai, Y.J. and Akiyama, H. 2019. A review of indirect N$_2$O emission factors from agricultural nitrogen leaching and runoff to update of the default IPCC values. Environmental Pollution 245: 300-306.

Tosha, T. and Shiro, Y. 2013. Crystal structures of nitric oxide reductases provide key insights into functional conversion of respiratory enzymes. IUBMB Life 65: 217-226.

Umana, N.H.-N. and Wanek, W. 2010. Large canopy exchange fluxes of inorganic and organic nitrogen and preferential retention of nitrogen by epiphytes in a tropical lowland rainforest. Ecosystems 13: 367-381.

Uscola, M., Villar-Salvador, P., Oliet, J. and Warren, C.R. 2014. Foliar absorption and root translocation of nitrogen from different chemical forms in seedlings of two Mediterranean trees. Environmental and Experimental Botany 104: 34-43.

Veraart, A., Klein, J. and Scheffer, M. 2011. Warming can boost denitrification disproportionately due to altered oxygen dynamics. PLoS One 6(3). e18508.

Vitousek, P.M., Porder, S., Houlton, B.Z. and Chadwick, O.A. 2010. Terrestrial phosphorus limitation: mechanisms, implications, and nitrogen–phosphorus interactions. Ecological Applications 20: 5-15.

Vitousek, P.M., Menge, D.N.L., Reed, S.C. and Cleveland, C.C. 2013. Biological nitrogen fixation: rates, patterns and ecological controls in terrestrial ecosystems. Philosophical Transactions of the Royal Society B. 368: 20130119.

Voss, M., Bange, H.W., Dippner, J.W., Middelburg, J.J., Montoya, J.P. and Ward, B. 2013. The marine nitrogen cycle: recent discoveries, uncertainties and the potential relevance of climate change. Philosophical Transactions of the Royal Society B: Biological Sciences 368: 20130121-20130121.

Wang, C., Liu, D. and Bai, E. 2018. Decreasing soil microbial diversity is associated with decreasing microbial biomass under nitrogen addition. Soil Biology and Biochemistry 120: 126-133.

Welten, B.G., Ledgard, S. and Luo, J. 2014. Administration of dicyandiamide to dairy cows via drinking water reduces nitrogen losses from grazed pastures. Journal of Agricultural Science 152: 150-158.

Wimalasekera, R., Tebart, F. and Scherer, G.F.E. 2011. Polyamines, polyamine oxidases and nitric oxide in development, abiotic and biotic stresses. Plant Science 181: 593-603.

WMO 2019. WMO statement on the state of the global climate in 2018. https://library.wmo.int/doc_num.php?explnum_id=5789.

WPP 2019. United Nations, Department of Economic and Social Affairs, Population Division. World Population Prospects 2019: Highlights (ST/ESA/SER.A/423).

Wrage-Mönnig, N., Horn, M.A., Well, R., Müller, C., Velthof, G. and Oenema, O. 2018. The role of nitrifier denitrification in the production of nitrous oxide revisited. Soil Biology and Biochemistry 123: A3-A16.

Wunsch, P., Herb, M., Wieland, H., Schiek, U.M. and Zumft, W.G. 2003. Requirements for Cu(A) and Cu-S center assembly of nitrous oxide reductase deduced from complete periplasmic enzyme maturation in the nondenitrifier Pseudomonas putida. Journal of Bacteriology 185: 887-896.

Zakir, H.A.K.M., Subbarao, G.V., Pearse, S.J., Gopalakrishnan, S., Ito, O., Ishikawa, T., et al. 2008. Detection, isolation and characterization of a root-exuded compound, methyl 3-(4-hydroxyphenyl) propionate, responsible for biological nitrification inhibition by sorghum (Sorghum bicolor). New Phytologist 180: 442-451.

Zaman, M. and Blennerhassett, J.D. 2010. Effects of the different rates of urease and nitrification inhibitors on gaseous emissions of ammonia and nitrous oxide, nitrate leaching and pasture production from urine patches in an intensive grazed pasture system. Agriculture Ecosystem and Environment 136: 236-246.

Zhang, Y., Song, L., Liu, X., Li, W., Lü, S., Zheng, L., et al. 2012. Atmospheric organic nitrogen deposition in China. Atmospheric Environment 46: 195-204.

Zhang, Y., Ding, H., Zheng, X., Cai, Z., Misselbrook, T., Carswell, A., et al. 2018. Soil N transformation mechanisms can effectively conserve N in soil under saturated conditions compared to unsaturated conditions in subtropical China. Biology and Fertility of Soils 54: 495-507.

Zhao, X., Liu, S.L., Pu, C., Zhang, X.Q., Xue, J.F., Zhang, R., et al. 2016. Methane and nitrous oxide emissions under no-till farming in China: a meta-analysis. Global Change Biology 22: 1372-1384.

Zhou, Y., Shuiyuan, C., Lang, J., Chen, D., Zhao, B., Liu, C., et al. 2015. A comprehensive ammonia emission inventory with high-resolution and its evaluation in the Beijing–Tianjin–Hebei (BTH) region, China. Atmospheric Environment 106: 305-317.

Zhou, Z., Takaya, N., Sakairi, M.A.C. and Shoun, H. 2001. Oxygen requirement for denitrification by the fungus Fusarium oxysporum. Archives of Microbiology 175: 19-25.

Zhou, Z., Wang, C., Zheng, M., Jiang, L. and Luo, Y. 2017. Patterns and mechanisms of responses by soil microbial communities to nitrogen addition. Soil Biology and Biochemistry 115: 433-441.

Zhu, L., Ding, W., Feng, L.J., Kong, Y., Xu, J. and Xu, X.Y. 2012. Isolation of aerobic denitrifiers and characterization for their potential application in the bioremediation of oligotrophic ecosystem. Bioresource Technology 108(Supplement C): 1-7.

Zhu, X., Burger, M., Doane, T.A. and Horwath, W.R. 2013. Ammonia oxidation pathways and nitrifier denitrification are significant sources of N and NO under low oxygen availability. Proceedings of the National Academy of Sciences 110: 6328-6333.

Zumft, W.G. 1997. Cell biology and molecular basis of denitrification. Microbiology and Molecular Biology Reviews 61: 533-616.

Zumft, W.G. 2005. Nitric oxide reductases of prokaryotes with emphasis on the respiratory, heme-copper oxidase type. Journal of Inorganic Biochemistry 99: 194-215.

# 4

# Nitrification and Denitrification Processes: Environmental Impacts

Jessica Purswani[*] and Clementina Pozo Llorente

## 4.1 INTRODUCTION

Nitrogen (N) is one of the most important nutrients for living organisms. It is a fundamental component of amino acids and nucleic acids. The element is highly demanded by plants and microorganisms and is indispensable for crop production. N is the most abundant element in our atmosphere. The majority of all N (78% of the global total) exists in the form of $N_2$ (as an inert gas) and is continuously released to the atmosphere from volcanic and thermal eruptions.

Nitrogen cycles between organic and inorganic forms. The biogeochemical cycle of N is one of the most studied and relevant cycles in which diazotrophs, nitrifying microorganisms (ammonium oxidiser bacteria [AOB], ammonium oxidiser archaea [AOA], nitrite oxidiser bacteria [NOB], commamox bacteria or anammox bacteria), and denitrifying microorganisms catalyse diverse processes such as nitrogen fixation, ammonium assimilation, ammonification, ammonium oxidation (nitrification and anammox reactions in aerobic and anaerobic conditions, respectively), and nitrate reduction.

Nitrate reduction yields ammonium which can be incorporated into biomass (assimilatory nitrate reduction), exist as free ammonium (dissimilatory nitrate reduction to ammonium [DNRA]), or reduce to $N_2$ via the sequential reduction of various inorganic forms of nitrogen, including nitrite, nitric oxide, and nitrous oxide ($N_2O$) as part the denitrification process (Kuypers et al. 2018). A more detailed description of the N-cycle has been included in Chapter 1 (Nitrogen Cycle: An Overview).

Due to anthropogenic activities, the biochemical cycle involving nitrogen has been drastically altered, which has produced markedly negative consequences both for the environment and human health (Vitousek et al. 1997). Excessive and irrational use of nitrogenous fertilisers in agriculture, discharges of sewage to soils, livestock activities, percolation of leachates from landfills, and the increased biological fixation of atmospheric $N_2$, industrial activities, wastewater treatment programmes, and the combustion of biomass and fossil fuels have enhanced the levels of several reactive N forms (Nr), including $N_2O$, nitrate ($NO_3^-$), nitrite ($NO_2^-$), ammonia ($NH_3$), and ammonium ($NH_4^+$), both in terrestrial and aquatic ecosystems and atmosphere (Mosier et al. 2001, Galloway et al. 2003, 2008, Fowler et al. 2013).

As a result of our need to feed the continuously growing human population, the application of N-fertilisers to cropping systems has increased markedly in recent decades. The application of

Department of Microbiology. Faculty of Sciences. Institute of Water Research. Av. Fuentenueva. The University of Granada. 18071. Granada (Spain).
[*] Corresponding author: jessicapurswani@ugr.es

fertilisers often has replaced N-biological fixation and manure amendments. Synthetic nitrogen fertilisers used in agricultural soils represent about 35% of the total nitrogen of anthropogenic origin and has been reported to be an important contributor to the observed disturbances in the soil's N-cycle (Spiertz et al. 2010). In fact, several studies have evaluated the effects of N-fertilisation on the abundance of microbial communities and genes involved in steps of the N-cycle (N-fixation, nitrification, and denitrification) (Gubry-Rangin et al. 2010, Wessen et al. 2011, Carey et al. 2016, Ouyang et al. 2018). The majority of these studies concluded that ammonium oxidising bacteria (AOB) and ammonium oxidising archaea (AOA) communities were affected by N-fertilisation. The AOB were determined to be more sensitive than AOA to N inputs, and genes involved in the nitrification and denitrification processes were modified, which resulted in increased levels of nitrates and emissions of Nr gasses emitted to the atmosphere. However, the effect of N-fertilisation on the structural and functional behaviour of diazotrophs has not been determined by performing short or long-term experiments (Reardon et al. 2014, Feng et al. 2018, Ouyang et al. 2018). No matter what type of N is applied to agricultural soil (livestock residues or synthetic fertilisers), cropping systems typically utilise the added N inefficiently (no more than 60% is used). Although denitrification removes some of the accumulated N (retained or unconsumed by plants or microorganisms), the remaining N may be lost via leaching or runoff processes, which are affected by soil characteristics, crop management practices, and climatic conditions. Hence, nitrate causes groundwater contamination and eutrophication of surface waters. Nitrate contamination renders these water sources inadequate for consumption. Drinking contaminated water can cause human health hazards, including methemoglobinemia and several cancers (Follet and Hatfield 2001).

In addition to nitrates, other forms of Nr that occur as a result of N-cycle disturbance have been detected. Nr gasses, including ammonia and $N_2O$, are released from the soil in a process that is largely induced by the presence of elevated levels of N inputs from synthetic N-fertiliser and manure within the soil (Davidson 2009). Changes in agronomic practices have the potential to reduce rates of $NH_3$ volatilisation. Huang et al. (2016) carried out a study in diverse types of croplands of China. The researchers were able to report how the deep placement of fertiliser significantly reduced the levels of $NH_3$ emissions by around 60%. Other strategies such as the use of urease inhibitors or other N fertilisers have also been effective in reducing emissions (Ju and Zhang 2017).

The concentration of $N_2O$ in the atmosphere is continually increasing, and the increases observed have been particularly acute as a result of increased levels of agricultural and industrial activities (IPCC 2013). $N_2O$ emissions contribute to global greenhouse gas (GHG) accumulation and stratospheric ozone depletion, which releases two molecules of nitric oxide (NO) per molecule of ozone destroyed. Although the levels of $N_2O$ emissions are lower compared with other greenhouse gases (such as methane and $CO_2$), its global warming power is still about 300 times higher than $CO_2$ (European Chemicals Agency, UE. https://echa.europa.eu/).

The terrestrial ecosystems release a greater proportion of nitrous oxide than aquatic ecosystems (two-thirds of the total released) (IPCC 2013). $N_2O$ emission to the atmosphere originates from both natural (forests and oceans) and anthropogenic sources. Anthropogenic sources of $N_2O$ include agricultural activities (widespread use of livestock residues or synthetic N-fertilisers), wastewater treatment plants, biomass/fossil fuel combustion, and industrial activities. Among the anthropogenic sources listed, agricultural activities contribute to the greatest percentage of $N_2O$ emitted (Firestone et al. 1980).

Denitrification, ammonium oxidation to nitrite, nitrifier denitrification, and to a lesser extent dissimilatory nitrate reduction to ammonium (DNRA) are microbial processes involved in the biochemical N-cycle. These processes contribute to diverse proportions of $N_2O$ released (Figure 4.1). Among the chemical process listed, denitrification processes release the greatest proportion of $N_2O$ molecule since it is an intermediate of the reduction of nitrate to $N_2$. Levels of the nitrous oxide reductase (Nos) enzyme, which is involved in the reduction of $N_2O$ to dinitrogen gas ($N_2$) and its inhibition by certain environmental/operational factors (pH, $O_2$, C/N ratio, or temperature), determines the levels of $N_2O$ released into the atmosphere (Philippot 2007).

The microbial nitrification processes also release $N_2O$ into the atmosphere. Ammonia is added to the soil through the use of inorganic, synthetic fertiliser, and manure (organic fertiliser). The compound can be oxidised to form hydroxylamine ($NH_2OH$), which is further oxidised to nitric oxide (NO). The NO can be either further oxidised to nitrite ($NO_2^-$) or reduced to $N_2O$.

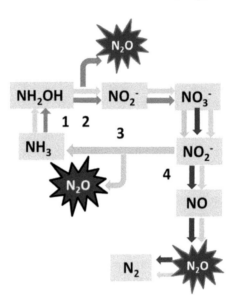

**FIGURE 4.1** Nitrous oxide sources from the N-cycle. 1: Nitrifier denitrification. 2: Nitrification. 3: Dissimilatory Nitrate Reduction to Ammonium (DNRA). 4: Denitrification.

## 4.2 NITRIFICATION

### 4.2.1 General Aspects

Nitrification involves the conversion of ammonium (via nitrite) to nitrate by microorganisms living in habitats, such as soil, freshwater, marine water, and WWTPS. The main habitats affected by anthropogenic addition of N are agricultural systems, which receive N via fertilisation. However, the environmental impact of increasing N levels extends beyond the soil, affecting the surface freshwater and groundwater of surrounding areas. Nitrate levels within European countries have remained constant throughout the last 30 years (https://www.eea.europa.eu/), even though large campaigns have been conducted to reduce anthropogenic use of N-based fertilisers. The biological factors that impact nitrification and their environmental effects are discussed below.

### 4.2.2 Enzymes

The main enzymatic steps involved in nitrification pathways have been previously elucidated (Figure 4.2). The microbial oxidation of ammonia to hydroxylamine is performed by ammonia monooxygenase (AMO) (Hooper et al. 1997). The AMO enzyme contains 3 polypeptides (encoded by genes *amoA*, *amoB*, and *amoC*) and is present in both aerobic oxidising bacteria and archaea. Although the AMO enzyme is closely related to methane monooxygenase (MMO), which is found in methanotrophs, the efficiency of the oxidation of ammonia to hydroxylamine by MMO is very low (Stein et al. 2011).

Hydroxylamine is further oxidised by the enzyme hydroxylamine oxidoreductase (HAO), which was initially thought to oxidise hydroxylamine to nitrite. However, researchers were subsequently able to show that nitric oxide was produced by the enzyme (Caranto et al. 2017). Nitric oxide (NO) is further oxidised to nitrate by nitric oxide oxidase (NOD) or to nitrite by an unknown enzyme (possibly a copper-containing nitrite reductase: Cu-NIR). Additionally, NO can be reduced to $N_2O$ by nitric oxide reductases (NOR: cytochrome c-dependant, cNOR; quinol-dependent, qNOR; Copper-containing quinol-dependent, $Cu_ANOR$; NADH-dependent cytochrome $P_{450}$, $P_{450}NOR$). Nitrite oxidation is the main biochemical pathway used to produce nitrate from nitrite and is catalysed by nitrite oxidoreductase (NXR) enzymes.

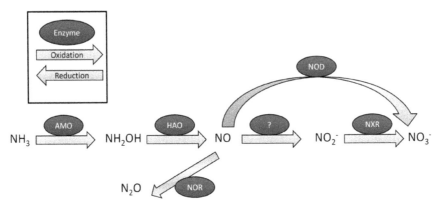

**FIGURE 4.2** Enzymatic transformations within the nitrification pathway. The following enzymes perform the nitrogen transformations: ammonia monooxygenase (AMO, *amoCBA*), hydroxylamine oxidoreductase (HAO, *hao*), nitrite oxidoreductase (NXR, *nxrAB*), nitric oxide reductase (NOR, *cnorG/norZ/norVW/p450nor*), nitric oxide oxidase (NOD, *hmp*).

## 4.2.3  Nitrifying Microorganisms

AOB oxidise ammonia to hydoxylamine using ammonia monooxygenase. Most AOB are chemolithotrophs that belong to the Betaproteobacteria and Gammaproteobacteria classes (Arp et al. 2003). One of the first microorganisms identified to be capable of performing this reaction was *Nitrosomonas europaea* by Sergei Winogradsky in 1892 (Watson 1971). Many other Nitrosomonas bacteria have been identified as AOB, including *N. communis*, *N. ureae*, *N. aestuarii*, *N. marina*, *N. nitrosa*, *N. eutropha*, *N. oligotropha*, and *N. halophila* (Koops et al. 1991). To date, other AOB that have been described include *Nitrosospira multiformis*, *N. briensis*, *Nitrisiciccus oceanus*, *Methylosinus trichosporium*, and *Methylococcus capsulatus* (Biller et al. 2012). Cold-adapted *Nitrosospira* spp. have been suggested to be potential ammonia oxidisers, which are present in the cryosols of permafrost within Siberian landscapes (Sanders et al. 2019).

AOA, such as *Nitrosopumilus maritimus,* that belong to the *Thaumarchaeota* phylum can also oxidise ammonia to hydroxylamine (Könneke et al. 2005). AOA were first reported in wastewater treatment plants, and the abundance of AOB and AOA have been monitored continually, although there is no consensus regarding whether AOA or AOB are more abundant in specific types of wastewater treatment schemes. However, the adaptability of thermophilic nitrifier archaea, including *Nitrosocaldus yellowstonii*, have increased the upper-temperature limit of nitrification (de la Torre et al. 2008). A large number of distinct aquatic habitats have previously been sampled to assess the presence of the archaeal *amoA* gene, which encodes the AMO enzyme. The study was able to determine that *amoA* sequences were specific to habitat (salinity, depth in the water column, temperature, aquatic environment) and identified only a few sequences in more than one habitat type. Therefore, the authors concluded that AOA populations thrive in specific aquatic environments (Biller et al. 2012). Whether AOA are specific to particular soil environments remains unknown.

Ammonia oxidising bacteria and AOA also contain hydroxylamine oxidoreductase (HAO) and the candidate Cu-NIR enzyme. Thus, the microorganisms are able to oxidise hydoxylamine to nitric oxide (NO) and nitrite. NOB can produce nitrate from nitrite using NXR enzymes. The diversity of NOB is large from Alphaproteobacteria, Betaproteobacteria, and Gammaproteobacteria to Chloroflexi, Nitrospirae, and Nitrospinae phyla (Daims et al. 2016). Anaerobic NOB, such as *Thiocapsa* sp. and *Rhodopseudomonas* sp., can oxidise nitrite anaerobically by coupling nitrite oxidation with phototrophy. Planctomycetes, such as *Jettenia*, *Brocadia*, and *Anammoxoglobus* genera, are anaerobic ammonium oxidising bacteria (annamox) that have been shown to convert ammonia to nitrite and reduce it to nitrogen gas without using a carbon source (van de Graaf et al. 1997).

Nitrifying microorganisms have the capacity to partially oxidise ammonia to nitrate but cannot perform all of the steps of the pathway. Additionally, AOB, AOA, and NOB are not phylogenetically related. Recently, it was determined that *Candidatus Nitrospira inopinata* (Daims et al. 2015) and other *Nitrospira* candidates (van Kessel et al. 2015) are capable of completely oxidising ammonia to nitrate. The complete oxidation of ammonia is referred to as "comammox". *Candidatus Nitrospira inopinata* contains all enzymes required for nitrification, including AMO, HAO, and NXR.

### 4.2.4 Nitrification and Its Impact on the Environment

The continuous addition of synthetic ammonium-based fertilisers in the soil, at rates and concentrations greater than the capacity of soil microbiome and plants to transform the compounds to biomass and nitrogen gas, largely results in N imbalance and the presence of excess nitrogen-based intermediate compounds. The microbial nitrification activities of AOB, AOA, NOB, and commamox bacteria result in the oxidation of ammonium to nitrate ions. When excess ammonium is present, both ammonium and nitrate ion concentrations reach aquatic systems via leaching and water runoff. This leads to eutrophication events in freshwater systems in which flow rates are low and stagnant.

The farming industry is also a large contributor to ammonia amendments to the soil. Fertilisers comprised of animal manure contain purines, which are degraded to uric acid and subsequently to urea and ammonia. High concentrations of organic ammonia produce the same types of problems as those of synthetic fertilisers when the soil capacity to transform ammonia and organic materials disrupt the balanced flow of N-assimilation and nitrogen gas production.

Natural sources of ammonia present in aquatic freshwater systems are predominantly derived from freshwater fish. Ammonia is toxic to all vertebrates, and its excretion is a necessity, whether in the form of ammonia or urea. The production of urea is a more metabolically expensive process than ammonia production and is mainly utilised by terrestrial vertebrates and marine fish (Olson et al. 1971). Hence, the cycling of nitrogen from ammonia to nitrate in freshwater via nitrification makes nitrates available to organisms, including microscopic algae and freshwater plants. The same is true for marine life, i.e., nitrification makes nitrates available to marine organisms.

The activity of nitrite- and ammonia-oxidising microorganisms in wastewater treatment plants is the first step of conventional nitrogen removal and is a key feature of the reduction of ammonium. In the event of incomplete nitrification, which occursunder low oxygen levels, the production of $N_2O$ increases.

#### 4.2.4.1 $N_2O$ Emissions and the Presence of $NO_3^-$ in Natural and Engineered Ecosystems

##### 4.2.4.1.1 $N_2O$ Emissions and $NO_3^-$ in Terrestrial Ecosystems

Nitrification in terrestrial ecosystems is fundamental for the recycling of nitrogen from ammonium to nitrate. Plants are capable of utilising both $NO_3^-$ and $NH_4^+$ ions as sources of N-assimilation. Ammonium is directly used to synthesise glutamine and subsequently other amino acids. Nitrate is assimilated to ammonium before it can be used (Salsac et al. 1987). Ammonium is the preferred form of N used for plant assimilation; however, under oxic conditions, AOB and AOA compete with plants for ammonium and thus nitrate assimilation will also occur. Mixtures of both types of N ions stimulate either root branching or root elongation. This increases the surface areas of the root network and increases nutrient uptake.

Up to 50% of the N available for plant assimilation is lost via leaching, which increases $NO_3$ levels in freshwater and the incidence of water eutrophication that occurs when the N compounds do not have sufficient metabolic flow and oxygen transfer rates. Partial inhibition of this process is essential for reducing nitrogen loss and reduce the fertiliser application in agricultural systems.

Inhibitors of nitrification include DCD (dicyandiamide), N-serve (nitrapyrin), and ENTEC (DMPP; 3,4-dimethylpyrazolephosphate), which act on the AmoB subunit of ammonia monooxygenase by chelating copper present at the active site of the enzyme (Beeckman et al. 2018). Nitrification is also affected by soil pH. The assumption that acidic soil conditions decrease nitrification is currently being challenged by new research, which has shown that equal levels of nitrification activity can be achieved at pH levels as low as 3.9 (Norton and Stark 2011).

Optimal temperature conditions for nitrification in terrestrial systems have been reported to be between 30°C and 40°C in soils for ammonia oxidisers (Avrahami et al. 2003). Among factors that affect the abundance and structure of archaeal and bacterial communities, the temperature was the most important. In fact, AOB abundance was shown to be stimulated by increases in temperature (Dai et al. 2013). The top layer of the soil (0-5 cm depth) in many areas is exposed to extreme temperatures as a result of sun exposure, especially in areas in which vegetation is not present. Most nitrifying microorganisms studied have been mesophiles, yet thermophilic AOB, NOB, and AOA have also been identified (Edwards et al. 2013, Abby et al. 2018, Lopez-Vazquez et al. 2014).

During nitrification, the production of NO intermediates is capable of being reduced to $N_2O$ when oxygen concentrations are limited. Furthermore, in carbon-limited systems, increased nitrate concentrations enhanced the levels of $N_2O$ emissions as a result of incomplete denitrification.

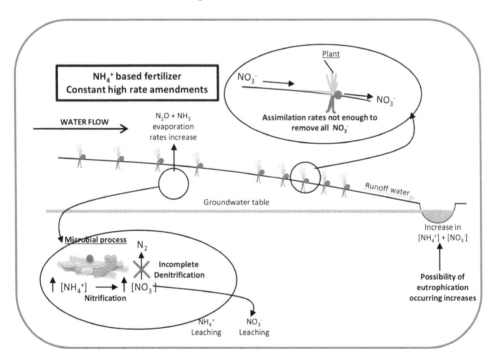

FIGURE 4.3 Environmental impacts of fertilizer overuse on freshwater. Constant amendments of ammonium-based fertilizers increase microbial nitrification rates. However, nitrate saturation is reached when plant assimilation and denitrification process are slower than the rate at which ammonium and nitrate uptake take place. With incomplete assimilation, runoff water will deposit high nitrate loads into the river. Additionally, nitric oxide production will increase due to the presence of high ammonium loads which inhibit complete denitrification. High loads of ammonium and nitrate ions will also leach into the groundwater table. Ultimately, an increase of nitrogen ions in freshwaters increases the probability of eutrophication events, especially when these waters areaccompanied by high phosphorous loads.

Overall, the overuse of ammonium-based fertilisers (Figure 4.3) leads to the saturation of N-assimilation rates and increases evaporation rates of ammonia. Nitrification rates of AOB and AOA are high, and microorganisms convert $NH_4^+$ to $NO_3^-$ steadily and yet high $NO_3^-$ loads inhibit complete denitrification and enhance the release of $N_2O$. The compound is a greenhouse gas and contributes highly to global warming. Additionally, the high concentration of $NO_3^-$ ions cannot be assimilated by plants and thus enter aquatic ecosystems via soil runoff or leaching to groundwater, enhancing the likelihood of the occurrence of eutrophication. Ammonia evaporation is also toxic to human health in areas with relatively high humidity. High levels of ammonia can cause skin and eye irritation as well as a burning sensation if inhaled. This is especially problematic when large amounts of manure are added as fertilisers.

### 4.2.4.1.2   $N_2O$ Emissions and $NO_3^-$ From Freshwater Ecosystems

Ammonium, nitrite, and nitrate are the most common ionic forms of dissolved inorganic nitrogen in aquatic ecosystems (Rabalais 2002). The main natural sources of these ions in freshwater are atmospheric deposition, dissolution of nitrogen-rich geological deposits, $N_2$ fixation, and biological degradation of organic matter (ammonia/urea from aquatic organisms or faeces deposits). Anthropogenic activities provide other sources of these nitrogen ions, which include purine leachates from farming activities, runoff, or groundwater containing leachates from ammonia-based agricultural fertilisers and contaminants from wastewater treatment plants.

The total quantity of nitrogen compounds leached depends on crop type, the dynamics of the nitrogen in the soil, soil characteristics, the level of organic matter available, and the pluviometry estimated for the year and climate of the region (Haygarth et al. 2003). The level of fertiliser leached into groundwater is variable and reported levels typically range from 5% to 25%, while some studies have reported higher values (30-50%) (Goderya et al. 1996, Shamrukh et al. 2001). Leaching was reported to be especially high in bare soils in which crops had been collected (Eugercios 2013).

The environmental impact of nitrification on freshwater systems has mainly been attributed to the accumulation of high nitrate concentrations (along with phosphorous pollution from households and industry) which lead to eutrophication, i.e., the overgrowth of algae results in rapid decreases in oxygen concentrations when the organisms die and decompose, and low oxygen levels are rapidly followed by the death of bottom-dwelling animals and fish. Fishes also emigrate from the affected area. Unbalanced ecosystems and alterations in the chemical composition of water bodies make them unsuitable for recreation, consumption, and other uses.

Under European law, nitrate concentrations are limited to 50 mg/L, yet 15% of the European countries have sources of groundwater that exceed these limits, and 3% have surface water bodies that exceed this limit (https://ec.europa.eu/environment/pubs/pdf/factsheets/nitrates.pdf). Decreased levels of nitrates (< 25 mg/L) within the EU were reported for 66% of European countries for groundwater, and 21% of countries contained < 2 mg/L nitrates in surface water. Eutrophication (or red tide) occurs when N concentrations in the water reach 300 µg/L and P concentrations reach 20 µg/L (Richardson et al. 2007). Hence, nitrate concentration cutoffs are monitored, and when they are exceeded, intervention strategies to reduce N concentrations are required to avoid eutrophication. Furthermore, countries that extract groundwater from deep depths for human consumption will be less affected by nitrate contamination from agricultural sources than those that extract groundwater from decreased depths.

High ammonium concentrations, under aerobic conditions, result in the release of $N_2O$ global greenhouse gas (GHG) to the atmosphere. In anaerobic aquatic environments and anoxic sediments, the facultative anaerobic bacteria utilise nitrite as terminal acceptors of electrons which results in the formation of $N_2O$ (Camargo and Alonso 2006). Hence, $N_2O$ is released under both aerobic and anaerobic conditions.

In reservoirs, nitrification is the main pathway of $N_2O$ production where seasonal dissolved oxygen stratification plays an important role in the regulation of its production. The reservoir of

$N_2O$ flux had a significant negative logarithmic relationship with hydraulic load, suggesting a role for the reservoir in the control of $N_2O$ emissions (Liang et al. 2019).

Furthermore, a global assessment by Camargo and Alonso (2006) indicates that inorganic nitrogen pollution increases the concentration of hydrogen ions in freshwater ecosystems without significant acid-neutralising capacity. This results in the acidification of the ecosystems when incomplete nitrification occurs (i.e., from NO accumulation) (Baker et al. 1991). In addition to eutrophication, inorganic nitrogen pollution induces the occurrence of toxic algae and also enhances the capacity of nitrogenous compounds, which impair the ability of aquatic animals to survive, grow, and reproduce (Camargo and Alonso 2006).

Ammonia in its ionised ($NH_4^+$) and un-ionised ($NH_3$) forms exist in chemical equilibrium that depends on the pH and temperature of the water. As pH and temperature increases, so do concentrations of $NH_3$, whilst ionised ammonia concentrations decrease (Emerson et al. 1975). Moreover, $NH_3$ causes toxicity to nitrifying bacteria, such as *Nitrosomonas* and *Nitrobacter*, which thereby inhibits nitrification and increases the accumulation of $NH_4^+$ in the aquatic environment. This causes toxicity in bacteria and aquatic animals (Russo 1985). Accumulation of nitrite ions ($NO_2^-$) and un-ionised nitrous acid ($HNO_2$) can also cause toxicity to *Nitrosomonas* and *Nitrobacter* and thus inhibits the nitrification process and affects aquatic organisms. Nitrite toxicity in organisms is caused by the entry of nitrite to red blood cells and prevents oxygen release to the body tissue due to its high dissociation constant (Jensen 2003). Levels of toxicity from nitrate ions are considered relatively low since branchial permeability to $NO_3$- is more limited than nitrite uptake. However, freshwater animals appear to be more sensitive to nitrate toxicity than seawater animals (Camargo et al. 2005); in Canada, a limit of 2 mg $NO_3$-N/L has been put in place.

The nitrification process can occur when levels of dissolved oxygen are as low as 1.0 mg $O_2$/L. However, at 0.5 mg $O_2$/L ammonia, oxidation was not affected and nitrite was accumulated under pure autotrophic conditions (Hanaki et al. 1990). Maximum levels of nitrification take place at 15-25°C in a hypereutrophic lake and arctic sediments (Pauer and Auer 2000, Thamdrup and Fleischer 1998). In experimentally acidified Canadian lakes, $NH_4^+$ nitrification was inhibited at pH values below 5.6 (Rudd et al. 1988, 1990). AOB growth within sediments or biofilms containing high concentrations of ammonium had been previously reported (Herrmann et al. 2011), while AOA grew under conditions in which ammonia concentrations were low, such as in oligotrophic or slightly eutrophic lakes (Wu et al. 2010, Wuchter et al. 2006).

### 4.2.4.1.3 $N_2O$ Emissions From Marine Ecosystems

The environmental impacts of $N_2O$ from rivers and groundwater cause the release of 35-50 Tg N per year to oceans, which in turn cause eutrophication, limited oxygen zones, and habitat degradation in coastal areas. The presence of AOB in coastal water with high levels of ammonium has previously been reported (Zhang et al. 2015). Seasonal changes between AOB and AOA have been observed in Rushan Bay, China, where surface sediments that accumulated in the summer were dominated by AOB, while AOA predominated during the winter (He et al. 2018).

However, nitrification has been estimated to have a flux of approximately 2,000 Tg N per year, which explains the high levels of marine-oxidising archaea observed, even when low levels of ammonia were present in the ocean (Kuypers et al. 2018). Additionally, nitrite-oxidising bacteria generate nitrate, the dominant form of biologically available nitrogen in the ocean that contributes to carbon fixation (Pachiadaki et al. 2017).

Nevertheless, in the near future, it is expected that a decrease in the levels of global oceanic nitrification will occur (Beman et al. 2011) as a result of oceanic acidification from the dissolution of anthropogenic $CO_2$ (Sabine et al. 2004). Beman et al. (2011) reported that modest pH decreases (from 7.99-8.09 to 7.85-7.99) induced modest decreases in $NH_4^+$ oxidation rates, yet doubled levels of $pCO_2$. In these studies, rates were reduced from 8-38%. The consequences of an imbalance in oceanic N cycling cannot be foretold since the capacities of microbes to adapt and evolve to these conditions are unknown.

The quantity of $N_2O$ released to the atmosphere via the reduction of NO is estimated to be 4 Tg N per year, and inhibition of complete nitrification would result in NO reduction and an increase in oceanic $N_2O$ emissions (Dore et al. 1998).

#### 4.2.4.1.4 $NH_4$ and $NO_3$ Concentrations in Engineered Ecosystems: Wastewater Treatment Plants

WWTPs have a high load of organic matter and are designed to achieve high nitrogen conversion rates, as well as removal rates, in order to safely return treated water to rivers or reuse the water for irrigation purposes. Biological elimination of nitrogen in WTTPs follows a two-step process of nitrification under strict aerobic conditions that is followed by a denitrification under anoxic conditions.

Until recently, nitrite ions were largely ignored during these processes due to their relatively high instability. However, variable capacities of $NO_2^-$ to be fixed as nitrite, oxidised further to $NO_3^-$, or to be reduced to NO, $N_2O$, or $N_2$ have multiple environmental implications. Therefore, recent WWTPs studies have focused on NOB activity since low levels of activity may increase nitrite leakage to freshwater systems (Daims et al. 2016) or enhance the production of $N_2O$ from incomplete hydroxylamine oxidation (Guo et al. 2018).

AOB microorganisms are present in conventional WWTPs and convert $NH_3$ to $NH_2OH$ and $NO_2^-$ using AMO and HAO enzymes. NOB converts $NO_2^-$ to $NO_3^-$. However, studies have shown that under anaerobic conditions, HAO converts $NH_2OH$ to a nitrosyl radical (NOH), which can lead to the formation of $N_2O$ (Guo et al. 2018). Additionally, during nitrifier denitrification, $NO_2^-$ is reduced to NO in the presence of nitrite reductase (NiR), and NO is then reduced to $N_2O$ in the presence of nitrite oxide reductase (NOR). Since AOB and NOB do not produce NOS enzyme, $N_2O$ is the end-product observed.

The different parameters that increase $N_2O$ production include low levels of oxygen (0.2-1.5 DO mg/L), which occur as a result of low levels of aeration and large quantities of organic matter and high levels of nitrite, which occur as a result of low dissolved oxygen levels, low pH, low temperature, high ammonium concentrations, high nitrite concentrations in WW, and low sludge retention times (Thakur and Medhi 2019, Frutos et al. 2018). Increasing levels of aeration are not always prudent since the stripping of $N_2O$ occurs at increased rates. Therefore, minimised stripping while providing indigenous microorganisms increased time to convert forms of N (Cui et al. 2015) or simultaneously enhancing nitrification-denitrification processes may limit $N_2O$ emissions (Kampschreur et al. 2009). For more details on this, see the discussion of wastewater treatment and the nitrogen cycle in Chapter 8.

## 4.3 DENITRIFICATION

### 4.3.1 General Aspects

Denitrification provides an alternative mechanism for respiration under oxygen limiting conditions in which microorganisms use soluble nitrate or nitrite as terminal electron acceptors. In a four-step process, nitrate is reduced to $N_2$ (Figure 4.4). This process has important effects on the global cycling of N because fixed nitrogen returns to the atmosphere in the $N_2$ form to complete the N-cycle (Knowles 1982, Zumft 1997, Philippot 2007).

$$NO_3^- + 2e^- + 2H^+ \rightarrow NO_2^- + H_2O \quad \textit{E'o + 420 mV}$$

$$NO_2^- + 2e^- + 4H^+ \rightarrow NO + H_2O + 2OH^- \quad \textit{E'o + 375 mV}$$

$$NO + 2e^- + 2H^+ \rightarrow N_2O + H_2O \quad \textit{E'o + 1175 mV}$$

$$N_2O + 2e^- + 2H^+ \rightarrow N_2 + H_2O \quad \textit{E'o + 1355 mV}$$

**FIGURE 4.4** Denitrification reactions. *E'o*: Relative redox potentials.

Heterotrophic microorganisms that perform the entire reduction reaction (nitrate to $N_2$ through $NO_2$, NO, and $N_2O$) are called classical denitrifiers. They use organic carbon compounds as carbon sources, and they are the most common denitrifiers found in nature (van Rijn et al. 2006). However, many other types of denitrifying microorganisms (bacteria, archaea, or fungi) lack the *nosZ* gene that encodes the last enzyme involved in denitrification, Nos. Furthermore, an imbalance in the activity of pooled enzymes involved in the denitrification process may inhibit Nos activity (Philippot et al. 2011, Pauleta et al. 2019), which determines the release of $N_2O$ to the atmosphere (incomplete denitrification).

Along with denitrification, other processes of the N-cycle are considered to be significant sources of atmospheric $N_2O$. These contribute to global $N_2O$ emissions through nitrification, nitrifier denitrification, and dissimilatory nitrate reduction to ammonium (DNRA).

Since denitrification is the main source of atmospheric NO and $N_2O$ and Nos is the enzyme (along with nitrogenase or multicopper oxidase under certain conditions) (Fernandes et al. 2010) most responsible for the reduction of $N_2O$ to $N_2$, denitrification processes may be considered both sources or sinks of $N_2O$. This occurs in contrast with the nitrification process that behaves like a net source of $N_2O$ (Robertson and Tiedje 1987). Mao et al. (2008) carried out a study of soil microcosms that showed how a microbial community with a low level of Nos activity relative to other reductases involved in denitrification process could be a strong $N_2O$ source, while other conditions in which high relative Nos activity may emit less $N_2O$ and could even work as a sink of $N_2O$ produced during nitrification.

## 4.3.2   Enzymes Involved

The denitrification proteome includes four main intracellular metallo-enzymes: nitrate- (Nar/Nap) reductase, nitrite- (NirK/NirS) reductase, NO-(cNor/qNor) reductase, and nitrous oxide-reductase (NosZI and NosZII). The enzymes are localised within periplasmic space or bound to membranes and are encoded by *narG/napA*, *nirK/nirS*, *c-norC/q-norC*, and *nosZI-nosZII* (clade I and clade II genes, respectively). In bacteria, multiple copies of genes involved in denitrification within a single organism and sometimes in plasmids. Therefore, the horizontal transfer of genes is a normal mechanism in which denitrification capacity can be acquired (Demanèche et al. 2009).

Nitrite reductase (Nir) is the crucial enzyme involved in denitrification since it catalyses the reaction that limits the rate of the denitrification process (Zumft 1997). Therefore, a subunit of nitrite reductase (NirS) is often used to determine denitrification enzyme activity (DEA), which is a measure of the denitrification potential of natural systems. Nir genes (*nirK/nirS*), along with *nosZ* genes, are widely used to describe the presence of denitrifier communities within an ecosystem (Demanèche et al. 2009, Philippot et al. 2011).

The enzymes involved in the denitrification process are sequentially induced when oxygen levels are depleted, and carbon sources are readily available. Different combinations of denitrification genes in the genomes of denitrifying microorganisms have been reported (Zumft 1997, Holtan-Hartwig et al. 2000, Philippot 2002, van Spanning 2011). Recently, several investigations have identified an atypical *nosZ* gene, which is present in non-denitrifying microorganisms capable of reducing $N_2O$. This clade is highly diverse and well-known in terrestrial environments (Sanford et al. 2012, Jones et al. 2013).

The lack of *nosZ* genes is widely distributed among denitrifiers and other organisms. Despite the presence of denitrifying machinery (Nar/Nap, Nir, Nor, Nos), NO and $N_2O$ are detected in some cases as a result of an imbalance in the activities of denitrifying enzymes (Lycus et al. 2017, Roco et al. 2017).

### 4.3.3 Denitrifying Microorganisms

Denitrification is a common feature of prokaryotes. There is a very high variety of microorganisms involved in denitrification reactions, which makes them ubiquitous in most natural environments. This abundance is due to the fact that most species of denitrifying organisms are facultative, which means that they are capable of using oxygen or nitrate and nitrite as final electron acceptors when needed. The quantity of ATP obtained from nitrite and nitrate reduction is less than that which is obtained from oxygen reduction and greater than that which is obtained from sulfate reduction. The efficiency of ATP production determines the preference of microorganisms to use a particular type of electron acceptor (Knowles 1982, Zumft 1997, Philippot 2007).

Denitrifying microorganisms are predominant in soils and account for up to 5% of the total microbial population. However, they have also been isolated from sediments and aquatic environments (Demanèche et al. 2009, Bru et al. 2011). Classical denitrifiers harbour the complete set of enzymatic machinery needed to produce $N_2$, and other microorganisms produce a mixture of $N_2$ and $N_2O$. However, some microorganisms lack the Nos gen (*nosZ*), which results in the release of $N_2O$. Some non-denitrifying bacteria and archaea are able to reduce $N_2O$ to $N_2$ (Graf et al. 2014). In addition to that, a few microorganisms cannot reduce $NO_3$ and use $NO_2$ as an electron acceptor.

Bacteria and archaea are the most prevalent denitrifiers, but fungal populations have also been shown to play a significant role in soil denitrification, even though incomplete denitrification is achieved because fungi lack the Nos enzyme that catalyses the last enzymatic step in the process (Payne 1981, Zumft 1992, Philippot et al. 2007, Zumft and Kroneck 2007).

Fungal $N_2O$-producing activity is comparable to that of bacteria in the soil, and it seems that both types of organisms participate equally in the release of $N_2O$ released from terrestrial ecosystems (Chen et al. 2014). Among the denitrifying fungi, members of *Sordariomycete* and *Eurotiomycete* classes (phylum *Ascomicota*) (especially several species of *Fusarium* genus), as well as several members of phylum *Basidiomycota*, are recognised as the most active $N_2O$ producers in several ecosystems (Mothapo et al. 2015). For more details on this, refer to Chapter 5 which details the role of fungi in the nitrogen cycle.

The majority of denitrifying bacteria identified have been facultative aerobic heterotrophs, although some organisms have been autotrophs, diazotrophs, and phototrophs (Zehr and Ward 2002). Many diazotrophic bacteria (such *Azospirillum* or *Bradyrhizobium* spp.) or ammonium oxidising bacteria (AOB) belonging to genera *Nitrosospira* or *Nitrosomonas* have the capacity to denitrify (Rösch et al. 2002, Shaw et al. 2006). Gram-negative bacteria are the most abundant bacterial denitrifiers, but gram-positive bacteria have also been identified. Most of these have been members of *Bacillales* and *Actinomycetales* families, and some strains belong to a non-spore-forming group of bacteria (Verbaendert et al. 2011).

### 4.3.4 Denitrification and Its Environmental Impacts

Denitrification is a process of global ecological importance because i) it eliminates dissolved inorganic forms of N from ecosystems (useful for the treatment of wastewater), ii) incomplete denitrification generates $N_2O$, a potent greenhouse gas, which actively participates in the destruction of the ozone layer, and iii) the last step of denitrification, which is catalysed by the Nos enzyme, allows for the reduction of $N_2O$ produced by other pathways (such as nitrification, nitrifier denitrification, and DNRA) to $N_2$.

The rate of $N_2O$ production from incomplete denitrification is controlled by biotic (microorganisms and enzymes involved) and environmental factors, including oxygen, organic carbon and nitrate availability, temperature, and pH. Under oxygen-limited and anaerobic environments with moderate organic carbon availability, high nitrate levels, and sub-neutral pH values, $N_2O$ production from denitrification has been detected (Quick et al. 2019).

### 4.3.4.1    N₂O Emissions from Natural and Engineered Ecosystems

According to the World Meteorological Organisation (WMO, Greenhouse Gas Bulletin 2017), the atmospheric concentration of $N_2O$ reached almost 330 ppb in 2016, which represented an increase of 122% from levels recorded in the pre-industrial era. Although the concentration of nitrous oxide is the third most important component of GHG detected in the atmosphere (after $CO_2$ and methane), its calorific power is 300 times greater that of $CO_2$, therefore its contribution to global warming is particularly high. Also, nitrous oxide actively participates in the destruction of the ozone layer by favouring the formation of NO in the stratosphere where $N_2O$ is fragmented by photolysis to $N_2$ and electronically excited oxygen atoms, and a portion of remaining $N_2O$ reacts with the molecules to form NO. A single NO molecule can destroy up to $10^5$ ozone molecules (Ravishankara et al. 2009).

Both natural (soils, oceans, and sediments) and anthropogenic sources of $N_2O$ emissions occur. Among these, agricultural activities (widespread use of livestock residues or application of synthetic N-fertilisers), wastewater treatment plants, biomass/fossil fuel combustion, and industrial activities are included. Of about 17Tg N-$N_2O$ is emitted to the atmosphere each year, approximately 60-70% of the emissions come from natural sources, and soil emission is the leading source of $N_2O$ emissions. Among the anthropogenic activities listed, agricultural practices are responsible for the majority of $N_2O$ emissions (Galloway 2003, Wuebbles 2009). The $N_2O$ emissions from natural and agricultural sources and wastewater treatment are caused by biological activities. It is common knowledge that $N_2O$ emissions are enhanced when concentrations of nitrogen forms (ammonia, nitrite, or nitrate) are high, and direct relationships between levels of nitrogen have been recorded (Galloway et al. 2008, Stein and Klotz 2016).

Along with incomplete denitrification, nitrification (ammonia oxidation to nitrite), nitrifier denitrification, and dissimilatory nitrate reduction to ammonium (DNRA) are the main biological processes that promote the emission of nitrous oxide to the atmosphere. These are processes in which $N_2O$ is not an end product but is instead a by-product of denitrification processes. In the following sections, only $N_2O$ emissions from the 'incomplete' denitrification process of several ecosystems will be addressed.

#### 4.3.4.1.1    N₂O Emissions From Terrestrial Ecosystems

In terrestrial ecosystems (forest soils, wetlands, deserts, grasslands, and cropped soils), the biological activity of the N-cycle is the main source of nitrous oxide to the atmosphere. Of processes involved in the N-cycle in the soil, incomplete heterotrophic denitrification contributes the highest proportion of nitrous oxide emissions and therefore nitrogen loss from the soil (Philippot et al. 2007, Park et al. 2012).

Of total $N_2O$ emitted from soil ecosystems ($\cong 10$ Tg/year), about 65% is released from agricultural land, mainly from soils of temperate grasslands. Saggar et al. (2013) reviewed and analysed data regarding N loss from grasslands around the world and concluded that each year 5.5 Tg of N could be lost from 'incomplete' denitrification (as $N_2O$ emissions) in temperate grasslands. On the other hand, soils of tropical forests have been considered to be the largest terrestrial natural source of $N_2O$ (about 3.0 Tg N year) (Werner et al. 2007a). Assessments of different types of forest soils have shown that incomplete denitrification contributes most highly to total volumes of $N_2O$ emitted (Werner et al. 2007a, Butterbach-Bahl et al. 2004).

In agricultural soils, the $N_2O$ emission rate is defined by abiotic factors, such as oxygen level, water content, soil pH value, organic carbon availability, $NO_2$ and $NO_3$ content (influenced by fertiliser types, climatic conditions, and agricultural practices), and biotic factors. Biotic factors and the relationships between microbial communities affect denitrifier density and the activity of enzymes that facilitate denitrification (Liu et al. 2010, Saggar et al. 2013).

Oxygen limited soils provide an adequate environment for the production of $N_2O$ via denitrification when nitrate is available. Oxygen inhibits nitrous oxide reductase activity, although other enzymes involved in the denitrification processes continue to function. This leads to the accumulation of $N_2O$

(Morley et al. 2008). In soil, water content strongly influences denitrification which is an effect that is partly due to its relationship with oxygen levels.

High levels of $N_2O$ release have been detected in soils with low to near-neutral pH values (Knowles 1982, Liu et al. 2010). This is maybe due to the irregular assembly of the nitrous oxide reductase enzyme in the periplasm of microorganisms (Hu et al. 2015). Rates of $N_2$ production from denitrification increase as levels of available carbon sources increase. In contrast, decreased levels of carbon sources enhance the levels of $N_2O$ emitted as a result of denitrification since the enzymes involved in the process compete for electrons, and Nos is the weakest competitor (Guo et al. 2018). Otherwise, a balance between carbon source availability and nitrate concentrations determines the ratio of $N_2O$ and $N_2$ produced (Robertson and Tiedje 1987).

The availability of soil nitrate to denitrifying microorganisms is required for denitrification to occur. Consequently, the concentrations of the ion have the potential to be one of the principal limiting factors of denitrification. On the other hand, high $NO_3$ and ammonium levels in soils are positively correlated with $N_2O$ emissions because nitrous oxide reductase (Nos) activity is inhibited as ammonium and $NO_3$ levels increase. As nitrate concentration decreases, an increased proportion of N is emitted as $N_2$ since Nos activity is stimulated and complete denitrification is achieved (Cai et al. 2002). However, a close relationship exists between $NO_3$, $NH_4$, carbon source, and $N_2O$ emissions. Therefore, within carbon limited systems, an increase in nitrate concentrations result in enhanced $N_2O$ emissions; in systems with abundant carbon, increased nitrate levels also increase denitrification rates but will not affect levels of $N_2O$ emissions (Quick et al. 2016). Available nitrate is either derived from the nitrification of ammonium or exogenously applied materials, such as nitrogen fertilisers.

Fertilisation management practices (fertiliser-type, nitrogen application rate, timing, and application method) affect denitrification rates in soils. Therefore, the rational use of nitrogen fertilisers appears to be an important way in which we can mitigate $N_2O$ emissions from agricultural soils. However, the increase in the global population that is predicted to occur in the coming decades and the intensive use of nitrogen fertilisers and manure in agriculture will make it difficult to reduce $N_2O$ emissions. $N_2O$ emissions from cultivated soils could make up the majority of total nitrous oxide emissions to the atmosphere in the near future (Hu et al. 2015).

### 4.3.4.1.2  *N₂O Emissions from River and Stream Ecosystems*

To date, the majority of studies of $N_2O$ production have focused on $N_2O$ release from soil environments. Several investigations have reported low levels of $N_2O$ emissions from lentic freshwater and marine ecosystems (Seitzinger and Kroeze 1998). Eutrophic systems have been reported to produce a higher level of $N_2O$ than oligo and mesophilic systems. However, recent investigations suggest that nitrous oxide emissions from streams and rivers may be as great as those that are emitted from agricultural soils (Rosamond et al. 2012, Gardner et al. 2016). In these ecosystems, $N_2O$ emissions can occur from saturated sediments, groundwater, or the water column (Quick et al. 2019). Incomplete denitrification is the pathway that most highly produces $N_2O$, which is followed by nitrification (Beaulieu et al. 2011). $N_2O$ may be accumulated as an intermediate under high and low oxygen levels, high nitrate availability, and low pH levels (Knowles 1996). Reisinger et al. (2016) assessed mesocosms in which water column denitrification rates were higher than those detected in sediments. This finding was contrary to reports of several studies that had been previously published.

The $N_2O$ production in aquatic ecosystems is complex and is affected by a wide variety of factors. Ammonium and nitrate concentrations in both surface and groundwater environments are positively correlated with nitrous oxide emissions. However, other factors like dissolved oxygen, carbon availability, temperature, or pH also affect the release of $N_2O$ (Cébron et al. 2005, Rosamond et al. 2012, Borges et al. 2015, Quick et al. 2019). Beaulieu et al. (2011) reported that high concentrations of $NO_3$ in streams stimulated the denitrification pathway and the production of $N_2O$ but did not increase $N_2O$ yield. These authors used a global river network model to estimate that microbial denitrification and nitrification processes in these aquatic systems account for up to 10% of the generation of global anthropogenic $N_2O$ emitted, which was much higher than estimates of other studies.

### 4.3.4.1.3  N₂O Emissions from Engineered Wastewater Treatment Plant Ecosystems

WWTPs provide unique environments for the production of $N_2O$ emissions. Emissions of the wastewater treatment sector account for up to 5% of the total anthropogenic $N_2O$ emitted globally (Tumendelger et al. 2019). Throughout the biological nitrogen removal processes, nitrous oxide is released via incomplete hydroxylamine ($NH_2OH$) oxidation, nitrifier denitrification, and incomplete heterotrophic denitrification (Kampschreur et al. 2009b). The pathways are carried out by both autotrophic and heterotrophic microorganisms, which include AOB, AOA, NOB, and denitrifying microorganisms under aerobic and oxic/anoxic conditions (Law et al. 2012, Guo et al. 2018). In the denitrification process, nitrite is reduced to NO and $N_2O$, which is considered its final product because AOB lack the nitrous oxide reductase enzyme.

Several operational and environmental parameters affect levels of $N_2O$ emissions from wastewater treatment processes. These include low dissolved oxygen (DO) levels, inadequate chemical oxygen demand (COD) to N ratios (COD/N), low temperature, nitrite accumulation in nitrifying zones, the availability of Cu ions, pH, and changes based on whether the conditions are anoxic or aerobic are also observed (Ghafari et al. 2008, Kampschreur et al. 2009b, Law et al. 2012, Zhu et al. 2013, Massara et al. 2017, Guo et al. 2018).

Along with operational conditions, reactor configurations affect $N_2O$ emissions from wastewater treatment plants (Wan et al. 2019). Sequencing batch reactor (SBR) configurations are assumed to promote high levels of $N_2O$ production (Pijuan et al. 2014). Furthermore, the lowest levels of nitrous oxide accumulation have been recorded in anaerobic-anoxic-oxic reactors ($A^2O$) using conventional activated sludge (CAS) configurations (Chen et al. 2016).

Numerous full-scale studies have been designed to discern the identity of the pathway responsible for $N_2O$ production in each configuration tested (Vasilaki et al. 2019). $NH_2OH$ oxidation has rarely been detected as the main pathway for $N_2O$ release. However, in alternating aerobic-anaerobic processes, the nitrifier denitrification pathway has been suggested to be dominant (Wang et al. 2016b). Castellano-Hinojosa et al. (2018) assessed a full-scale pre-denitrification-nitrification system (two sequential bioreactors, anoxic, and aerated) and using molecular and bioinformatics tools were able to report that nitrifier denitrification was the dominant pathway for $N_2O$ release. Regardless of the configuration used, systems in which high $N_2O$ concentrations detected also have displayed a positive and marked relationship between elevated concentrations of nitrite and $N_2O$ release.

In the biological nitrogen removal processes that take place in wastewater plants, the main challenge for the mitigation of $N_2O$ emissions involves reducing levels of $N_2O$ produced. Therefore, the application of optimal aeration intensities and DO concentrations have been shown to prevent the accumulation of high concentrations of nitrite. Furthermore, the presence of adequate concentrations of carbon sources in anoxic tanks has been reported to be essential for ensuring that the complete denitrification process is allowed to occur (Kampschreur et al. 2009b, Law et al. 2012, Wu et al. 2014). Another possible means to mitigate $N_2O$ production may involve the application of biovalorisation strategies (Vasilaki et al. 2019). The use of nitrous oxide in reactions involving oxygen atom transfer, metal-catalysed oxidation, or nitrogen atom incorporation reactions result in numerous products of importance (Severin 2015). Other recently proposed mitigation strategies include the formation of value-added products like polyhydroxialkanoates (homo- and heteropolymers, PHAs) (Frutos et al. 2017b), extracellular polymeric substances (EPS) (Lai et al. 2017), and rerouting $N_2O$ emission for energy production (Scherson et al. 2014). For more details on this topic, refer to the discussion of wastewater treatment and the nitrogen cycle explained in Chapter 8.

## 4.4 CONCLUSIONS AND FUTURE REMARKS

Ecosystem resilience and resistance mechanisms have limits to their capacities to mitigate the effects of environmental disturbances. More should be known in order to prevent key ecosystems from collapsing. When the natural biochemical recycling of nutrients is interrupted, reduced, or stopped completely, the consequences maybe that the known nutrient recycling process has reached its point of no return. Restoration of the nutrient recycling process will then only be possible if microbial communities are capable of adapting and evolving the ability to function under the new conditions. Little is known about the evolution of microorganisms under different environmental conditions, yet genetic exchange between mesophilic and thermophilic microorganisms involved in the N-cycle can arise as an adaptation to low pH and high temperatures in acidophiles. However, the rates of evolution may limit the effectiveness of adaptation, therefore, directed evolution strategies should be studied to determine the impact of evolution in the short and long term in different environmental systems.

Global population growth is accompanied by an increasing demand for wastewater treatments and crops. This increased demand is predicted to increase the anthropogenic contribution to global N. These increases will mainly come from the increased use of N-containing fertilisers (synthetic or organic) to maximise soil potential.

Levels of N (Bouwman et al. 2009) in agricultural soils have been estimated to increase to 154-231 Tg N/year by the year 2050. Nitrogen will be lost via ammonia volatilisation (24-30%), denitrification (46%-49%), and leaching (23%-26%). The introduction of 36-59 Tg N/year into aquatic systems is predicted to remain high even when the N cycling is highly efficient. Lixiviation will depend on soil type and the rainfall of each environmental zone will be considered. Eugercios et al. (2017) concluded that northern Europe will experience a reduction in the lixiviation of nitrates, while in countries, including Spain, a reduction of lixiviation will occur in the northern half of the country; the lixiviation of the bottom half will remain unpredictable, except in areas surrounding Spain's two main rivers (Ebro and Guadalquivir). In these areas, an increase in lixiviation is predicted. With increasing N, levels of $N_2O$ will also increase. This correspondingly results in increased levels of global warming, which occurs as a result of the nature of $N_2O$ as a GHG.

Urgent measures are needed to palliate ecosystem imbalances occurring in the N cycle globally, and the willingness to solve this problem will have to come from both political and agricultural sectors to limit the negative effects N imbalance. Additionally, water contaminated with high levels of nitrates has the potential to render fresh water undrinkable. Though we have the technology to remedy this issue, not every country will be economically capable of doing so. Thus, human health will be at risk if measures are not taken to decrease $NO_3$ levels.

## 4.5 REFERENCES

Abby, S.S., Melcher, M., Kerou M., Krupovic, M., Stieglmeier, M., Rossel C., et al. 2018. Candidatus *Nitrosocaldus cavascurensis*, an ammonia oxidizing, extremely thermophilic archaeon with a highly mobile genome. Frontiers in Microbiology 9: 28. doi: 10.3389/fmicb.2018.00028.

Arp, D.J. and Stein, L.Y. 2003. Metabolism of inorganic N compounds by ammonia-oxidizing bacteria. Critical Reviews in Biochemistry and Molecular Biology 38: 471-495. doi: 10.1080/10409230390267446.

Avrahami, S., Liesack, W. and Conrad, R. 2003. Effects of temperature and fertilizer on activity and community structure of soil ammonia oxidizers. Environmental Microbiology 5: 691-705. doi:10.1046/j.1462-2920.2003.00457.x.

Baker, L.A., Herlihy, A.T., Kaufmann, P.R. and Eilers, J.M. 1991. Acidic lakes and streams in the United States: the role of acidic deposition. Science 252: 1151-1154.

Beaulieu, J.J., Tank, J.L., Hamilton, S.K., Wollheim, W.M., Hall, R.O., Jr., Mulholland, P.J., et al. 2011. Nitrous oxide emission from denitrification in stream and river networks. Proceedings of the National Academy of Sciences 108(4): 214-219.

Beeckman, F., Motte, H. and Beeckman, T. 2018. Nitrification in agricultural soils: impact, actors and mitigation Current Opinion in Biotechnology 50: 166-173.

Beman, J.M., Chow, C.E., King, A.L., Feng, Y., Fuhrman, J.A., Andersson, A., et al. 2011. Global declines in oceanic nitrification rates as a consequence of ocean acidification. Proceedings of the National Academy of Sciences 108: 208-213. Doi: 10.1073/pnas.1011053108.

Biller, S.J., Mosier, A.C., Wells, G.F. and Francis, C.A. 2012. Global biodiversity of aquatic ammonia-oxidizing archaea is partitioned by habitat. Frontiers in Microbiology 3: 252. doi:10.3389/fmicb.2012.00252.

Borges, A.V., Darchambeau, F., Teodoru, C.R., Marwick, T.R., Tamooh, F., Geeraert, N., et al. 2015. Globally significant greenhouse-gas emissions from African inland waters. Nature Geoscience 8(8): 637-642. https://doi.org/10.1038/ngeo2486.

Bouwman, A.F., Beusen, A.H.W. and Billen, G. 2009. Human alteration of the global nitrogen and phosphorus soil balances for the period 1970-2050. Global Biochemical Cycles 23: 1-16.

Bru, D., Ramette, R., Saby, N.P.A., Dequiedt, S., Ranjard, L., Jolivet, C., et al. 2011. Determinants of the distribution of nitrogen-cycling microbial communities at the landscape scale. The ISME Journal 5: 532-542. https://doi.org/10.1038/ismej.2010.130.

Butterbach-Bahl, K., Kock, M., Willibald, G., Hewett, B., Buhagiar, S., Papen, H., et al. 2004. Temporal variations of fluxes of NO, $NO_2$, $N_2O$, $CO_2$, and $CH_4$ in a tropical rain forest ecosystem. Global Biogeochemical Cycles 18(3): 1-11. https://doi.org/10.1029/2004gb002243.

Cai, G., Chen, D., White, R.E., Fan, X.H., Pacholski, A., Zhu, Z.L., et al. 2002. Gaseous nitrogen losses from urea applied to maize on a calcareous fluvo-aquic soil in the North China Plain. Australian Journal of Soil Research 40(5): 737-748.

Camargo, J.A., Alonso, A. and Salamanca, A. 2005. Nitrate toxicity to aquatic animals: a review with new data for freshwater invertebrates. Chemosphere 58: 1255-1267.

Camargo, J.A. and Alonso, A. 2006. Ecological and toxicological effects of inorganic nitrogen pollution in aquatic ecosystems: a global assessment. Environment International 32: 831-849.

Caranto, J.D., Vilbert, A.C. and Lancaster, K.M. 2016. *Nitrosomonas europaea* cytochrome P460 is a direct link between nitrification and nitrous oxide emission. Proceedings of the National Academy of Sciences 113: 14704-14709.

Caranto, J.D. and Lancaster, K.M. 2017. Nitric oxide is an obligate bacterial nitrification intermediate produced by hydroxylamine oxidoreductase. Proceedings of the National Academy of Sciences 114: 8217-8222.

Carey, C.J., Dove, N.C., Beman, J.M., Hart, S.C. and Aronson, E.L. 2016. Meta-analysis reveals ammonia-oxidizing bacteria respond more strongly to nitrogen addition than ammonia-oxidizing archaea. Soil Biology and Biochemistry 99: 158-166. https://doi.org/10.1016/j.soilbio.2016.05.014.

Castellano-Hinojosa, A., Maza-Márquez, P., Melero-Rubio, Y., González-López, J. and Rodelas, B. 2018. Linking nitrous oxide emissions to population dynamics of nitrifying and denitrifying prokaryotes in four full-scale wastewater treatment plants. Chemosphere 200: 57-66. https://doi.org/10.1016/j.chemosphere.2018.02.102.

Cébron, A., Garnier, J. and Billen, G. 2005. Nitrous oxide production and nitrification kinetics by natural bacterial communities of the lower Seine river (France). Aquatic Microbial Ecology 41(1): 25-38.

Chen, H., Mothapo, N.V. and Shi, W. 2014. The significant contribution of fungi to soil $N_2O$ production across diverse ecosystems. Applied Soil Ecology 73: 70-77.

Chen, W.H., Yang, J.H., Yuan, C.S. and Yang, Y.H. 2016. Toward better understanding and feasibility of controlling greenhouse gas emissions from treatment of industrial wastewater with activated sludge. Environmental Science and Pollution Research 23: 20449-20461. https://doi.org/10.1007/s11356-016-7183-2.

Cui, M., Ma, A., Qi, H., Zhuang, X. and Zhuang, G. 2015. Anaerobic oxidation of methane: an "active" microbial process. Microbiologyopen 4: 1-11.

Dai, J.Y., Gao, G., Chen, D., Tang, X.M., Shao, K.Q. and Cai, X.L. 2013. Effects of trophic status and temperature on communities of sedimentary ammonia oxidizers in Lake Taihu. Geomicrobiology Journal 30: 886-896. doi: 10.1080/01490451.2013.791353.

Daims, H., Lebedeva, E.V., Pjevac, P., Han, P., Herbold, C., Albertsen, M., et al. 2015. Complete nitrification by Nitrospira bacteria. Nature 24: 504-509. doi: 10.1038/nature16461.

Daims, H., Lücker, S. and Wagner, M. 2016. A new perspective on microbes formerly known as nitrite-oxidizing bacteria. Trends in Microbiology 24: 699-712.

Davidson, E.A. 2009. The contribution of manure and fertilizer nitrogen to atmospheric nitrous oxide since 1860. Nature Geoscience 2(9): 659-662.

Demanèche, S., Philippot, L., David, M.M., Navarro, E., Vogel, T.M. and Simonet, P. 2009. Characterization of denitrification gene clusters of soil bacteria via a metagenomic approach. Applied Environmental Microbiology 75: 534-537. https://doi.org/10.1128/AEM.01706-08.

de la Torre, J.R., Walker, C.B., Ingalls, A.E., Könneke, M. and Stahl, D.A. 2008. Cultivation of a thermophilic ammonia oxidizing archaeon synthesizing crenarchaeol. Environmental Microbiology 10(3): 810-818. doi: 10.1111/j.1462-2920.2007.01506.x.

Dore, J.E., Popp, B.N., Karl, D.M. and Sansone, F.J. 1998. A large source of atmospheric nitrous oxide from subtropical North Pacific surface waters. Nature 396: 63-66.

Edwards, T.A., Calica, N.A., Huang, D.A., Manoharan, N., Hou, W., Huang, L., et al. 2013. Cultivation and characterization of thermophilic Nitrospira species from geothermal springs in the US Great Basin, China, and Armenia. FEMS Microbiology Ecology 85: 283-292. doi: 10.1111/1574-6941.12117.

Emerson, K., Russo, R.C., Lund, R.E. and Thurston, R.V. 1975. Aqueous ammonia equilibrium calculations: effect of pH and temperature. Journal of the Fisheries Board of Canada 32: 2379-2383.

Eugercios, A.R. 2013. Interacciones acuífero-lago y biogeoquímica del nitrógeno en ambientes kársticos. Tesis Doctoral. Universidad Complutense de Madrid, Madrid, España. http://eprints.ucm.es/21579/.

Eugercios Silva, A.R., Álvarez-Cobelas, M. and Montero González, E. 2017. Impactos del nitrógeno agrícola en los ecosistemas acuáticos. Ecosistemas 26: 37-44 [Enero-abril 2017] doi: 10.7818/ECOS.2017.26-1.06.

Feng, M., Adams, J.M., Fan, K., Shi, Y., Sun, R., Wang, D., et al. 2018. Long-term fertilization influences community assembly processes of soil diazotrophs. Soil Biology and Biochemistry 126: 151-158.

Fernandes, A.T., Damas, J.M., Todorovic, S., Huber, R., Baratto, M.C., Pogni, R., et al. 2010. The multicopper oxidase from the archaeon *Pyrobaculum aerophilum* shows nitrous oxide reductase activity. The FEBS Journal 277: 3176-3189. https://doi.org/10.1111/j.1742-4658.2010.07725.x.

Firestone, M.K., Firestone, R.B. and Tiedje, J.M. 1980. Nitrous-oxide from soil denitrification: factors controlling its biological production. Science 208: 749-751. https://doi.org/10.1126/science.208.4445.749.

Follett, R.F. and Hatfield, J.L. 2001. Nitrogen in the environment: sources, problems and management. Proceedings of the 2nd International Nitrogen Conference on Science and Policy. The Scientific World 1(S2): 920-926. https://doi.org/10.1100/tsw.2001.269.

Fowler, D., Coyle, M., Skiba, U., Sutton, M.A., Cape, J.N., Reis, S., et al. 2013. The global nitrogen cycle in the twenty-first century. Philosophical Transactions of the Royal Society B: Biological Sciences 368: 20130164.

Frutos, O.D., Cortes, I., Cantera, S., Arnaiz, E., Lebrero, R. and Muñoz, R. 2017. Nitrous oxide abatement coupled with biopolymer production as a model GHG biorefinery for cost-effective climate change mitigation. Environmental Science and Technology 51: 6319-6325.

Frutos, O.D., Quijano, G., Aizpuru, A. and Muñoz, R. 2018. A state-of-the-art review on nitrous oxide control from waste treatment and industrial sources. Biotechnology Advances 36: 1025-1037.

Galloway, J.N., Aber, J.D., Erisman, J.W., Seitzinger, S.P., Howarth, R.W., Cowling, E.B., et al. 2003. The nitrogen cascade. Bioscience 53(4): 341-356. https://doi.org/10.1641/0006-3568(2003)053[0341:TNC]2.0.CO;2.

Galloway, J.N., Townsend, A.R., Erisman, J.W., Bekunda, M., Cai, Z., Freney, J.R., et al. 2008. Transformation of the nitrogen cycle. Recent trends, questions, and potential solutions. Science 320: 889-892. https://doi.org/10.1126/science.1136674.

Gardner, J.R., Fisher, T.R., Jordan, T.E. and Knee, K.L. 2016. Balancing watershed nitrogen budgets: accounting for biogenic gases in streams. Biogeochemistry 127(2-3): 231-253. https://doi.org/10.1007/s10533-015-0177-1.

Ghafari, S., Hasan, M. and Aroua, M.K. 2008. Bio-electrochemical removal of nitrate from water and wastewater - a review. Bioresource Technology 99: 3965-3974.

Goderya, F.S., Dahab, M.F., Woldt, W.E. and Bogardi, I. 1996. Incorporation of spatial variability in modeling non-point source groundwater nitrate pollution. Water Science and Technology 33: 233-240.

Graf, D.R.H., Jones, C.M. and Hallin, S. 2014. Intergenomic comparisons highlight modularity of the denitrification pathway and underpin the importance of community structure for $N_2O$ emissions. PLoS One 9(12): e114118. https://doi.org/10.1371/journal.pone.0114118.

Gubry-Rangin, C., Nicol, G.W. and Prosser, J.I. 2010. Archaea rather than bacteria control nitrification in two agricultural acidic soils. FEMS Microbiology Ecology 74: 566-574. https://doi.org/10.1111/j.1574-6941.2010.00971.x.

Guo, G., Wang, Y., Hao, T., Wu, D. and Chen, G.H. 2018. Enzymatic nitrous oxide emissions from wastewater treatment. Frontiers of Environmental Science and Engineering 12(1): 10. https://doi.org/10.1007/s11783-018-1021-3.

Hanaki, K., Wantawin, C. and Ohgaki, S. 1990. Nitrification at low levels of dissolved oxygen with and without organic loading in a suspended-growth reactor. Water Research 24: 297-302.

Haygarth, P., Johnes, P., Butterfield, D., Foy, R. and Withers, P. 2003. Land use for achieving 'good ecological status' of waterbodies in England and Wales: a theoretical exploration for nitrogen and phosphorus. Supplementary report for Defra project PE0203. Department for Environment Food and Rural Affairs. Chelmsford, UK.

He, H., Zhen, Y., Mi, T.Z., Fu, L.L., Yu, Z.G. 2018. Ammonia-oxidizing archaea and bacteria differentially contribute to ammonia oxidation in sediments from adjacent waters of Rushan Bay, China. Frontiers in Microbiology 9: 116. doi: 10.3389/fmicb.2018.00116.

Herrmann, M., Scheibe, A., Avrahami, S. and Küsel, K. 2011. Ammonium availability affects the ratio of ammonia-oxidizing bacteria to ammonia-oxidizing archaea in simulated creek ecosystems. Applied Environmental Microbiology 77: 1896-1899. doi: 10.1128/AEM.02879-10.

Holtan-Hartwig, L., Dörsch, P. and Bakken, L.R. 2000. Comparison of denitrifying communities in organic soils: kinetics of $NO_3^-$ and $N_2O$ reduction. Soil Biology and Biochemistry 32: 833-843.

Hooper, A.B., Vannelli, T., Bergmann, D.J. and Arciero, D.M. 1997. Enzymology of the oxidation of ammonia to nitrite by bacteria. Antonie Leeuwenhoek 71: 59-67.

Hu, H.W., Chen, D. and He, J.Z. 2015. Microbial regulation of terrestrial nitrous oxide formation: understanding the biological pathways for prediction of emission rates. FEMS Microbiology Reviews 39: 729-749. https://doi.org/10.1093/femsre/fuv021.

Huang, S., Lv, W.S., Bloszies, S., Shi, Q.H., Pan, X.H. and Zeng, Y.J. 2016. Effects of fertilizer management practices on yield scaled ammonia emissions from croplands in China: a meta-analysis. Field Crops Research 192: 118-125. https://doi.org/10.1016/j.fcr.2016.04.023.

IPCC. 2013. Climate Change 2013: The Physical Science Basis. Contribution of Working Group I to the Fifth Assessment Report of the Intergovernmental Panel on Climate Change, Cambridge University Press.

Jensen, F.B. 2003. Nitrite disrupts multiple physiological functions in aquatic animals. Comparative Biochemistry and Physiology Part A: Molecular and Integrative Physiology 135: 9-24.

Jones, C.M., Graf, D.R.H., Bru, D., Philippot, L. and Hallin, S. 2013. The unaccounted yet abundant nitrous oxide-reducing microbial community: a potential nitrous oxide sink. The ISME Journal 7(2): 417-426. https://doi.org/10.1038/ismej.2012.125.

Ju, X. and Zhang, C. 2017. Nitrogen cycling and environmental impacts in upland agricultural soils in North China: a review. Journal of Integrative Agriculture 16(12): 2848-2862. https://doi.org/10.1016/S2095-3119(17)61743-X.

Kampschreur, M.J., Temmink, H., Kleerebezem, R., Jetten, M.S. and van Loosdrecht, M.C. 2009. Nitrous oxide emission during wastewater treatment. Water Research 43: 4093-4103. https://doi.org/10.1016/j.watres.2009.03.001.

Knowles, R. 1982. Denitrification. Microbiology Reviews 46: 43-70.

Knowles, R. 1996. Denitrification: microbiology and ecology. Life Support and Biosphere Science 3(1-2): 31-34.

Könneke, M., Bernhard, A.E., de la Torre, J.R., Walker, C.B., Waterbury, J.B. and Stahl, D.A. 2005. Isolation of an autotrophic ammonia-oxidizing marine archaeon. Nature 437: 543-546.

Koops, H.P., Böttcher, B., Möller, U.C., Pommerening-Röser, A. and Stehr, G. 1991. Classification of eight new species of ammonia-oxidizing bacteria: Nitrosomonas communis sp. nov., Nitrosomonas ureae sp. nov., Nitrosomonas aestuarii sp. nov., Nitrosomonas marina sp. nov., Nitrosomonas nitrosa sp. nov., Nitrosomonas eutropha sp. nov., Nitrosomonas oligotropha sp. nov. and Nitrosomonas halophila sp. nov. Microbiology. 137. https://doi.org/10.1099/00221287-137-7-1689.

Kuypers, M.M.M., Marchant, H.K. and Kartal, B. 2018. The microbial nitrogen-cycling network. Nature Reviews Microbiology 16: 263-276. https://doi.org/10.1038/nrmicro.2018.9.

Lai, Y.S., Ontiveros-Valencia, A., Ilhan, Z.E., Zhou, Y., Miranda, E., Maldonado, J., et al. 2017. Enhancing biodegradation of C16-alkyl quaternary ammonium compounds using an oxygen-based membrane biofilm reactor. Water Research 123: 825-833. https://doi.org/10.1016/j.watres.2017.07.003.

Law, Y., Ye, L., Pan, Y. and Yuan, Z. 2012. Nitrous oxide emissions from wastewater treatment processes. Philosophical Transactions of the Royal Society B: Biological Sciences 367: 1265-1277. https://doi.org/10.1098/rstb.2011.0317.

Liang, X., Xing, T., Li, J., Wang, B., Wang, F., He, C., et al. 2019. Control of the hydraulic load on nitrous oxide emissions from cascade reservoirs. Environmental Science and Technology 15: 11745-11754. doi: 10.1021/acs.est.9b03438.

Liu, B., Mørkved, P.T., Frostegård, Å. and Bakken, L.R. 2010. Denitrification gene pools, transcription and kinetics of NO, N$_2$O and N$_2$ production as affected by soil pH. FEMS Microbiology Ecology 72: 407-417. https://doi.org/10.1111/j.1574- 6941.2010.00856.x.

Lopez-Vazquez, C.M., Kubare, M., Saroj, D.P., Chikamba, C., Schwarz, J., Daims, H., et al. 2014. Thermophilic biological nitrogen removal in industrial wastewater treatment. Applied Microbiology and Biotechnology 98: 945-956. doi: 10.1007/s00253-013-4950-6.

Lycus, P., Lovise Bøthun, K., Bergaust, L., Shapleigh, J.P., Bakken, L.R. and Frostegård, Å. 2017. Phenotypic and genotypic richness of denitrifiers revealed by a novel isolation strategy. The ISME Journal 11(10): 2219-2232. https://doi.org/10.1038/ismej.2017.82.

Mao, Y., Bakken, L.R., Zhao, L. and Frostegård, Å. 2008. Functional robustness and gene pools of a wastewater nitrification reactor: comparison of dispersed and intact biofilms when stressed by low oxygen and low pH. FEMS Microbiology Ecology 66: 167-180. https://doi.org/10.1111/j.1574-6941.2008.00532.x.

Massara, T.M., Malamis, S., Guisasola, A., Baeza, J.A., Noutsopoulos, C. and Katsou, E. 2017. A review on nitrous oxide (N$_2$O) emissions during biological nutrient removal from municipal wastewater and sludge reject water. Science of the Total Environment 596: 106-123. https://doi.org/10.1016/j.scitotenv.2017.03.191.

Morley, N., Baggs, E.M., Dorsch, P. and Bakken, L. 2008. Production of NO, N$_2$O and N$_2$ by extracted soil bacteria, regulation by NO$_2$ and O$_2$ concentrations. FEMS Microbiology Ecology 65: 102-112. https://doi.org/10.1111/j.1574-6941.2008.00495.x.

Mosier, A.R., Blenken, M.A., Chaiwanakupt, P., Ellis, E.C., Freney, J.R., Horwarth, R.B., et al. 2001. Policy implications of human-accelerated nitrogen cycling. Biogeochemistry 52: 281-320.

Mothapo, N., Chen, H., Cubeta, M.A., Grossman, J.M., Fuller, F. and Shi, W. 2015. Phylogenetic, taxonomic and functional diversity of fungal denitrifiers and associated N$_2$O production efficacy. Soil Biology and Biochemistry 83: 160-175. https://doi.org/10.1016/j.soilbio.2015.02.001.

Norton, J.M. and Stark, J.M. 2011. Regulation and measurement of nitrification in terrestrial systems. Methods in Enzymology 486: 343-368.

Olson, K.R. and Fromm, P.O. 1971. Excretion of urea by two teleosts exposed to different concentrations of ambient ammonia. Comparative Biochemistry and Physiology Part A: Physiology 40: 999-1007.

Ouyang, Y., Evans, S.E., Friesen, M.L. and Tiemann, L.K. 2018. Effect of nitrogen fertilization on the abundance of nitrogen cycling genes in agricultural soils. A meta-analysis of field studies. Soil Biology and Biochemistry 127: 71-78. https://doi.org/10.1016/j.soilbio.2018.08.024.

Pachiadaki, M.G., Sintes, E., Bergauer, K., Brown, J.M., Record, N.R., Swan, B.K., et al. 2017. Major role of nitrite-oxidizing bacteria in dark ocean carbon fixation. Science 358: 1046-1051.

Park, S., Croteau, P., Boering, K.A., Etheridge, D., Ferretti, D., Fraser, P., et al. 2012. Trends and seasonal cycles in the isotopic composition of nitrous oxide since 1940. Nature Geoscience 5(4): 261-265. https://doi.org/10.1038/ngeo1421.

Pauer, J.J. and Auer, M.T. 2000. Nitrification in the water column and sediment of a hypereutrophic lake and adjoining river system. Water Research 34: 1247-1254. doi: 10.1016/S0043-1354(99)00258-4.

Pauleta, S.R., Carepo, M.S.P. and Moura, I. 2019. Source and reduction of nitrous oxide. Coordination Chemistry Reviews 387: 436-449. https://doi.org/10.1016/j.ccr.2019.02.005.

Payne, W.J. 1981. Denitrification. John Wiley and Sons, New York, NY, USA.

Philippot, L. 2002. Denitrifying genes in bacterial and archaeal genomes. Biochimica et Biophysica Acta (BBA) – Gene Structure and Expression 1577: 355-376. https://doi.org/10.1016/s0167-4781(02)00420-7.

Philippot, L., Hallin, S. and Schloter, M. 2007. Ecology of denitrifying prokaryotes in agricultural soil. Advances in Agronomy 96: 249-305.

Philippot, L., Andert, J., Jones, C.M., Bru, D. and Hallin, S. 2011. Importance of denitrifiers lacking the genes encoding the nitrous oxide reductase for N$_2$O emissions from soil. Global Change Biology 17: 1497-1504. https://doi.org/10.1111/j.1365-2486.2010.02334.x.

Pijuan, M., Tora, J., Rodríguez-Caballero, A., Cesar, E., Carrera, J. and Pérez, J. 2014. Effect of process parameters and operational mode on nitrous oxide emissions from a nitritation reactor treating reject wastewater. Water Research 49: 23-33. https://doi.org/10.1016/j.watres.2013.11.009.

Quick, A.M., Reeder, W.J., Farrell, T.B., Tonina, D., Feris, K.P. and Benner, S.G. 2016. Controls on nitrous oxide emissions from the hyporheic zones of streams. Environmental Science and Technology 50: 11491-11500. https://doi.org/10.1021/acs.est.6b02680.

Quick, A.M., Reeder, W.J., Farrell, T.B., Tonina, D., Feris, K.P. and Benner, S.G. 2019. Nitrous oxide from streams and rivers: a review of primary biogeochemical pathways and environmental variables. Earth-Science Reviews 191: 224-262. https://doi.org/10.1016/j.earscirev.2019.02.021.

Rabalais, N.N. 2002. Nitrogen in aquatic ecosystems. AMBIO A Journal of the Human Environment 31: 102-112.

Ravishankara, A.R., Daniel, J.S. and Portmann, R.W. 2009. Nitrous oxide ($N_2O$): the dominant ozone-depleting substance emitted in the 21st century. Science 326(5949): 123-125.

Reardon, C.L., Gollany, H.T. and Wuest, S.B. 2014. Diazotroph community structure and abundance in wheat–fallow and wheat–pea crop rotations. Soil Biology and Biochemistry 69: 406-412. https://doi.org/10.1016/j.soilbio.2013.10.038.

Reisinger, A.J., Groffman, P.M. and Rosi-Marshall, E.J. 2016. Nitrogen-cycling process rates across urban ecosystems, FEMS Microbiology Ecology 92: fiw198. https://doi.org/10.1093/femsec/fiw198.

Richardson, C.J., King, R.S., Qian, S.S., Vaithiyanathan, P., Qualls, R.G., Stow, C.A. 2007. Estimating ecological thresholds for phosphorus in the Everglades. Environmental Science and Technology 41(23): 8084-8091. doi: 10.1021/es062624w.

Robertson, G.P. and Tiedje, J.M. 1987. Nitrous oxide sources in aerobic soils: nitrification, denitrification and other biological processes. Soil Biology and Biochemistry 19(2): 187-193.

Roco, C.A., Bergaust, L.L., Bakken, L.R., Yavitt, J.B. and Shapleigh, J.P. 2017. Modularity of nitrogen-oxide reducing soil bacteria: linking phenotype to genotype. Environmental Microbiology 19: 2507-2519. https://doi.org/10.1111/1462-2920.13250.

Rosamond, M.S., Thuss, S.J. and Schiff, S.L. 2012. Dependence of riverine nitrous oxide emissions on dissolved oxygen levels. Nature Geoscience 5(10): 715-718. https://doi.org/10.1038/ngeo1556.

Rösch, C., Mergel, A. and Bothe, H. 2002. Biodiversity of denitrifying and dinitrogen-fixing bacteria in an acid forest soil. Applied Environmental Microbiology 68(8): 3818-3829. https://doi.org/10.1128/AEM.68.8.3818-3829.2002.

Rudd, J.W.M., C.A. Kelly, Schindler D.W., Turner M.A. 1988. Disruption of the nitrogen cycle in acidified lakes. Science 240: 1515-1517.

Rudd, J.W.M., Kelly, C.A., Schindler, D.W. and Turner, M.A. 1990. A comparison of the acidification efficiencies of nitric and sulfuric acids by two whole-lake addition experiments. Limnology and Oceanography 35: 663-679.

Russo, R.C. 1985. Ammonia, nitrite and nitrate. pp. 455-471. *In*: Rand, G.M. and Petrocelli, S.R. (eds.). Fundamentals of aquatic toxicology. Washington DC: Hemisphere Publishing Corporation.

Sabine, C.L., Feely, R.A., Gruber, N., Key, R.M., Lee, K., Bullister, J.L., et al. 2004. The oceanic sink for anthropogenic $CO_2$. Science 305: 367-371.

Saggar, S., Jha, N., Deslippe, J., Bolan, N.S., Luo, J., Giltrap, D.L., et al. 2013. Denitrification and $N_2O$: $N_2$ production in temperate grasslands: processes, measurements, modelling and mitigating negative impacts. Science of the Total Environment 465: 173-195. http://dx.doi.org/10.1016/j.scitotenv.2012.11.050.

Salsac, L., Chaillou, S., Morot-Gaudry, J.F., Lesaint, C. and Jolivet, E. 1987. Nitrate and ammonium nutrition in plants. Advances in Botanical Research 25: 805-812.

Sanders, T., Fiencke, C., Hüpeden, J., Pfeiffer, E.M. and Spieck, E. 2019. Cold adapted nitrosospira sp.: a potential crucial contributor of ammonia oxidation in cryosols of permafrost-affected landscapes in Northeast Siberia. Microorganisms 14: 7(12): 699. doi: 10.3390/microorganisms7120699.

Sanford, R.A., Darlene, D., Wagner, Q.W., Chee-Sanford, J.C., Thomas, S.H., Cruz-García, C., et al. 2012. Unexpected nondenitrifier nitrous oxide reductase gene diversity and abundance in soils. Proceedings of the National Academy of Sciences 109(48): 19709-19714. https://doi.org/10.1073/pnas.1211238109.

Scherson, Y.D., Woo, S.G. and Criddle, C.S. 2014. Production of nitrous oxide from anaerobic digester centrate and its use as a co-oxidant of biogas to enhance energy recovery. Environmental Science and Technology 48: 5612-5619. https://doi.org/10.1021/es501009j.

Seitzinger, S.P. and Kroeze, C. 1998. Global distribution of nitrous oxide production and N inputs in freshwater and coastal marine ecosystems. Global Biogeochemical Cycles 12: 93-113.

Severin, K. 2015. Synthetic chemistry with nitrous oxide. Chemical Society Reviews 44: 6375-6386.

Shamrukh, M., Corapcioglu, M. and Hassona, F. 2001. Modeling the effect of chemical fertilizers on ground water quality in the Nile Valley Aquifer, Egypt. Ground Water 39: 59-67.

Shaw, L.J., Nicol, G.W., Smith, Z., Fear, J., Prosser, J.I. and Baggs, E.M. 2006. *Nitrosospira* spp. can produce nitrous oxide via a nitrifier denitrification pathway. Environmental Microbiology 8(2): 214-222.

Spiertz, J.H.J. 2010. Nitrogen, sustainable agriculture and food security. A review. Agronomy for Sustainable Development 30: 43-55. https://doi.org/10.1051/agro:2008064ff.ffhal-00886486f.

Stein, L.Y. and Klotz, M.G. 2011. Nitrifying and denitrifying pathways of methanotrophic bacteria. Biochemical Society Transactions 39: 1826-1831.

Stein, L.Y. and Klotz, M.G. 2016. The nitrogen cycle. Current Biology 26(3): 94-98. https://doi.org/10.1016/j.cub.2015.12.021.

Thakur IS1 and Medhi K2. 2019. Nitrification and denitrification processes for mitigation of nitrous oxide from waste water treatment plants for biovalorization: challenges and opportunities. Bioresource Technology 282: 502-513. doi: 10.1016/j.biortech.2019.03.069.

Thamdrup, B. and Fleischer, S. 1998. Temperature dependence of oxygen respiration, nitrogen mineralization, and nitrification in Arctic sediments. Aquatic Microbial Ecology 15: 191-199. doi: 10.3354/ame015191.

Tumendelger, A., Alshboul, Z. and Lorke, A. 2019. Methane and nitrous oxide emission from different treatment units of municipal wastewater treatment plants in Southwest Germany. PLoS One 14(1): e0209763. https://doi.org/10.1371/journal.pone.0209763.

van de Graaf, A.A., de Bruijn, P., Robertson, L.A., Jetten, M.S.M. and Kuenen, J.G. 1997. Metabolic pathway of anaerobic ammonium oxidation on the basis of 15N studies in a fluidized bed reactor. Microbiology 143: 2415-2421.

van Kessel, M.A., Speth, D.R., Albertsen, M., Nielsen, P.H., Op den Camp, H.J., Kartal, B., et al. 2015. Complete nitrification by a single microorganism. Nature 24; 528(7583): 555-9. doi: 10.1038/nature16459.

van Rijn, J., Tal, Y. and Schreier, H.J. 2006. Denitrification in recirculating systems: theory and applications. Aquacultural Engineering 34: 364-376.

van Spanning, R. 2011. Structure, function, regulation and evolution of the nitrite and nitrous oxide reductases: denitrification enzymes with a beta-propeller fold. pp. 135-161. *In:* Moir, J.W.B. (ed.). Nitrogen Cycling in Bacteria: Molecular Analysis. Caister Academic Press, UK.

Vasilaki, V., Massara, T.M., Stanchev, P., Fatone, F. and Katsou, E. 2019. A decade of nitrous oxide ($N_2O$) monitoring in full-scale wastewater treatment processes: a critical review. Water Research 161: 392-412. https://doi.org/10.1016/j.watres.2019.04.022.

Verbaendert, I., De Vos, P., Boon, N. and Heylen, K. 2011. Denitrification in Gram-positive bacteria: an underexplored trait. Biochemical Society Transactions 39(1): 254-258. https://doi.org/10.1042/BST0390254.

Vitousek, P.M., Aber, J.R., Howarth, W.G., Likens, E., Matson, P.A., Schindler, D.W., et al. 1997. Human alteration of the global nitrogen cycle: causes and consequences. Ecological Applications 1: 1-15.

Wan, X., Baeten, J.E. and Volcke, E.I.P. 2019. Effect of operating conditions on $N_2O$ emissions from one-stage partial nitritation-anammox reactors. Biochemical Engineering Journal 143: 24-33. https://doi.org/10.1016/j.bej.2018.12.004.

Wang, Y., Lin, X., Zhou, D., Ye, L., Han, H. and Song, C. 2016. Nitric oxide and nitrous oxide emissions from a full-scale activated sludge anaerobic/anoxic/oxic process. Chemical Engineering Journal 289: 330-340. https://doi.org/10.1016/j.cej.2015.12.074.

Watson, S.W. 1971. Taxonomic considerations of the family nitrobacteraceae buchanan. International Journal of Systematic Bacteriology 21: 254–270.

Werner, C., Butterbach-Bahl, K., Haas, E., Hickler, T. and Kiese, R. 2007. A global inventory of $N_2O$ emissions from tropical rain forest soils using a detailed biogeochemical model. Global Biogeochemical Cycles 21: GB3010. https://doi.org/10.1029/2006GB002909.

Wessen, E., Soderstrom, M., Stenberg, M., Bru, D., Hellman, M., Welsh, A., et al. 2011. Spatial distribution of ammonia-oxidizing bacteria and archaea across a 44-hectare farm related to ecosystem functioning. The ISME Journal 5: 1213-1225.

Wu, G., Zheng, D. and Xing, L. 2014. Nitritation and $N_2O$ emission in a denitrification and nitrification two-sludge system treating high ammonium containing wastewater. Water 6: 2978-2992.

Wu, Y.C., Xiang, Y., Wang, J.J., Zhong, J.C., He, J.Z. and Wu, Q.L. 2010. Heterogeneity of archaeal and bacterial ammonia-oxidizing communities in Lake Taihu, China. Environmental Microbiology Reports 2: 569-576. doi:10.1111/j.1758-2229.2010.00146.x.

Wuchter, C., Abbas, B., Coolen, M.J.L., Herfort, L., van Bleijswijk, J., Timmers, P., et al. 2006. Archaeal nitrification in the ocean. Proceedings of the National Academy of Sciences 103: 12317-12322. doi: 10.1073/pnas.0600756103.

Wuebbles, D.J. 2009. Nitrous oxide: no laughing matter. Science 326(5949): 56-57. https://doi.org/10.1126/science.1179571.

Zehr, J.P. and Ward, B.B. 2002. Nitrogen cycling in the ocean: new perspectives on processes and paradigms. Applied Environmental Microbiology 68: 1015-1024. https://doi.org/10.1128/AEM.68.3.1015-1024.2002.

Zhang, Y., Chen, L.J., Dai, T.J., Sun, R.H. and Wen, D.H. 2015. Ammonia manipulates the ammonia-oxidizing archaea and bacteria in the coastal sediment-water microcosms. Applied Microbiology and Biotechnology 99: 6481-6491. doi: 10.1007/s00253-015-6524-2.

Zhu, X., Chen, Y., Chen, H., Li, X., Peng, Y. and Wang, S. 2013. Minimizing nitrous oxide in biological nutrient removal from municipal wastewater by controlling copper ion concentrations. Applied Microbiology and Biotechnology 97(3): 1325-1334.

Zumft, W.G. 1992. The denitrifying prokaryotes. pp. 554-582. *In:* Balows, A., Trüper, H.G., Dworkin, M. Harder, W. and Schleifer, K.H. (eds.). The prokaryotes. Springer-Verlag, New York, NY, USA.

Zumft, W.G. 1997. Cell biology and molecular basis of denitrification. Microbiology and Molecular Biology Reviews 61: 533-616.

Zumft, W.G. and Kroneck, P.M.H. 2007. Respiratory transformation of nitrous oxide ($N_2O$) to dinitrogen by bacteria and archaea. Advances in Microbial Physiology 52: 107-227. https://doi.org/10.1016/S0065-2911(06)52003-X 9.2002.

# 5

# The Contribution of Fungi and Their Lifestyle in the Nitrogen Cycle

Mercedes García Sánchez[1], Sergio Saia[2] and Elisabet Aranda[3,*]

## 5.1 INTRODUCTION

The cycling of N in the soil, plant, atmosphere and water is of paramount importance on Earth, and over several decades, it has been thought to be exclusively mediated by prokaryotes. However, fungi were recently found to play an important role in the different reactions and processes of the N-cycle.

Archaea, bacteria, fungi and plants interact in the reactions of oxidation and reduction within the N-cycle on Earth. Among these reactions, biological $N_2$ fixation involves the conversion of $N_2$ into $NH_3$, which is mediated exclusively by $N_2$ fixing microorganisms, including, up-to-date, only bacteria and archaea, free-living or in symbiosis (Offre et al. 2013). The few fungal species discovered until now and the lack of fungal studies suggest that $N_2$ fixation may also exist in this kingdom. The nitrification process includes the oxidation of $NH_4^+$ to $NO_3^-$ producing hydroxylamine ($NH_2OH$), nitrite ($NO_2^-$) and nitrate ($NO_3^-$). Nitrification is performed by a broad spectrum of microorganisms, including ammonia-oxidizing archaea and bacteria (AOA and AOB, respectively), nitrite-oxidizing bacteria (NOB) and complete ammonia oxidizers (Comammox) (Hayatsu et al. 2008, Prosser and Nicol 2012). The reduction of $NO_3^-$ to $NO_2^-$, nitric oxide (NO), nitrous oxide ($N_2O$) and $N_2$ through denitrification is performed by archaea, bacteria and fungi (Hayatsu et al. 2008). Under anaerobic or microaerophilic conditions, the dissimilatory nitrate reduction to ammonia (DNRA) is performed by bacteria and fungi (Kamp et al. 2015), and the ammonia fermentation process that couples the dissimilatory and assimilatory pathways of nitrate and ethanol is found in some representatives of the fungal kingdom (Takasaki et al. 2004). Finally, some bacteria can form $N_2$ from $NO_2^-$ and $NH_3$ through NO and hydrazine ($N_2H_4$) via anaerobic ammonium oxidation (Anammox) (van Niftrik and Jetten 2012).

The kingdom of fungi is one of the largest groups of eukaryotes. Fungi play a pivotal role in nutrient and carbon cycling, in both terrestrial and aquatic ecosystems. The plasticity of the lifestyle of some fungal species (free-living saprotrophs, pathogens and mutualists) also leads to high diversity in the roles in the biogeochemical cycles. In addition to that, fungal lifestyle plasticity is often related to their hosts and environments rather than their genetic makeup. For instance, fungal endophytes can be latent saprotrophs or pathogens and mutualists. Also, some fungi which are symbiotic with plants, such as some mycorrhizal fungi, can have a free-living style and rarely

[1] INRA, UMR Eco and Sols, 2 Place Pierre Viala, F-34060 Montpellier, France [ORCID: 0000-0001-5967-4091].
[2] Department of Veterinary Sciences, University of Pisa via delle Piagge 2, Pisa 56129, IT.
[3] Institute of Water Research, University of Granada. Department of Microbiology. Ramón y Cajal, 4. Bldg Fray Luis. ZIP 18071, Granada, Spain. [ORCID: 0000-0001-5915-2445].
* Corresponding author: earanda@ugr.es

act as saprotrophs or as antagonists of host plants. Nonetheless, the saprotrophic and antagonistic effects against plants are under debate and appear to be determined by the fitness benefits conferred on or from their hosts, the nutrients' availability in the environment, the production of secondary metabolites and/or their colonization strategies (Kuo et al. 2014). Since endophytism is not a stable trophic state and many fungi frequently convert into saprophytic lineages, we will show the role of fungi in the N-cycle separately by investigating free lifestyles and strict symbionts (i.e., arbuscular mycorrhizal (AMF) and ectomycorrhizal fungi [EMF]).

Stimulation of the plant growth by the beneficial fungi has a direct and strong impact on N-cycling. Plant biomass per unit area or unit soil weight is various orders of magnitude higher than fungal biomass. For example, many annual crops can produce several tons of above-ground biomass per hectare per year and an amount of root biomass that is around the 20% and 40% of the above-ground biomass with higher fractions in monocot compared to dicot herbs and angiosperm compared to gymnosperm woody species (Poorter et al. 2012). Perennial crops have comparable sizes of both above- and below-ground biomass that are not completely renewed yearly.

Estimation of the plant below-ground biomass per unit area and depth span from 0.8 t C ha$^{-1}$ (in a depth of 1 m) in agricultural systems to 840 t C ha$^{-1}$ in forest systems (Robinson 2007). Estimation of the root N can also be dramatically variable due to the variability in root N concentration, the root C amounts and root C and N turnover rates (Zhou et al. 2019). In general, the root C:N ratio may change from 7 to up 170 (Zhou et al. 2019). In annual field crops, root N concentration at maturity rarely exceeds the 2% of the dry biomass of which at least 50% is made up of C. Living or dead root and litter C:N ratios were estimated close to 100 (Zechmeister-Boltenstern et al. 2015).

Also, total plant litter (including above- and below-ground litter) spans from values close to 0 to 30 t ha$^{-1}$ yr$^{-1}$ (Zhou et al. 2019) in forest soils and may be of similar magnitude in annual field crops.

This leads to a brute estimation of the amount of N stored in roots and root litter that is below 0.2-0.3 t N ha$^{-1}$.

In contrast, the whole microbial or the sole fungal biomass in the soil is several orders of magnitude lower than the plant biomass, the plant litter or the sole root litter. In Australia, Gonzalez-Quiñones et al. (2011) estimated a soil microbial C (SMC) spanning from 0.14 to 0.52 g C kg$^{-1}$ soil in the first 10 cm of soil (with median close to 0.25 g kg$^{-1}$). If considering a soil bulk density close to 1 t m$^{-3}$, such an estimation leads to values of 14-52 g of microbial C m$^{-2}$ (i.e., 0.14-0.52 t microbial C ha$^{-1}$). The C:N ratio of the soil organic carbon (SOC) and the soil microbes is dramatically lower than that of plants. In general, SOC has a C:N ratio close to 12-15, soil fungi close to 8-10 and soil bacteria close to 5-8 (Zechmeister-Boltenstern et al. 2015). This leads to an estimation of less than 0.028-0.104 t soil microbial N (SMN) ha$^{-1}$ (in the first 10 cm of soil), and if considering that around 70% of the SMN is located in the first 10 cm of soil (Murphy et al. 1998), it is extremely difficult to find more than 0.04-0.15 t SMN ha$^{-1}$ in the first 100 cm of soil. In forest soils, the first, second and third quartile of the SMN were found to be 0.043, 0.089 and 0.186 mg g$^{-1}$ soil, respectively (Li et al. 2014). Applying these values to the same estimation process, soil bulk density, depth and SMN distribution by depth used for the agricultural soil would correspond to 0.06, 0.13 and 0.27 t SMN ha$^{-1}$. In addition to this, plant biomass has a slower turnover than the whole soil microbial or the sole soil fungal biomass. Such computation can, however, be affected by high variability due to each term of the estimation. Nonetheless, estimation to a fraction of the microbial biomass or N from other reports can yield similar results. For example, applying the same process to the data provided by Baldrian et al. (2013) on fungal biomass in forest soils and assuming an N concentration of the fungi of 70 mg N g$^{-1}$ biomass leads to estimations of 0.02 to 0.07 t soil fungal N ha$^{-1}$, although this method of fungal biomass estimation can lead to estimations that are almost double the previous value. Similarly, Wallander et al. (2001) estimated an amount of EMF mycelium between 0.7 and 0.9 t ha$^{-1}$, which would correspond to 0.049-0.063 t soil fungal N ha$^{-1}$.

When coupling these data with the mean SOC concentration (which is usually lower than 5%, especially in a cropped field, but see Barré et al. (2017) and Schillaci et al. (2019), this leads to an estimation of the soil organic N that is usually below the 5‰ and frequently below 2‰. In a 0-100 cm deep layer, when considering a soil bulk density close to 1 t m$^{-3}$, such an estimation leads to amounts of soil organic N spanning from 20 to 50 t N ha$^{-1}$.

These data suggest that the whole SMN or the soil fungal N can be one or two orders of magnitude lower than root N and that root N is one or two orders of magnitude lower than soil organic N.

## 5.2 FREE LIFESTYLES

Saprophytes are known to be responsible for the destructive decay of organic matter, promoting the carbon and nitrogen cycles in the soil. As saprotrophs, fungi can participate in different chemical reactions related to N.

### 5.2.1 Fungal Denitrification

Denitrification is responsible for the formation of molecular N$_2$ through the anoxic reduction of nitrate to nitrite nitric oxide and nitrous oxide (N$_2$O) under limited conditions and 70% water-filled soil pore space (Butterbach-Bahl et al. 2013). Denitrification is widespread in the fungal kingdom and plays an important role in N$_2$O emissions since fungi are recalcitrant to convert nitrous oxide (N$_2$O) to dinitrogen (N$_2$) due to the high energy demand process (Chen et al. 2014) (Figure 5.1). Therefore, this consists of the formation of N$_2$O as the end product, which has roughly a 300-fold radiative forcing compared to carbon dioxide and thus is a potent greenhouse gas and ozone-depleting compound (Novinscak et al. 2016).

The denitrification process in fungi was discovered in the fungus *Aspergillus nidulans* in 1963 (Cove and Pateman 1963) and since this moment, several studies have been performed in order to get new insights on the biochemistry of this process. *Fusarium oxysporum* and *F. solani* were tested for their capability to release nitrous oxide in 1972; however, the authors did not make any clear conclusions since the nitrate was not detected during growth under aerobic or partially anaerobic conditions (Bollag and Tung 1972). Today, it is well known that denitrification reaction in fungi involves the participation of nitrate reductase (Nar), nitrite reductase (Nir), nitric oxide reductase and nitrous oxide reductase (Nor) (Morozkina and Kurakov 2007).

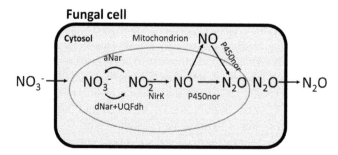

**FIGURE 5.1** Denitrification pathway of fungi, aNar assimilatory NO$_3^-$ reductase, dNAR, dissimilatory NO$_3^-$ reductase, UQFdh: ubiquinone dependent formate dehydrogenase, Nirk: copper containing NO$_2^-$ reductase, P450nor and heme-cytochrome P450 NO reductase. Modified from Mothapo et al. (2015).

Under limiting oxygen conditions, a denitrifying system is coupled with the mitochondrial electron transport chain, allowing anaerobic respiration associated with ATP synthesis. The experiment performed by Novinscak et al. (2016) isolated 492 facultative anaerobic fungi from 15 soil ecosystems. They found 27 fungal denitrifiers belonging to ten genera, including *Aspergillus*, *Penicillium* and some lesser-known species—such as *Byssochlamys nivea*, *Volutella ciliata*, *Chloridium* spp. and *Trichocladium* spp.—with the presence of the P450nor and the nirK genes, which can produce N$_2$O from nitrite (Novinscak et al. 2016).

Several studies have focused on the isolation of denitrifier fungi from agricultural soils and compost samples. However, the production of nitric oxide (NO) in some fungal species has been related to pathogenic activity, being that NO is an endogenous signaling molecule related to defense reactions under aerobic conditions (Arasimowicz-Jelonek and Floryszak-Wieczorek 2016). NOx synthesis has been related to the development, pathogen virulence and survival in the host, and it is closely regulated depending on the species, environmental conditions and development stage.

### 5.2.1.1   Co-Denitrification

Genes encoding nitrous oxide reductase, which reduces $N_2O$ to $N_2$, are not found in fungi (Shoun et al. 2012); for this reason, it is considered a prokaryotic process. However, fungal denitrification is often accompanied by co-denitrification—an enzymatically mediated nitrosation process in which a hybrid $N_2$ or $N_2O$ species is formed upon a combination of nitrogen atoms from nitrite and other nitrogen compounds, such as amines (nitrogen donor), under denitrifying conditions. However, the specific pathway is still unknown (Phillips et al. 2016). There are three fungal species belonging to the order Hypocreales that have demonstrated to perform co-denitrification: *Fusarium oxysporum*, *Fusarium solani* and *Cylindrocarpon tonkinense* (Spott et al. 2011). Co-denitrification has an important role when organic nitrogen is present in the soil. It has been estimated that fungal co-denitrification contributes to up to 92% of the $N_2$ produced in grassland soils (Laughlin and Stevens 2002).

### 5.2.1.2   Dissimilatory Nitrate Reduction (DNRA)

$NO_3^-$ can be stored intracellularly in living prokaryotic cells at concentrations which, by far, exceed ambient concentrations—a trait that is also known to occur in eukaryotic cells as was reviewed by Kamp et al. (2015). The extra- and intracellular-nitrate can be used for both assimilation and dissimilation processes. Thus, the assimilation of nitrate into ammonia is incorporated into the biomass or to build-up proteins and/or nucleic acids. Conversely, the dissimilation nitrate is proposed as a mechanism for energy conversion in which the $NO_3^-$ acts as an electrons acceptor with really low $O_2$ conditions (Kraft et al. 2011, Thamdrup 2012). Therefore, the dissimilation nitrate and in particular the intracellular nitrate store are physiological life traits that provide microbes with environmental flexibility (metabolic activity under oxidant or anoxic conditions) and resource independence (anaerobic metabolism without immediate nitrate supply). Such life traits are especially important in environments that are temporarily anoxic and/or nitrate-free, and they may have developed as a 'life strategy'. The dissimilation of nitrate can occur through different pathways (Shoun and Tanimoto 1991, Takaya 2009): i) denitrification in which the nitrate is reduced to nitrous oxide ($NO_3^- \rightarrow NO_2^- \rightarrow NO \rightarrow N_2O$), (see 5.2.1); ii) dissimilatory nitrate reduction to ammonia ($NO_3^- \rightarrow NO_2^- \rightarrow NH_4^+$) in which the nitrate is reduced through nitrite to ammonia through respiratory (DNRA) or fermentative (ammonia fermentation) metabolism using organic/inorganic and organic electron donors, respectively.

The dissimilatory nitrate reduction to ammonia (DNRA) is a mechanism well known in prokaryotes; however, in fungi, the ability to switch from oxygen to intracellular nitrate for anaerobic energy metabolism under temporarily oxic or anoxic conditions is still in its infancy. To date, the fungus *Aspergillus terreus*, isolated from sediment in the seasonal oxygen minimum zone of the Arabian Sea, has shown abilities to dissimilate nitrate to ammonia through the DNRA pathway (Stief et al. 2014). A conclusive experiment with $^{15}NO_3^-$ labeling was made for only one single strain of *A. terreus*. Interestingly, this work revealed that the intracellular nitrate concentration in this strain reached up to 0.4 mM. Additionally, it was shown that the main product of its nitrate reduction activity was ammonium, indicating the ability of this fungus to use the dissimilatory nitrate pathway. However, it is not easy to disseminate if the mechanism of this fungus to dissimilate the nitrate was carried out through the DNRA or ammonia fermentation since the spreading oxygen-deficient zones in the ocean could activate one or the other mechanism. Nevertheless, it was presumably assumed that the metabolic pathway, termed ammonia fermentation, has apparently evolved in this fungus as well as within individual soil fungi (Zhou et al. 2002). The prevalence of either pathway is controlled by ambient oxygen levels with hypoxic and anoxic levels triggering the different dissimilatory nitrate pathways (Takaya 2009).

### 5.2.1.3 Ammonia Fermentation

$NO_3^-$ is generally metabolized by organisms in assimilatory and dissimilatory reductive pathways. A dissimilatory reduction in which $NO_3^-$ is used as an alternative electron acceptor for respiration when $O_2$ is not available is a strategy extensive in facultative anaerobic microorganisms. As previously explained, the denitrification in eukaryotes ($NO_3^- \rightarrow NO_2^- \rightarrow NO \rightarrow N_2O$) is the form in which fungi can produce energy under anoxic conditions. However, another metabolic system found in fungi is known as ammonia fermentation and is based on the dissimilatory nitrate reduction to ammonium ($NO_3^- \rightarrow NO_2^- \rightarrow NH_4^+$) (Berks et al. 1995, Zumft 1997).

Although the ammonia fermentation is thought to only occur among prokaryotic organisms, there are some studies reporting this system in fungi (Shoun and Tanimoto 1991, Shoun et al. 1992, Tsuruta et al. 1998). The main importance of the dissimilatory pathways in relation to denitrification is that the $NO_3^-$ is transferred into another mineral form of N, which is less mobile and may thus, in contrast to denitrification, conserve the N in the ecosystem (Lu et al. 2013) and support cell growth under conditions more anoxic than those of denitrification (Zhou et al. 2002, Hawkins and Kranabetter 2017). Therefore, the ammonia fermentation has been called a "short circuit in the biological N cycle" (Cole and Brown 1980) as the direct transfer of $NO_3^-$ and $NO_2^-$ to $NH_4^+$. This might be the second pathway of dissimilatory nitrate metabolism in denitrifying fungi and for the alternative expression of ammonia fermentation and denitrification under anaerobic conditions in response to the $O_2$ supply.

Bacteria usually regulate these pathways independently and produce respective enzymes that catalyze these reactions (Lin and Stewart 1997, Zumft 1997). However, in the case of fungi, there is some evidence showing that assimilatory and dissimilatory pathways through ammonia fermentation might share similarities in regards to enzymes. Unlike denitrification, in which *dNar* and *dNir* are involved, assimilatory $NO_3^-$ reductase (*aNar*) and $NO_2^-$ reductase (*aNir*) are probably used for reducing $NO_3^-$ to $NH_4^+$ for ammonia fermentation. This was the first indication that the assimilatory $NO_3^-$ reducing system, which is generally distributed among plants and microorganisms, is utilized in a dissimilatory function producing ATP. Thus, the nitrate-assimilating pathway in the fungus *A. nidulans*, which has been extensively studied (Cove 1979), assimilatory $NO_3^-$ reductase (*aNar*) and $NO_2^-$ reductase (*aNir*) located in the cytosol use NADH or NADPH as an electron donor (Campbell and Kinghorn 1990, Johnstone et al. 1990). However, these properties are quite different from those involved in nitrate respiration (Kobayashi et al. 1996, Zumft 1997) but are similar to those in ammonia fermentation (Zhou et al. 2002).

The fungal denitrifying system is localized to mitochondria where it acts as a mechanism for anaerobic respiration similar to that of bacteria (Kobayashi et al. 1996). The novel nitrate metabolism of eukaryotes consists of the reduction of $NO_3^-$ to $NH_4^+$ coupled with the catabolic oxidation of electron donors (i.e., ethanol) to acetate and substrate-level phosphorylation that support growth under axenic conditions. This process has been described for the common soil fungus *F. oxysporum* (Zhou et al. 2002). Thus, the metabolism of this fungus depending on $O_2$ status would be classified as ammonia fermentation under anoxic conditions, denitrification when $O_2$ supply would be limited and aerobic respiration under sufficient $O_2$ supply. This study also reported that this mechanism would be extended to other fungal species such as *Talaromyces rotundus* (IFO; 9142), *Trichophyton rubrum* (IFO; 5807), *Hyalodendron sp.* (IFO; 31243), *Penicillium abeanum* (IFO; 6239), *Petriella guttulata* (IFO; 8613), *Calonectria kyotensis* (IFO; 8962), *Hypocrea nigricans* (IFO; 31290), *Hypomyces trichothecoides* (IFO; 6892), *Neurospora crassa* (IFO; 6067), *F. oxysporum* (IFO; 30710), *F. oxysporum* (IFO; 9968), *Cylindrocarpon tonkinense* (IFO; 30561), *Gibberella fujikuroi* (IFO; 6349), *F. oxysporum* (Institute of Applied Microbiology, University of Tokyo; 5009) and *F. lini* (Institute of Applied Microbiology, University of Tokyo; 5008), which have been considered aerobic organisms and should be classified as facultative anaerobes. However, the identification of the genes involved in ammonia fermentation in *A. nidulans* revealed that the assimilatory $NO_3^-$ reductase (*aNar*) and $NO_2^-$ reductase (*aNir*) were essential for reducing nitrate and for anaerobic cell growth during ammonia fermentation (Takasaki et al. 2004). Additionally, the ethanol oxidation was also coupled with nitrate reduction, as previously described for *F. oxysporum*, with the exception that *A. nidulas* requires functional gen (*facA*) that encodes the acetyl-CoA synthetase responsible of ATP synthesis. Therefore, this mechanism represents the fungi adaptation to anaerobiosis conditions by using assimilating enzymes for dissimilatory purposes, indicating that both reactions are intimately associated (Zhou et al. 2002, Takasaki et al. 2004).

## 5.2.2   Heterotrophic Nitrification

Many saprotrophic fungi can oxidize organic nitrogen compounds to $NO_3^-$. This process is known as heterotrophic nitrification (Prosser 2007). Nitrification is an acidifying process that has a greater impact on acidification of forest soils. This is also a very important process for N-cycling because the nitrate formed is readily available for plants and more easily leached from the soil than ammonium.

Different studies have been performed using $^{15}N$ tracing studies (Zhu et al. 2013) and antibiotics for bacterial and fungal populations, cycloheximide (fungal inhibitor) and streptomycin (bacterial inhibitor) to analyze the contribution of both populations to the nitrification processes in these environments (Figure 5.2). Zhu et al (2015), by using these experiments, demonstrated that fungi rather than bacteria dominate heterotrophic nitrification in subtropical coniferous forest soils. Other studies have confirmed this fact under acidic conditions (Stroo et al. 1986). Genera such as *Aspergillus*, *Penicillium* and *Absidia* (*A. cylindrospora*) have been isolated from acidic soils. *A. cylindrospora* was examined for production of $NO_2^-$ and $NO_3^-$ in media containing $(NH_4)_2SO_4$ with positive results when sterile soils were added to the medium, confirming that nitrification in acidic soil is a heterotrophic process catalyzed by acid-tolerant fungi and not by autotrophs or heterotrophs in non-acid microsites (Stroo et al. 1986).

The mechanism involved in heterotrophic nitrification remains unclear; nonetheless, some studies indicated that oxidation of organic nitrogen compounds involves reactions with hydroxyl radicals produced in the presence of hydrogen peroxide and superoxide or by oxidation forms of Mn.

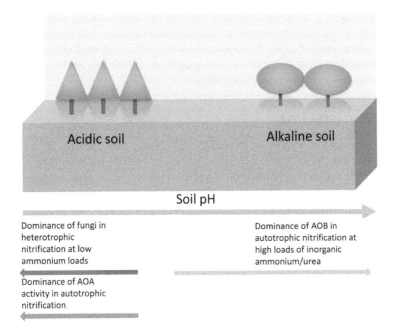

**FIGURE 5.2** Dominance of microbial populations in heterotrophic nitrification according to the soil pH.

### 5.2.3 Ammonification

Nitrogen is a major element controlling soil fertility and biomass production, and it plays an important role in organism metabolism and synthesis (Vitousek and Howarth 1991). The nitrogen dynamics are often linked to the decomposition of the organic matter because all N transformations and uptake processes are mediated by enzymatic systems that require carbon and energy for their synthesis and expression (Geisseler et al. 2010, Burke et al. 2011). In natural environments, such as soils, filamentous fungi are an important key element in transforming the organic matter by producing some extracellular enzymes (Chigineva et al. 2011, Chigineva et al. 2011, Fontaine et al. 2011). A study showed that when nutrient availability is low, soil fungi predominantly mediate the nutrient release through degrading recalcitrant organic matter (Fontaine et al. 2011). Thus, the decomposition of the organic matter by fungi might contribute to the N mineralization, increasing N plant availability. Some studies have observed that a soil fungi decomposer can bridge the soil-litter interface and determine the inorganic N distribution in the soil profile. This finding revealed that fungal hyphal translocation might be suggested as a mechanism for N input and can account for the N immobilized in the residues decomposing in the field (Frey et al. 2000, Frey et al. 2003). On the other hand, some fungal endophytes may play a role in N-cycling from organic matter decomposition. This is the case of the fungus *Phomopsis liquidambari* that appeared to promote the growth of rice (*Oryza sativa*) while allowing the marked reduction in the amount of N fertilizer soil application (Li et al. 2009). The mechanisms of *P. liquidambari* improving the N mineralization have been proposed by Chen et al. (2013). In this latter study, it was evaluated the priming effect might induce the ability of this fungus to N mineralize either *in vitro* or in soil conditions. It was thus found that fungus was able to release the $NH_4^+$-N as the result of the litter decomposition. Likewise, this effect was also found in soils, thus the increase in $NH_4^+$-N stimulated the soil ammonia oxidizer bacteria (AOB) communities and enhanced nitrification, leading to an increase in soil $NO_3^-$-N with an increased possibility of soil N loss. Later on, Yang et al. (2015) conducted a similar experiment where they deeply analyzed the composition and structure of AOB and diazotrophs as well as the profile of root exudates. They found that the presence of *P. liquidambari* positively affected the potential of nitrification rates, modifying the abundance and structure of AOA, AOB and diazotrophs, especially under low N inputs. Likewise, the profile of plant root exudates was altered, increasing the release of soluble saccharides, total free amino acids and organic acids. This finding suggested that the plant-soil feedback mechanisms might be mediated by the interaction of root-endophyte interactions, especially under nutrient-limitation. Afterward, it was observed as the interaction between plant roots, *P. liquidambari* might drive the below-ground decomposition affecting the N transformation. Nevertheless, this finding is limited to environmental N levels and plant growth stage-development (Sun et al. 2019).

### 5.3 SYMBIOTIC LIFESTYLE

A wealth of fungi have a symbiotic behavior with plants and bacteria in the soil. In some circumstances, this symbiosis is obligate for either one or both of partners. The role of symbiotic fungi on soil N-cycle is thus tightly interconnected with the role of its symbiotic partner on both the C and N-cycles.

Among the symbiotic fungi, two groups are particularly widespread: arbuscular mycorrhizal fungi (AMF) and ectomycorrhizal fungi (EMF).

These fungi interact in a mutualistic symbiosis with most of the land plants. In particular, EMF are usually non-obligate symbionts and interact with the roots of many trees, especially in forest soils. EMF can form a mycelium structure all-around root, bearing some invasion structures into the root cortex without entering the root cells. In addition to this, they grow a dense mycelium in the soil.

AMF are obligate symbionts of plants. AMF live in a quiescent status as a spore or non-active mycelium when not engaging in the mycorrhizal symbiosis. In contrast to EMF, AMF are polynucleate haploid organisms and can also form specific exchange structures, the arbuscules, in the root cells. Arbuscules are responsible for a high surface and tight connection with the plant cytoplasm. The symbiosis between plants and ERM is mostly confined to tree species, whereas the symbiosis between plants and AMF is confined to both annual and perennial herbaceous tree species.

Other mycorrhizal symbioses are less widespread and thus are of limited importance for the terrestrial N-cycling. These symbioses are engaged by the *Ericaeae* and *Orchidacea* species with the Ericoid and Orchid mycorrhizal fungi, respectively.

AMF are presently classified as members of the subphylum *Glomeromycotina*, into the phylum *Mucoromycota* (Spatafora et al. 2016), and were previously classified as members of the phylum *Glomeromycota*. Apart from the classification, the relationship between genetic traits or phylogenetic position of AMF and their functionality in the environment and within the symbiosis is under debate (Bruns et al. 2018). In contrast, the EMF, Ericoid and Orchid mycorrhizal fungi and other fungi engaging in mutualistic relationships with plants are from the phyla *Ascomycota* and *Basidiomycota* (Tedersoo et al. 2018).

Other fungi, from either *Ascomycota* or *Basidiomycota*, engage in mutualistic symbioses with plants but are not strictly mycorrhizal. The most known fungi comprise species from the families *Sebacinaceae* (classes *Sebacinales*; e.g., *Piriformospora indica*) living as endophytes and *Hypocreaceae* (classes *Sordariomycete*; e.g., *Trichoderma spp.*) in the soil.

AMF are usually present in environments with limited P supply for the plant and SOC; EMF are present in environments with high SOC and thus high organic N but limited N availability for the plants. All of these beneficial fungi stimulate plant growth through a wealth of mechanisms depending on both the plant and fungal species and their environmental conditions. The most important way of plant biostimulation usually entails a direct uptake of the nutrient by the fungus and its transfer to the plant, especially in mycorrhizal symbiosis.

In these symbioses, the fungus also offers a range of additional benefits, which encompass the following:

1. An indirect increase of the nutrient solubility in the soil through the release of exudates and decomposition of their biomass since hyphae are dramatically finer than roots, and root hairs are thus able to better explore the soil and help the soil bacteria spread.
2. An improvement of the soil status through both the entanglement of the particles with the hyphae and for the AMF, the release of glomalin, a glycoprotein that is strongly implied in the binding of the mineral particles (Rillig and Mummey 2006).
3. Protection against pathogens (Pozo and Azcón-Aguilar 2007).
4. Stimulation of the plant physiology through the release in the roots of compounds with hormonal activity (Chanclud and Morel 2016).

Soil fungi with a symbiotic lifestyle play two contrasting roles in the N-cycling; on the one hand, they can reduce the amount of mineral nitrogen in the soil by increasing the yield potential of the plant and thus the plant demand of N from the soil or directly taking up soil mineral N. The uptake of soil mineral N is tightly linked to the availability of photosynthates for the fungi. Also, the sequestration of the mineral nitrogen in the soil can depend on the effect of the soil fungi on the stabilization of the soil organic matter of which biomass roughly needs 10% of N (Yue et al. 2016).

On the other hand, fungi acting in symbioses with plants can both play a role in the early stages of mineralization, especially the depolymerization of the N-containing molecules (Schimel and Bennett 2004). Such a role can depend on both the direct activity of the fungi on the mineralization rate and the stimulation of the bacterial mineralization (Hodge and Storer 2015). In the next sections, we will deal with peculiarities of the symbiotic fungi with special emphasis on AMF and EMF and their roles on both plant and soil N transformations. In addition to this, we will give hints on the relationship between soil fungi and soil bacteria in the framework of N-cycling in the soil.

### 5.3.1  The Role of Arbuscular Mycorrhizal Fungi on N-Cycling in the Plant

AMF act as an extension of the root system. The AMF hyphae spreading in the soil hyphae are several folds thinner and longer than the roots and capable of penetrating soil pores inaccessible to the finer roots or the root hairs. Such a trait gives them the ability to efficiently scavenge the soil for mineral elements with scarce solubility and at one time, AMF have a certain ability to take up nutrients from the organic sources. Such a trait, coupled with the extensive penetration into the root cortex and the formation of the arbuscules into the root cells, give AMF the ability to strongly sustain the plant demand for some mineral elements and mostly P, especially when the mineral element availability for the plant is scarce (Smith and Read 2010). In return to P and other minerals delivered from the AMF to the plant, the plant can deliver to the AMF up to 20% of its fixed carbon and generally around 5-10% (Bryla and Eissenstat 2005). Under increasing P availability, the benefit from the AMF to the plant progressively declines (Janos 2007).

The role of the AMF on plant P nutrition has received attention for a long time, despite, ironically, AMF being earlier believed to sustain the plant N nutrition (Koide and Mosse 2004). The role of AMF on the N-cycling in the soil-plant system is receiving attention in the last 10-15 years, despite results *in vivo* being provided earlier (Ames et al. 1983, Corrêa et al. 2015).

A wealth of studies have shown that AMF are capable of taking up both organic and inorganic forms of N and has also elucidated the mechanisms of uptake and transport into the hyphae (see [Hodge and Storer 2015] and reference therein). Also, mycorrhizal-specific ammonium transporters were found in plants to take up N in the root cells released by the arbuscule, which suggested that the plant-AMF trade-off of nutrients also directly targeted the N exchange (Guether et al. 2009).

The transfer of N taken up as organic or inorganic forms from the AMF to monoxenic culture of root organs was shown. Such a transfer may also imply a net contribution to the whole plant N content under *in vivo* condition (Ames et al. 1983, Leigh et al. 2011, Saia et al. 2014b). However, the higher uptake of N of the arbuscular mycorrhizal plant from the soil if compared to the non-mycorrhizal counterpart may, on the whole, depend on two co-occurring mechanisms: the direct uptake from the AMF and its further transfer to the host plant and the higher uptake from the plant; thanks to its biomass increase and thus the demand for N of the plant from the soil.

The AMF extraradical mycelium can take up N in both organic and inorganic forms. In general, AMF seem to prefer organic and ammonium N since these forms do not have a high energy content (if compared to nitrate) and thus require less energy for the organication from the energy-limited AMF. It seems that AMF have the ability of N taking up nitrate, but its transfer to the plant is limited (Tanaka and Yano 2005) especially when no stress for the plants or the mass flow of solutes occurs (Tobar et al. 1994).

The transfer of N from the AMF to the plant seems, however, to occur only if the C supply to the AMF is adequate, irrespective of the C donor (Walder et al. 2012). On the whole, the AMF have a high demand for N to sustain its growth given the high N concentration of the chitin.

In cropped environments, non-N depleted conditions (such as the cultivation of $N_2$-fixing species or the N fertilization) can increase the benefit from the AMF to the plants, especially if the AMF contribute to the relieving of other stresses (such as lack of other nutrients, abiotic or biotic stresses) (Saia et al. 2020).

The amount of N that the AMF can transfer to the plant is, however, a little fraction of the whole plant N content and is usually below 10% (Atul-Nayyar et al. 2009, Saia et al. 2014b). AMF increased the plant P concentration and the amount of N in the plant coming from an organic patch compared to the non-inoculated control, and the N from the organic source was one-fifth of the plant N (Leigh et al. 2009). However, when AMF reduced the N concentration of the plant, the plant capture of N from an organic material spread in the soil was almost halved and at the one time, the number of bacteria in the soil increased (Saia et al. 2014b). On the whole, the direct stimulus of the AMF to the plant growth and its augmented uptake of minerals apart from N and especially P—which further increased the plant demand for N—is very likely the strongest effect of the AMF on the availability of mineral N in the soil. Such a reduction of the N availability in soil and increase for the plant can strongly affect the soil N-cycle (but see further).

AMF can also play an important role in the increase of the total N supply in both the plant and soil; thanks to their role in the nitrogen fixation with special emphasis on the symbiotic N fixation (SNF) of the Rhizobia-legumes symbiosis (Hayman 1986). The stimulation of the SNF by the AMF is variable. Under controlled conditions, a benefit was frequently seen (Kaschuk et al. 2009, Schütz et al. 2018) of which the occurrence is not constant under field conditions (Chalk et al. 2006, Kaschuk et al. 2009). In particular, it seems that the AMF may benefit the net amount of N derived by the SNF when the legume plants experience some stresses directly relieved by the AMF (Saia et al. 2014a, Püschel et al. 2017).

The symbiosis between plants and AMF also has an altered N physiology compared to that of the non AMF colonized plants. Such an alteration can also occur irrespective of a change in the N concentration in the plant tissue (Bücking and Kafle 2015). In particular, AMF seem to directly affect the organication activity of the plant in both the root and leaves. In such plant organs, the formation of non-structural amino acids, especially the gamma amino butyric acid (GABA) was increased in contrast to the glutamine formation through the glutamine synthetase (GS) and glutamate synthase (glutamine:2-oxoglutarate aminotransferase; GOGAT) (Saia et al. 2015b, Rivero et al. 2015), and GABA can be implied in the N transfer from the plant to the AMF more than the structural amino acids. Saia et al. (2019) also found a reduction in the plant of nitrate when an AMF-based inoculum was added in the soil. At the one time, AMF can also stimulate the expression of the N uptake genes in roots (Cappellazzo et al. 2008, Tian et al. 2010, Saia et al. 2015a).

## 5.3.2 The Role of Arbuscular Mycorrhizal Fungi on N-Cycling in the Soil

AMF also have direct and strong effects on N-cycling in the soil. AMF are not able to derive N from the atmosphere, but there are indications that N-fixing bacteria engage in bacterial-fungal relationships with AMF in which the bacteria may provide N derived from the atmosphere and the AMF may provide organic C. However, such a possibility would imply a limitation of the C for the fungus (Bianciotto et al. 1996, Bianciotto and Bonfante 2002) and thus likely to occur under severe N limitation for the AMF. Nonetheless, such a trait of a fungal-bacterial symbiosis was observed and was important for the rice N nutrition when engaging in a relationship with a symbiotic endophytic fungus harboring its own symbiotic $N_2$-fixing bacteria (Paul et al. 2020).

The role of the AMF in N transformation in the soil can be both direct or mediated by the relationship between AMF and bacteria involved in the N transformation. In any case, AMF have no saprotrophic ability so that the direct effects mostly regard the uptake of mineral and inorganic N forms or the release of N forms as mycorrhizal exudates.

In a wealth of studies, the effects of AMF on soil bacterial mediating N transformation or the activity of these bacteria were evident; the latter of which is most notably due to an alteration of the substrate availability through both a direct uptake from the AMF or an augmented uptake from the mycorrhizal plant compared to the non-mycorrhizal control.

Indeed, it was shown that AMF subtracted ammonium or arginine from the growing medium (Johansen et al. 1992, Frey and Schüepp 1993, Jin et al. 2005). Along with the uptake of nutrients by the plant, such a reduction of the substrate may be responsible for both a reduction of the nitrification and for the N losses of the systems as seen elsewhere (Bender and van der Heijden 2015, Köhl et al. 2014), despite an increase of the nitrate losses being also observed in one circumstance (Köhl et al. 2014). Other indirect effects on the reduction of N losses in the AMF treatments compared to a non-AMF counterpart can also be due to the effects of the AMF on soil aggregation, which is improved by AMF, aeration and change in soil pH (Marschner and Baumann 2003, Rillig and Mummey 2006). Lastly, AMF are responsible for a massive release of glomalin (Rillig 2004), by far the most abundant and stable protein in the soil of which abundance is directly related to aggregate water stability and implies a strong stabilization of huge amounts of N.

In addition to this, it was shown that AMF can directly influence the bacterial community in the soil, and these effects can be either directly or indirectly mediated by the effects of the AMF on the soil condition, availability of pabulum for the bacteria and change of the N forms in the soils.

Amora-Lazcano et al. (1998) found that the AMF decreased the ammonifier and nitrifier bacteria and increased the ammonium oxidizers compared to the non-inoculated control. Such an effect can be due to both the presence of the AMF hyphae or its exudates (Toljander et al. 2007). Other authors also showed that AMF can change the bacterial community in the soil (Walley and Germida 1997, Marschner et al. 2001, Marschner and Baumann 2003, Singh et al. 2008).

Notably, the relationship between soil bacteria and AMF can also consist of competition for resources as shown by Leigh et al. (2011). In an experiment where both the root systems, the AMF hyphae and soil bacteria were allowed to access organic material; the presence of the AMF was associated with an increase of the bacterial activity in the short term (7-9 weeks) (Saia et al. 2014b).

Finally, AMF may be indirectly responsible for a generalized increase of bacterial activity in the soil, thanks to their ability to help the bacterial spread in nutrient-rich patches (Bianciotto et al. 2001).

On the whole, the effects of the AMF on the N availability and forms in the soil and on the bacterial population and activity are likely to be masked by the effects of the AMF on the soil status or the plant (Veresoglou et al. 2012).

### 5.3.3 Ectomycorrhizal Fungi in Contrast to AMF

Few species of plants (ca. the 2%) can engage in symbiosis with EMF, and these species are mostly tree-species living in forest ecosystems (van der Heijden et al. 2015), so the ectomycorrhizal symbiosis is more relevant for forest areas than for cropped areas especially when N is limited (Courty et al. 2010).

In contrast to AMF, EMF can have saprotrophic abilities and can assume a high relevance for the N nutrition of the host plant (Read and Perez-Moreno 2003). However, such traits are under debate and appear to be highly variable among EMF lineages (Lindahl and Tunlid 2015, Pellitier and Zak 2018). EMF can increase the access of the host to the labile organic N (Michelsen et al. 1996), whereas their role in plant N nutrition when the organic N source is recalcitrant can be minimal (Wu et al. 2003). Also, EMF are able to mobilize high amounts of inorganic N from the soil (Hawkins and Kranabetter); at one time, their C release is more important than the C release from the roots in stimulating the N-cycling in the soil (Zhang et al. 2019), and this may be due to the high turnover rate of some necromass of the ECM especially when its melanin concentration is low (Fernandez et al. 2016). Similarly, AMF necromass is highly decomposable but not the glomalin released by the AMF (Steinberg and Rillig 2003).

Similar to that of AMF, the biomass of ECM can represent a high fraction of the microbial biomass in the ECM-dominated environments (Högberg and Högberg 2002), which should imply a high contribution of their organic N to the whole organic nitrogen in the living biomass of the soil. At the one time, ECM are able to favor the release of organic N from the soil organic matter (SOM) (Trap et al. 2017, Makarov 2019) and transfer N to the host plant (Müller et al. 2007, Näsholm et al. 2009).

However, the net release of N from organic sources, thanks to the activity of the ECM, can also consist of competition for N and sequestration for other microbes (Averill et al. 2014), as also seen for AMF (Bukovská et al. 2018), and this may further limit the N mineralization from the soil. Also, EMF have a relatively low production of proteases compared to the whole exoenzyme production (Botton and Chalot 1999, Smith and Read 2008). Protease activity, especially protein depolymerization, is a key point of the N mineralization in the soil (Schimel and Bennett 2004), so the low protease release of the ECM pose a question on their role in N availability for the microbes in the soil and the plants. Direct comparisons between plants interacting with AMF or EMF showed that the latter has a higher ability to deplete N from the SOM and increase the C:N ratio of the SOM (Lin et al. 2017), which could also depend on the litter traits of the species used (Lin et al. 2017, Badagliacca et al. 2017) and the N demands of each plant species and its microbial symbiont. Also, it was shown that the higher the EMF mycelial biomass, the lower the $^{15}N$ signature of the pine needles and that EMF retain high amounts of N for their own needs (Hobbie et al. 2008, Hobbie and Högberg 2012).

The N delivery from the EMF to the host plant, as for AMF, should occur when both the plant and the EMF are not C limited and when the plant, but not the EMF, is N limited (Makarov 2019). Such a condition is more likely to occur when the N available in the system is mostly organic, and the EMF also have some ability to decompose the organic C and N (Hobbie et al. 2008, Lilleskov et al. 2011, Näsholm et al. 2013, Lindahl and Tunlid 2015). In such conditions, the benefit of the organic matter decomposition for the EMF is likely to occur, thanks to the release of N from the organic matter rather than the release of C (Lindahl and Tunlid 2015) and this especially occurs when the EMF interact with N-depleted, non $N_2$-fixing host for which the EMF may act as a bridge for N transfer among plant species (Simard et al. 2015).

AMF and EMF fungi play contrasting roles in the N-cycling in both soil and plants. Ericoid mycorrhizal fungi frequently act similarly to EMF.

The most important differences between AMF and ERM are the higher ability of the ERM to retain N for their own need and thus deplete the soil N pool in favor of the fungal N pool. In such conditions, the delivery of N from the fungal partner to the plant strongly depends on the plant or plant system ability to feed the fungus with organic C for which some EMF have saprotrophic abilities that AMF do not have. In turn, AMF massively release glomalin, a protein with a low decomposition rate, thus consisting of a higher release of stable and organic N in the soil.

Both fungal groups are able to feed the plant only when their own N needs are satisfied, but such conditions are more likely to occur in natural environments, especially forest, where most of the ECM live compared to cropped environments, which are often highly depleted in N and are more colonized by the plant species getting into the symbiosis with AMF.

## 5.4  FUTURE PERSPECTIVE AND CHALLENGES

Fungi clearly play an important role in global N-cycle through different reactions. However, many fungal species need to be studied to understand their specific interactions in the N- (and C-) cycle in both natural and cropped environments. We still lack a basic understanding of many N-related chemical pathways, genes encoding proteins that are the key steps in denitrification, co-denitrification and ammonia fermentation and all of the genomic changes associated with these pathways. Furthermore, *omics* studies are crucial to understanding the implications of such reactions in the environmental network. Moreover, this information will be important to predict how fungi will respond to challenges created by anthropogenic actions, such as global warming, greenhouse gas emissions or pollution-related to soil fertilization.

## 5.5  CONCLUSIONS

In conclusion, fungi are pivotal components of the N-cycle in both natural and anthropic systems and should be carefully taken into account in the ecosystems or agro-ecosystems. It is of great interest to understand the specific role of fungi in each ecological niche, the origin of the presence and impact of their metabolic pathways in eukaryotes, whether these genes were acquired by gene transference and how they were evolved in eukaryotes. The question of if $N_2$ fixation is exclusively a prokaryotic trait is under debate but hints at the common genes between rhizobia—the most important $N_2$ fixers—and some fungal species pave the way to future use of the fungal technology in the environment. In contrast, fungal DNRA and ammonification represent efficient mechanisms for obtaining energy at low oxygen concentrations, which could represent an advantageous trait in some environments, such as extreme conditions, and could be transferred to other microbes. Finally, the contribution of symbiotic fungi in the global net of N is of main concern, especially if considering their role in changing the plant-soil microbiomes relationship, which can shape nutrient cycles, contributing to the mitigation of N pollution from the agro-ecosystems to the natural ecosystems. Lastly, proper management of the fungal activity in the plant-soil system can strongly help mitigate climate change, especially by improving the stabilization of the SOM, which is the largest carbon reservoir in terrestrial ecosystems and requires large amounts of N to be stabilized.

## 5.6 ACKNOWLEDGMENTS

We would like to thank the Ministry of Economy and Competitiveness (MINECO) and the European Regional Development Fund (ERDF) for their funds CTM2017-84332-R (MINECO/AEI/FEDER/UE)].

## 5.7 REFERENCES

Ames, R.N., Reid, C.P.P., Porter, L.K. and Cambardella, C. 1983. Hyphal uptake and transport of nitrogen from two 15n-labelled sources by *Glomus mosseae*, a vesicular-arbuscular mycorrhizal fungus. New Phytologist 95(3): 381-396. doi: 10.1111/j.1469-8137.1983.tb03506.x.

Amora-Lazcano, E., Vazquez, M. and Azcon, R. 1998. Response of nitrogen-transforming microorganisms to arbuscular mycorrhizal fungi. Biology and Fertility of Soils 27(1): 65-70.

Arasimowicz-Jelonek, M. and Floryszak-Wieczorek, J. 2016. Nitric oxide in the offensive strategy of fungal and oomycete plant pathogens. Frontiers in Plant Science 7(252). doi: 10.3389/fpls.2016.00252.

Atul-Nayyar, A., Hamel, C., Hanson, K. and Germida, J. 2009. The arbuscular mycorrhizal symbiosis links N mineralization to plant demand. Mycorrhiza 19(4): 239-246. doi: 10.1007/s00572-008-0215-0.

Averill, C., Turner, B.L. and Finzi, A.C. 2014. Mycorrhiza-mediated competition between plants and decomposers drives soil carbon storage. Nature 505(7484): 543.

Badagliacca, G., Ruisi, P., Rees, R.M. and Saia, S. 2017. An assessment of factors controlling $N_2O$ and $CO_2$ emissions from crop residues using different measurement approaches. Biology and Fertility of Soils 53(5): 547-561. doi: 10.1007/s00374-017-1195-z.

Baldrian, P., Větrovský, T., Cajthaml, T., Dobiášová, P., Petránková, M., Šnajdr, J., et al. 2013. Estimation of fungal biomass in forest litter and soil. Fungal Ecology 6(1): 1-11. doi: https://doi.org/10.1016/j.funeco.2012.10.002.

Barré, P., Durand, H., Chenu, C., Meunier, P., Montagne, D., Castel, G., et al. 2017. Geological control of soil organic carbon and nitrogen stocks at the landscape scale. Geoderma 285: 50-56. doi: https://doi.org/10.1016/j.geoderma.2016.09.029.

Bender, S.F. and van der Heijden, M.G.A. 2015. Soil biota enhance agricultural sustainability by improving crop yield, nutrient uptake and reducing nitrogen leaching losses. Journal of Applied Ecology 52(1): 228-239. doi: 10.1111/1365-2664.12351.

Berks, B.C., Ferguson, S.J., Moir, J.W. and Richardson, D.J. 1995. Enzymes and associated electron transport systems that catalyse the respiratory reduction of nitrogen oxides and oxyanions. Biochimica. et Biophysica. Acta. (BBA)-Bioenergetics 1232(3): 97-173.

Bianciotto, V., Bandi, C., Minerdi, D., Sironi, M., Tichy, H.V. and Bonfante, P. 1996. An obligately endosymbiotic mycorrhizal fungus itself harbors obligately intracellular bacteria. Applied and Environmental Microbiology 62(8): 3005-3010. doi: https://aem.asm.org/content/aem/62/8/3005.full.pdf.

Bianciotto, V., Andreotti, S., Balestrini, R., Bonfante, P. and Perotto, S. 2001. Extracellular polysaccharides are involved in the attachment of Azospirillum brasilense and *Rhizobium leguminosarum* to arbuscular mycorrhizal structures. European Journal of Histochemistry 45(1): 39-50.

Bianciotto, V. and Bonfante, P. 2002. Arbuscular mycorrhizal fungi: a specialised niche for rhizospheric and endocellular bacteria. Antonie van Leeuwenhoek 81(1): 365-371. doi: 10.1023/a:1020544919072.

Bollag, J.-M. and Tung, G. 1972. Nitrous oxide release by soil fungi. Soil Biology and Biochemistry 4(3): 271-276.

Botton, B. and Chalot, M. 1999. Nitrogen assimilation: enzymology in ectomycorrhizas. pp. 333-372. *In*: Varma, A. and Hock, B. (eds.). Mycorrhiza: Structure, Function, Molecular Biology and Biotechnology. Berlin, Heidelberg: Springer Berlin Heidelberg.

Bruns, T.D., Corradi, N., Redecker, D., Taylor, J.W. and Öpik, M. 2018. Glomeromycotina: what is a species and why should we care? New Phytologist 220(4): 963-967. doi: 10.1111/nph.14913.

Bryla, D.R. and Eissenstat, D.M. 2005. Respiratory costs of mycorrhizal associations. pp. 207-224. *In*: Lambers, H. and Ribas-Carbo, M. (eds.). Plant Respiration: From Cell to Ecosystem. Dordrecht: Springer Netherlands.

Bücking, H. and Kafle, A. 2015. Role of arbuscular mycorrhizal fungi in the nitrogen uptake of plants: current knowledge and research gaps. Agronomy 5(4): 587-612.

Bukovská, P., Bonkowski, M., Konvalinková, T., Beskid, O., Hujslová, M., Püschel, D., et al. 2018. Utilization of organic nitrogen by arbuscular mycorrhizal fungi—is there a specific role for protists and ammonia oxidizers? Mycorrhiza 28(3): 269-283. doi: 10.1007/s00572-018-0825-0.

Burke, D.J., Weintraub, M.N., Hewins, C.R. and Kalisz, S. 2011. Relationship between soil enzyme activities, nutrient cycling and soil fungal communities in a northern hardwood forest. Soil Biology and Biochemistry 43(4): 795-803.

Butterbach-Bahl, K., Baggs Elizabeth, M., Dannenmann, M., Kiese, R. and Zechmeister-Boltenstern, S. 2013. Nitrous oxide emissions from soils: how well do we understand the processes and their controls? Philosophical Transactions of the Royal Society B. Biological Sciences 368(1621): 20130122. doi: 10.1098/rstb.2013.0122.

Campbell, W.H. and Kinghorn, J.R. 1990. Functional domains of assimilatory nitrate reductases and nitrite reductases. Trends in Biochemical Sciences 15(8): 315-319.

Cappellazzo, G., Lanfranco, L., Fitz, M., Wipf, D. and Bonfante, P. 2008. Characterization of an amino acid permease from the endomycorrhizal fungus *Glomus mosseae*. Plant Physiology 147(1): 429-437. doi: 10.1104/pp.108.117820.

Chalk, P.M., Souza, R.d.F., Urquiaga, S., Alves, B.J.R. and Boddey, R.M. 2006. The role of arbuscular mycorrhiza in legume symbiotic performance. Soil Biology and Biochemistry 38(9): 2944-2951. doi: https://doi.org/10.1016/j.soilbio.2006.05.005.

Chanclud, E. and Morel, J.-B. 2016. Plant hormones: a fungal point of view. Molecular Plant Pathology 17(8): 1289-1297. doi: 10.1111/mpp.12393.

Chen, H., Mothapo, N.V. and Shi, W. 2014. The significant contribution of fungi to soil $N_2O$ production across diverse ecosystems. Applied Soil Ecology 73: 70-77. doi: https://doi.org/10.1016/j.apsoil.2013.08.011.

Chen, Y., Ren, C.-G., Yang, B., Peng, Y. and Dai, C.-C. 2013. Priming effects of the endophytic fungus *Phomopsis liquidambari* on soil mineral N transformations. Microbial Ecology 65(1): 161-170.

Chigineva, N., Aleksandrova, A., Marhan, S., Kandeler, E. and Tiunov, A. 2011. The importance of mycelial connection at the soil–litter interface for nutrient translocation, enzyme activity and litter decomposition. Applied Soil Ecology 51: 35-41.

Cole, J. and Brown, C. 1980. Nitrite reduction to ammonia by fermentative bacteria: a short circuit in the biological nitrogen cycle. FEMS Microbiology Letters 7(2): 65-72.

Corrêa, A., Cruz, C. and Ferrol, N. 2015. Nitrogen and carbon/nitrogen dynamics in arbuscular mycorrhiza: the great unknown. Mycorrhiza 25(7): 499-515. doi: 10.1007/s00572-015-0627-6.

Courty, P.-E., Buée, M., Diedhiou, A.G., Frey-Klett, P., Le Tacon, F., Rineau, F., et al. 2010. The role of ectomycorrhizal communities in forest ecosystem processes: new perspectives and emerging concepts. Soil Biology and Biochemistry 42(5): 679-698. doi: https://doi.org/10.1016/j.soilbio.2009.12.006.

Cove, D. 1979. Genetic studies of nitrate assimilation in *Aspergillus nidulans*. Biological Reviews 54(3): 291-327.

Cove, D.J. and Pateman, J.A. 1963. Independently segregating genetic loci concerned with nitrate reductase activity in *Aspergillus nidulans*. Nature 198(4877): 262-263. doi: 10.1038/198262a0.

Fernandez, C.W., Langley, J.A., Chapman, S., McCormack, M.L. and Koide, R.T. 2016. The decomposition of ectomycorrhizal fungal necromass. Soil Biology and Biochemistry 93: 38-49. doi: https://doi.org/10.1016/j.soilbio.2015.10.017.

Fontaine, S., Henault, C., Aamor, A., Bdioui, N., Bloor, J., Maire, V., et al. 2011. Fungi mediate long term sequestration of carbon and nitrogen in soil through their priming effect. Soil Biology and Biochemistry 43(1): 86-96.

Frey, B. and Schüepp, H. 1993. Acquisition of nitrogen by external hyphae of arbuscular mycorrhizal fungi associated with *Zea mays* L. New Phytologist 124(2): 221-230. doi: 10.1111/j.1469-8137.1993.tb03811.x.

Frey, S., Elliott, E., Paustian, K. and Peterson, G. 2000. Fungal translocation as a mechanism for soil nitrogen inputs to surface residue decomposition in a no-tillage agroecosystem. Soil Biology and Biochemistry 32(5): 689-698.

Frey, S., Six, J. and Elliott, E. 2003. Reciprocal transfer of carbon and nitrogen by decomposer fungi at the soil–litter interface. Soil Biology and Biochemistry 35(7): 1001-1004.

Geisseler, D., Horwath, W.R., Joergensen, R.G. and Ludwig, B. 2010. Pathways of nitrogen utilization by soil microorganisms–a review. Soil Biology and Biochemistry 42(12): 2058-2067.

Gonzalez-Quiñones, V., Stockdale, E., Banning, N., Hoyle, F., Sawada, Y., Wherrett, A., et al. 2011. Soil microbial biomass—Interpretation and consideration for soil monitoring. Soil Research 49(4): 287-304.

Guether, M., Neuhäuser, B., Balestrini, R., Dynowski, M., Ludewig, U. and Bonfante, P. 2009. A mycorrhizal-specific ammonium transporter from *Lotus japonicus* acquires nitrogen released by arbuscular mycorrhizal fungi. Plant Physiology 150(1): 73-83.

Hawkins, B. and Kranabetter, J. 2017. Quantifying inorganic nitrogen uptake capacity among ectomycorrhizal fungal species using MIFE microelectrode ion flux measurements: theory and applications. Botany 95(10): 963-969.

Hayatsu, M., Tago, K. and Saito, M. 2008. Various players in the nitrogen cycle: diversity and functions of the microorganisms involved in nitrification and denitrification. Soil Science and Plant Nutrition 54(1): 33-45. doi: doi:10.1111/j.1747-0765.2007.00195.x.

Hayman, D. 1986. Mycorrhizae of nitrogen-fixing legumes. MIRCEN Journal of Applied Microbiology and Biotechnology 2(1): 121-145.

Hobbie, E.A., Colpaert, J.V., White, M.W., Ouimette, A.P. and Macko, S.A. 2008. Nitrogen form, availability, and mycorrhizal colonization affect biomass and nitrogen isotope patterns in *Pinus sylvestris.* Plant and Soil 310(1): 121. doi: 10.1007/s11104-008-9637-x.

Hobbie, E.A. and Högberg, P. 2012. Nitrogen isotopes link mycorrhizal fungi and plants to nitrogen dynamics. New Phytologist 196(2): 367-382. doi: 10.1111/j.1469-8137.2012.04300.x.

Hodge, A. and Storer, K. 2015. Arbuscular mycorrhiza and nitrogen: implications for individual plants through to ecosystems. Plant and Soil 386(1): 1-19. doi: 10.1007/s11104-014-2162-1.

Högberg, M.N. and Högberg, P. 2002. Extramatrical ectomycorrhizal mycelium contributes one-third of microbial biomass and produces, together with associated roots, half the dissolved organic carbon in a forest soil. New Phytologist 154(3): 791-795. doi: 10.1046/j.1469-8137.2002.00417.x.

Janos, D.P. 2007. Plant responsiveness to mycorrhizas differs from dependence upon mycorrhizas. Mycorrhiza 17(2): 75-91. doi: 10.1007/s00572-006-0094-1.

Jin, H., Pfeffer, P.E., Douds, D.D., Piotrowski, E., Lammers, P.J. and Shachar-Hill, Y. 2005. The uptake, metabolism, transport and transfer of nitrogen in an arbuscular mycorrhizal symbiosis. New Phytologist 168(3): 687-696. doi: 10.1111/j.1469-8137.2005.01536.x.

Johansen, A., Jakobsen, I. and Jensen, E.S. 1992. Hyphal transport of 15N-labelled nitrogen by a vesicular—arbuscular mycorrhizal fungus and its effect on depletion of inorganic soil N. New Phytologist 122(2): 281-288. doi: 10.1111/j.1469-8137.1992.tb04232.x.

Johnstone, I., McCabe, P., Greaves, P., Gurr, S., Cole, G., Brow, M., et al. 1990. Isolation and characterisation of the crnA-niiA-niaD gene cluster for nitrate assimilation in *Aspergillus nidulans.* Gene. 90(2): 181-192.

Kamp, A., Høgslund, S., Risgaard-Petersen, N. and Stief, P. 2015. Nitrate storage and dissimilatory nitrate reduction by eukaryotic microbes. Frontiers in Microbiology 6: 1492.

Kaschuk, G., Kuyper, T.W., Leffelaar, P.A., Hungria, M. and Giller, K.E. 2009. Are the rates of photosynthesis stimulated by the carbon sink strength of rhizobial and arbuscular mycorrhizal symbioses? Soil Biology and Biochemistry 41(6): 1233-1244. doi: https://doi.org/10.1016/j.soilbio.2009.03.005.

Kobayashi, M., Matsuo, Y., Takimoto, A., Suzuki, S., Maruo, F. and Shoun, H. 1996. Denitrification, a novel type of respiratory metabolism in fungal mitochondrion. Journal of Biological Chemistry 271(27): 16263-16267.

Köhl, L., Oehl, F. and van der Heijden, M.G.A. 2014. Agricultural practices indirectly influence plant productivity and ecosystem services through effects on soil biota. Ecological Applications 24(7): 1842-1853. doi: 10.1890/13-1821.1.

Koide, R.T. and Mosse, B. 2004. A history of research on arbuscular mycorrhiza. Mycorrhiza 14(3): 145-163. doi: 10.1007/s00572-004-0307-4.

Kraft, B., Strous, M. and Tegetmeyer, H.E. 2011. Microbial nitrate respiration – genes, enzymes and environmental distribution. Journal of Biotechnology 155(1): 104-117. doi: https://doi.org/10.1016/j.jbiotec.2010.12.025.

Kuo, H.-C., Hui, S., Choi, J., Asiegbu, F.O., Valkonen, J.P.T. and Lee, Y.-H. 2014. Secret lifestyles of *Neurospora crassa*. Scientific Reports 4(1): 5135. doi: 10.1038/srep05135.

Laughlin, R.J. and Stevens, R.J. 2002. Evidence for fungal dominance of denitrification and codenitrification in a grassland soil. Soil Science Society of America Journal 66(5): 1540-1548. doi: 10.2136/sssaj2002.1540.

Leigh, J., Hodge, A. and Fitter, A.H. 2009. Arbuscular mycorrhizal fungi can transfer substantial amounts of nitrogen to their host plant from organic material. New Phytologist 181(1): 199-207. doi: 10.1111/j.1469-8137.2008.02630.x.

Leigh, J., Fitter, A.H. and Hodge, A. 2011. Growth and symbiotic effectiveness of an arbuscular mycorrhizal fungus in organic matter in competition with soil bacteria. FEMS Microbiology Ecology 76(3): 428-438.

Li, P., Yang, Y., Han, W. and Fang, J. 2014. Global patterns of soil microbial nitrogen and phosphorus stoichiometry in forest ecosystems. Global Ecology and Biogeography 23(9): 979-987. doi: 10.1111/geb.12190.

Li, X., Wang, C. and Ren, C. 2009. Effect of endophytic fungus $B_3$ and different amounts of nitrogen applied on growth and yield in rice (Oryza sativa L.). Jiangsu Journal of Agricultural Sciences 25(6): 1207-1212.

Lilleskov, E.A., Hobbie, E.A. and Horton, T.R. 2011. Conservation of ectomycorrhizal fungi: exploring the linkages between functional and taxonomic responses to anthropogenic N deposition. Fungal Ecology 4(2): 174-183. doi: https://doi.org/10.1016/j.funeco.2010.09.008.

Lin, G., McCormack, M.L., Ma, C. and Guo, D. 2017. Similar below-ground carbon cycling dynamics but contrasting modes of nitrogen cycling between arbuscular mycorrhizal and ectomycorrhizal forests. New Phytologist 213(3): 1440-1451. doi: 10.1111/nph.14206.

Lin, J.T. and Stewart, V. 1997. Nitrate assimilation by bacteria. pp. 1-30. *In*: Poole, R.K. (ed.). Advances in Microbial Physiology. Academic Press, Elsevier. doi: https://doi.org/10.1016/S0065-2911(08)60014-4.

Lindahl, B.D. and Tunlid, A. 2015. Ectomycorrhizal fungi–potential organic matter decomposers, yet not saprotrophs. New Phytologist 205(4): 1443-1447.

Lu, W.-W., Zhang, H.-L. and Shi, W.-M. 2013. Dissimilatory nitrate reduction to ammonium in an anaerobic agricultural soil as affected by glucose and free sulfide. European Journal of Soil Biology 58: 98-104.

Makarov, M.I. 2019. The role of mycorrhiza in transformation of nitrogen compounds in soil and nitrogen nutrition of plants: a review. Eurasian Soil Science 52(2): 193-205. doi: 10.1134/s1064229319020108.

Marschner, P., Crowley, D. and Lieberei, R. 2001. Arbuscular mycorrhizal infection changes the bacterial 16 S rDNA community composition in the rhizosphere of maize. Mycorrhiza 11(6): 297-302. doi: 10.1007/s00572-001-0136-7.

Marschner, P. and Baumann, K. 2003. Changes in bacterial community structure induced by mycorrhizal colonisation in split-root maize. Plant and Soil 251(2): 279-289. doi: 10.1023/a:1023034825871.

Michelsen, A., Schmidt, I.K., Jonasson, S., Quarmby, C. and Sleep, D. 1996. Leaf 15N abundance of subarctic plants provides field evidence that ericoid, ectomycorrhizal and non-and arbuscular mycorrhizal species access different sources of soil nitrogen. Oecologia 105(1): 53-63. doi: 10.1007/bf00328791.

Morozkina, E. and Kurakov, A. 2007. Dissimilatory nitrate reduction in fungi under conditions of hypoxia and anoxia: a review. Applied Biochemistry and Microbiology 43(5): 544-549.

Mothapo, N., Chen, H., Cubeta, M.A., Grossman, J.M., Fuller, F. and Shi, W. 2015. Phylogenetic, taxonomic and functional diversity of fungal denitrifiers and associated $N_2O$ production efficacy. Soil Biology and Biochemistry 83: 160-175. doi: https://doi.org/10.1016/j.soilbio.2015.02.001.

Müller, T., Avolio, M., Olivi, M., Benjdia, M., Rikirsch, E., Kasaras, A., et al. 2007. Nitrogen transport in the ectomycorrhiza association: the Hebeloma cylindrosporum–Pinus pinaster model. Phytochemistry 68(1): 41-51. doi: https://doi.org/10.1016/j.phytochem.2006.09.021.

Murphy, D.V., Sparling, G.P. and Fillery, I.R.P. 1998. Stratification of microbial biomass C and N and gross N mineralisation with soil depth in two contrasting Western Australian agricultural soils. Soil Research 36(1): 45-56. doi: https://doi.org/10.1071/S97045.

Näsholm, T., Kielland, K. and Ganeteg, U. 2009. Uptake of organic nitrogen by plants. New Phytologist 182(1): 31-48. doi: 10.1111/j.1469-8137.2008.02751.x.

Näsholm, T., Högberg, P., Franklin, O., Metcalfe, D., Keel, S.G., Campbell, C., et al. 2013. Are ectomycorrhizal fungi alleviating or aggravating nitrogen limitation of tree growth in boreal forests? New Phytologist 198(1): 214-221. doi: 10.1111/nph.12139.

Novinscak, A., Goyer, C., Zebarth, B.J., Burton, D.L., Chantigny, M.H. and Filion, M. 2016. Novel *P450nor* gene detection assay used to characterize the prevalence and diversity of soil fungal denitrifiers. Applied and Environmental Microbiology 82(15): 4560-4569. doi: 10.1128/aem.00231-16.

Offre, P., Spang, A. and Schleper, C. 2013. Archaea in biogeochemical cycles. Annual Review of Microbiology 67(1): 437-457. doi: 10.1146/annurev-micro-092412-155614.

Paul, K., Saha, C., Nag, M., Mandal, D., Naiya, H., Sen, D., et al. 2020. A tripartite interaction among the basidiomycete Rhodotorula mucilaginosa, $N_2$-fixing endobacteria, and rice improves plant nitrogen nutrition. The Plant Cell 32(2): 486-507.

Pellitier, P.T. and Zak, D.R. 2018. Ectomycorrhizal fungi and the enzymatic liberation of nitrogen from soil organic matter: why evolutionary history matters. New Phytologist 217(1): 68-73. doi: 10.1111/nph.14598.

Phillips, R.L., Song, B., McMillan, A.M.S., Grelet, G., Weir, B.S., Palmada, T., et al. 2016. Chemical formation of hybrid di-nitrogen calls fungal codenitrification into question. Scientific Reports 6(1): 39077. doi: 10.1038/srep39077.

Poorter, H., Niklas, K.J., Reich, P.B., Oleksyn, J., Poot, P. and Mommer, L. 2012. Biomass allocation to leaves, stems and roots: meta-analyses of interspecific variation and environmental control. New Phytologist 193(1): 30-50. doi: 10.1111/j.1469-8137.2011.03952.x.

Pozo, M.J. and Azcón-Aguilar, C. 2007. Unraveling mycorrhiza-induced resistance. Current Opinion in Plant Biology 10(4): 393-398. doi: https://doi.org/10.1016/j.pbi.2007.05.004.

Prosser, J.I. 2007. Chapter 15 - The ecology of nitrifying bacteria. pp. 223-243. *In*: Bothe, H., Ferguson, S. and Newton, W.E. (eds.). Biology of the Nitrogen Cycle. Amsterdam: Elsevier.

Prosser, J.I. and Nicol, G.W. 2012. Archaeal and bacterial ammonia-oxidisers in soil: the quest for niche specialisation and differentiation. Trends in Microbiology 20(11): 523-531. doi: https://doi.org/10.1016/j.tim.2012.08.001.

Püschel, D., Janoušková, M., Voříšková, A., Gryndlerová, H., Vosátka, M. and Jansa, J. 2017. Arbuscular mycorrhiza stimulates biological nitrogen fixation in two *Medicago* spp. through improved phosphorus acquisition. Frontiers in Plant Science 8(390). doi: 10.3389/fpls.2017.00390.

Read, D.J. and Perez-Moreno, J. 2003. Mycorrhizas and nutrient cycling in ecosystems – a journey towards relevance? New Phytologist 157(3): 475-492. doi: 10.1046/j.1469-8137.2003.00704.x.

Rillig, M.C. 2004. Arbuscular mycorrhizae, glomalin, and soil aggregation. Canadian Journal of Soil Science 84(4): 355-363.

Rillig, M.C. and Mummey, D.L. 2006. Mycorrhizas and soil structure. New Phytologist 171(1): 41-53. doi: 10.1111/j.1469-8137.2006.01750.x.

Rivero, J., Gamir, J., Aroca, R., Pozo, M.J. and Flors, V. 2015. Metabolic transition in mycorrhizal tomato roots. Frontiers in Microbiology 6(598). doi: 10.3389/fmicb.2015.00598.

Robinson, D. 2007. Implications of a large global root biomass for carbon sink estimates and for soil carbon dynamics. Proceedings of the Royal Society B. Biological Sciences 274(1626): 2753-2759.

Saia, S., Amato, G., Frenda, A.S., Giambalvo, D. and Ruisi, P. 2014a. Influence of arbuscular mycorrhizae on biomass production and nitrogen fixation of berseem clover plants subjected to water stress. PLOS One 9(3): e90738. doi: 10.1371/journal.pone.0090738.

Saia, S., Benítez, E., García-Garrido, J., Settanni, L., Amato, G. and Giambalvo, D. 2014b. The effect of arbuscular mycorrhizal fungi on total plant nitrogen uptake and nitrogen recovery from soil organic material. The Journal of Agricultural Science 152(3): 370-378.

Saia, S., Rappa, V., Ruisi, P., Abenavoli, M.R., Sunseri, F., Giambalvo, D., et al. 2015a. Soil inoculation with symbiotic microorganisms promotes plant growth and nutrient transporter genes expression in durum wheat. Frontiers in Plant Science 6(815). doi: 10.3389/fpls.2015.00815.

Saia, S., Ruisi, P., Fileccia, V., Di Miceli, G., Amato, G. and Martinelli, F. 2015b. Metabolomics suggests that soil inoculation with arbuscular mycorrhizal fungi decreased free amino acid content in roots of durum wheat grown under N-limited, P-rich field conditions. Plos One 10(6): e0129591. doi: 10.1371/journal.pone.0129591.

Saia, S., Colla, G., Raimondi, G., Di Stasio, E., Cardarelli, M., Bonini, P., et al. 2019. An endophytic fungi-based biostimulant modulated lettuce yield, physiological and functional quality responses to both moderate and severe water limitation. Scientia Horticulturae 256: 108595. doi: https://doi.org/10.1016/j.scienta.2019.108595.

Saia, S., Aissa, E., Luziatelli, F., Ruzzi, M., Colla, G., Ficca, A.G., et al. 2020. Growth-promoting bacteria and arbuscular mycorrhizal fungi differentially benefit tomato and corn depending upon the supplied form of phosphorus. Mycorrhiza. doi:10.1007/s00572-019-00927-w.

Schillaci, C., Acutis, M., Vesely, F. and Saia, S. 2019. A simple pipeline for the assessment of legacy soil datasets: an example and test with soil organic carbon from a highly variable area. Catena 175: 110-122. doi: https://doi.org/10.1016/j.catena.2018.12.015.

Schimel, J.P. and Bennett, J. 2004. Nitrogen mineralization: challenges of a changing paradigm. Ecology 85(3): 591-602. doi: 10.1890/03-8002.

Schütz, L., Gattinger, A., Meier, M., Müller, A., Boller, T., Mäder, P., et al. 2018. Improving crop yield and nutrient use efficiency via biofertilization—a global meta-analysis. Frontiers in Plant Science 8(2204). doi: 10.3389/fpls.2017.02204.

Shoun, H. and Tanimoto, T. 1991. Denitrification by the fungus *Fusarium oxysporum* and involvement of cytochrome P-450 in the respiratory nitrite reduction. Journal of Biological Chemistry 266(17): 11078-11082. doi: http://www.jbc.org/content/266/17/11078.abstract.

Shoun, H., Kim, D.-H., Uchiyama, H. and Sugiyama, J. 1992. Denitrification by fungi. FEMS Microbiology Letters 94(3): 277-281.

Shoun H., Fushinobu, S., Jiang, L., Kim, S.W. and Wakagi, T. 2012. Fungal denitrification and nitric oxide reductase cytochrome P450nor. Philosophical Transactions of the Royal Society of London. Series B. Biological Sciences vol. 367(1593): 1186-1194. doi: 10.1098/rstb.2011.0335.

Simard, S., Asay, A., Beiler, K., Bingham, M., Deslippe, J., He, X., et al. 2015. Resource transfer between plants through ectomycorrhizal fungal networks. pp. 133-176. *In*: Horton, T.R. (ed.). Mycorrhizal Networks.Dordrecht: Springer Netherlands.

Singh, B.K., Nunan, N., Ridgway, K.P., McNicol, J., Young, J.P.W., Daniell, T.J., et al. 2008. Relationship between assemblages of mycorrhizal fungi and bacteria on grass roots. Environmental Microbiology 10(2): 534-541. doi: 10.1111/j.1462-2920.2007.01474.x.

Smith, S.E. and Read, D. 2008. pp. 1-9. *In*: Smith, S.E. and Read, D. (eds.). Mycorrhizal Symbiosis (Third Edition) London: Academic Press.

Smith, S.E. and Read, D.J. 2010. Mycorrhizal Symbiosis. Academic Press, London, UK.

Spatafora, J.W., Chang, Y., Benny, G.L., Lazarus, K., Smith, M.E., Berbee, M.L., et al. 2016. A phylum-level phylogenetic classification of zygomycete fungi based on genome-scale data. Mycologia 108(5): 1028-1046. doi: 10.3852/16-042.

Spott, O., Russow, R. and Stange, C.F. 2011. Formation of hybrid $N_2O$ and hybrid $N_2$ due to codenitrification: first review of a barely considered process of microbially mediated N-nitrosation. Soil Biology and Biochemistry 43(10): 1995-2011.

Steinberg, P.D. and Rillig, M.C. 2003. Differential decomposition of arbuscular mycorrhizal fungal hyphae and glomalin. Soil Biology and Biochemistry 35(1): 191-194. doi: https://doi.org/10.1016/S0038-0717(02)00249-3.

Stief, P., Fuchs-Ocklenburg, S., Kamp. A., Manohar, C.-S., Houbraken, J., Boekhout, T., et al. 2014. Dissimilatory nitrate reduction by *Aspergillus terreus* isolated from the seasonal oxygen minimum zone in the Arabian Sea. BMC Microbiology 14(1): 35.

Stroo, H.F., Klein, T.M. and Alexander, M. 1986. Heterotrophic nitrification in an acid forest soil and by an acid-tolerant fungus. Applied and Environmental Microbiology 52(5): 1107-1111. doi: https://aem.asm.org/content/aem/52/5/1107.full.pdf.

Sun, K., Cao, W., Hu, L., Fu, W., Gong, J., Kang, N., et al. 2019. Symbiotic fungal endophyte Phomopsis liquidambari-rice system promotes nitrogen transformation by influencing below-ground straw decomposition in paddy soil. Journal of Applied Microbiology 126(1): 191-203.

Takasaki, K., Shoun, H., Yamaguchi, M., Takeo, K., Nakamura, A., Hoshino, T., et al. 2004. Fungal ammonia fermentation, a novel metabolic mechanism that couples the dissimilatory and assimilatory pathways of both nitrate and ethanol role of acetyl coa synthetase in anaerobic atp synthesis. Journal of Biological Chemistry 279(13): 12414-12420.

Takaya, N. 2009. Response to hypoxia, reduction of electron acceptors, and subsequent survival by filamentous fungi. Bioscience, Biotechnology, and Biochemistry 73(1): 1-8.

Tanaka, Y. and Yano, K. 2005. Nitrogen delivery to maize via mycorrhizal hyphae depends on the form of N supplied. Plant, Cell and Environment 28(10): 1247-1254. doi: 10.1111/j.1365-3040.2005.01360.x.

Tedersoo, L., Sánchez-Ramírez, S., Kõljalg, U., Bahram, M., Döring, M., Schigel, D., et al. 2018. High-level classification of the Fungi and a tool for evolutionary ecological analyses. Fungal Diversity 90(1): 135-159. doi: 10.1007/s13225-018-0401-0.

Thamdrup, B. 2012. New pathways and processes in the global nitrogen cycle. Annual Review of Ecology, Evolution, and Systematics 43: 407-428.

Tian, C., Kasiborski, B., Koul, R., Lammers, P.J., Bücking, H. and Shachar-Hill, Y. 2010. Regulation of the nitrogen transfer pathway in the arbuscular mycorrhizal symbiosis: gene characterization and the coordination of expression with nitrogen flux. Plant Physiology 153(3): 1175-1187. doi: 10.1104/pp.110.156430.

Tobar, R., Azcón, R. and Barea, J. 1994. Improved nitrogen uptake and transport from 15N-labelled nitrate by external hyphae of arbuscular mycorrhiza under water-stressed conditions. New Phytologist 126(1): 119-122.

Toljander, J.F., Lindahl, B.D., Paul, L.R., Elfstrand, M. and Finlay, R.D. 2007. Influence of arbuscular mycorrhizal mycelial exudates on soil bacterial growth and community structure. FEMS Microbiology Ecology 61(2): 295-304.

Trap, J., Akpa-Vinceslas, M., Margerie, P., Boudsocq, S., Richard, F., Decaëns, T., et al. 2017. Slow decomposition of leaf litter from mature Fagus sylvatica trees promotes offspring nitrogen acquisition by interacting with ectomycorrhizal fungi. Journal of Ecology 105(2): 528-539. doi: 10.1111/1365-2745.12665.

Tsuruta, S., Takaya, N., Zhang, L., Shoun, H., Kimura, K., Hamamoto, M., et al. 1998. Denitrification by yeasts and occurrence of cytochrome P450nor in *Trichosporon cutaneum*. FEMS Microbiology Letters 168(1): 105-110.

van der Heijden, M.G.A., Martin, F.M., Selosse, M.-A. and Sanders, I.R. 2015. Mycorrhizal ecology and evolution: the past, the present, and the future. New Phytologist 205(4): 1406-1423. doi: 10.1111/nph.13288.

van Niftrik, L. and Jetten, M.S.M. 2012. Anaerobic ammonium-oxidizing bacteria: unique microorganisms with exceptional properties. Microbiology and Molecular Biology Reviews 76(3): 585-596. doi: 10.1128/mmbr.05025-11.

Veresoglou, S.D., Chen, B. and Rillig, M.C. 2012. Arbuscular mycorrhiza and soil nitrogen cycling. Soil Biology and Biochemistry 46: 53-62.

Vitousek, P.M. and Howarth, R.W. 1991. Nitrogen limitation on land and in the sea: how can it occur? Biogeochemistry 13(2): 87-115.

Walder, F., Niemann, H., Natarajan, M., Lehmann, M.F., Boller, T. and Wiemken, A. 2012. Mycorrhizal networks: common goods of plants shared under unequal terms of trade. Plant Physiology 159(2): 789-797. doi: 10.1104/pp.112.195727.

Wallander, H., Nilsson, L.O., Hagerberg, D. and Bååth, E. 2001. Estimation of the biomass and seasonal growth of external mycelium of ectomycorrhizal fungi in the field. New Phytologist 151(3): 753-760.

Walley, F. and Germida, J. 1997. Response of spring wheat (*Triticum aestivum*) to interactions between Pseudomonas species and Glomus clarum NT4. Biology and Fertility of Soils 24(4): 365-371.

Wu, T., Sharda, J.N. and Koide, R.T. 2003. Exploring interactions between saprotrophic microbes and ectomycorrhizal fungi using a protein–tannin complex as an N source by red pine (Pinus resinosa). New Phytologist 159(1): 131-139.

Yang, B., Wang, X., Ma, H., Yang, T., Jia, Y., Zhou, J., et al. 2015. Fungal endophyte *Phomopsis liquidambari* affects nitrogen transformation processes and related microorganisms in the rice rhizosphere. Frontiers in Microbiology 6: 982.

Yue, K., Peng, Y., Peng, C., Yang, W., Peng, X. and Wu, F. 2016. Stimulation of terrestrial ecosystem carbon storage by nitrogen addition: a meta-analysis. Scientific Reports 6(1): 19895. doi: 10.1038/srep19895.

Zechmeister-Boltenstern, S., Keiblinger, K.M., Mooshammer, M., Peñuelas, J., Richter, A., Sardans, J., et al. 2015. The application of ecological stoichiometry to plant–microbial–soil organic matter transformations. Ecological Monographs 85(2): 133-155.

Zhang, Z., Phillips, R.P., Zhao, W., Yuan, Y., Liu, Q. and Yin, H. 2019. Mycelia-derived C contributes more to nitrogen cycling than root-derived C in ectomycorrhizal alpine forests. Functional Ecology 33(2): 346-359.

Zhou, G., Xu, S., Ciais, P., Manzoni, S., Fang, J., Yu, G., et al. 2019. Climate and litter C/N ratio constrain soil organic carbon accumulation. National Science Review 6(4): 746-757.

Zhou, Z., Takaya, N., Nakamura, A., Yamaguchi, M., Takeo, K. and Shoun, H. 2002. Ammonia fermentation, a novel anoxic metabolism of nitrate by fungi. Journal of Biological Chemistry 277(3): 1892-1896. doi: 10.1074/jbc.M109096200.

Zhu, T., Meng, T., Zhang, J., Yin, Y., Cai, Z., Yang, W., et al. 2013. Nitrogen mineralization, immobilization turnover, heterotrophic nitrification, and microbial groups in acid forest soils of subtropical China. Biology and Fertility of Soils 49(3): 323-331.

Zhu, T., Meng, T., Zhang, J., Zhong, W., Müller, C. and Cai, Z. 2015. Fungi-dominant heterotrophic nitrification in a subtropical forest soil of China. Journal of Soils and Sediments 15(3): 705-709. doi: 10.1007/s11368-014-1048-4.

Zumft, W.G. 1997. Cell biology and molecular basis of denitrification. Microbiology and Molecular Biology Reviews 61(4): 533-616.

# Biotechnological Applications of Bioeffectors Derived From the Plant Microbiome to Improve Plant's Physiological Response for a Better Adaptation to Biotic and Abiotic Stress: Fundamentals and Case Studies

Ana García-Villaraco[*], Jose Antonio Lucas,
Beatriz Ramos-Solano and Francisco Javier Gutierrez-Mañero

## 6.1 INTRODUCTION

Plants are sessile organisms. This condition involves a continuous adaptation to environmental changes to overcome all stressing situations that happen throughout their life until seeds are formed, ensuring the survival of the species. Unlike other organisms that can run away from danger, plants are hooked to the place where seeds germinate for life. Their only chance to survive is to synthesize chemical molecules for each specific situation. The set of reactions arranged in metabolic pathways that lead to this enormous array of different chemical molecules is known as secondary metabolism. The term 'secondary' is an excluding definition from primary metabolism that involves metabolic pathways dealing with sugar, lipid and protein synthesis and degradation, which are present in all living beings and share high similarities. The carbon scaffoldings that supply these secondary metabolism pathways come from photosynthesis, which requires water and nutrients, so plants have depicted complex physiological mechanisms to ensure proper fitness; in addition, they establish intimate relationships with beneficial microorganisms in the underground and aboveground levels, which are known as the plant microbiome. Within this microbiome are beneficial bacteria, known as Plant Growth Promoting Rhizobacteria (PGPR), which represent a great tool to boost plant fitness in different aspects. Further progress has been made on this topic, including bacterial derived molecules able to mimic the bacterial response or trigger a different plant response. As both PGPR and derived molecules trigger plant metabolism, they are known as bioeffectors.

The present work describes the physiological mechanisms involved in plant adaptation to water stress, nutrient absorption and adaptative responses to biotic stress and how bioeffectors are able to modulate these responses, focusing on the mechanisms involved in plant adaptation to water stress

Universidad San Pablo-CEU Universities. Facultad de Farmacia. Ctra. Boadilla del Monte km 5.300, Boadilla del Monte 28668 Madrid, Spain.
[*] Corresponding author: anabec.fcex@ceu.es

(salinity and water shortage), plant innate immunity and general mechanisms involved in plant protection to pathogen outbreaks. The final section ends with several examples about modulation of secondary metabolism with bioeffectors, addressing beneficial molecules for human health in specific species with interest for the pharmaceutical industry (opium poppy, foxglove, St.John's Wort) and in specific agronomic crops (blackberry, soybean).

## 6.2   PLANT MECHANISMS TO IMPROVE ADAPTATION TO WATER STRESS

In the last decade, factors that usually limit crop productivity have seen an extreme increase driven by climate change. As water and nutrient availability are the most limiting factors, water stress is critical, not only to strict water limitation itself, but also due to salinity. In addition to these factors, the availability of fertile soils is low, limiting the high demand of fertile surface to feed an increasing population requires using poor soils: saline and arid soils.

### 6.2.1   The Physiological Problem

Effects of high salinity in soil result in decreased fertility, mainly associated to two reasons: i) on one hand, upon low values of soil water potential, there is a concomitant increase of osmotic pressure that results in an impaired capacity of the plant to absorb water; ii) on the other hand, the toxicity of sodium ions upon accumulation (Munns and Tester 2008). In either case, a number of plant metabolic targets are affected in a non-specific way. Hyperaccumulation of sodium or potassium ions usually takes place on leaves causing necrosis of different extent. The negative effect of salinity and water stress results from unbalancing many physiological processes, such as normal functioning of enzymes or those that require a precise structural arrangement as photosynthesis (Hanin et al. 2016).

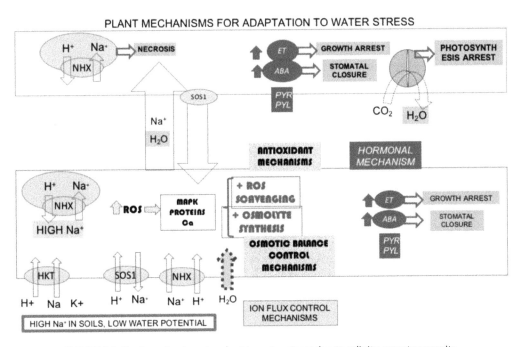

**FIGURE 6.1** Plant mechanisms to adapt to water stress due to salinity or water scarcity.

When water stress starts, the immediate physiological response is to control plant water potential by closing stomata, which leads to photosynthesis arrest (Fricke et al. 2006), compromising plant fitness and therefore decreasing primary production and crop yield. Secondary to the urgent control of water loss, molecular responses follow to fix the damage and to start changes leading to adaptation to the settled stress situation (Figure 6.1).

Despite the tolerance of most plant species to moderate salinity, probably due to the origin of first cells on salty water, only 1-2% of plant species survive on high saline environments as seawater (470 mM NaCl). As a general rule, 80 mM is the salt threshold tolerance for most plants, while halophytes tolerate up to 200 mM NaCl (Volkov and Beilby 2017).

Sodium ions ($Na^+$) enter the plant symplast through subfamily 1-HKT transport proteins, located in epidermal cells of the root system. The concomitant increase in ion concentration in the symplast is detected by hyperosmosis sensors that activate the signaling cascade of molecular systems involved in protection and adaptation: $Ca_2^+$-sensing proteins (CDPKs, CBLs and CIPKs) and ROS-sensing kinases (MAPK) (Hanin et al. 2016). Reactive oxygen species (ROS) modulate response through MAPK, so they are necessary to start responses. However, under stress conditions, there is an overproduction of ROS that causes dramatic cell damage due to phospholipid and protein peroxidation. In addition to structural damage, peroxidation of cell membranes dramatically affects ionic homeostasis as ionic fluxes are compromised and so are plant fitness and developmental and growth processes (Xing et al. 2008, Jammes et al. 2009). Consequently, plant survival depends on its ability to cope with this unbalance. Plant's endowment to overcome a ROS burst consists on an enzymatic system for ROS scavenging, such as superoxide dismutase (SOD) and ascorbate peroxidase (APX) and the associated ascorbate-glutation cycle (García-Cristobal et al. 2015), and a non-enzymatic system constituted by several antioxidant metabolites (phenols, flavonols, ascorbate and gluthation); in addition to this, a pool of organic molecules (proline, soluble sugars) contribute to overcoming osmotic unbalance.

### 6.2.2 The Multifactor Solution: PGPR

Plants develop their life cycle in soils, coexisting with native soil microorganisms that colonize plant root surface due to the carbon-rich exudates released by the plant, such as phenolic compounds, terpenes, organic acids, sugars, amino acids, proteins and fatty acids among others (Narasimhan et al. 2003); exudates are key to support all microorganisms growing therein, the high specific ecosystem termed rhizosphere, where only selected beneficial microorganisms' survive. Microorganisms colonizing the rhizosphere are specific to the plant species and therefore specific microbiomes are described for each plant species and even plant organ within the same species (Wintermans et al. 2016). Consistent with this statement, rhizosphere bacterial strains that are able to improve plant adaptative capacity have been identified, and benefits of this interaction are revealed upon stress challenge; for example, pathogen outbreaks, herbivores, high or low temperatures, salinity, water deficit. As along life's way, these stress situations take place at different time points and an array of bacterial mechanisms involved in improving plant fitness may be activated depending on the stress type; furthermore, they may not necessarily be activated one by one but be simultaneously activated, triggering different plant targets to overcome the stressful situation. In order to be successful when using PGPR, a deep knowledge of basic physiological processes affected is a must. Given the many bacterial mechanisms and plant targets, it is very unlikely that a single PGPR would trigger all plant targets simultaneously. Therefore, the first step is to focus on one of the affected processes, to select those strains that are able to help the plant to overcome it.

### 6.2.3 Mechanisms of PGPR to Ameliorate Water Stress

In the case of water stress, either for lack of water or salinity, the mechanisms that PGPR may use to improve plant fitness are regulation/modulation of water exchange, regulation/modulation of nutrient exchange, photosynthesis, regulation of plant hormonal balance and boosting ROS scavenging systems at different target points. Specific PGPR have demonstrated an enormous benefit to improve plant fitness under these stress conditions, up to the point that the term induced systemic tolerance (IST) has been coined to define the plant metabolic situation triggered by these PGPR.

Regulation/modulation of water and nutrient exchange are tightly related. Ion exchange and water relations belong to an integrated and dynamic system comprising different levels from gene expression to histology and anatomy: ion transport pumps, signal transduction molecules and pathways or stomatal closure. More precisely, among the genes affected by salt stress are aquaporins ZmPIP (Marulanda et al. 2010), $Na^+/H^+$ antiport pumps (SOS1) and HKT $Na^+$, $K^+$ in the plasma membrane and $Na^+$, $K^+/H^+$ (NHX) pumps in the tonoplast (Apse et al. 1999, Sunarpi et al. 2005).

Among PGPR strains that have been able to improve adaptation to water stress, *Bacillus subtilis* (Zhang et al. 2008a) and *Bacillus amyloliquefaciens* strain SN13 (Nautiyal et al. 2013) have reported to be very efficient triggering expression of genes encoding ion pumps to help the plant achieve a healthy ion/water balance. Although not as frequent, some *Pseudomonas* strains have achieved similar results conferring resistance to salinity by activating gene transcription (Kasotia et al. 2016). The ability of some other gram (–) strains to protect plants against water stress has been described, although the mechanisms involved are associated with hormonal balance and stomatal closure rather than gene expression, contrary to the response triggered by *Bacillus* strains.

The systemic adaptative response to water stress involving modification of plant hormonal balance is probably the most frequent and therefore most studied response. The mechanism involved in this response consists of activation of endogenous ethylene synthesis leading to growth arrest in the root in the first instance followed by shoot growth inhibition in order to overcome stress conditions, which limit productivity. Certain PGPR strains are able to degrade the ethylene precursor ACC, creating a gradient of ACC from the plant to the rhizosphere such that as ACC concentration in the plant root decreases, ethylene synthesis is impaired and as a consequence, the inhibitory effect of ethylene disappears; then plant growth is recovered, therefore, productivity is improved or maintained under water stress (Glick et al. 2007). ACC degrading strains are present among gram (+) and gram (–) strains (Shrivastava and Kumar 2015, Kang et al. 2014). Nevertheless, lowering plant ethylene levels results in plants more exposed to the negative effects of drought and salinity and in this case, ABA is released to trigger stomatal closure, compromising growth. Therefore, in order to maintain photosynthetic capacity and the concomitant yield, ethylene-lowering strains should be combined with other strains that are able to target osmotic homeostasis mechanisms, increasing the chances of success by tackling both mechanisms simultaneously. Despite the relevant role of the plant growth regulator ABA in stomatal closure, few studies report the ability of PGPR to increase endogenous ABA concentration and a simultaneous decrease in ethylene (Porcel et al. 2014). Irrespective of the low number of reports, there is a wide representation of bacterial genera (*Pseudomonas, Burkholderia, Acinetobacter* and *Bacillus*) with a great potential to trigger this ITS mechanism (Ilangumaran and Smith 2017). In summary, to improve plant's adaptation to water stress while keeping photosynthetic capacity and productivity high, a combination of strains with ITS-triggering capacity and osmotic homeostasis-triggering capacity should be used as plants become more exposed to negative effects of salinity and drought.

**Osmolite accumulation** is one of the most effective mechanisms plants have depicted to protect themselves from osmotic stress. Osmolite is synthesized to increase endogenous osmotic pressure, favouring hidric homeostasis. Among these molecules are sugars, amino acids and derived molecules, such as proline, that derive from amino acids or the sugars threalose, manitol and sorbitol or glycine betaine, a quaternary ammonium compound. As biosynthesis of most of these metabolites depends on a single or few genes, this seems to be a quite easy mechanism for strains to activate. For example, the main enzyme for proline biosynthesis is 1-pyrroline-5-carboxylate synthetase (P5CS), which catalyzes the conversion of glutamate to 1-pyrroline-5-carboxylate that is further reduced to proline. Several studies have described that the overexpression of P5CS gene rendered tolerance for salt stress in transgenic *Arabidopsis*. This evidences that modification of a single gene is a good strategy to obtain plants with enhanced resistance to salinity. Nevertheless, this effect has been described for

*Bacillus amyloliquefaciens*, which not only triggers proline biosynthetic pathway (Chauhan et al. 2019) but also triggers expression of ion pumps achieving, therefore, a more efficient response than the genetic manipulation and of course, at a lower cost.

## 6.3 EFFECT OF BIOEFFECTORS ON GROWTH PROMOTION AND IMPROVEMENT OF PLANT'S FITNESS IMPROVEMENT OF PLANT NUTRITION

Both developed and developing countries are currently facing the major challenge of maintaining high agricultural productivity due, on the one hand, to a growing global demand for food for a growing population and, on the other, to the effects of climate change, such as desertification of soils, salinization of aquifers and high temperatures among others. Traditional nutrient management is primarily based on external fertilizer inputs to crops, especially in the case of intensive agricultural practices, which guarantee high yield and quality; the use of chemical fertilizers, which are expensive and create environmental problems, is highly extended. However, in the recent decades, increase in fertilizer inputs have not shown a correlation with crop yields increases, revealing a low efficiency in nutrient use and a greater environmental risk (Zhang et al. 2010). Therefore, there has been a resurgence of interest in environmentally friendly, sustainable and organic farming practices (Esitken et al. 2005). It is known that the use of biofertilizers composed by beneficial microorganisms instead of synthetic chemicals, improves plant growth and can help maintain environmental health and soil productivity (O'Connell 1992). In fact, the use of biofertilizers has become a feasible production practice and many commercial biofertilizers based primarily on plant growth-promoting rhizobacteria (PGPR) that have beneficial effects on plant development often related to increased nutrient availability for the host plant (Vessey 2003).

However, not all PGPR strains exert their positive effect on plant growth by increasing the nutrient status of host plants. Some PGPR can promote growth by suppressing plant diseases (Zehnder et al. 2001) or by producing phytohormones and peptides that act as biostimulants (Glick et al. 1998, Jimenez-Delgadillo 2004). The production of plant growth regulators such as auxin, cytokinin and gibberellin by PGPR has also been suggested as a possible mechanism of action that affects plant growth. Numerous studies have shown an improvement in plant growth and development in response to seed or root inoculation with several microbial inoculants capable of producing plant growth regulators (Zahir et al. 2004). Likewise, PGPR can improve the general fitness of the plant by increasing the rate or efficiency of photosynthesis (García-Cristobal et al. 2015).

Thus, it seems that the effect on growth promotion and improvement of plant's fitness could be due to more than one factor induced by the PGPR. In fact, there seems to be a relationship between the improvement in the acquisition of nutrients and auxins. Plant growth promotion induced by PGPR by increasing the number of nutrients absorbed and accumulated in plant tissues can be achieved not only exclusively via increased availability of nutrients but also through the functionality of plasma membrane transport proteins involved in nutrient uptake at the root level (Pii et al. 2015). The transmembrane electrochemical gradient is maintained by the plasma membrane $H^+$-ATPase, which expels $H^+$ outside the root epidermal cells. It has been demonstrated that the electrochemical gradient across the membrane favors nutrients' movement through the membranes since $H^+$ entrance is coupled with the transport of several nutrients (e.g., Pi and $NO_3^-$) (White 2003). Stimulation of plasma membrane $H^+$-ATPase activity by PGPR has been reported by several authors (Bertrand et al. 2000, Canellas et al. 2013) and it has been attributed to an IAA-derived effect (Canellas et al. 2013). However, the mechanisms through which PGPRs are able to stimulate the plasma membrane $H^+$-ATPase activity (both at the transcriptional and posttranscriptional level) are still unknown (Pii et al. 2015).

## 6.3.1 Nitrogen

Apart from bacteria being able to fix atmospheric nitrogen (free-living or symbiotic), PGPR can improve nitrogen nutrition to the plant. The absorption of nitrate ($NO_3^-$) is a symport $H+/NO_3^-$ and the energy is provided by a proton gradient maintained by the activity of the plasma membrane $H^+$-ATPase (Touraine and Glass 1997). Bertrand et al. (2000) hypothesized an effect of the bacteria on the cHATS (High Affinity Transport System) as responsible for an enhanced uptake of $NO_3^-$ and consequent higher content of this anion in the plant tissues. Regarding $NH_4^+$ and urea uptake, there are not any evidence of microbial effects on these forms of organic nitrogen acquisition by the plant (Pii et al. 2015).

Soil microorganism plays a key role in soil organic matter mineralization, atmospheric $N_2$ fixation and denitrification and therefore in the biogeochemical cycle of nitrogen and consequently in soil fertility (Jetten 2008). Around 90% of the nitrogen in the soil is organic nitrogen (organic matter), so nitrification and ammonification processes (mineralization) carried out by bacteria are key for plant's nutrition. In fact, microorganisms such as mycorrhizal fungi and PGPR mineralize organic matter by releasing hydrolytic enzymes and therefore improving nutrient availability in the soil (Ollivier et al. 2011).

The denitrification generates $NO_2^-$ which can be then transformed into nitrogen oxides ($NO_2$ and NO), which can be absorbed by plants. These nitrogen oxides could play an important role in the rhizosphere as the second messenger in the indol-3-acetic acid (IAA) signaling pathway, participating in the induction pathway of both adventitious and lateral roots (Correa-Aragunde et al. 2004). Moreover, different PGPR are able to synthesize NO, which have been reported to induce changes in plant root architecture, inducing the development of lateral and adventitious roots (Creus et al. 2005) and root hairs (Hadas and Okon 1987) and promoting activity on tomato root branching (Molina-Favero et al. 2008).

## 6.3.2 Phosphorus

Phosphorous mineralization is required prior plants can use it. Phosphatases released by bacteria can mineralize organic P (like phosphoesters and phosphotriesters) (Pii et al. 2015).

Monobasic $H_2PO_4^-$ (Pi) is the form of P that plants preferably uptake (Marschner 2011) through a phosphate/$H^+$ symporter from PHT1 gene family, which is considered to be the most important for Pi in the plasma membrane (Karthikeyan et al. 2002, Schünmann et al. 2004).

Both inorganic and organic P forms are very insoluble compounds and in general presents an availability problem of for the plants, even in soils with high P content. Most of the inorganic P in soils is bound to Fe, Al or Ca reducing its solubility, by precipitation and adsorption processes (Gyaneshwar et al. 2002).

Bacteria from genera *Azospirillum*, *Azotobacter*, *Bacillus*, *Burkholderia*, *Pseudomonas*, *Rhizobium* and *Serratia* have been reported to solubilize P releasing of organic acids (Bhattacharyya and Jha 2012), such as acetate, oxalate, succinate, citrate, and gluconate, which can desorb Pi from soil adsorption sites by ligand exchange and thus solubilizing Pi (Tomasi et al. 2008).

## 6.3.3 Iron

Two different mechanisms for the acquisition of iron have been described according to the type of plant. In dicotyledons, the strategy I has been described (Marschner and Römheld 1994). The acquisition of Fe is based on a mechanism at the plasma membrane level that involves the reduction of FeIII to FeII and the absorption of FeII, thanks to the transmembrane electrochemical gradient produced by the activity of the plasma membrane H+ -ATPase. When these plants need to take iron, they increase protons release of to the rhizosphere and increase the FeIII reduction activity carried out by ferric reductase oxidase (FRO) and the transport of FeII through the membranes, which is produced through an iron transporter (IRT) (Connolly et al. 2003). In monocotyledons, on the other hand, another mechanism has been described; it is called strategy II and consists of biosynthesis and release of phytosiderophores (PS) that have a strong affinity for chelating FeIII (Schaaf et al. 2004). The Fe-Siderophore complex is water soluble and in this form, they are able to approximate to the reduction-transport systems in the membrane for absorption and assimilation.

In order to improve Fe availability in soil, plants and microorganisms have evolved similar strategies for the Fe mobilization relying on the exudation of a huge variety of organic compounds (e.g., organic acids, phenolic compounds and siderophores) that are able to complex Fe (Mimmo et al. 2014). Rhizosphere microorganisms can synthesize and release microbial siderophores (MS) (Lemanceau et al. 2009) that show very high affinity for FeIII (Guerinot 1994).

Microorganisms also contribute to Fe availability in soil solution by acidification of soil through respiration due to the increase in carbonic acid in the surrounding soil (Hinsinger et al. 2003). Plant exudates along with microbial activity can lead to a decrease of 1 or 2 units of pH of the rhizosphere as compared to bulk soil with a consequent increase of Fe mobilization (Tomasi et al. 2009).

Therefore, it is clear that PGPR are able to improve Fe bioavailability by different mechanisms but up till now no evidence is available concerning the capability of microorganisms to influence the mechanisms responsible for Fe uptake in strategy I and strategy II plants (Pii et al. 2015).

It has been shown that the capability of plants to recover from nitrate scarcity is based on the increased of genes encoding proteins involved in $NO_3^-$ uptake (Nacry et al. 2013). Similar responses have been also described for P and Fe (Nacry et al. 2013, Liang et al. 2014). Thus, it is possible to think that PGPR could be capable to influence the expression of genes involved in nutrient uptake. However, it is necessary to continue moving forward to test this hypothesis.

Beyond inorganic nutrients, PGPR can improve the entry of carbon in the form of energy-rich sugar molecules into plants by optimizing photosynthesis. Photosynthesis can be optimized either by increasing the concentration of photosynthetic pigments of reaction centers of photosystems or pigments that dissipate excess energy as well as improving other antioxidant systems in order to reduce the effect of light excess resulting in free radicals' formation. On the other hand, optimization of $CO_2$ fixation (Galicia-Campos et al. 2020), allowing stomata opening or reducing photorespiration, maybe another point where PGPR could act.

The majority of works that report an improvement of photosynthesis by PGPR do so based on parameters that measure the response of photosystems, the number of photosynthetic pigments or the amount of $CO_2$ fixed. All are easily measurable parameters with different types of equipment. However, it is difficult to determine what is exactly the effect of these PGPR on the plant by which light reactions or $CO_2$ fixation are optimized. The majority of studies that report an improvement in plant photosynthesis were carried out in stressful situations, especially in water stress (Liu et al. 2019, Galicia-Campos et al. 2020). Under these circumstances, the plant sends a signal (Abcisic Acid) to close the stomata, which will limit the entry of $CO_2$ into the plant and therefore the production of trioses. Therefore, beyond the direct effect on the number of pigments and proteins of the photosystems, it is clear that PGPR improve plants adaptation to stressful situations. PGPR can alleviate the effects of excess of radiation by increasing the number of pigments that dissipate energy and in the case of water stress, allowing the plant to open stomata to a greater extent. To do so, plants have different mechanisms that can be reinforced by PGPR, such as the synthesis of compatible solutes (Galicia-Campos et al. 2020), the reactive oxygen species scavenging or an improvement in the efficiency of water use (WUE) (Galicia-Campos et al. 2020).

Certain PGPR increase photosynthesis rate through the modulation of endogenous sugar/ABA signaling leading to sugar accumulation as well as the suppression of classic glucose signaling responses. Sugars produced in photosynthesis are also signaling molecules for plant growth, development and stress responses. High sugar levels induce storage processes and feedback inhibition of photosynthesis. To date, Arabidopsis hexokinase (HXK1) is the only identified sugar sensor in plants, Hexokinase-dependent glucose signaling requires ABA signal transduction as the two signaling pathways positively interact with each other (Zhang et al. 2008b).

Based on the foregoing, we can conclude that PGPR improve plant nutrition by different mechanisms, some of which are not described, so more studies are needed to progress. The same PGPR can favor the nutrition of the plant by making a nutrient more bioavailable to the plant, increasing the activity of ATP-H$^+$ase (which increases the uptake of the nutrient by the simporters) or favoring the development or modifying the architecture of radical nutrient absorption surface (Figure 6.2). The effect of promoting plant growth can be produced by an improvement in plant nutrition but also by other factors that PGPR affect, such as an improvement in photosynthesis and water use efficiency (García-Cristobal et al. 2015, Galicia-Campos et al. 2020). This is why, regarding PGPRs' effects, usually an improvement in the general state or the fitness of the plant is considered.

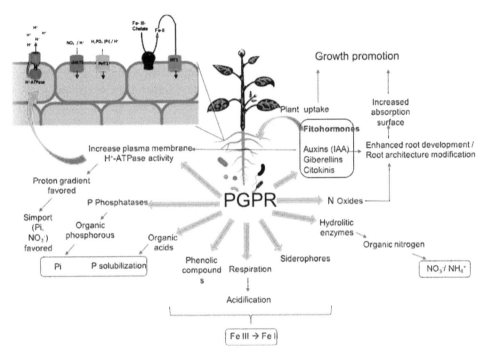

**FIGURE 6.2** Schematic representation of the different mechanisms used by PGPR to improve plant nutrition and growth promotion.

Therefore, the use of PGPR to improve plant nutrition is a complete approach, more convenient than the application of chemical fertilizers that do not always solve the bioavailability problem. PGPR are an excellent biotechnology tool to improve crop production with healthier and protected plants from other biotic and abiotic stresses and capable of optimizing water and nutrients.

## 6.4 PLANT MECHANISMS TO IMPROVE ADAPTATION TO BIOTIC STRESS

Living beings depend on an efficient immune system for survival, being it especially relevant in those organisms with a long life span. This immune system has a few requirements: high specificity—has to be able to differeniate own structures from foreign molecules—and have a memory. The most sophisticated immune system is that of vertebrates constituted by an innate immune system with a broad spectrum constitutively expressed and non-specific and an adaptive immune system, which is strictly regulated and highly specific. The latter system involves circulating cells (T and B cells) which are able to use a vast array of structural receptors and immunoglobulin-type secreted proteins, creating along time an enormous library of receptors and immunoglobulins that treasures the memory of immune system (Spoel and Dong 2012).

As plants lack a circulatory system, they lack the corresponding adaptative immune system of vertebrates. However, they have developed an extremely specific immunitary system, though with a limited self-reactivity are able to create a life-long memory with pathogens to which they have survived (Spoel and Dong 2012).

Plant innate immunity is based on receptors able to detect invasion. Traditionally, plant innate immunity has been described as a two-stage process. The first stage involves plant membrane receptors able to recognize specific molecular patterns from pathogens (PAMPS; Pathogen Associated Molecular Patterns), nemathods (NAMPs; Nematode Associated Molecular Patterns), danger signals started by the plant itself (DAMPs; Danger Associated Molecular Patterns) or any microbe irrespective of their pathogenic capacity (MAMPs; Microbe Associated Molecular Patterns). Upon completing this first stage, the second stage takes place with the onset of the response known as PTI or MTI (PAMP-triggered Immunity or MAMP-triggered immunity). However, many pathogenic microbes are able to inject effectors in the plant's cytosol to block the PTI/MTI response, enhancing the chances of a successful invasion. To overcome this secondary attack, plants have developed specific receptors to these effectors that are able to trigger a new defensive response, way more intense known as ETI (Effector-triggered immunity) (Jones and Dangl 2006, Cook et al. 2015, Gust et al. 2018, Alhoraibi et al. 2019) (Figure 6.3).

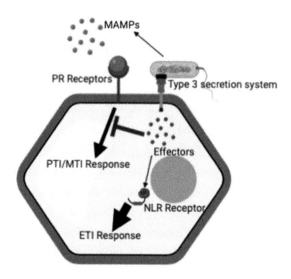

**FIGURE 6.3** Plant innate immunity responses.

Plant membrane receptors involved in PTI response are known as pattern recognition receptors (PRR). There are two types of PRRs depending on the presence or absence of a kinase domain: those showing a kinase domain are termed RLK and those without are termed RLP showing instead leucine rich areas in the reception area (LRR). Among the latter type are two well-known receptors the FLS2 and the EFR, which specifically bind bacterial flagellin (epitope flg22) or the Tu elongation factor (EF-Tu; epitope elf18/elf26), respectively. Within the LRK group are lysine rich receptors (LysM) with kinase domain (RLK) that specifically bind chiting oligomers from fungi and bacterial peptidoglycans (Saijo and Po-iian Loo et al. 2019).

Although plant and animal PRRs share a few traits in their general structure, as leucine rich repeated domains (LRR) for ligand binding, a convergent evolution process is generally accepted rather than a divergent process. This is supported by comparing the analogous transmembrane flagellin receptors FLAGELLIN-SENSITIVE 2 (FLS2) in *Arabidopsis thaliana* and Toll-like receptor 5 (TLR5) in humans, which only share similarities in the LRR domain. Despite both recognized conserved epitopes of bacterial flagellin, each one binds a specific epitope therein (Couto and Zipfel 2016, Alhoraibi et al. 2019, Yu et al. 2019).

Receptors involved in the ETI response are termed nucleotide-binding leucine-rich domains (NLRs). ETI response results in programmed cell death, a process known as a hypersensitive response. Animals also have NLRs determining cellular apoptosis. However, the process involves different mechanisms in each organism; in animals, proinflammatory cytokines activated by caspases are involved, while plants lack caspases, so the process is mediated by a different mechanism (Spoel and Dong 2012, Saijo and Po-iian Loo 2019).

In view of the above, it is evident that one of the plants' responses is started outside the plant and the second starts from the inside. Research carried out in the last years has shown the complexity of these two processes, leading to a new classification based on the origin of the response being extracellular or intracellular. Consequently, all molecular patterns are able to trigger an immune response in the plant (PAMPs, MAMPs, DAMPs) will be in the danger signal group (DA; Danger signal); extracellular immunogenic pattern (ExIP) and intracellular immunogenic pattern (InIP) are the terms coined to describe where the danger signal is perceived, and extracellular triggered immunity (ExTI) and intracellular triggered immunity (InTI) to refer to the place where the response is initiated (Burgh and Joosten 2019).

Activation of PRRs result inside the cell results in increased $Ca_2^+$ levels and the concomitant activation of Ca dependent kinases (CDPK) and mitogen-activated protein kinase cascades, which will finally result in activation of transcription factors (TF) and start transcription of defense-related genes. Ca-dependent protein kinases trigger membrane NADPH oxidases, which generate superoxide ions ($O_2^-$) in the first instance, activate in turn superoxide-dismutase (SOD) releasing $H_2O_2$ and creating a ROS burst in the apoplast; all these reactions comprise the PTI triggered response. The ETI response amplifies the previous response creating an internal ROS burst leading to programmed cell apoptosis designed to block pathogen invasion. Salicylic acid plays a key role in this performance as well as three intracellular SA-receptors termed NPR1, NPR3 and NPR4 that function as cotranscriptional regulators (Liu et al. 2016, Alhoraibi et al. 2019).

Cells that are being invaded generate alert signals that are quickly sent throughout the plant delivered and switching non-attacked cells to alert status. This phenomenon is known as acquired systemic resistance (SAR) to which several molecules have been associated as methyl salicylic acid, azelaic acid, picolinic acid, gliceraldehide 3 P. These molecules activate SA synthesis in distant cells that in turn interacts with NPR1, NPR3 and NPR4 resulting in the synthesis of resistance proteins (PR) before pathogen invasion (Backer et al. 2019, Maruri-López et al. 2019).

Plant growth regulators play a key role in innate immunity mechanisms. Not only SA intervenes in innate immunity but also Jasmonic acid (JA) and ethylene (ET). Salicylic acid is involved in response toward biotrophic pathogens while JA/Et are involved in necrotrophic pathogens; a few of the earlier studies on the interconnection of these two pathways suggested an antagonism, but there is increasing evidence of the opposite (Liu et al. 2016, Backer et al. 2019, Martin-Rivilla et al. 2019).

As early as the 1990s, the ability of specific bacterial strains from the rhizosphere to trigger PTI response in plants has been reported in many studies. The mechanism to trigger the systemic protection is similar to SAR and has been termed Induced Systemic Resistance (ISR) in which the plant shows an enhanced capacity to overcome a pathogen attack. ISR inducing strains trigger a special metabolic situation termed 'priming' which is evidenced upon pathogen challenge with a more effective and intense defensive response, preventing pathogen infection; this mechanism is similar to vaccines for animals (Mauch-Mani et al. 2017, Sukanya et al. 2018, Mhlongo et al. 2018) (Figure 6.4).

The use of bioeffectors to trigger plant defensive metabolism is currently an excellent alternative to phytosanitary products. From the first description of ISR by Van Loon et al. (1998) when only the jasmonic/ethylene pathway was associated to beneficial microorganisms, evidence has shown that the signal transduction pathway is special for each bioeffector and they can be simultaneously triggered (Ramos Solano et al. 2008, Martin-Rivilla et al. 2019).

On the grounds of the above descriptions, our group's contributions to these topics include the use of bioeffectors (bacterial strains) and derived metabolic elicitors (secreted molecules to the culture media) to trigger plant immune system and understand the mechanisms involved. In earlier works, we have shown the ability of some bioeffectors to improve plant defensive capacity affecting ROS scavenging enzymes so that treated plants were better able to cope with ROS burst upon stress and significant increases on PR2 (glucanases) and PR3 (chitinases) activity (Lucas et al. 2014, Garcia-Cristobal et al. 2015).

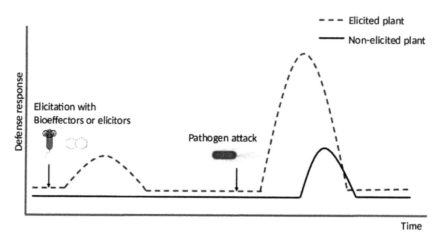

**FIGURE 6.4** Defense response of plants elicited with bioeffectors or elicitors.

However, these data did not reveal the signal transduction pathway to innate immunity triggered by bioeffectors. This objective was undertaken by analyzing the differential expression of marker genes of each signal transduction pathway like *NPR1*, *isocorismate synthase* (ICS, involved in SA synthesis), genes encoding resistance proteins *PR1* and *PR2* as markers of SA mediated pathway and PDF1, Lox2 (involved in JA synthesys), MYC2 and PR3 as markers of the JA/Et mediated pathway (Martin-Rivilla et al. 2019). Firstly, 25 strains isolated from the rhizosphere of *N. glauca* were characterized for their PGPR putative traits and their taxonomic classification based on 16s sequence was carried out. Most were able to trigger defense mechanisms on *Arabidopsis thaliana* seedlings. The most effective in protection (4 Gram-negative and 2 Gram-positive) were studied to unravel the mechanism involved in signal transduction triggered by each strain. As the main outcomes, no common pattern was detected, and at least one strain was able to trigger both pathways, ruling out the proposed antagonisms between pathways. Strains that are able to trigger simultaneously both pathways are of great agronomic interest as a broader protective response is achieved (Martin-Rivilla personal communication).

Current research on bioeffectors is switching toward using bacterial molecular elicitors rather than the bacterial strain as it involves several advantages. On one hand, the effective concentration range is much lower which results in lower environmental impact and an easier agronomic and commercial management practices since no attention need to be paid to viability of commercial products; finally, it is very unlikely that a bacterial molecule can dramatically affect the plant's microbiome. Fatima and Anjum (2017) and Wu et al. (2018) have reported the effectiveness of elicitors from *Pseudomonas* and *Bacillus,* respectively, to protect tomato and arabidopsis. We have also shown that Pseudomonas fluorescens (N21.4) that holds at least two elicitors soluble in hexane (Sumayo et al. 2013, Fatima and Anjum 2017) were able to simultaneously trigger both defensive pathways, increasing PR activity while lowering oxidative stress (Martin-Rivilla et al. 2019).

We have evidenced the ability of bioeffectors (bacterial strains and derived elicitors) to trigger plant metabolism, showing enormous biotechnological potential. Any product developed from these raw materials will contribute to an environmentally friendly agronomic production, sustainable in which the plant plays the main role in the production system by modulation of its innate immune system.

## 6.5 PLANT SECONDARY METABOLISM: FROM ITS ROLE IN PLANTS TO APPLICATIONS IN THE AGRIFOOD AND PHARMACOLOGICAL ENVIRONMENTS THROUGH BIOTIC ELICITATION

Plants are sessile organisms. This condition involves a continuous adaptation to environmental changes to overcome all stressing situations that happen throughout their life until seeds are formed, ensuring the survival of the species. Unlike other organisms that can run away from danger, plants are hooked to the place seeds germinate for life. Their only chance to survive is to synthesize chemical molecules for each specific situation. The set of reactions arranged in metabolic pathways that lead to this enormous array of different chemical molecules is known as secondary metabolism. The term 'secondary' is an excluding definition from primary metabolism that involves metabolic pathways dealing with sugar, lipid and protein synthesis and degradation which are present in all living beings and share high similarities. Moreover, this secondary metabolism is not only exclusive of sessile organisms but is at some points present only in some groups of plants being taxonomic criteria.

As plants colonize most environments on Earth, a tremendous catalogue of different molecules can be anticipated if we consider the different environmental conditions determined by latitude, light hours, light intensity, soil composition, available water and temperature to which plants are adapted.

Interestingly, the chemical arsenal created by plants for their interest is not limited to the plant kingdom, but it includes a good number of molecules that extends to other kingdoms. In line with this, plants design secondary metabolism for different purposes addressing survival. Some are related to direct communication with other kingdoms, for example, inhibiting neighbor plants growth, pollinator attractants, the establishment of symbiotic relationships or molecules contained in the plant directly related to protection against abiotic factors, like sunscreens (UV protection), antioxidants and osmoprotectants, and biotic factors, like food deterrents, insecticides, bactericides, fungicides, etc. Plant immunity has been described in a different heading and involves active enzymes like chitinases (PR3) or glucanases (PR2) to mention a few; the protection described herein refers to that achieved with metabolites that target receptors of other living beings.

The underlying reason for such a broad spectrum of activities relies on evolution: we all share a Last Unique Common Ancestor (LUCA) from which the different kingdoms evolved. To illustrate this assessment within this context, at some point in evolution, plants diverged from animals; both groups carry on the information for a determined molecule that is used with a different purpose in each. Hence, plants synthesize molecules for their survival that target the receptors still present in mammals. For example, morphine the most effective analgesic molecule nowadays is obtained from *Papaver somniferum* and binds our endorphin receptors; morphine and endorphin are different molecules but share the same structure in the molecule area that fits in the same receptor.

Secondary metabolism is present in all groups of plants, and it is more relevant in phanerogams. Three great groups of compounds have been defined according to biosynthetic origin and structure: terpenes, phenolic compounds and alkaloids (Figure 6.5). All of them are present in all plants since they lead to i) structural molecules basic for plant growth like phytosterols or aromatic amino acids or ii) molecules with a primary physiological role, like carotenoids in light-harvesting, or plant growth regulators, like abscisic acid or gibberellins. However, plants express some secondary metabolic pathways with a greater intensity, which transforms them into medicinal plants with a given activity, functional foods or raw materials to obtain fibres, latex, etc.

The diversity of plant species is great, but they all have a common point that is their growth. The purpose of secondary metabolism to adapt itself, and here comes the biotechnological challenge for any of the above purposes: to manipulate secondary metabolism in order to keep the metabolites of interest high and within a certain range, so that edible foods have a consistent effect on health and medicinal plants provide a constant yield of bioactives to sustain the industry.

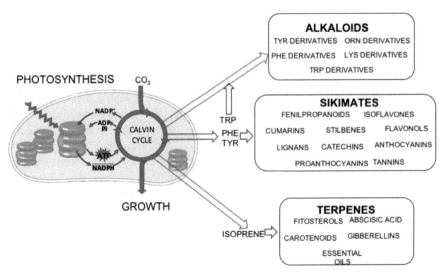

**FIGURE 6.5** Connections of primary and secondary metabolism. Overview of general groups of secondary metabolites.

In view of the above, plants appear as an enormous and unlimited source of molecules with a great potential for multiple purposes: food, functional foods, raw materials for industry, active molecules for the pharmaceutical and food industry and active molecules for environmentally safe and sustainable agronomic practices among others. As described in the introduction, plants inhabit most soils where they survive, thanks to their adaptive secondary metabolism and the best-fitted microbiome within their rhizosphere. The selection of bacteria that has taken place along evolution has resulted in effective isolates to trigger plant metabolism. Within this context, beneficial bacteria and bacterial derived molecules, known as elicitors along with this chapter, appear like a challenging and effective alternative to trigger plant secondary metabolism, increasing bioactives and stabilizing its contents. In the present chapter, five examples will be presented to illustrate functional foods, functional ingredients and pharmacologically active molecules, under field production, in controlled conditions or *in vitro* culture using beneficial bacteria and/or elicitors: fox glove, opium poppy, St. Johns wort, soybean and blackberries.

## 6.5.1 *Digitalis lanata* L (Scrophulariaceae) Foxglove (Gutierrez-Mañero et al. 2003)

Agronomic production of *Digitalis lanata* is the main source of cardenolides as these molecules accumulate preferentially in the rosette leaves at the end of the first year prior to flowering the second year (Figure 6.6).

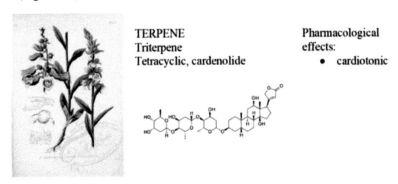

**FIGURE 6.6** *Digitalis lanata*. Botany, structure of main secondary bioactive and pharmacological effects.

This article describes the isolation of bacterial strains from the rhizosphere of wild populations of two digitalis species, as potential candidates to trigger terpene biosynthesis in *D. lanata*, the high yield species used in the industry. The 480 isolates obtained in the screening from those pre-selected by plants were characterized at the genus level, *Bacillus* being the most abundant genus. After PCR-RAPDs analysis, 12 groups appeared in *D. thapsi* rhizosphere and 18 from *D. parviflora* sharing 85% similarity. One strain from each group was selected for biological assay evaluating growth promotion and cardenolide contents in the rosette leaves, under the influence of each isolate, delivered by soil drench. Among strains tested, 17 increased at least one of the evaluated parameters and only 12 stimulated growth and enhanced cardenolide contents, simultaneously. The most interesting facet of this research is i) the simultaneous enhancement of growth and secondary metabolism since, according to cost-benefit balance, there is a negative correlation between both and ii) specialized bacterial strains are able to trigger cardenolide synthesis in a selected high-yield variety. This is of major economic importance for the pharmaceutical industry because it would further increase field production, which is the current way to obtain cardiotonics.

### 6.5.2 *Papaver somniferum* (Papaveraceae) Opium poppy (Bonilla et al. 2014)

Agronomic production of *Papaver somniferum* is the only source of morphinanes as these molecules only accumulate in its capsules upon fruiting (Figure 6.7).

ALKALOID
Morphinanes

Pharmacological effects:

• analgesic.

**FIGURE 6.7** *Papaver somniferum.* Botany, structure of main secondary bioactive and pharmacological effects.

Different to the *digitalis* case presented above in which bacteria were isolated from the same genus, in opium poppy 10 PGPR from different backgrounds were assayed on *Papaver somniferum* to evaluate their potential as biotic elicitors to increase alkaloid. As some specificity exists between plants and microorganisms, the ability of the ten strains and their culture media to trigger seed germination was tested resulted in only three strains being effective. Then, a biological assay to increase seedling growth and alkaloid levels under greenhouse conditions was carried out with these three strains: N5.18 *Stenotrophomonas maltophilia*, Aur9 *Chryseobacterium balustinum* and N21.4 *Pseudomonas fluorescens*. After two applications, by soil drench or foliar spray, only N5.18 delivered by foliar spray significantly increased plant height, total alkaloids and straw dry weight; these increases were supported by a better photosynthetic efficiency. The relative contents of morphinanes (morphine, thebaine, codeine and oripavine) were affected by this treatment recording a significant increase in morphine coupled to a decrease in thebaine, demonstrating the effectivity of elicitors from N5.18, isolated from *Nicotiana glauca* (Solanaceae) in this plant species (Papaveraceae). This shows that some strains are 'wide spectrum' showing interspecies effectiveness. Considering the increase in capsule biomass and alkaloids together with the acceleration of germination, strain N5.18 appears as a good candidate to elicit plant metabolism and consequently to increase the productivity of *Papaver somniferum*.

Further field trials have shown that leaf spraying of this strain in the early stages of flowering increases total morphine concentration, targeting codeinone-O-demethylase that catalyzes the transformation of thebaine and codeine into morphine (unpublished). However, the intensity of biotic modulation on alkaloid biosynthesis is further subjected to environmental conditions, and morphinane increases the change depending on location and season (unpublished).

### 6.5.3 *Hypericum perforatum* (Guttaceae) St. John's Wort (Gutierrez-Mañero et al. 2012)

*Hypericum perforatum* L. (Guttiferae) appears as an alternative treatment to mild and moderate depression and been traditionally used as a health enhancer based on the phytochemicals hyperforin and the naphtodiantrone hypericin (Figure 6.8). As the patent for the antidepressant Prozac expired in 2001, obtaining natural sources of antidepressant molecules from *H. perforatum* is a challenge. Hypericin can be obtained either from field-grown plants or *in vitro* shoot cultures with good yields; however, as levels of phytopharmaceuticals fluctuate due to environmental conditions, elicitation is a good strategy to trigger secondary metabolism.

Phenolic compound
Poliketide
Naftodiantrone

Pharmacological effects:

• Antidepressive.

**FIGURE 6.8** *Hypericum perforatum* shoot cultures and wild plant. Botany, structure of main secondary bioactive and pharmacological effects.

The aim of this study is to identify rhizobacterial strains to trigger secondary metabolism in *H. perforatum* seedlings among six from different origins and to obtain the molecular elicitors from the most effective strain to test them in shoot cultures. As for opium poppy, strain *Stenotrophomonas maltophilia* N5.18 significantly increased hypericin up to 1.2 ppm and pseudohypericin up to 3.4 ppm, over controls (0.3 and 2.5 ppm, respectively) when delivered to seedlings, confirming its wide spectrum ability. In shoot cultures, three bacterial fractions of different molecular weight were obtained from the supernatants and delivered to shoot cultures at different concentrations. Only pseudohypericin was detected (168.9 ppm), and significant increases were observed under the different elicitors, reaching values of 3,164.8 ppm with small elicitors in the intermediate concentration. Once more, inducibility of plant secondary metabolism by bacteria and derived elicitors is shown even in shoot cultures; a plant material where secondary metabolism is usually depressed due to the lack of environmental pressure. Therefore, these elicitors have great potential to enhance phytopharmaceutical production.

### 6.5.4 *Glycine max* (Fabaceae) Soybean (Ramos-Solano et al. 2010, Algar et al. 2013, Algar et al. 2014)

Soybean is a legume with high nutritional value due to several components: proteins, lipids and isoflavones (daidzein and genistein) mostly contained in the seeds. The goal set for this plant species ranged from an increase in plant yield to increase in isoflavones for their interest as phytopharmaceuticals because they are defensive molecules for the plant, which use beneficial bacteria and derived elicitors (Figure 6.9).

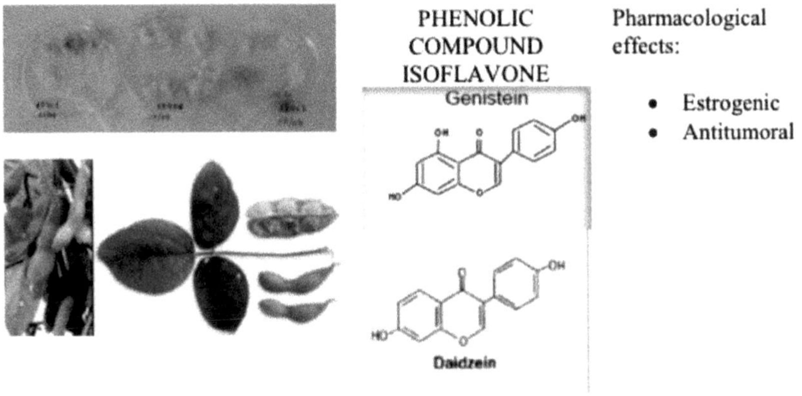

**FIGURE 6.9** *Glycine max.* Calli and crop, structure of main secondary bioactive and pharmacological effects.

As in previous cases, a screening of effective strains to trigger soybean metabolism was carried out among nine strains from different origins. Strains were delivered to soybean seedlings and four different patterns were detected according to changes in daidzein and genistein with an outstanding increase in both IF by *Pseudomonas fluorescens* N21.4 and outstanding reduction by *Curtobacterium sp.* M84; N5.18 (*Stenotrophomonas maltophila*) and Aur9 (*Chryseobacterium balustinum*) increase one IF family over the other in opposite way (Ramos Solano et al. 2010).

These four strains were assayed on a plant protection assay in older seedlings against the leaf pathogen *Xanthomonas axonopodis* pv. *Glycines* resulting in enhanced protection irrespective of changes on IF: in M84 treated seedlings, a marked increase in IF was detected associated to protection suggesting a role of phytoalexins, while in N5.18 treated plants, IF decreased significantly suggesting their transformation in other defensive molecules, gliceolins, therefore playing a role as phytoanticipins (Algar et al. 2014).

The next step was to evaluate the effects of metabolic elicitors from N21.4 on cell lines from soybean with different IF contents. Three fractions of different molecular weight were obtained and assayed on the three lines (high intermediate and low IF contents). Results revealed that N21.4 releases at least two different metabolic elicitors that are able to trigger significant increases in all cell lines, even in the high yield line so these elicitors appear as an alternative to obtain isoflavones for food supplements, leaving soybean crop for feeding (Algar et al. 2012).

The last biotechnological application is to obtain functional foods, like nutritious soybean sprouts enriched in IF. Elicitors from N21.4 and N5.18 were obtained as described above and were delivered to soybean seeds before germination, confirming that either the inactivated bacterial strain and the metabolic elicitors are able to trigger IF biosynthesis in the germinating seed. This is the first step to establish a biotechnological process to obtain enriched functional foods (Algar et al. 2013).

### 6.5.5   *Rubus cv Loch ness* (Rosaceae) Blackberry (García-Seco et al. 2012, Ramos Solano et al. 2014)

The ability of *Pseudomonas fluorescens* N21.4 isolated from the rhizosphere of *N. glauca*, which was able to trigger secondary metabolism related to the flavonoid pathway in soybean, was tested on blackberries. Blackberries are an excellent source of bioactive compounds as compared to other marketable berries (Figure 6.10). Their production in counter season provides fruit to the market from October through June, representing a high income through export to other countries. Biotic elicitation with plant growth-promoting rhizobacteria has been shown to improve biomass production and to trigger secondary metabolism in other plant species in controlled conditions, but few have confirmed effects in field conditions under continuous environmental changes.

 PHENOLIC
COMPOUND
Flavonoids
Anthocyanins

Pharmacological effects:
- antioxidant,
- prevents onset of metabolic syndrome.

**FIGURE 6.10** Blackberry. Phenolic compounds and pharmacological effects.

We have demonstrated the ability of N21.4 to increase blackberry fitness, fruit quality and protection against natural pests under field conditions when delivered through the root or sprayed in leaves along its entire production period (July through November). Our results showed an average increase up to 800 g per plant in total production, directly related to the increase in flowering buds. Protection against *Spodoptera littoralis* in inoculated plants was higher than in control plants, hence contributing to increase in yield. Fruits from inoculated plants showed significant increases of up to 3 °Brix, total phenolics of up to 18% and flavonoids of up to 22%. We conclude that *P. fluorescens* N21.4 enhances plant defense and fruit quality together with increased productivity as compared to current management practices, already obtaining high yields with economic profit (García-Seco et al. 2012).

In view of these results, we aimed to i) characterize blackberry fruits from *Rubus* sp. var. Lochness along the year to confirm the composition of bioactives throughout the year and ii) to evaluate the ability of N21.4 to improve fruit yield and quality in production greenhouses throughout the year in southwestern Spain (Huelva). Blackberry plants were leaf or root inoculated and fruits were harvested in each season. Nutritional parameters, antioxidant potential and bioactive contents were analyzed and accumulated fruit yield was recorded. Blackberries that were grown under short-day conditions (autumn and winter) showed significantly lower Brix values than fruits grown under long-day conditions. Interestingly, an increase in fruit Brix, relevant for quality, was detected after bacterial challenge, together with significant and sustained increases in total phenolics and flavonoids; anthocyanins were not affected as the plant already accumulates huge amounts of cyanidin derivatives. Improvements in inoculated fruits were more evident when environmental conditions are worse (October through early March). In summary, N21.4 is an effective agent to increase fruit quality and production along the year in blackberry; this is an environmentally friendly approach to increase fruit quality (Ramos Solano et al. 2014).

Further studies have been conducted to unravel the points of the flavonoid-anthocyanin pathway tackled by this strain as well as the transcription factors triggered. The first step was to study the expression of genes involved in the flavonoid-anthocyanin pathway at three developmental stages of maturation (green, red and black) as well as transcription factors of MYB type. The transcriptomic analysis allowed us to identify the genes for blackberry and to find the midripen stage as the most sensitive for elicitation with PGPR (García-Seco et al. 2015a) and then compare the expression of these genes on inoculated plants. The most relevant results were the ability of N21.4 to trigger overexpression of CHS (chalcone synthase), CHI (Chalcone isomerase) and F3H (Flavonol-3-hydroxylase), under the control of MYB1, which was consistently overexpressed in red fruits (García-Seco et al. 2015b).

In view of the above, it is evidenced that PGPR are able to trigger specific metabolic pathways. On one hand, the metabolic pathways involved in plant defense and adaptation to the everchanging situations along with their life and therefore introducing PGPR in current agronomic management will contribute to environmental sustainability. On the other hand, the metabolic pathways lead to bioactive molecules that improve human health, therefore, management of PGPRs will result in high-quality fruits, enriched in bioactives.

## 6.6  ACKNOWLEDGEMENTS

The authors would like to acknowledge all entities that have provided funding for these studies: Ministerio de Ciencia y Tecnologia, Fundacion Universitaria San Pablo-CEU Universities, Federación de Arroceros de Sevilla, Alcaliber S.A., Agricola el Bosque as well as all collaborators in field trials from FAS, Alcaliber, Agricola el Bosque, and students that have participated in all studies.

## 6.7  REFERENCES

Algar, E., Ramos-Solano, B., García-Villaraco, A., Saco Sierra, M.D., Martín Gómez, M.S. and Gutiérrez-Mañero, F.J. 2013. Bacterial bioeffectors modify bioactive profile and increase isoflavone content in soybean sprouts (*Glycine max* var Osumi). Plant Foods for Human Nutrition 68: 299-305. doi: 10.1007/s11130-013-0373-x.

Algar, E., Gutierrez-Mañero, F.J., Garcia-Villaraco, A., García-Seco, D., Lucas, J.A. and Ramos-Solano, B. 2014. The role of isoflavone metabolism in plant protection depends on the rhizobacterial MAMP that triggers systemic resistance against Xanthomonas axonopodis pv. glycines in *Glycine max* (L.) Merr. cv. Osumi. Plant Physiology and Biochemistry 82: 9e16. doi: http://dx.doi.org/10.1016/j.plaphy.2014.05.001.

Alhoraibi, H., Bigeard, J., Rayapuram, N., Colcombet, J. and Hirt, H. 2019. Plant immunity: the MTI-ETI model and beyond. Current Issues Molecular Biology 30: 39-58.

Apse, M.P., Aharon, G.S., Snedden, W.A. and Blumwald, E. 1999. Salt tolerance conferred by overexpression of a vacuolar $Na^+/H^+$ antiport in Arabidopsis. Science 285: 1256-1258. doi: 10.1126/science.285.5431.1256.

Backer, R., Naidoo, S. and van den Berg, N. 2019. The Nonexpressor of Pathogenesis-Related Genes 1 (NPR1) and related family: mechanistic insights in plant disease resistance. Frontiers in Plant Science 10: 412.

Bertrand, H., Plassard, C., Pinochet, X., Touraine, B., Normand, P. and Cleyet-Marel, J.C. 2000. Stimulation of the ionic transport system in *Brassica napus* by a plant growth-promoting rhizobacterium (*Achromobacter* sp.). Canadian Journal of Microbiology 46: 229-236. doi: 10.1139/w99-137.

Bhattacharyya, P.N. and Jha, D.K. 2012. Plant growth-promoting rhizobacteria (PGPR): emergence in agriculture. World Journal of Microbiology and Biotechnology 28: 1327-1350. doi: 10.1007/s11274-011-0979-9.

Bonilla, A, Sarria, A.L.F., Algar, E., Muñoz Ledesma, F.J., Ramos Solano, B., Fernandes, J.B., et al. 2014. Microbe associated molecular patterns from rhizosphere bacteria trigger germination and Papaver somniferum metabolism under greenhouse conditions. Plant Physiology and Biochemistry 74: 133e140.

Canellas, L., Balmori, D., Médici, L., Aguiar, N., Campostrini, E., Rosa, R., et al. 2013. A combination of humic substances and *Herbaspirillum seropedicae* inoculation enhances the growth of maize (*Zea mays* L.). Plant and Soil. doi: 10.1007/s11104-012-1382-5.

Chauhan, P.S., Lata, C., Tiwari, S., Chauhan, A.S., Mishra, S.K., Agrawal, L., et al. 2019. Transcriptional alterations reveal *Bacillus amyloliquefaciens*-rice cooperation under salt stress. Scientific Reports 9: 11912. doi: 10.1038/s41598-019-48309-8.

Connolly, E., Campbell, N., Grotz, N., Prichard, C. and Guerinot, M. 2003. Overexpression of the FRO2 ferric chelate reductase confers tolerance to growth on low iron and uncovers posttranscriptional control. Plant Physiology 133: 1102-1110. doi: 10.1104/pp. 103.025122.

Cook, D.E., Mesarich, C.H. and Thomma, B.P.H.J. 2015. Understanding plant immunity as a surveillance system to detect invasion. Annual Review of Phytopathology 53(1): 541-563.

Correa-Aragunde, N., Graziano, M. and Lamattina, L. 2004. Nitric oxide plays a central role in determining lateral root development in tomato. Planta 218: 900-905. doi: 10.1007/s00425-003-1172-7.

Couto, D. and Zipfel, C. 2016. Regulation of pattern recognition receptor signalling in plants. Nature Reviews Immunology 16: 537-552.

Creus, C.M., Graziano, M., Casanovas, E.M., Pereyra, M.A., Simontacchi, M., Puntarulo, S., et al. 2005. Nitric oxide is involved in the Azospirillum brasilense-induced lateral root formation in tomato. Planta 221: 297-303.

Esitken, A., Ercisli, S., Karlidag, H. and Sahin, F. 2005. Potential use of plant growth promoting rhizobacteria (PGPR) in organic apricot production. *In*: Proceedings of the International Scientific Conference of Environmentally Friendly Fruit Growing, Tartu-Estonia, September 7-9, 2005, pp. 90-97.

Fatima, S. and Anjum, T. 2017. Identification of a potential ISR determinant from *Pseudomonas aeruginosa* PM12 against Fusarium wilt in Tomato. Frontiers in Plant Science 8: 69.

Fricke, W., Akhiyarova, G., Wei, W.X., Alexandersson, E., Miller, A., Kjellbom, P.O., et al. 2006. The short-term growth response to salt of the developing barley leaf. Journal of Experimental Botany 57: 1079-1095. doi: 10.1093/jxb/erj095.

Galicia-Campos, E., Ramos-Solano, B., Montero-Palmero, M.B., Gutierrez-Mañero, F.J. and Garcia-Villaraco, A. 2020. Management of Plant Physiology with Beneficial Bacteria to Improve Leaf Bioactive Profiles and Plant Adaptation under Saline Stress in Olea europea L. Foods 9(1): 57-56. doi: 10.3390/foods9010057.

García-Cristobal, J., García-Villaraco, A., Ramos, B., Gutierrez-Manero, F.J. and Lucas, J.A. 2015. Priming of pathogenesis related-proteins and enzymes related tooxidative stress by plant growth promoting rhizobacteria on riceplants upon abiotic and biotic stress challenge. Journal of Plant Physiology 188: 72-79. doi: http://dx.doi.org/10.1016/j.jplph.2015.09.011.

García-Seco, D., Bonilla, A., Algar, E., García-Villaraco, A., Gutierrez Mañero, F.J. and Ramos-Solano, B. 2012. Enhanced blackberry production using *Pseudomonas fluorescens* as elicitor. Agronomy for Sustainable Development doi: 10.1007/s13593-012-0103-z.

García-Seco, D., Zhang, Y., Gutierrez-Mañero, F.J., Martin, C. and Ramos-Solano, B. 2015a. RNA-seq analysis and transcriptome assembly for blackberry (Rubus sp. Var. Lochness) fruit. BMC Genomics 16: 5. doi: 10.1186/s12864-014-1198-1.

García-Seco, D., Zhang, Y., Gutierrez-Mañero, F.J., Martin, C. and Ramos-Solano, B. 2015b. Application of *Pseudomonas fluorescens* to blackberry under field conditions improves fruit quality by modifying flavonoid metabolism. PLoS One 10(11): e0142639. doi: 10.1371/journal.pone.0142639.

Glick, B.R., Penrose, D.M. and Li, J.P. 1998. A model for the lowering of plant ethylene concentrations by plant growth-promoting bacteria. Journal of Theoretical Biology 190: 63-68.

Glick, B.R., Cheng, Z., Czarny, J. and Duan, J. 2007. Promotion of plant growth by ACC deaminase-producing soil bacteria. European Journal of Plant Pathology 119: 329-339. doi: 10.1007/s10658-007-9162-4.

Guerinot, M.L. 1994. Microbial iron transport. Annual Review of Microbiology 48: 743-772. doi: 10.1146/annurev.mi.48.100194.003523.

Gust, A.A., Pruitt, R. and Nürnberger, T. 2017. Sensing danger: key to activating plant immunity. Trends in Plant Science 22(9): 779-791.

Gutiérrez Mañero, F.J., Ramos, B., Lucas Garcia, J.A., Probanza, A. and Barientos Casero, M.L. 2003. Systemic induction of the biosynthesis of terpenic compounds in *Digitalis lanata*. Journal of Plant Physiology 160(2): 105-109.

Gutiérrez Mañero, F.J., Algar, E., Martín Gómez, M.S., Saco Sierra, M.D. and Beatriz Ramos Solano. 2012. Elicitation of secondary metabolism in *Hypericum perforatum* by rhizosphere bacteria and derived elicitors in seedlings and shoot cultures. Pharmaceutical Biology doi: 10.3109/13880209.2012.664150.

Gyaneshwar, P., Naresh Kumar, G., Parekh, L.J. and Poole, P.S. 2002. Role of soil microorganisms in improving P nutrition of plants. Plant and Soil 245: 83-93. doi: 10.1023/A:1020663916259.

Hadas, R. and Okon, Y. 1987. Effect of *Azospirillum brasilense* inoculation on root morphology and respiration in tomato seedlings. Biology and Fertility of Soils 5: 241-247. doi: 10.1007/BF00256908.

Hanin, M., Ebel, C., Ngom, M., Laplaze, L. and Masmoudi, K. 2016. New insights on plant salt tolerance mechanisms and their potential use for breeding. doi: 10.3389/fpls.2016.01787.

Hinsinger, P., Plassard, C., Tang, C. and Jaillard, B. 2003. Origins of root-mediated pH changes in the rhizosphere and their responses to environmental constraints: a review. Plant and Soil 248: 43-59. doi: 10. 1023/A:1022371130939.

Ilangumaran, G. and Smith, D.L. 2017. Plant growth promoting rhizobacteria in amelioration of salinity stress: a systems biology perspective. Frontiers in Plant Science 23 doi: https://doi.org/10.3389/fpls.2017.01768.

Jammes, F., Song, C., Shin, D., Munemasa, S., Takeda, K., Gu, D., et al. 2009. MAP kinases MPK9 and MPK12 are preferentially expressed in guard cells and positively regulate ROS-mediated ABA signaling. Proceedings of the National Academy of Sciences 106: 20520-20525. doi: 10.1073/pnas.0907205106.

Jetten, M.S.M. 2008. The microbial nitrogen cycle. Environmental Microbiology 10: 2903-2909. doi: 10.1111/j.1462-2920.2008.01786.x.

Jimenez-Delgadillo, M.R. 2004. Peptidos Secretados por *Bacillus* Subtilis que Codifican la Arquitectura de la Raiz de *Arabidopsis thaliana*. PhD Dissertation. CINVESTAV, Unidad Irapuato, MX.

Kang, S.M., Khan, A.L., Waqas, M., You, Y.-H., Kim, J.-H., Kim, J.-G., et al. 2014. Plant growth-promoting rhizobacteria reduce adverse effects of salinity and osmotic stress by regulating phytohormones and antioxidants in *Cucumis sativus*. Journal of Plant Interactions 9: 673-682. doi: 10.1080/17429145.2014.894587.

Karthikeyan, A.S., Varadarajan, D.K., Mukatira, U.T., D'Urzo, M.P., Damsz, B. and Raghothama, K.G. 2002. Regulated expression of Arabidopsis phosphate transporters. Plant Physiology 130: 221-233. doi: 10.1104/pp. 020007.

Kasotia, A., Varma, A., Tuteja, N. and Kumar Choudhary, D. 2016. Amelioration of soybean plant from saline-induced condition by exopolysaccharide producing *Pseudomonas*-mediated expression of high affinity K+-transporter (HKT1) gene. 2016. Current Science 111(12): 1961-1967.

Lemanceau, P., Expert, D., Gaymand, F., Bakker, P.A.H.M. and Briat, J.F. 2009. Role of iron in plant–microbe interactions. Advances in Botanical Research 51: 491-549. doi: 10.1016/S0065-2296(09)51012-9.

Liu, F., Hailin, M., Peng, L., Zhenyu, D., Bingyao, M. and Xinghong, L. 2019. Effect of the inoculation of plant growth-promoting rhizobacteria on the photosynthetic characteristics of Sambucus williamsii hance container seedlings under drought stress. AMB Express 9: 169.

Liu, L., Sonbol, F.-M., Huot, B., Gu, Y., Withers, J., Mwimba, M., et al. 2016. Salicylic acid receptors activate jasmonic acid signalling through a non-canonical pathway to promote effector-triggered immunity. Nature Communications 7: 13099.

Lucas, J.A., Garcia-Cristobal, J., Bonilla, A., Ramos, B. and Gutiérrez-Mañero, F.J. 2014. Beneficial rhizobacteria from rice rhizosphere confers high protection against biotic and abiotic stress inducing systemic resistance in rice seedlings. Plant Physiology and Biochemistry 82: 44-53.

Marschner, H. and Römheld, V. 1994. Strategies of plants for acquisition of iron. Plant and Soil 165: 261-274. doi: 10.1007/BF00008069.

Marschner, P. 2011. Marschner's Mineral Nutrition of Higher Plants, 3rd ed. London.

Martin-Rivilla, H., García-Villaraco, A., Ramos Solano, B., Gutierrez Mañero, F.J. and Lucas, J.A. 2019. Extracts from cultures of *Pseudomonas fluorescens* induce defensive patterns of gene expression and enzyme activity while depressing visible injury and reactive oxygen species in *Arabidopsis thaliana* challenged with pathogenic *Pseudomonas syringae*. AoB Plants 11(5): 1-9.

Marulanda, A., Azcon, R., Chaumont, F., Ruiz-Lozano, J.M. and Aroca, R. 2010. Regulation of plasma membrane aquaporins by inoculation with a *Bacillus megaterium* strain in maize (*Zea mays* L.) plants under unstressed and salt-stressed conditions. Planta 232: 533-543. doi: 10.1007/s00425-010-1196-8.

Maruri-López, I., Aviles-Baltazar, N.Y., Buchala, A. and Serrano, M. 2019. Intra and extracellular journey of the phytohormone salicylic acid. Frontiers in Plant Science 10: 423.

Mhlongo, M.I., Piater, L.A., Madala, N.E., Labuschagne, N. and Dubery, I.A. 2018. The chemistry of plant–microbe interactions in the rhizosphere and the potential for metabolomics to reveal signaling related to defense priming and induced systemic resistance. Frontiers in Plant Science 9: 1908-1917.

Mimmo, T., Del Buono, D., Terzano, R., Tomasi, N., Vigani, G., Crecchio, C., et al. 2014. Rhizospheric organic compounds in the soil-microorganism-plant system: their role in iron availability. European Journal of Soil Science 65: 629-642. doi: 10.1111/ejss.12158.

Molina-Favero, C., Creus, C.M., Simontacchi, M., Puntarulo, S. and Lamattina, L. 2008. Aerobic nitric oxide production by *Azospirillum brasilense* Sp245 and its influence on root architecture in tomato. Molecular Plant Microbe Interactions 21: 1001-1009. doi: 10.1094/MPMI-21-7-1001.

Munns, R. and Tester, M. 2008. Mechanisms of salinity tolerance. Annual Review of Plant Biology 59: 651-681. doi: 10.1146/annurev.arplant.59.032607.092911.

Nacry, P., Bouguyon, E. and Gojon, A. 2013. Nitrogen acquisition by roots: physiological and developmental mechanisms ensuring plant adaptation to a fluctuating resource. Plant and Soil. doi: 10.1007/s11104-013-1645-9.

Narasimhan K., Basheer, C., Bajic, V. and Swarup, S. 2003. Enhancement of plant–microbe interactions using a rhizosphere metabolomics-driven approach and its application in the removal of polychlorinated biphenyls. Plant Physiology 132: 146-153.

Nautiyal, C.S., Srivastava, S., Chauhan, P.S., Seem, K., Mishra, A. and Sopory, S.K. 2013. Plant growth-promoting bacteria *Bacillus amyloliquefaciens* NBRISN13 modulates gene expression profile of leaf and rhizosphere community in rice during salt stress. Plant Physiology and Biochemistry 66: 1-9. doi: 10.1016/j.plaphy.2013.01.020.

O'Connell, P.F. 1992. Sustainable agriculture—a valid alternative. Outlook on Agriculture 21: 5-12.

Ollivier, J., Töwe, S., Bannert, A., Hai, B., Kastl, E.M., Meyer, A., et al. 2011. Nitrogen turnover in soil and global change. FEMS Microbiology Ecology 78: 3-16. doi: 10.1111/j. 1574-6941.2011.01165.x.

Pii, Y., Mimmo, T., Tomasi, N., Terzano, R., Cesco, S. and Crecchio, C. 2015. Microbial interactions in the rhizosphere: beneficial influences of plant growth-promoting rhizobacteria on nutrient acquisition process. A review. Biology and Fertility of Soils 51(4): 403-415. doi: 10.1007/s00374-015-0996-1.

Porcel, R., Zamarreño, A.M., García-Mina, J.M. and Aroca, R. 2014. Involvement of plant endogenous ABA in *Bacillus megaterium* PGPR activity in tomato plants. BMC Plant Biology 14: 36.

Ramos Solano, B., Barriuso Maicas, J., Pereyra De La Iglesia, M.T., Domenech, J. and Gutiérrez Mañero, F.J. 2008. Systemic disease protection elicited by plant growth promoting rhizobacteria strains: relationship between metabolic responses, systemic disease protection, and biotic elicitors. Phytopathology 98(4): 451-457.

Ramos-Solano, B., Algar, E., García-Villaraco, A., García-Cristobal, J., Lucas Garcia, J.A. and Gutierrez-Mañero, F.J. 2010. Biotic elicitation of isoflavone metabolism with plant growth promoting rhizobacteria in early stages of development in *Glycine max* var. Osumi. Journal of Agricultural and Food Chemistry 58: 1484-1492. doi: 10.1021/jf903299a.

Ramos Solano, B., Garcia-Villaraco, A., Gutierrez-Mañero, F.J., Lucas, J.A., Bonilla, A. and García-Seco, D. 2014. Annual changes in bioactive contents and production in field-grown blackberry after inoculation with *Pseudomonas fluorescens*. Plant Physiology and Biochemistry 74: 1e8.

Saijo Y. and Loo, P. 2019. Plant immunity in signal integration between biotic and abiotic stress responses. New Phytologist 204: 273-318.

Schaaf, G., Ludewig, U., Erenoglu, B.E., Mori, S., Kitahara, T. and von Wirén, N. 2004. ZmYS1 functions as a proton-coupled symporter for phytosiderophore- and nicotianamine-chelated metals. Journal of Biological Chemistry 279: 9091-9096. doi: 10.1074/jbc.M311799200.

Schünmann, P.H.D., Richardson, A.E., Smith, F.W. and Delhaize, E. 2004. Characterization of promoter expression patterns derived from the Pht1 phosphate transporter genes of barley (*Hordeum vulgare* L.). Journal of Experimental Botany 55: 855-865. doi:10.1093/jxb/erh103.

Shrivastava P. and Kumar, R. 2015. Soil salinity: a serious environmental issue and plant growth promoting bacteria as one of the tools for its alleviation. Saudi Journal of Biological Sciences 22(2): 123-131. doi: https://doi.org/10.1016/j.sjbs.2014.12.001.

Spoel, S.H. and Dong, X. 2012. How do plants achieve immunity? Defence without specialized immune cells. Nature Reviews Immunology 12(2): 89-100.

Sukanya, V., Patel, R.M., Suthar, K.P. and Singh, D. 2018. An overview: mechanism involved in bio-priming mediated plant growth promotion. International Journal of Pure and Applied Bioscience 6(5): 771-783.

Sumayo, M., Hahm, M.-S. and Ghim, S.-Y. 2013. Determinants of plant growth-promoting *Ochrobactrum lupini* KUDC1013 involved in induction of systemic resistance against *Pectobacterium carotovorum* subsp. carotovorum in tobacco leaves. Plant Pathology Journal 29(2): 174-181.

Sunarpi, H., Horie, T., Motoda, J., Kubo, M., Yang, H., Yoda, K., et al. 2005. Enhanced salt tolerance mediated by AtHKT1 transporter-induced Na unloading from xylem vessels to xylem parenchyma cells. Plant Journal 44: 928-938. doi: 10.1111/j.1365-313X.2005.02595.x.

Tomasi, N., Weisskopf, L., Renella, G., Landi, L., Pinton, R., Varanini, Z., et al. 2008. Flavonoids of white lupin roots participate in phosphorus mobilization from soil. Soil Biology and Biochemistry 40: 1971-1974. doi: 10.1016/j.soilbio.2008. 02.017.

Tomasi, N., Kretzschmar, T., Espen, L., Weisskopf, L., Fuglsang, A.T., Palmgren, M.G., et al. 2009. Plasma membrane $H^+$-ATPase-dependent citrate exudation from cluster roots of phosphate-deficient white lupin. Plant, Cell and Environment 32: 465-475. doi: 10.1111/j.1365-3040.2009.01938.x.

Touraine, B. and Glass, A. 1997. NO3- and ClO3- Fluxes in the chl1-5 Mutant of Arabidopsis thaliana (Does the CHL1-5 Gene Encode a Low-Affinity NO3- Transporter?). Plant Physiology 114(1): 137-144. doi: 10.1104/pp.114.1.137.

van der Burgh, A.M. and Joosten, M.H.A.J. 2019. Plant immunity: thinking outside and inside the box. Trends in Plant Science 24(7): 587-601.

Vessey, J.K. 2003. Plant growth promoting rhizobacteria as biofertilizers. Plant and Soil 255: 571-586.

Volkov, V. and Beilby, M.J. 2017. Salinity tolerance in plants: mechanisms and regulation of ion transport. Frontiers in Plant Science doi: https://doi.org/10.3389/fpls.2017.01795.

White, P.J. 2003. Ion transport. pp. 625-634. *In*: Thomas, B., Murphy, D.J. and Murray, B.G. (eds.). Encyclopedia of Applied Plant Sciences. Academic Press, London.

Wintermans, P.C., Bakker, P.A. and Pieterse, C.M. 2016. Natural genetic variation in *Arabidopsis* for responsiveness to plant growth-promoting rhizobacteria. Plant Molecular Biology 90: 623-634. doi: 10.1007/s11103-016-0442-2.

Wu, G., Liu, Y., Xu, Y., Zhang, G., Shen, Q. and Zhang, R. 2018. Exploring elicitors of the beneficial rhizobacterium *Bacillus amyloliquefaciensSQR9* to induce plant systemic resistance and their interactions with plant signaling pathways. Molecular Plant Microbe Interactions 31(5): 560-567.

Xing, Y., Jia, W. and Zhang, J. 2008. AtMKK1 mediates ABA-induced CAT1 expression and $H_2O_2$ production via AtMPK6-coupled signaling in *Arabidopsis*. Plant Journal 54: 440-451. doi: 10.1111/j.1365-313X.2008.03433.x.

Youry, P., Mimmo, T., Tomasi, N., Terzano, R., Cesco, S. and Crecchio, C. 2015. Microbial interactions in the rhizosphere: beneficial influences of plant growth-promoting rhizobacteria on nutrient acquisition process. A review. Biology and Fertility of Soils 4: 403-415.

Yu, K., Pieterse, C.M.J., Bakker, P.A.H.M. and Berendsen, R.L. 2019. Beneficial microbes going underground of root immunity. Plant, Cell and Environment 6: 906-911.

Zahir, A.Z., Arshad, M. and Frankenberger, Jr., W.T. 2004. Plant growth promoting rhizobacteria: applications and perspectives in agriculture. Advances in Agronomy 81: 97-168.

Zehnder, G.W., Murphy, I.F., Sikora, E.J. and Kloepper, J.W. 2001. Application to rhizobacteria for induced resistance. European Journal of Plant Pathology 107: 39-50.

Zhang, F., Shen, J., Zhang, J., Zuo, Y., Li, L. and Chen, X. 2010. Rhizosphere processes and management for improving nutrient use efficiency and crop productivity: implications for China, 1st ed. Advances in Agronomy 107: 1-32. doi: 10.1016/S0065-2113(10)07001-X.

Zhang, H., Kim, M.S., Sun, Y., Dowod, S.E., Shi, H. and Pare, P.W. 2008a. Soil bacteria confer plant salt tolerance by tissue-specific regulation of the sodium transporter HKT1. Molecular Plant Microbe Interactions 21: 731-734.

Zhang, H., Xie, X., Mi-Seong, K., Kornyeyev, D.A., Holaday, S. and Pare, P.W. 2008b. Soil bacteria augment Arabidopsis photosynthesis by decreasing glucose sensing and abscisic acid levels in planta. The Plant Journal 56: 264-273 doi: 10.1111/j.1365-313X.2008.03593.

# Anammox Process:
# Biotechnological Impact

Alejandro Gonzalez-Martinez[1,*], Irina Levchuk[2] and Riku Vahala[2]

## 7.1 INTRODUCTION

More than 50% of the global population lives in urban areas, which occupy about 3% of the global land (Grimm et al. 2008). Such imbalance makes urban areas one of the main sources of environmental pollution. Urban wastewater is a source of various contaminants, which damage the environment (Rizzo et al. 2019). Among these contaminants, reactive nitrogen species play an important role. Often, reactive nitrogen species are often not eliminated during urban wastewater treatment (Holmes et al. 2019). As a result, Large amounts of reactive nitrogen are discharged to the aquatic environment globally. Moreover, according to the literature (van Drecht et al. 2009), the release of nitrogen from wastewater treatment plants (WWTPs) will only increase in the future. Thus, van Drecht et al. (2009) estimated that global nitrogen discharge from WWTPs could reach 12.0-15.5 Tg of nitrogen annually in 2050, which is significantly higher than 6.4 Tg of nitrogen released annually in 2000. Factors such as global growth of population, urbanization and development of sewage treatment play a key role in the expected increase of nitrogen discharge from urban WWTPs (van Drecht et al. 2009).

Over-enrichment of water bodies with reactive nitrogen poses a serious threat to the environment, economy and human health. Reactive nitrogen significantly stimulates the growth of algae in the aquatic environment, which in turn leads to anthropogenic eutrophication (Figure 7.1) (Holmes et al. 2019) and hypoxic (Funkey et al. 2014) or anoxic conditions (so-called dead zones) (Guo 2007). Anthropogenic eutrophication usually leads to negative socio-economic consequences, such as for fishing, tourism, drinking water supply, etc. For instance, algae bloom in the Lake Erie (North America), which took place in 2014, led to a generation of microcystin toxin and as a result, about 400,000 people were left without drinking water supply (Smith et al. 2015). Reactive nitrogen species occurring in drinking water sources (Wu et al. 2019) endanger public health (Holmes et al. 2019). Thus, a high concentration of nitrate in drinking water may provoke the development of serious and often mortal disease (methemoglobinemia) in infants (Holmes et al. 2019). Presence of cyanobacteria (often appear during algae bloom) in drinking water can cause adverse effects for human health and be potentially lethal due to the generation of toxic compounds (Guo 2007, Metcalf and Souza 2019).

---
[1] Department of Microbiology and Institute of Water Research, University of Granada, Granada, Spain.
[2] Aalto University, P.O. Box 15200, FI-00076 AALTO, Tietotie 1E, Espoo, Finland.
* Corresponding author: agon@ugr.es

**FIGURE 7.1** Example of eutrophication in Southern Finland during summer 2018.

To prevent undesired environmental, socio-economical and health-related issues caused by the discharge of reactive nitrogen species from WWTPs, nitrogen removal from wastewater should be conducted. Discharge limits for nitrogen in urban wastewater are established in many countries. For instance, in Europe minimun requirement is set in Urban Waste Water Directive 91/271/EEC in which established discharge limits for total nitrogen (TN) is 10-15 mg/L of TN and/or a minimum of 70-80% TN removal should be achieved during urban wastewater treatment in sensitive areas, which are subject to eutrophication. However, many European Member States have set more stringent effluent standards for wastewater effluent.

Widely applied conventional treatment methods for nitrogen elimination in urban WWTPs include biological nitrification and denitrification. It should be noted that aeration and the addition of organic matter are required for nitrification and denitrification, respectively, making this process energy quite demanding. Thus, according to one estimation, about 25% of energy during conventional wastewater treatment is consumed for the elimination of nitrogen (Siegrist et al. 2008). However, in recent decades, an alternative biological method for nitrogen removal from wastewater, namely Anaerobic Ammonium Oxidation (Anammox), was discovered (Kuenen 2008). Interestingly, only a few decades ago, it was generally accepted that ammonium was chemically inert and anaerobic oxidation of ammonium was not feasible (Kuenen 2008). However, observations of anaerobic oxidation of ammonium in natural environments were already reported a few decades before the discovery of Anammox bacteria (Hamm and Thompson 1941, Richards 1965, Allgeier et al. 1932). For instance, in 1932, Allgeier et al. (1932) reported the generation of nitrogen after fermentation of lake sediments. In an attempt to explain the origin of this phenomenon, authors assumed that one of the possible reasons could be simultaneous nitrification and denitrification (Allgeier et al. 1932). Ammonium loss in anoxic areas of oceanic environment, which was not explained at that time, was also observed and reported (Hamm and Thompson 1941, Richards 1965). Later, the existence of microorganisms, which was able to oxidize ammonium to $N_2$, was predicted (Broda 1977). Already in the 1990s, Anammox was discovered in the denitrifying pilot plant, which was operated in the Gist Brocades company for wastewater treatment (Kuenen 2008, Mulder et al. 1995). The discovery of Anammox played a crucial role in our understanding of the nitrogen cycle.

## 7.2 NITROGEN CYCLE IN WASTEWATER TREATMENT

A schematic nitrogen cycle in wastewater treatment is presented in Figure 7.2. As it was mentioned earlier, the conventional path for nitrogen removal from wastewater consists of two steps, namely biological nitrification and denitrification. During biological nitrification in activated sludge, i) oxidation of ammonia to nitrite is followed by ii) oxidation of nitrite to nitrate that takes place (Tchobanoglous et al. 2003). Oxidation of ammonia to nitrite occurs due to the activity of aerobic autotrophic bacteria, namely *Nitrosomonas*, according to Reaction 1.

$$2NH_4^+ + 3O_2 \rightarrow 2NO_2^- + 4H^+ + 2H_2O \tag{1}$$

Other groups of aerobic autotrophic bacteria, such as *Nitrobacter, Nitrococcus*, etc., are responsible for subsequent oxidation of nitrite to nitrate (Tchobanoglous et al. 2003), as shown in Reaction 2.

$$2NO_2^- + O_2 \rightarrow 2NO_3^- \tag{2}$$

The overall reaction of nitrification is shown in Reaction 3.

$$NH_4^+ + 2O_2 \rightarrow NO_3^- + 2H^+ + H_2O \tag{3}$$

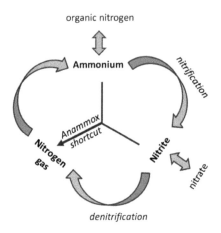

**FIGURE 7.2** Schematic representation of Anammox shortcut in the nitrogen cycle.

Nitrification is a sensitive process. Nitrifiers (*Nitrobacter* spp. and *Nitrococcus* spp.) use inorganic carbon (carbon dioxide, carbonates and bicarbonates) as a source of carbon. A characteristic feature of nitrifiers is a relatively slow growth rate. Hence, in case of malfunctioning of the nitrification process, a long time is required for the regeneration of nitrifiers. Successful nitrification occurs during wastewater treatment when parameters such as oxygen concentration, temperature, pH and alkalinity are optimized. The amount of oxygen required per g of ammonia is 4.25 g (Tchobanoglous et al. 2003). Therefore, continuous oxygen supply is needed during nitrification, which leads to significant energy consumption as mentioned earlier. The pH has a significant influence on the efficiency of nitrification. The optimal pH range for nitrification is between 7.5 and 8.0 (Tchobanoglous et al. 2003). Other environmental factors, such as toxic organic compounds in wastewater (Inglezakis et al. 2017), metals (Zhang et al. 2019a) and free (un-ionized) ammonia (Jiang et al. 2019), can inhibit nitrification.

Nitrate formed during nitrification is converted to nitric oxide, nitrous oxide and nitrogen gas during biological denitrification process (Tchobanoglous et al. 2003). Denitrification is a heterotrophic process, which means that organic carbon is consumed. In comparison with nitrification, bacteria responsible for denitrification have higher growth rates, which make denitrification more robust as compared to nitrification. Denitrifying bacteria can be heterotrophic and autotrophic (Tchobanoglous et al. 2003). A variety of heterotrophic denitrifiers was reported earlier, including *Xanthomonas*, *Thiobacillus*, *Spirillum*, *Pseudomonas*, *Propionibacterium*, *Paracoccus*, *Moraxella*, *Halobacterium*, *Corynebacterium*, *Chromobacterium*, *Bacillus*, *Alcaligenes* and *Achromobacter* (Payne, 1981). Typical denitrifiers available in activated sludge are *Pseudomonas*, *Alcaligenes* and *Bacillus*.

The optimal concentration of dissolved oxygen for denitrification is below 0.5 mg/L when the anaerobic respiration occurs. At a higher level of dissolved oxygen, denitrifiers prefer oxygen over nitrite and switch to aerobic respiration, which leads to the absence of denitrification reactions. The overall reaction of denitrification is shown below.

$$NO_3^- + COD \rightarrow N_{2(gas)} + OH^- \tag{4}$$

Anammox process represents a shortcut in the nitrogen cycle by direct oxidation of ammonium to gaseous nitrogen using nitrite in three steps (Kartal et al. 2011, Strous et al. 2006), starting with nitrite reduction to nitric oxide as shown in Reaction 5.

$$NO_2^- + 2H^+ \rightarrow NO + 2H_2O \tag{5}$$

After that nitric oxide reacts with ammonium leading to the generation of hydrazine (Reaction 6).

$$NO + NH_4^+ + 2H^+ \rightarrow N_2H_4 + H_2O \tag{6}$$

The final step of the Anammox process is the oxidation of hydrazine to a gaseous form of nitrogen as demonstrated in Reaction 7.

$$N_2H_4 \rightarrow N_{2(gas)} + 4H^+ \tag{7}$$

An overall Anammox reaction is shown below (Reaction 8).

$$NH_4^+ + NO_2^- \rightarrow N_{2(gas)} + 2H_2O \tag{8}$$

## 7.3   CELL STRUCTURE AND METABOLISM OF ANAMMOX BACTERIA

Anammox bacteria belong to phylum Planctomycetes, class of Planctomycetia and order Brocadiales (Jetten et al. 2015). A plethora of studies devoted to phylum Planctomycetes has been published (Fuerst and Sagulenko 2011, Fuerst 2017), which is not surprising when taking their unique features into consideration. Planctomycetes can be considered an exception among prokaryotic organisms (Fuerst 2005) as they possess a unique feature of intracellular compartmentalization (Fuerst and Sagulenko 2011, Lindsay et al. 2001), which is not typical for prokaryotes. Planctomycetes are classified as Gram-negative bacteria. The majority of Planctomycetes are classified as aerobic chemoorganoheterotrophic bacteria (van Niftrik et al. 2004). Anammox bacteria hold a specific place in the phylum of Planctomycetes, mainly due to their unique property of anaerobic ammonium oxidation (Kartal et al. 2012). Anammox bacteria belong to the group of chemolithoautotrophic organisms (van Niftrik et al. 2004). As shown in Figure 7.3, Anammox bacteria have a spherical form and a diameter below 1 μm (van Niftrik et al. 2004). The cell of Anammox bacteria consists of three compartments, namely paryphoplasm, riboplasm and anammoxosome (Figure 7.3), and each compartment is separated by a lipid membrane layer. About 50-70% of the volume of the Anammox cell is occupied by the anammoxosome organelle (Lindsay et al. 2001, van Niftrik et al. 2004).

Anammoxosome is responsible for energy metabolism or, in other words, for Anammox reaction (Peeters and van Niftrik 2019). Interestingly, isolated anammoxosome was reported to function independently (Neumann et al. 2014). Due to unique features the anammoxosome organelle, such as the presence of iron-rich nanoparticles (Ferousi et al. 2017) and tubule-like structures (de Almeida et al. 2015), etc., it has been a subject of numerous studies (Peeters and van Niftrik 2019, van Niftrik et al. 2004).

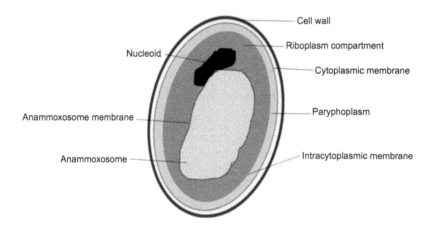

**FIGURE 7.3** Schematic representation of Anammox planctomycetes. Adapted from (Lindsay et al. 2001).

Anammox bacteria possess another unique feature—ladderane lipids (Kartal et al. 2012)—which were reported to be present in all membranes of the Anammox cell (Neumann et al. 2014). Remarkably, ladderane lipids were found only in Anammox bacteria (Sinninghe Damste et al. 2002) and play the role of lipid biomarker (Lanekoff and Karlsson 2010). Ladderane lipids, discovered in 2002 (Sinninghe Damste et al. 2002), possess a unique structure, combining a hydrocarbon chain with three or five cyclobutane rings concatenated linearly. Based on molecular modeling, it was suggested that presence of ladderane lipids cause the formation of a tightly packed membrane (Sinninghe Damste et al. 2002), which is considered to play a significant role in protecting the Anammox cell from toxic compounds formed in anammoxosome during metabolism such as hydrazine.

The pathway of Anammox metabolism is a fascinating research topic with many open questions (Peeters and van Niftrik 2019). It was suggested that Anammox process can be divided into three steps (Kartal and Keltjens 2016). Firstly, the reduction of nitrite ($NO_2^-$) to nitric oxide (NO) occurs with the help of enzyme nitrite reductase (NirS) and one electron (Figure 7.4). After that hydrazine ($N_2H_4$) is generated through a reaction of nitric oxide and ammonium by hydrazine synthase (HZS) and three electrons. It should be mentioned that hydrazine formed during Anammox metabolism is an extremely reactive compound and it is widely used on an industrial scale—for instance, in rocket fuels. Hydrazine possesses extremely high toxicity for living organisms (Shi et al. 2018). For example, inactivation of bacterial spores by hydrazine was reported (Schubert et al. 2008). Hence, the anammoxosome membrane (especially ladderane lipids) plays a crucial role in the confinement of intermediates, such as hydrazine within the anammoxosome (van Niftrik et al. 2004). The final step of Anammox metabolism is the oxidation of hydrazine to nitrogen gas ($N_2$) by hydrazine dehydrogenase (HDH) during which four electrons are released. These electrons are used for earlier redox reactions within the pathway (Strous et al. 2006). During Anammox metabolism, an accumulation of protons in anammoxosome occurs, which leads to the formation of proton motive force (PMF) in the membrane of this organelle (van Der Star et al. 2010). It was reported that PMF plays a crucial role in ATP synthesis (Karlsson et al. 2009). It should be noted that the pathway of Anammox metabolism may vary depending on bacteria species (Peeters and van Niftrik 2019).

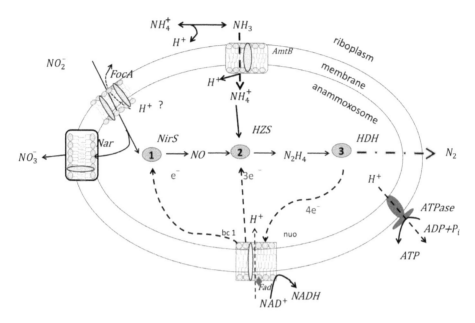

**FIGURE 7.4** Schematic representation of Anammox bacteria metabolism. Numbers show the main reactions taking place during metabolism: 1) reduction of nitrite to nitric oxide by nitrite reductase (NirS); 2) reaction of nitric oxide and ammonium by hydrazine synthase (HZS) leading to the formation of hydrazine ($N_2H_4$); 3) oxidation of hydrazine to dinitrogen gas ($N_2$) by hydrazine dehydrogenase (HDH). Nar—nitrate reductase; nuo, NADH—ubiquinone oxidoreductase; NAD$^+$—ferredoxin oxidoreductase; bc1—cytochrome bc1 complex; FocA—nitrite transporter; AmtB—protein transporting ammonium; adapted and modified from (Hu et al. 2012).

## 7.4 SPECIES OF ANAMMOX BACTERIA

A characteristic feature of Anammox bacteria is relatively slow growth rate. The doubling time reported for Anammox bacteria is between 11 and 20 days (Kartal et al. 2012, Jetten et al. 2009). Recently, some authors have reported that doubling time for Anammox bacteria can be significantly shorter (Lotti et al. 2015a, Lotti et al. 2014). Due to the slow growth rate, for a relatively long time, it was challenging to culture and study Anammox bacteria. Until today pure cultures of Anammox bacteria were not obtained (Peeters and van Niftrik 2019). All known species of Anammox bacteria have 'Candidatus' status, which means that they have not been cultivated in pure culture yet. There have been five genera of Anammox bacteria discovered to date, namely *Brocadia*, *Scalindua*, *Kuenenia*, *Anammoxoglobus* and *Jettenia* (Peeters and van Niftrik 2019). More than 20 species of Anammox bacteria are currently known. It should be mentioned that known Anammox bacteria were found not only in the wastewater environment but also in natural anoxic ecosystems—for instance, in marine or freshwater sediments (Ali et al. 2015). Different species of Anammox bacteria known to date are presented in Table 7.1.

**TABLE 7.1** Different species of Anammox bacteria. Adapted from (Holmes et al. 2019)

| Specie | Reference |
|--------|-----------|
| Candidatus Brocadia anammoxidans | (Jetten et al. 2001) |
| Candidatus Brocadia sinica | (Oshiki et al. 2011) |
| Candidatus Brocadia sapporoensis | (Narita et al. 2017) |
| Candidatus Brocadia fulgida | (Kartal et al. 2008) |
| Candidatus Brocadia caroliniensis | (Rothrock et al. 2011) |
| Candidatus Brocadia brasiliensis | (Araujo et al. 2011) |
| Candidatus Scalindua arabica | (Woebken et al. 2008) |
| Candidatus Scalindua brodae | (Schmid et al. 2003) |
| Candidatus Scalindua wagneri | (Schmid et al. 2003) |
| Candidatus Scalindua pacifica | (Dang et al. 2013) |
| Candidatus Scalindua profunda | (van de Vossenberg et al. 2013) |
| Candidatus Scalindua sinooilfield | (Li et al. 2010) |
| Candidatus Scalindua zhenghei | (Hong et al. 2011) |
| Candidatus Scalindua sorokinii | (Kuypers et al. 2003) |
| Candidatus Scalindua richardsii | (Fuchsman et al. 2012) |
| Candidatus Scalindua rubra | (Speth et al. 2017) |
| Candidatus Scalindua marina | (Jiang, X. et al. 2017) |
| Candidatus Jettenia asiatica | (Quan et al. 2008) |
| Candidatus Jettenia moscovienalis | (Nikolaev et al. 2015) |
| Candidatus Jettenia caeni | (Ali et al. 2015) |
| Candidatus Jettenia ecosi | (Botchkova et al. 2018) |
| Candidatus Anammoxoglobus propionicus | (Kartal et al. 2007) |
| Candidatus Kuenenia stuttgartiensis | (Strous et al. 2006) |

## 7.5 IMPACT OF ANAMMOX ON THE ENVIRONMENT AND WASTEWATER TREATMENT

Anammox bacteria, which only a few decades ago were considered to be an 'impossible' microorganism, are extremely interesting from the point of view of both fundamental microbiology and potential applications. Despite the fact that Anammox was first discovered in a wastewater treatment system, different species of these bacteria were also found in more than 30 natural ecosystems (including both freshwater and marine environment) around the world. It was suggested that Anammox bacteria can be responsible for production from 30 to 50% of nitrogen gas in the oxygen minimum zones of the oceans (Devol 2003), thus becoming a major player in the nitrogen cycle.

Anammox stands head and shoulders above other biological methods applied for nitrogen removal from wastewater. Significant energy and cost reduction can be achieved when Anammox is applied as compared to conventional treatment. Thus, it was suggested that the amount of energy consumed can be reduced by about 60% when Anammox is applied (van Dongen et al. 2001). According to some estimation, the operational costs of nitrogen elimination during wastewater treatment can be reduced by about 90% if Anammox process is used (Jetten et al. 2001).

## 7.6 ANAMMOX PROCESS IN WATER TREATMENT TECHNOLOGIES

### 7.6.1 Autotrophic Nitrogen Technologies

Novel biological technologies based on autotrophic nitrogen removal bacteria have been developed in the last years for the treatment of effluents with low C/N concentration. Their most prominent advantages are the reduction of energy demand, the absence of external carbon addition and the lower production of sludge compared to conventional treatment. However, the main weaknesses of these technologies are their slow growth and metabolism of the Anammox bacteria (Gonzalez-Martinez et al. 2011). To cope with these problems, several kinds of autotrophic nitrogen removal technologies have been designed for enhancing the efficiency of N removal (Table 7.2).

**TABLE 7.2** Comparison of different autotrophic nitrogen removal technologies (van Dongen et al. 2001; Wyffels et al 2004; Van Hulle et al. 2010; Gonzalez-Martinez et al. 2011; 2015a; 2015b; Zhang et al. 2015; Garcia-Ruiz et al. 2018; Yang et al. 2019)

| | Partial Nitritation-Anammox | Single Anammox Bioreactor | CANON Process | IFAS Process | DEMON Process | Submerged bed-Anammox Process | OLAND Process |
|---|---|---|---|---|---|---|---|
| Number of bioreactors | 2 | 1 | 1 | 1 | 1 | 1 | 1 (or 2) |
| Biomass configurations | Flocs/granules | Granules | Granules | Moving bed carriers | Microgranules | Fixed carriers | Granules |
| Oxygen conditions | Oxic/anoxic | Anoxic | Oxygen limited | Oxygen limited | Oxygen limited | Oxygen limited | Oxygen limited |
| Temperature (°C) | 35 | 35 | 35 | 27–29 | 35 | 35 | 35 |
| Dissolved oxygen requirement (mg $O_2$/L) | 1.72 | 0 | 1.5–2 | 0.1–0.35 | 0–0.3 | 1–1.5 | 1.94 |
| pH | 7–8 | 7–8 | 7–8 | 7.6–8.0 | 6.71–6.69 | 7–8 | 7–8 |
| Nitrogen removal performance (%) | 90 | 90 | 90 | 80 | 90 | 92 | 85 |
| Types of bacteria | AOB and Anammox | Anammox | AOB and Anammox | AOB and Anammox | AOB and Anammox | AOB and Anammox | AOB and Anammox |

Anammox technologies can be divided into two different groups: single bioreactor systems and two-stage systems. In two-stage systems, there is a physical separation of both steps in two different bioreactors. In this way, the partial-nitritation process is followed by subsequent anaerobic ammonium oxidation, using nitrite as the terminal electron acceptor, (Sri Shalini et al. 2012). Thus, the ammonia contained in the partial nitrification effluent is fed to the anaerobic bioreactor where the Anammox process can be done. However, in the single-stage system, the partial nitritation and the Anammox process take place at the same time and in the same bioreactor. The main advantages of the single-stage Anammox system are the low construction cost, low nitrite and nitrate accumulation and the small space requirements (Li et al. 2014). However, the two-stage Anammox systems have a short start-up phase, high process stability and easy control (Veys et al. 2010). For these reasons, both configurations have been widely used in full-scale wastewater treatment plants. However, these new technologies meet some challenges, such as the strict operational conditions, for the optimal application in a full-scale wastewater treatment plant.

## 7.6.2 Effect of Operational Conditions on Anammox Process in Autotrophic Nitrogen Technologies

According to Ali and Okabe (2015), 88 full-scale reactors based on the Anammox process have been built in Europe. Nevertheless, it is well known that the slow growth rate, the low cell yield and the high sensitivity to changing environmental conditions are the main challenges for an efficient application of these autotrophic nitrogen removal systems. In this way, the Anammox process application has been restricted by the control of the influent composition and the control of the physico-chemical conditions. Thus, for optimal functioning, several authors have reported that strict control of different parameters (such as temperature, pH, dissolved oxygen demand (DO), carbon source concentration, etc. is needed in order to obtain high-efficiency rates in these technologies (Gonzalez-Martinez et al. 2016, Sobotka et al. 2016, Lu et al. 2017, Wang et al. 2019).

### 7.6.2.1 Temperature Effect

The range of temperatures in which the growth and activity of Anammox bacteria have been reported is very wide (Lotti et al. 2014, Sobotka et al. B 2016; Gonzalez-Martinez et al. 2016). Several Anammox species were found in extreme temperature conditions as high as 60-80°C and as low as 2°C. This suggests that Anammox bacteria could be used to operate over a wide range of temperatures in the wastewater treatment process. However, in wastewater treatment, the Anammox nitrogen removal performance does not offer excellent temperature flexibility with optimal temperature values between 30°C and 40°C. (Zhu et al. 2015).

Under such conditions, the growth and the biological activity of the Anammox bacteria are strongly favored. This is particularly important in single-stage reactors where this temperature range triggers the development of ammonium oxidation bacteria (AOB) and facilitates the partial nitrification process with significant inhibition of nitrite-oxidizing bacteria (NOB). Thus, nitrite oxidation activity could easily be suppressed without nitrate formation. This means that the required water temperature is much higher than that found in conventional wastewater treatment plants. This highlights the need to develop Anammox technologies to ensure a long-term and sustainable nitrogen removal process under low-temperature conditions. For these reasons, several authors have studied the effect of temperature influence on the Anammox process over different biotechnological configurations and scales. In this sense, Lotti et al. (2014) showed a proof of concept for the application of the Anammox process in a gas-lift sequencing batch single reactor under low-temperature operation. This experiment showed nitrogen removal rates of 0.44 and 0.40 g-$N_{Tot}$ $L^{-1}d^{-1}$ at 20°C and 15°C with optimal nitrogen removal efficiencies of 86% and 73%, respectively. These data correlate with other studies, where a successful performance of the Anammox process was also possible at 15°C, with nitrogen removal rates between 1.1 and 2.44 g N $L^{-1}d^{-1}$ (Sobotka et al. 2016, Gonzalez-Martinez et al. 2016). Hu et al. (2013) reported that the lab-scale nitritation/ Anammox bioreactor was operated for over 300 days without nitrite accumulation and was able to remove over 90% of the supplied nitrogen at temperatures as low as 12°C. In the same way, Isaka et al. (2008) found that high nitrogen removal could be maintained when the temperature decreased from 20°C to 12°C and that effluent ammonia and nitrite concentrations increased gradually as the operation temperature fell below 10°C. However, an important number of new studies suggest that with temperatures lower than 13°C, an unstable growth, performance and stability are observed with nitrogen removal rates between 0.04-0.029 g N $L^{-1}d^{-1}$. (Lotti et al. 2014, Ma et al. 2016). In this way, when the temperature was decreased to 11°C, several attempts to obtain high nitrogen removal efficiency while operating at low temperatures have been made with important nitrite accumulation (Lotti et al. 2015b, Sobotka et al. 2016). However, it is well known that NOB had a lower affinity to nitrite than Anammox bacteria, which allows for the modification of other operational conditions to reduce the nitrite oxidation bacteria activity (Hu et al. 2013). These results show it remains a challenge to achieve good nitrogen removal performance from urban wastewater at low-temperature conditions. Either way, the vast majority of experimental research shows that the adapted Anammox biomass has a lower decrease in activity at lower temperatures in comparison with non-adapted biomass. Thus, a sudden drop in temperature has a more deleterious effect on the Anammox process than a stepwise decrease in temperature.

### 7.6.2.2 Dissolved Oxygen Demand (DO) Effect

The Anammox process converts ammonium and nitrite directly to dinitrogen gas under anaerobic conditions; this process does not require aeration and other electron donors (Gonzalez-Martinez et al. 2011). In this sense, the optimum DO concentration for Anammox growth is lower than 0.4 mg/l of oxygen to obtain a good performance. Nevertheless, oxygen is still required for the production of nitrite by AOB. For this reason, in a single Anammox technology in which the partial nitritation and the Anammox process can be done in the same bioreactor, the DO control is of crucial importance in the Anammox enrichment process. This is not only because a high oxygen concentration may be inhibitive to Anammox but also because it could negatively impact stability on AOB and NOB microorganisms, which compete with Anammox bacteria for the substrate. The strong limitation of oxygen allows the $NH_4^+$ oxidation to $NO_2^-$; however, it avoids the oxidation of $NO_2^-$ to $NO_3^-$, achieving optimal environmental conditions for the Anammox reaction (Nielsen et al. 2005, Gonzalez-Martinez et al. 2018). For this reason, in the Anammox technologies, DO concentration significantly affects the biological activity of the nitritation process control. It is based on different oxygen affinities of AOB and NOB. The half-saturation constant value of AOB (0.3 mg/L) is lower than that of NOB (1.1 mg/L), which means that AOB will outcompete under limited DO condition (Ma et al. 2016). Some researchers have used a DO concentration around 1.0 mg/L as a suitable concentration for nitritation in order to inhibit nitratation and enhancing partial nitritation and Anammox (Ma et al. 2016, Ruiz et al. 2006). Several authors have been reported that reported that DO concentration of 0.06-1.3 mg/L is a good design condition for the Anammox process; however, this concentration is depended on the thickness of biofilm, density and thickness of the boundary layer (Liu et al. 2017, Yue et al. 2018).

In the last years, intermittent aeration has been considered a promising novel strategy to inhibit the nitratation process due to the lag phase under aerobic conditions (Ma et al. 2015). The key of this strategy is to shut down the aeration before the NOB activity is restored; nitrite can accumulate at the end of the aeration phase and then be consumed by Anammox bacteria. The possibility of implementing this method is based on the fact that Anammox activity was reversibly inhibited by oxygen. In this sense, Egli et al. (2001) reported that the Anammox process could be inhibited reversibly by DO at a low level (<1% air saturation) but an irreversible inhibition happened at higher oxygen concentrations (>18% air saturation). Therefore, intermittent aeration can successfully inhibit NOB and enhance the performance of nitrogen removal (Miao et al. 2018).

### 7.6.2.3 The pH Effect

Activity in natural environments has been observed in a wider pH range from 4 to 9 Zhu et al. (2015). However, it is well known that in wastewater treatment the Anammox process is very sensitive to pH changes (Lotti et al. 2015b). In this sense, several authors have reported that the overall optimum pH for growth and Anammox activity is within the range 6 to 8 (Table 7.3). Lu et al. (2017) revealed that the estimated optimal pH value for Anammox was 7.6 and that the maximum specific growth rates of these microorganisms under this pH was 0.36-0.38, day 1, at 30°C. In this way, pH is an important control parameter during the operation of Anammox bioreactors, which must be kept at near optimum values because this directly affects the process (Carvajal-Arroyo et al. 2014, Puyol et al. 2014, Wang and Gu 2014).

**TABLE 7.3** Optimal pH and temperature in autotrophic nitrogen removal bioreactors

| Optimal pH Range | Optimal Temperature Range (°C) | References |
|---|---|---|
| 7.4 | 30±2 | Carvajal-Arroyo et al. (2014) |
| 7.5–7.7 | 35 | Gonzalez-Martinez et al. (2016) |
| 8 | 37 | Egli et al. (2001) |
| 7.2–7.6 | 30±2 | Puyol et al. (2014a) |
| 7–8 | 35–40 | Dosta et al. (2008) |
| 6.8 | 35 | Xiong et al. (2013) |
| 8 | 35 | Wang et al. (2012) |
| 7.8 | 30 | Fernandez et al. (2012) |
| 7.8 | 40 | Sobotka et al. (2015) |

In the Anammox reaction process, the energy obtained from the reaction is used for the fixation of $CO_2$ that allows the growth of cells. At the same time, the use of nitrite leads to the $H^+$ consumption that causes an increase in pH. Therefore, the bioreactors must be buffered because this reaction produces a weak buffering capacity in the bioreactor (Yin et al. 2016). Different strategies can be used effectively to maintain the optimal pH in Anammox technology bioreactors, such as by adding $H_2SO_4$ (Gonzalez-Martinez et al. 2016), by using Argon and $CO_2$ gas (Lotti et al. 2014), by adding HCl solution (Tang et al. 2009) or by using acid/base dosing systems (Lackner et al. 2014). In a single partial nitration-Anammox bioreactor, the pH has an important effect on the nitrification process. In the partial nitration process, 7.07 mg $CaCO_3$ is needed per mg of ammonia oxidized. Thus, without an optimal buffer addition, the nitrite production causes a pH depletion in the Anammox technology. Moreover, changes in the pH values can cause the appearance of other compounds with inhibitory capacity, such us free ammonium (FA) and free nitrous acid (FNA).

The FA concentration increases at high pH values and decreases at low pH, while the FNA concentration increases at low pH values and decreases at high pH. This effect must be controlled because the inhibitory effects over the Anammox process of these compounds have been reported as one of the most important parameters that would be taken into account (Fernández et al. 2012). In this way, Van Hulle et al. (2010) reported the direct correlation between FA and FNA with the pH by using mathematical equations:

$$[NH_3] = \frac{[TAN] \times 10^{pH}}{e^{-\frac{2300}{(T+273)}} + 10^{pH}} \qquad [NHO_2] = \frac{[TNO_2] \times 10^{-pH}}{e^{-\frac{2300}{(T+273)}} + 10^{-pH}}$$

For these reasons, it is very important to reach an optimal pH of around 7 to avoid the accumulation of nitrite and ammonium in the bioreactor.

### 7.6.2.4 Carbon Source Effect

Anammox bacteria are chemoautotrophic microorganisms that use $CO_2$ or $HCO_3^-$ as a carbon source to grow and maintain their activity. Moreover, the Anammox process has no need for an external organic carbon source, making it a key technology for the treatment of streams with a low carbon/nitrogen ratio and a high ammonia concentration, such as the leachate from the anaerobic digestion (Ma et al. 2016). For these reasons, it is very important in the Anammox process to distinguish between inorganic carbon sources and organic carbon sources. The organic carbon source effect has been heavily investigated in the last decade. In this way, different studies have shown that a high concentration of organic carbon has a clear negative impact on the Anammox process (Tang et al. 2013, Pearsson et al. 2017). This effect could be explained because a high concentration of organic carbon source fosters the growth of heterotrophic denitrifying bacteria, which use nitrite and organic matter, competing against Anammox bacteria. On the other hand, Jin et al. (2012) suggested that the inhibitory effect of the organic matter can produce microbial poisoning or enzyme inactivation. Nevertheless, it is well known that low concentrations of an organic carbon source did not have a significant negative impact. It may even produce some positive changes in the Anammox biomass under a low carbon source concentration. Also, in a single partial-nitration Anammox, the partial nitration process can be aided by the presence of a low organic carbon source. In this way, Mosquera-Corral et al. (2005) reported that the acetate could encourage AOB development and activity under concentration <300 mg acetate-C/L. In this way, the conclusions of Milia et al. (2016) are similar, reporting an optimal partial nitration performance with the presence of organic carbon source due to the coexistence with heterotrophic microorganisms. That could be the reason why the Anammox process could achieve high performance in effluents with abundant organic matter, such as digested animal waste monosodium glutamate wastewater.

On the other hand, it is well known that Anammox bacteria oxidize ammonium with nitrite as the electron acceptor and with $CO_2$ as the main carbon source. However, Ma et al. (2015) reported that Anammox bacteria are far more vulnerable than AOB and NOB when they compete for the use of inorganic carbon as the preferred assimilative carbon source. For this reason, the need to use a sufficient amount of inorganic carbon is a key factor in order to grow them and maintain the Anammox process activity. However, the presence of this compound could be influenced by different operational conditions. In the partial nitration process, a reduction in the pH occurs which is accompanied by an inorganic carbon consumption to neutralize the pH. From the Anammox reaction equation, it can be observed that the oxidation of 0.066 mol inorganic carbon is needed per mole of ammonia. For this reason, it is particularly necessary to maintain a high concentration of inorganic carbon source with the addition of bicarbonate or $CO_2$ for the cultivation of Anammox bacteria (Gonzalez-Martinez et al. 2015a).

### 7.6.2.5 Salinity Effect

Salinity is acknowledged by several authors as an important operational factor in the wastewater treatment process (Lu et al. 2019, Wang et al. 2019). The salinity has been shown to impact biological wastewater treatment processes, and it may affect the microbial population metabolism, inhibition of degradation processes, accumulation of harmful metabolites, which could persist in the bioreactor, and higher cell death rates due to higher differential osmotic pressure across cell membranes (Rodriguez-Sanchez et al. 2018). Moreover, although autotrophic nitrogen removal technologies are widely used for effluents at high temperatures and high-salinity concentrations due to their achievement of high rates of nitrogen removal, it is important to consider the salinity impact as a potential inhibitory parameter when employing the Anammox process to treat salts-containing wastewater (Gonzalez-Martinez et al. 2018). In this way, Gonzalez-Silva et al. (2017) found that *Candidatus* Brocadia dominated under 0-3 g NaCl $L^{-1}$, while *Candidatus* Kuenenia dominated in Anammox communities acclimated to salinity condition up to 30 g NaCl $L^{-1}$. These results suggest that although the Anammox communities adapt to different saline conditions, the increase in salinity determines changes in bacterial diversity populations, including the Anammox bacteria. It was reported that salinity in certain concentrations (3-15 g NaCl $L^{-1}$) promoted the Anammox granular sludge aggregation and enhanced the yield of the Anammox population in the autotrophic nitrogen removal bioreactors (Speth et al. 2017). In this sense, Dapena-Mora et al. (2010) reported that Anammox reached a maximum activity value at NaCl concentration of about 3 g/L. This favorable effect could be explained by the enzyme stimulation of specific ions under these conditions. However, this effect vanished for high salinity concentrations with a strong inhibition and a long-time harmful effect over the Anammox process.

Although autotrophic nitrogen removal technologies are widely used for industrial WWT at high temperatures and high-salinity concentrations due to their achievement of high rates of nitrogen removal, the capacity to treat up to 30 g NaCl $L^{-1}$ and the specific Anammox activity are lower than under low-salt concentrations (Dapena-Mora et al. 2010). The salinity effect could be increased by other parameters, such as the temperature. In this way, Wang et al. (2019) reported that Anammox activity was inhibited by 70% when the NaCl concentration was increased from 2 to 10 g/L at 20°C. The activity was lost when the reactor reached a stable state. Besides that, low salinity adapted Anammox sludge could adapt to high salinity quickly, and, currently the acclimatization of Anammox bacteria to saline condition is considered to be one of the best methods to undermine salt inhibition (Liu et al. 2014, Gonzalez-Silva et al. 2017). In this sense, Windey et al. (2005) reported that acclimatized Anammox microorganisms in comparison with non-acclimated Anammox microorganisms achieved a 36% higher activity under 30 g/L of salinity. In the same way, Kartal et al. (2006) reported that Anammox activity can be reduced up to 85% under 45 g/L of salinity with a total inhibition under 60 g/L of salinity. Moreover, this study shows that Anammox bacteria acclimated to salinity can tolerate levels up to 75 g $L^{-1}$ of NaCl. However, even with long-term acclimatization, Anammox sludge cannot be applied in autotrophic nitrogen removal bioreactors at salt concentrations higher than 30 g $L^{-1}$ (Liu et al. 2014).

### *7.6.2.6 Antibiotics Effect*

Indeed, most of the wastewater effluents with high antibiotic concentrations are treated through a biological treatment process. Among these, human and veterinary wastewater constitutes the major discharge of antibiotics since they are discharged in the form of urine and feces in wastewater (Zuccato et al. 2010). Antibiotics are just barely metabolized by humans and animals; nearly 30-90% of the parent compound is excreted as a mixture of metabolites and unchanged or conjugated substances. The anaerobic digestion process has been regarded as an efficient treatment for livestock manure effluents due to the reduction of its environmental pollution and its high performance at low cost. However, this technology has been shown to not have the capacity to remove antibiotics, such as fluoroquinolones, sulphonamides and tetracyclines (Van Epps et al. 2016). In this way, given the extended application of Anammox technologies for the treatment of the supernatant of the anaerobic digester, a high amount of antibiotics and different metabolites of these antibiotic compounds can be found in Anammox bioreactors (Rodriguez-Sanchez et al. 2017). To date, several studies have shown that antibiotics produce an important inhibitory effect on the Anammox process (Shi et al. 2017, Rodriguez-Sanchez et al. 2017). In this sense, Fernández et al. (2009) reported that when an Anammox technology has been exposed to 10 mg/L of tetracycline hydrochloride, a strong inhibition (decreased by 60%) of the specific Anammox activity was observed. Moreover, tetracycline led to a breakup of granules into small fractions and also to the nitrite accumulation in Anammox bioreactors (Shi et al. 2013). In this context, Shi et al. (2017) showed similar results with 2 mg/L of oxytetracycline with a sharp activity decrease of 81.3%. Similar results were observed with Erythromycin. Thus, it has been observed that the Anammox performance is not affected by low Erythromycin concentrations (≤1 mg $L^{-1}$), thanks to the EPS layer protection. However, significant inhibition was observed when the concentration reached 10 mg $L^{-1}$ (Zhang et al. 2019b).

A similar effect was seen when the Anammox bioreactor was amended with chloramphenicol. In this sense, it showed a 25% decrease in nitrogen removal efficiency with a concentration of 20 mg $L^{-1}$ (Fernández et al. 2009). On the other hand, Rodriguez-Sanchez et al. (2017) studied the removal rate and the effect of a mixture of antibiotics in an Anammox bioreactor. This researcher reported that azithromycin, norfloxacin and trimethoprim antibiotics were efficiently removed in the Anammox bioreactor, reducing its concentration 77.9 ± 11.2%, 51.7 ± 10.7% and 57.8 ± 8.1%, respectively. However, when the Anammox bioreactor was amended with these antibiotics, they caused a loss of Anammox performance, coupled with a deep change in the bacterial community. For these reasons, the presence of antibiotics directly affect Anammox activity, leading to process failures, and must be removed before being discharged into the Anammox bioreactor (Shi et al. 2017).

## 7.7  ANAMMOX FUTURE PROSPECTS

Over the past few decades, research and application of Anammox process showed exponential growth. The reveal of anammox metabolic pathway and intercellular structure have led to significant progress concerning nitrogen cycle knowledge and the applications of biotechnological processes. In this way, the Autotrophic nitrogen removal technologies have been proved as a low cost and sustainable nitrogen removal processes for wastewater containing high ammonia concentration.

Despite its fast progress and success, there are important issues that engineers and scientists still have to overcome in order to increase the chances to apply anammox systems in different wastewater treatment plants. Specifically, the slow Anammox double-time, the high-temperature requirements and the complexity of the real wastewater composition provide different challenges to the Anammox process that must be carried out in the next future. Therefore, future studies should be carried out with a view to discovering faster start-up and sustainable operational control methods to enhance the nitrogen removal performance under adverse operational conditions. On the other hand, another important objective in the future could be the recovery resources from these autotrophic nitrogen removal technologies. In this way, a deeper exploration of different compounds, such as ladderane lipids or the EPS, should be regarded as an important target in the anammox research studies to achieve new recovery resources strategies in the wastewater treatment plants.

Finally, more anammox researches should be more deeply focused on the anammox population response under emerging pollutants which are designed essentially to be resistant in biodegradation processes. For these reasons, it would be necessary to develop new operational strategies and bioreactors designs to allow the adaptation of the anammox population under these influent conditions.

## 7.8  REFERENCES

Ali, M. and Okabe, S. 2015. Anammox-based technologies for nitrogen removal: advances in process start-up and remaining issues. Chemosphere 141: 144-153.

Ali, M., Oshiki, M., Awata, T., Isobe, K., Kimura, Z., Yoshikawa, H., et al. 2015. Physiological characterization of anaerobic ammonium oxidizing bacterium 'Candidatus J ettenia caeni'. Environmental Microbiology 17: 2172-2189.

Allgeier, R., Peterson, W., Juday, C. and Birge, E. 1932. The anaerobic fermentation of lake deposits. Internationale Revue der Gesamten Hydrobiologie und Hydrographie 26: 444-461.

Araujo, J., Campos, A., Correa, M., Silva, E., Matte, M., Matte, G., et al. 2011. Anammox bacteria enrichment and characterization from municipal activated sludge. Water Science and Technology 64: 1428-1434.

Botchkova, E., Litti, Y.V., Novikov, A., Grouzdev, D., Bochkareva, E., Beskorovayny, A., et al. 2018. Description of "Candidatus Jettenia ecosi" sp. nov., a new species of anammox bacteria. Microbiology 87: 766-776.

Broda, E. 1977. Two kinds of lithotrophs missing in nature. Zeitschrift Für Allgemeine Mikrobiologie 17: 491-493.

Carvajal-Arroyo, J.M., Puyol, D., Li, G., Sierra-Alvarez, R. and Field, J.A. 2014. The role of pH on the resistance of resting- and active anammox bacteria to $NO_2$-inhibition. Biotechnology and Bioengineering 111(10): 1949-1956.

Chen, Y.-P., Li, S., Fang, F., Guo, J.-S., Zhang, Q. and Gao, X. 2012. Effect of inorganic carbon on the completely autotrophic nitrogen removal over nitrite (canon) process in a sequencing batch biofilm reactor. Environmental Technology 33(23): 2611-2617.

Dang, H., Zhou, H., Zhang, Z., Yu, Z., Hua, E., Liu, X., et al. 2013. Molecular detection of Candidatus Scalindua pacifica and environmental responses of sediment anammox bacterial community in the Bohai Sea, China. PLoS One 8: e61330.

Dapena-Mora, A., Vázquez-Padín, J.R., Campos, J.L., Mosquera-Corral, A., Jetten, M.S.M. and Méndez, R. 2010. Monitoring the stability of an Anammox reactor under high salinity conditions. Biochemical Engineering Journal 51(3): 167-171.

de Almeida, N.M., Neumann, S., Mesman, R.J., Ferousi, C., Keltjens, J.T., Jetten, M.S., et al. 2015. Immunogold localization of key metabolic enzymes in the anammoxosome and on the tubule-like structures of kuenenia stuttgartiensis. Journal of Bacteriology 197: 2432-2441.

Devol, A.H. 2003. Nitrogen cycle: solution to a marine mystery. Nature 422: 575.

Dosta, J., Fernández, I., Vázquez-Padín, J.R., Mosquera-Corral, A., Campos, J.L. and Mata-Álvarez Méndez, R. 2008. Short- and long-term effects of temperature on the Anammox process. Journal of Hazardous Material 154: 688-693.

Egli, K., Fanger, U., Alvarez, P.J.J., Siegrist, H., Van Der Meer, J.R. and Zehnder, A.J.B. 2001. Enrichment and characterization of an anammox bacterium from a rotating biological contactor treating ammoniumrich leachate. Archives of Microbiology 175: 198-207.

Fernández, I., Mosquera-Corral, A., Campos, J.L. and Méndez, R. 2009. Operation of an Anammox SBR in the presence of two broad-spectrum antibiotics. Process Biochemistry 44(4): 494-498.

Fernández, I., Dosta, J., Fajardo, C., Campos, J.L., Mosquera-Corral, A. and Méndez, R. 2012. Short- and long-term effects of ammonium and nitrite on the Anammox process. Journal of Environmental Management 95(SUPPL.): S170-S174.

Ferousi, C., Lindhoud, S., Baymann, F., Kartal, B., Jetten, M.S. and Reimann, J. 2017. Iron assimilation and utilization in anaerobic ammonium oxidizing bacteria. Current Opinion in Chemical Biology 37: 129-136.

Fuchsman, C.A., Staley, J.T., Oakley, B.B., Kirkpatrick, J.B. and Murray, J.W. 2012. Free-living and aggregate-associated Planctomycetes in the Black Sea. FEMS Microbiology Ecology 80: 402-416.

Fuerst, J.A. 2005. Intracellular compartmentation in planctomycetes. Annu. Rev. Microbiol. 59: 299-328.

Fuerst, J.A. and Sagulenko, E. 2011. Beyond the bacterium: planctomycetes challenge our concepts of microbial structure and function. Nature Reviews Microbiology 9: 403.

Fuerst, J.A. 2017. Planctomycetes—new models for microbial cells and activities. pp. 1-27. *In*: Ipek, Kurtböke (ed.). Mircobial Resources From Functional Existence in Nature to Applications. Academic Press, London, United Kingdom.

Funkey, C.P., Conley, D.J., Reuss, N.S., Humborg, C., Jilbert, T. and Slomp, C.P. 2014. Hypoxia sustains cyanobacteria blooms in the Baltic Sea. Environmental Science and Technology 48: 2598-2602.

García-Ruiz, M.J., Maza-Márquez, P., González-Martínez, A., Campos, E., González-López, J. and Osorio, F. 2018. Performance and bacterial community structure in three autotrophic submerged biofilters operated under different conditions. Journal of Chemical Technology and Biotechnology 93(8): 2429-2439.

Gonzalez-Martinez, A., Morillo, J.A., Garcia-Ruiz, M.J., Gonzalez-Lopez, J., Osorio, F., Martinez-Toledo, M.V., et al. 2015a. Archaeal populations in full-scale autotrophic nitrogen removal bioreactors operated with different technologies: CANON, DEMON and partial nitritation/anammox. Chemical Engineering Journal 277: 194-201.

Gonzalez-Martinez, A., Osorio, F., Morillo, J.A., Rodriguez-Sanchez, A., Gonzalez-Lopez, J., Abbas, B.A., et al. 2015b. Comparison of bacterial diversity in full scale anammox bioreactors operated under different conditions. Biotechnology Progress 31(6): 1464-1472.

Gonzalez-Martinez, A., Poyatos, J.M., Hontoria, E., Gonzalez-Lopez, J. and Osorio, F. 2011. Treatment of effluents polluted by nitrogen with new biological technologies based on autotrophic nitrification-denitrification processes. Recent Patents on Biotechnology 5: 74-84.

Gonzalez-Martinez, A., Rodriguez-Sanchez, A., Garcia-Ruiz, M.J., Muñoz-Palazon, B., Cortes-Lorenzo, C., Osorio, F., et al. 2016. Performance and bacterial community dynamics of a CANON bioreactor acclimated from high to low operational temperatures. Chemical Engineering Journal 287: 557-567.

Gonzalez-Martinez, A., Muñoz-Palazon, B., Rodriguez-Sanchez, A. and Gonzalez-Lopez, J. 2018. New concepts in anammox processes for wastewater nitrogen removal: recent advances and future prospects. FEMS Microbiology Letters 365(6): fny031.

Gonzalez-Silva, B.M., Rønning, A.J., Andreassen, I.K., Bakke, I., Cervantes, F.J., Østgaard, K., et al. 2017. Changes in the microbial community of an anammox consortium during adaptation to marine conditions revealed by 454 pyrosequencing. Applied Microbiology and Biotechnology 101(12): 5149-5162.

Grimm, N.B., Faeth, S.H., Golubiewski, N.E., Redman, C.L., Wu, J., Bai, X., et al. 2008. Global Change and the Ecology of Cities. Science 319: 756-760.

Guo, L. 2007. Ecology. Doing battle with the green monster of Taihu Lake. Science 317: 1166.

Hamm, R.E. and Thompson, T.G. 1941. Dissolved Nitrogen in the Sea Water of the Northeast Pacific with Notes on the Total Carbon Dioxide and the Dissolved Oxygen. University of Washington.

Holmes, D.E., Dang, Y. and Smith, J.A. 2019. Chapter Four - nitrogen cycling during wastewater treatment. Advances in Applied Microbiology 106: 113-192.

Hong, Y., Li, M., Cao, H. and Gu, J. 2011. Residence of habitat-specific anammox bacteria in the deep-sea subsurface sediments of the South China Sea: analyses of marker gene abundance with physical chemical parameters. Microbial Ecology 62: 36-47.

Hu, Z., Speth, D.R., Francoijs, K., Quan, Z. and Jetten, M. 2012. Metagenome analysis of a complex community reveals the metabolic blueprint of anammox bacterium "Candidatus Jettenia asiatica". Frontiers in Microbiology 3: 366.

Hu, Z., Lotti, T., de Kreuk, M., Kleerebezem, R., van Loosdrecht, M., Kruit, J., et al. 2013. Nitrogen removal by a nitritation-anammox bioreactor at low temperature. Applied and Environmental Microbiology 79(8): 2807-2812.

Inglezakis, V.J., Malamis, S., Omirkhan, A., Nauruzbayeva, J., Makhtayeva, Z., Seidakhmetov, T., et al. 2017. Investigating the inhibitory effect of cyanide, phenol and 4-nitrophenol on the activated sludge process employed for the treatment of petroleum wastewater. Journal of Environmental Management 203: 825-830.

Isaka, K., Date, Y., Kimura, Y., Sumino, T. and Tsuneda, S. 2008. Nitrogen removal performance using anaerobic ammonium oxidation at low temperatures. FEMS Microbiology Letters 282(1): 32-38.

Jetten, M.S., Niftrik, L.v., Strous, M., Kartal, B., Keltjens, J.T., Op den Camp H.J.O., et al. 2009. Biochemistry and molecular biology of anammox bacteria. Critical Reviews in Biochemistry and Molecular Biology 44: 65-84.

Jetten, M.S., Op den Camp, H.J.O., Gijs Kuenen, J. and Strous, M. 2015. "Candidatus Brocadiales" ord. nov, In Bergey's Manual of Systematics of Archaea and Bacteria, John Wiley and Sons, Inc., in association with Bergey's Manual Trust. p 1. https://doi.org/10.1002/9781118960608.obm00068.

Jetten, M.S.M., Wagner, M., Fuerst, J., van Loosdrecht, M., Kuenen, G. and Strous, M. 2001. Microbiology and application of the anaerobic ammonium oxidation ('anammox') process. Current Opinion in Biotechnology 12: 283-288.

Jiang, X., Hou, L., Zheng, Y., Liu, M., Yin, G., Gao, J., et al. 2017. Salinity-driven shifts in the activity, diversity, and abundance of anammox bacteria of estuarine and coastal wetlands. Physics and Chemistry of the Earth, Parts A/B/C 97: 46-53.

Jiang, Y., Poh, L.S., Lim, C., Pan, C. and Ng, W.J. 2019. Effect of free ammonia inhibition on process recovery of partial nitration in a membrane bioreactor. Bioresource Technology Reports 6: 152-158.

Jin, R.-C., Yu, J.-J., Ma, C., Yang, G.-F., Hu, B.-L. and Zheng, P. 2012. Performance and robustness of an ANAMMOX anaerobic baffled reactor subjected to transient shock loads. Biores. Technol. 114: 126-136.

Karlsson, R., Karlsson, A., Bäckman, O., Johansson, B.R. and Hulth, S. 2009. Identification of key proteins involved in the anammox reaction. FEMS Microbiology Letters 297: 87-94.

Kartal, B., Koleva, M., Arsov, R., van der Star, W., Jetten, M.S.M. and Strous, M. 2006. Adaptation of a freshwater anammox population to high salinity wastewater. Journal of Biotechnology 126(4): 546-553.

Kartal, B., Rattray, J., van Niftrik, L.A., van de Vossenberg, J., Schmid, M.C., Webb, R.I., et al. 2007. Candidatus "Anammoxoglobus propionicus" a new propionate oxidizing species of anaerobic ammonium oxidizing bacteria. Systematic and Applied Microbiology 30: 39-49.

Kartal, B., van Niftrik, L., Rattray, J., Van De Vossenberg Jack, L.C.M., Schmid, M.C., Damsté, J.S., et al. 2008. Candidatus 'Brocadia fulgida': an autofluorescent anaerobic ammonium oxidizing bacterium. FEMS Microbiology Ecology 63: 46-55.

Kartal, B., Maalcke, W.J., de Almeida, N.M., Cirpus, I., Gloerich, J., Geerts, W., et al. 2011. Molecular mechanism of anaerobic ammonium oxidation. Nature 479: 127.

Kartal, B., van Niftrik, L., Keltjens, J.T., Op den Camp, H.J.M. and Jetten, M.S.M. 2012. Chapter 3 - anammox—growth physiology, cell biology, and metabolism. Advances in Microbial Physiology 60: 211-262.

Kartal, B. and Keltjens, J.T. 2016. Anammox biochemistry: a tale of heme c proteins. Trends in Biochemical Sciences 41: 998-1011.

Kuenen, J.G. 2008. Anammox bacteria: from discovery to application. Nature Reviews Microbiology 6: 320.

Kuypers, M.M., Sliekers, A.O., Lavik, G., Schmid, M., Jørgensen, B.B., Kuenen, J.G., et al. 2003. Anaerobic ammonium oxidation by anammox bacteria in the Black Sea. Nature 422: 608.

Lackner, S., Gilbert, E.M., Vlaeminck, S.E., Joss, A., Horn, H. and van Loosdrecht, M.C.M. 2014. Full-scale partial nitritation/anammox experiences - an application survey. Water Research 55: 292-303.

Lanekoff, I. and Karlsson, R. 2010. Analysis of intact ladderane phospholipids, originating from viable anammox bacteria, using RP-LC-ESI-MS. Analytical and Bioanalytical Chemistry 397: 3543-3551.

Li, H., Chen, S., Mu, B. and Gu, J. 2010. Molecular detection of anaerobic ammonium-oxidizing (anammox) bacteria in high-temperature petroleum reservoirs. Microbial Ecology 60: 771-783.

Li, Z., Xu, X., Shao, B., Zhang, S. and Yang, F. 2014. Anammox granules formation and performance in a submerged anaerobic membrane bioreactor. Chemical Engineering Journal 254: 9-16.

Lindsay, M.R., Webb, R.I., Strous, M., Jetten, M.S., Butler, M.K., Forde, R.J., et al. 2001. Cell compartmentalisation in planctomycetes: novel types of structural organisation for the bacterial cell. Archives of Microbiology 175: 413-429.

Liu, C., Yamamoto, T., Nishiyama, T., Fujii, T. and Furukawa, K. 2009. Effect of salt concentration in anammox treatment using non woven biomass carrier. Journal of Bioscience and Bioengineering 107(5): 519-523.

Liu, M., Peng, Y., Wang, S., Liu, T. and Xiao, H. 2014. Enhancement of anammox activity by addition of compatible solutes at high salinity conditions. Bioresource Technology 167: 560-563.

Liu, W., Yang, D., Chen, W. and Gu, X. 2017. High-throughput sequencing-based microbial characterization of size fractionated biomass in an anoxic anammox reactor for low-strength wastewater at low temperatures. Bioresource Technology 231: 45-52.

Lotti, T., Kleerebezem, R., Hu, Z., Kartal, B., Jetten, M.S.M. and van Loosdrecht, M.C.M. 2014a. Simultaneous partial nitritation and anammox at low temperature with granular sludge. Water Research 66: 111-121.

Lotti, T., Kleerebezem, R., Lubello, C. and van Loosdrecht, M.C.M. 2014 2014b. Physiological and kinetic characterization of a suspended cell anammox culture. Water Research 60: 1-14.

Lotti, T., Kleerebezem, R., Abelleira-Pereira, J.M., Abbas, B. and van Loosdrecht, M.C.M. 2015a. Faster through training: the anammox case. Water Research 81: 261-268.

Lotti, T., Kleerebezem, R. and van Loosdrecht, M.C.M. 2015b. Effect of temperature change on Anammox activity. Biotechnology and Bioengineering 122(1): 98e103.

Lu, H., Li, Y., Shan, X., Abbas, G., Zeng, Z., Kang, D. et al. 2019. holistic analysis of ANAMMOX process in response to salinity: from adaptation to collapse. Separation and Purification Technology 215: 342-350.

Lu, X., Yin, Z., Sobotka, D., Wisniewski, K., Czerwionka, K., Xie, L., et al. 2017. Modeling the pH effects on nitrogen removal in the anammox-enriched granular sludge. Water Science and Technology 75(2): 378-386.

Ma, B., Wang, S., Cao, S., Miao. Y., Jia, F., Du, R., Peng, Y., et al. 2016. Biological nitrogen removal from sewage via anammox: recent advances. Bioresource Technology 200: 981-990.

Ma, C., Jin, R.C., Yang, G.F., Yu, J.J., Xing, B.S. and Zhang, Q.Q. 2012. Impacts of transient salinity shock loads on Anammox process performance. Bioresource Technology 112: 124-130.

Ma, Y., Sundar, S., Park, H. and Chandran, K. 2015. The effect of inorganic carbon on microbial interactions in a biofilm nitritation–anammox process. Water Research 70: 246-254.

Meng, Y., Sheng, B. and Meng, F. 2019. Changes in nitrogen removal and microbiota of anammox biofilm reactors under tetracycline stress at environmentally and industrially relevant concentrations. Science of the Total Environment 668: 379-388.

Metcalf, J.S. and Souza, N.R. 2019. Chapter 6 - cyanobacteria and their toxins. Separation Science and Technology 11: 125-148.

Miao, Y., Peng, Y., Zhang, L., Li, B., Li, X., Wu, L., et al. 2018. Partial nitrification-anammox (PNA) treating sewage with intermittent aeration mode: effect of influent C/N ratios (2018). Chemical Engineering Journal 334: 664-672.

Milia, S., Perra, M., Cappai, G. and Carucci, A. 2016a. SHARON process as preliminary treatment of refinery wastewater with high organic carbon-to-nitrogen ratio. Desalination and Water Treatment 57: 17935-17943.

Mosquera-Corral, A., González, F., Campos, J.L. and Méndez, R. 2005. Partial nitrification in a SHARON reactor in the presence of salts and organic carbon compounds. Process Biochemistry 40: 3109-3118.

Mulder, A., van de Graaf, A.A., Robertson, L.A. and Kuenen, J.G. 1995. Anaerobic ammonium oxidation discovered in a denitrifying fluidized bed reactor. FEMS Microbiology Ecology 16: 177-183.

Narita, Y., Zhang, L., Kimura, Z., Ali, M., Fujii, T. and Okabe, S. 2017. Enrichment and physiological characterization of an anaerobic ammonium-oxidizing bacterium 'Candidatus Brocadia sapporoensis'. Systematic and Applied Microbiology 40: 448-457.

Neumann, S., Wessels, H.J., Rijpstra, W.I.C., Sinninghe Damsté, J.S., Kartal, B., Jetten, M.S.M., et al. 2014. Isolation and characterization of a prokaryotic cell organelle from the anammox bacterium K uenenia stuttgartiensis. Molecular Microbiology 94: 794-802.

Nielsen, M., Bollmann, A., Sliekers, O., Jetten, M., Schmid, M., Strous, M., et al. 2005. Kinetics, diffusional limitation and microscale distribution of chemistry and organisms in a CANON reactor. FEMS Microbiology Ecology 51(2): 247-256.

Nikolaev, Y.A., Kozlov, M., Kevbrina, M., Dorofeev, A., Pimenov, N., Kallistova, A.Y., et al. 2015. Candidatus "Jettenia moscovienalis" sp. nov., a new species of bacteria carrying out anaerobic ammonium oxidation. Microbiology 84: 256-262.

Oshiki, M., Shimokawa, M., Fujii, N., Satoh, H. and Okabe, S. 2011. Physiological characteristics of the anaerobic ammonium-oxidizing bacterium 'Candidatus Brocadia sinica'. Microbiology 157: 1706-1713.

Payne, W.J. 1981. Denitrification. New York: John Wiley and Sons Inc.

Peeters, S.H. and van Niftrik, L. 2019. Trending topics and open questions in anaerobic ammonium oxidation. Current Opinion in Chemical Biology 49: 45-52.

Persson, F., Suarez, C., Hermansson, M., Plaza, E., Sultana, R. and Wilén, B.-M. 2017. Community structure of partial nitritation-anammox biofilms at decreasing substrate concentrations and low temperature- Microbial Biotechnology 10(4): 761-772.

Puyol, D., Carvajal-Arroyo, J.M., Sierra-Alvarez, R. and Field, J.A. 2014a. Nitrite (not free nitrous acid) is the main inhibitor of the anammox process at common pH conditions. Biotechnology Letters 36(3): 547-551.

Quan, Z., Rhee, S., Zuo, J., Yang, Y., Bae, J., Park, J.R., et al. 2008. Diversity of ammonium-oxidizing bacteria in a granular sludge anaerobic ammonium-oxidizing (anammox) reactor. Environmental Microbiology 10: 3130-3139.

Richards, F.A. 1965. Anoxic basins and fjords. Chemical oceanography Vol. 1. Academic Press, New York, USA, 611-645.

Rizzo, L., Malato, S., Antakyali, D., Beretsou, V.G., Đolić, M.B., Gernjak, W., et al. 2019. Consolidated vs new advanced treatment methods for the removal of contaminants of emerging concern from urban wastewater. Science of the Total Environment 655: 986-1008.

Rodriguez-Sanchez, A., Margareto, A., Robledo-Mahon, T., Aranda, E., Diaz-Cruz, S., Gonzalez-Lopez, J., et al. 2017. Performance and bacterial community structure of a granular autotrophic nitrogen removal bioreactor amended with high antibiotic concentrations. Chemical Engineering Journal 325: 257-269.

Rodriguez-Sanchez, A., Leyva-Diaz, J.C., Muñoz-Palazon, B., Rivadeneyra, M.A., Hurtado-Martinez, M., Martin-Ramos, D., et al. 2018. Biofouling formation and bacterial community structure in hybrid moving bed biofilm reactor-membrane bioreactors: influence of salinity concentration. Water 10(9): 1133.

Rothrock, M.J., Vanotti, M.B., Szögi, A.A., Gonzalez, M.C.G. and Fujii, T. 2011. Long-term preservation of anammox bacteria. Applied Microbiology and Biotechnology 92: 147.

Ruiz, G., Jeison, D., Rubilar, O., Ciudad, G. and Chamy, R. 2006. Nitrification-denitrification via nitrite accumulation for nitrogen removal from wastewaters. Bioresource Technology 97(2): 330-335.

Schmid, M., Walsh, K., Webb, R., Rijpstra, W.I., van de Pas-Schoonen, K., Verbruggen, M.J., et al. 2003. Candidatus "Scalindua brodae", sp. nov., Candidatus "Scalindua wagneri", sp. nov., two new species of anaerobic ammonium oxidation bacteria. Systematic and Applied Microbiology 26: 529-538.

Schubert, W., Plett, G., Yavrouian, A. and Barengoltz, J. 2008. Viability of bacterial spores exposed to hydrazine. Advances in Space Research 42: 1144-1149.

Shi, X., Yin, C., Wen, Y., Zhang, Y. and Huo, F. 2018. A probe with double acetoxyl moieties for hydrazine and its application in living cells. Spectrochimica Acta Part A: Molecular and Biomolecular Spectroscopy. 203: 106-111.

Shi, Y., Xing, S., Wang, X. and Wang, S. 2013. Changes of the reactor performance and the properties of granular sludge under tetracycline (TC) stress. Bioresource Technology 139: 170-175.

Shi, Z.-J., Hu, H.-Y., Shen, Y.-Y., Xu, J.-J., Shi, M.-L. and Jin, R.-C. 2017. Long-term effects of oxytetracycline (OTC) on the granule-based anammox: process performance and occurrence of antibiotic resistance genes. Biochemical Engineering Journal 127: 110-118.

Siegrist, H., Salzgeber, D., Eugster, J. and Joss, A. 2008. Anammox brings WWTP closer to energy autarky due to increased biogas production and reduced aeration energy for N-removal. Water Science and Technology 57: 383-388.

Sinninghe Damste, J.S., Strous, M., Rijpstra, W.I.C., Hopmans, E.C., Geenevasen, J.A., Van Duin, A.C., et al. 2002. Linearly concatenated cyclobutane lipids form a dense bacterial membrane. Nature 419: 708.

Smith, D.R., King, K.W. and Williams, M.R. 2015. What is causing the harmful algal blooms in Lake Erie? Journal of Soil and Water Conservation 70: 27A-29A.

Sobotka, D., Czerwionka, K. and Makinia, J. 2016. Influence of temperature on the activity of anammox granular biomass. Water Science and Technology 73: 2518-2525.

Speth, D.R., Guerrero-Cruz, S., Dutilh, B.E. and Jetten, M.S. 2016. Genome-based microbial ecology of anammox granules in a full-scale wastewater treatment system. Nature Communications 7: 11172.

Speth, D.R., Lagkouvardos, I., Wang, Y., Qian, P., Dutilh, B.E. and Jetten, M.S. 2017. Draft genome of Scalindua rubra, obtained from the interface above the discovery deep brine in the Red Sea, sheds light on potential salt adaptation strategies in anammox bacteria. Microbial Ecology 74: 1-5.

Sri Shalini, S. and Joseph, K. 2012. Nitrogen management in landfill leachate: application of SHARON, ANAMMOX and combined SHARON-ANAMMOX process. Waste Management 32(12): 2385-2400.

Strous, M., Pelletier, E., Mangenot, S., Rattei, T., Lehner, A., Taylor, M.W., et al. 2006. Deciphering the evolution and metabolism of an anammox bacterium from a community genome. Nature 440: 790.

Tang, C.-J., Zheng, P., Mahmood, Q. and Chen, J.-W. 2009. Start-up and inhibition analysis of the Anammox process seeded with anaerobic granular sludge. Journal of Industrial Microbiology and Biotechnology 36: 1093-1100.

Tang, C.-J., Zheng, P., Chai, L.-Y. and Min, X.-B. 2013. Thermodynamic and kinetic investigation of anaerobic bioprocesses on ANAMMOX under high organic conditions. Chemical Engineering Journal 230: 149-157.

Tchobanoglous, G., Burton, F.L. and Stensel, H.D. 2003. Wastewater Engineering. Treatment and Reuse, Boston Mcgraw-Hill. 1819.

van de Vossenberg, J., Woebken, D., Maalcke, W.J., Wessels, H.J., Dutilh, B.E., Kartal, B., et al. 2013. The metagenome of the marine anammox bacterium 'Candidatus Scalindua profunda' illustrates the versatility of this globally important nitrogen cycle bacterium. Environmental Microbiology 15: 1275-1289.

van Der Star, W., Dijkema, C., de Waard, P., Picioreanu, C., Strous, M. and van Loosdrecht, M.C. 2010. An intracellular pH gradient in the anammox bacterium Kuenenia stuttgartiensis as evaluated by 31 P NMR. Applied Microbiology and Biotechnology 86: 311-317.

van Dongen, U., Jetten, M.S. and Van Loosdrecht, M. 2001. The SHARON®-Anammox® process for treatment of ammonium rich wastewater. Water Science and Technology 44: 153-160.

van Drecht, G., Bouwman, A., Harrison, J. and Knoop, J. 2009. Global nitrogen and phosphate in urban wastewater for the period 1970 to 2050. Global Biogeochem. Cycles 23, GB0A03.

Van Epps, A. and Blaney, L. 2016. Antibiotic residues in animal waste: occurrence and degradation in conventional agricultural waste management practices. Current Pollution Reports 2(3): 135-155.

Van Hulle, S.W.H., Vandeweyer, H.J.P., Meesschaert, B.D., Vanrolleghem, P.A., Dejans, P. and Dumoulin, A. 2010. Engineering aspects and practical application of autotrophic nitrogen removal from nitrogen rich streams. Chemical Engineering Journal 162(1): 1-20.

van Niftrik, L.A., Fuerst, J.A., Damsté, J.S.S., Kuenen, J.G., Jetten, M.S. and Strous, M. 2004. The anammoxosome: an intracytoplasmic compartment in anammox bacteria. FEMS Microbiology Letters 233: 7-13.

Veys, P., Vandeweyer, H., Audenaert, W., Monballiu, A., Dejans, P., Jooken, E., et al. 2010. Performance analysis and optimization of autotrophic nitrogen removal in different reactor configurations: a modelling study. Environmental Technology 31(12): 1311-1324.

Wang, G., Dai, X. and Zhang, D. 2019. Effects of NaCl and phenol on anammox performance in mainstream reactors with low nitrogen concentration and low temperature. Biochemical Engineering 147: 72-80.

Wang, T., Zhang, H.M., Gao, D.W., Yang, F.L. and Zhang, G.Y. 2012. Comparison between MBR and SBR on Anammox start-up process from the conventional activated sludge. Bioresource Technology 122: 78-82.

Wang, Y.-F. and Gu, J.-D. 2014. Effects of allylthiourea, salinity, and pH on ammonium/ammonium-oxidizing prokaryotes in mangrove sediment incubated in laboratory microcosms. Applied Microbiology and Biotechnology 98: 3257–3274.

Windey, K., De Bo, I. and Verstraete, W. 2005. Oxygen-limited autotrophic nitrification denitrification (OLAND) in a rotating biological contactor treating high-salinity wastewater. Water Research 39(18): 4512-4520.

Woebken, D., Lam, P., Kuypers, M.M., Naqvi, S.W.A., Kartal, B., Strous, M., et al. 2008. A microdiversity study of anammox bacteria reveals a novel Candidatus Scalindua phylotype in marine oxygen minimum zones. Environmental Microbiology 10: 3106-3119.

Wu, P., Cheng, Y., Chen, H., Chueh, T., Chen, H., Huang, L., et al. 2019. Using the entrapped bioprocess as the pretreatment method for the drinking water treatment receiving eutrophic source water. Environmental Pollution 248: 57-65.

Wyffels, S., Boeckx, P., Pynaert, K., Zhang, D., Van Cleemput, O., Chen, G., et al. 2004. Nitrogen removal from sludge reject water by a two-stage oxygen-limited autotrophic nitrification denitrification process. Water Science and Technology 49: 57-64.

Xiong, L., Wang, Y.Y., Tang, C.J., Chai, L.Y., Xu, K.Q., Song, Y.X., et al. 2013. Start-up characteristics of a granule-based Anammox UASB reactor seeded with anaerobic granular sludge. Biomed Research International 2013: 9.

Yang, S., Peng, Y., Zhang, L., Zhang, Q., Li, J. and Wang, X. 2019. Autotrophic nitrogen removal in an integrated fixed-biofilm activated sludge (IFAS) reactor: Anammox bacteria enriched in the flocs have been overlooked. Bioresource Technology 288: 121512.

Yin, Z., Santos, C.E.D.D., Vilaplana, J.G., Sobotka, D., Czerwionka, K., Damianovic, M.H.R.Z., et al. 2016. Importance of the combined effects of dissolved oxygen and pH on optimization of nitrogen removal in anammox-enriched granular sludge (2016). Process Biochemistry 51(9): 1274-1282.

Yue, X., Yu, G., Lu, Y., Liu, Z., Li, Q., Tang, J., et al. 2018. Effect of dissolved oxygen on nitrogen removal and the microbial community of the completely autotrophic nitrogen removal over nitrite process in a submerged aerated biological filter. Bioresource Technology 254: 67-74.

Zhang, L., Fan, J., Nguyen, H.N., Li, S. and Rodrigues, D.F. 2019a. Effect of cadmium on the performance of partial nitrification using sequencing batch reactor. Chemosphere 222: 913-922.

Zhang, L., Zhang, S., Peng, Y., Han, X. and Gan, Y. 2015a. Nitrogen removal performance and microbial distribution in pilot- and full-scale integrated fixed-biofilm activated sludge reactors based on nitritation-anammox process. Bioresource Technology 196: 448-453.

Zhang, X., Chen, Z., Ma, Y., Chen, T., Zhang, J., Zhang, H., et al. 2019b. Impacts of erythromycin antibiotic on Anammox process: performance and microbial community structure. Biochemical Engineering Journal 143: 1-8.

Zhu, G., Xia, C., Shanyun, W., Zhou, L., Liu, L., and Zhao, S. 2015. Occurrence, activity and contribution of anammox in some freshwater extreme environments. Environmental Microbiology Reports 7: 961-969.

Zuccato, E., Castiglioni, S., Bagnati, R., Melis, M. and Fanelli, R. 2010. Source, occurrence and fate of antibiotics in the Italian aquatic environment. Journal of Hazardous Materials 179(1-3): 1042-1048.

# 8

# Nitrogen Cycle in Wastewater Treatment Plants

Belén Rodelas

## 8.1 INTRODUCTION: N IN WASTEWATERS AND THE NEED FOR ITS REMOVAL

### 8.1.1 Wastewater: Definition, Types and Composition

Wastewater comprises all the liquid wastes generated by human communities after using clean water supplies in a variety of applications (Tchobanoglous et al. 2003). It has been defined as "a combination of liquid or water-carried wastes removed from residences, institutions, and commercial and industrial establishments, together with ground water, surface water and storm water" (Sonune and Gathe 2004). Wastewaters can be classified into three main groups, according to their major input source: 1) domestic (mainly generated in households), 2) commercial and industrial, and 3) municipal, which is composed of domestic wastewater mixed with varying proportions of effluents from commercial/industrial origins, either pretreated or not (Wiessman et al. 2007).

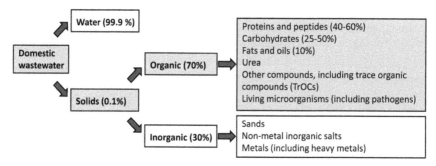

**FIGURE 8.1** Average composition of domestic wastewater (Tchobanoglous et al. 2003, Bitton 2011).

Domestic wastewaters typically derive from toilet flushing, bathing, washing, cooking, laundry, and other household activities (Bitton 2011). The quality and volumes generated are subjected to seasonal, geographic, and socio-cultural variations being influenced by weather conditions and depending strongly on the patterns of human activity resulting from different habits, cultures, lifestyles, and levels of environmental awareness (Seviour 2010, Friedler et al. 2013). Despite these dissimilarities, all domestic wastewaters mostly carry organic compounds, the bulk of which are easily biodegradable (Bitton 2011). Average composition of domestic wastewaters is reflected in Figure 8.1. In contrast, industrial wastewaters widely vary in their composition, strength, flow, and volume depending on the specific activities and can contribute organic compounds recalcitrant

Department of Microbiology and Institute of Water Research, University of Granada, Granada, Spain.
E-mail: mrodelas@ugr.es

to biodegradation or even highly toxic wastes, which often require specific *in situ* treatment at the source facilities before being discharged into municipal wastewater treatment (WWT) plants (Sonune and Gathe 2004, von Sperling 2007, Bitton 2011).

## 8.1.2 Occurrence of N Compounds in Wastewater

N is a major component of most wastewaters from either domestic or industrial origins, occurring in both organic and inorganic forms. The range of N concentrations and the major compounds contributing to the N load of wastewater widely vary depending on the source. N concentrations in domestic/municipal wastewater usually range between 20-85 mg/L, while industrial wastewater may bear much higher loads (Table 8.1).

**TABLE 8.1** Nitrogen and organic matter content of wastewaters. *Total N Kjeldahl (TNK)

| Typical Concentrations of N and C Compounds | | | |
|---|---|---|---|
| Source of Wastewater | Total N (mg/L) | Total Ammonia N (TAN, mg/L) | COD (mg/L) | References |
| Domestic | 20-85 | 12-50 | 250-1000 | Tchobanoglous et al. (2003) |
| Agriculture and farming: | | | | |
| Poultry | 49-140 | 47-115 | 320-919 | Zheng et al. (2019) |
| Piggery | 200-2055 | 110-1650 | 2000-30000 | Cheng et al. (2019) |
| Food making: | | | | |
| Brewery | 25-450 | 5-22 | 750-80000 | Arantes et al. (2017), Vymazal (2014) |
| Dairy and cheese | 14-5600 | 20-184 | 2000-95000 | Vymazal (2014), Finnegan et al. (2018) |
| Distillery | 7000* | | 4000-21000 | Vymazal (2014) |
| Fish processing | 77-3000* | 1-860 | 325-90000 | Vymazal (2014) |
| Meat processing | 530-810* | 65-740 | 400-11200 | Vymazal (2014) |
| Olive oil milling | 9-1540* | | 37000-318000 | Niaounakis and Halvadakis (2006), Vymazal (2014) |
| Winery | 0-176 | | 320-296119 | Bustamante et al. (2005), Welz et al. (2016) |
| Other: | | | | |
| Coke plants | 200-550* | 50-562 | 525-10000 | Vymazal (2014) |
| Fertilizers | | 2-1710 | 44-125000 | Bhandari et al. (2016) |
| Landfill leachate | 163-4230* | 91-3550 | 1300-69018 | Carrera et al. (2003), Iskander et al. (2016), Akgul et al. (2018) |
| Papermaking | 0.8-600 | | 500-40000 | Slade et al. (2004), Vymazal (2014) |
| Tannery | 144-370* | 33-335 | 2155-11159 | Durai and Rajasimman (2011) |

| Nitrate-rich Wastewaters | | | |
|---|---|---|---|
| Source of Wastewater | Nitrate (mg/L) | COD (mg/L) | References |
| Effluent of hydroponic cultures | >300 | | Park et al. (2009) |
| Manufacturing of cellophane, pectin, fertilizers | >1000 | | Ghafari et al. (2008) |
| Manufacturing of explosives | 3500-3700 | 60-80 | Shen et al. (2009) |
| Nuclear industries | 500-50000 | | Glass and Silverstein (1999) |

In fresh, raw domestic wastewaters, most of the N input (60%) has a physiological origin and is in the form of organic molecules bearing amino ($R-NH_2$) groups, typically proteins (which are often insoluble macromolecules that remain in suspension), soluble free amino acids resulting from protein hydrolysis, and urea, which is a neutral molecule highly soluble in water and a major component of urine in both humans and animals (Gerardi 2006, von Sperling 2007, Wiessman et al. 2007). All the organic forms of N are subsequently transformed into ammonia during biodegradation by microorganisms occurring in the wastewater by the process called ammonification (Daims and Wagner 2010) through which proteins are converted to shorter peptides and single amino acids by extracellular proteolytic enzymes, and amino acids are then deaminated (Bitton 2011). In particular, urea is fastly decomposed into bicarbonate and ammonia by hydrolysis primarily mediated by bacterial ureases, being seldom detectable in wastewater (Gerardi 2006, von Sperling 2007). Consequently, ammonia is the most abundant form of dissolved inorganic N in domestic wastewater, while very low concentrations (0-2 mg/L) of either nitrate or nitrite occur (von Sperling 2007, Wiessman et al. 2007). Free ammonia ($NH_3$) or ammonium ion ($NH_4^+$) are in dynamic equilibrium in water and interconvert depending on both temperature and pH (von Sperling 2007). Free ammonia becomes more abundant with lower temperature and pH (Emerson et al. 1975). At temperatures ranging 15-25°C, over 95% ammonia occurs as $NH_4^+$ at pH < 8, while virtually all ammonia is present as $NH_3$ at pH > 11 (von Sperling 2007). Since domestic and municipal wastewaters are usually in the 7-8 pH range, ammonia is mainly present in the ionized form (von Sperling 2007). For an ammonium concentration of 30 mg N/L and at pH= 7-8 at 20°C, the concentration of free ammonia in a WWT line ranges 0.14-1.38 mg $NH_3$-N/L (Liu et al. 2019). Typically, the influents of municipal WWT plants carry concentrations of 15-25 mg organic N/L and 20-35 mg $NH_4^+$/L (von Sperling 2007).

In industrial wastewaters, ammonia concentrations are often 1-2 orders of magnitude higher than in domestic wastewaters, and nitrate may be the prevalent form of N in wastewaters generated from some specific activities, such as hydroponic cultures, nuclear industries, and manufacturing of cellophane, pectin, fertilizers, or explosives. Nonetheless, some industrial waters like those derived from papermaking, wineries, or olive oil mills may carry high loads of organic matter (measured as chemical oxygen demand or COD) but very low concentrations of N (Table 8.1).

### 8.1.3 The Need for Nutrient Removal in Wastewater Treatment

Although sewage collection systems for the disposal of wastewater were already introduced by ancient civilizations, such as the Babylonians, Assyrians, and Romans, WWT did not become a general practice until the beginning of the twentieth century (Seviour 2010). The conventional activated sludge (CAS) process was patented by Arden and Lockett in 1914 (Arden and Locket 1914) and was subsequently developed as a combination of biological processes and physical separation methods. The biological treatment relays on the ability of microbes to either assimilate or oxidize the organic matter in influent wastewater when a source of aeration is provided. Subsequently, clarification of the effluent is achieved by decantation of the suspended biomass, the so-called 'activated sludge' consisting of tridimensional cell aggregates or 'flocs', which is then partly reused to further inoculate fresh untreated wastewater (Seviour 2010). The excess waste sludge must be additionally treated as solid waste, more often through anaerobic digestion (Bitton 2011). A flow diagram showing the basics of a CAS treatment is depicted in Figure 8.2. Besides the suspended growth approach, early WWT methods, such as trickling filters, relied on microorganisms growing in biofilms (Cooper 2001).

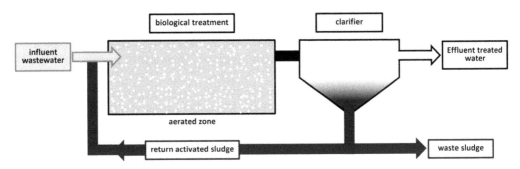

**FIGURE 8.2** Flow diagram of the conventional activated sludge processes. Adapted from Seviour et al. (1999).

Reducing organic matter and suspended solids were for many years the main goals of WWT (Seviour 2010). More recently, additional objectives became of importance in order to improve the quality of the treated water, such as elimination of toxic metals, odors, pathogens, and N and P compounds, leading to the design of many modifications of the CAS process and the development of new, more advanced, and cost-effective biological treatment technologies (Ferrera and Sánchez 2016). The removal of N and P from wastewater is of particular importance for several reasons, related to the deleterious effects of these nutrients in the aquatic environments which receive discharges from WWT plants, and the toxicity for humans of some N compounds (Seviour et al. 1999, Wiessman et al. 2007, Bitton 2011):

- Nitrate and phosphates are naturally present at low concentrations in aquatic environments and limit the growth of photoautotrophic microorganisms, such as algae and cyanobacteria. Uncontrolled discharges of these N and P compounds promote their overgrowth, leading to eutrophication of water bodies and damaging the aquatic ecosystem. In addition to this, contact with and intake of eutrophied water causes a variety of medical conditions, including reaction to toxins produced by cyanobacteria.
- Ammonia, particularly the free $NH_3$ form, is very toxic to fishes even at low concentrations. Nitrite is toxic for both eukaryotes and bacteria.
- Nitrate and nitrite may reach drinking waters and pose risks for human health. Intake of nitrate and nitrite is directly linked to methemoglobinemia in babies. Also, upon ingestion, these compounds undergo chemical reactions with secondary amines present in the diet and generate nitrosamines regarded as carcinogens.

Due to these adverse effects, N and P concentration limits for WWT plant effluents are currently regulated in most developed countries in order to protect the environment and public health. In the European Union, the requisites are particularly restrictive when the effluents are discharged into the so-called sensitive areas, i.e., those water bodies found to be eutrophic or which in the near future may become eutrophic if protective action is not taken (Council Directive 91/271/EEC 1991). Advanced technologies designed to improve N and P removal from wastewater are termed 'biological nutrient removal (BNR) processes' (Wiessman et al. 2007). Technologies designed for the enhanced biological P removal (EBPR) fall outside of the scope of the present chapter and will not be discussed here, but the readers can refer to previously published reviews on the topic (Oehmen et al. 2007, Nielsen et al. 2012, 2019, Esfahani et al. 2019).

## 8.2 BIOLOGICAL PERSPECTIVE OF THE N-CYCLE

Nitrogen (N) is a major element for all living organisms, is the fourth most abundant in cellular biomass in both eukaryotes and prokaryotes (the N:C ratio incorporated into cells ranges 1:50-1:5 depending on the life form), and is also required for the synthesis of proteins and nucleic acids, two essential macromolecules for life (Canfield et al. 2010). Most N in Earth is in the form of $N_2$, which is the most abundant gas in the atmosphere (79%). However, $N_2$ is a very stable molecule largely inaccessible to most living organisms since it can be used as the N source only by a few prokaryotes (diazotrophic bacteria and archaea), making it a limiting macronutrient in many aquatic and soil environments (Canfield et al. 2010).

Like all the elements which are part of organic matter, N undergoes oxidations and reductions in a cyclic process, which allows its circulation between the atmospheric (inert $N_2$ gas) and terrestrial compartments (reactive N forms) (Stein and Klotz 2016). N-cycle is schematically depicted in Figure 8.3. To put it simply, atmospheric $N_2$ is reduced to ammonia throughout the cycle, which is incorporated into the cellular organic macromolecules; after this, organic matter is degraded generating ammonia again, which is re-oxidized to $N_2$ through different oxidation states ($NO_x$: $NO_3^-$, $NO_2^-$, NO, and $N_2O$; nitrate, nitrite, nitric oxide, and nitrous oxide, respectively).

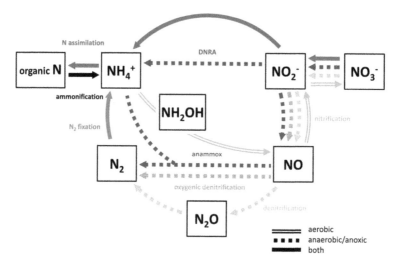

**FIGURE 8.3** Main processes of the N-cycle. DNRA: Dissimilatory nitrate/nitrite reduction to ammonium.

A large part of the transformations in N-cycle are biologically driven, and microorganisms are key players of the interchange of N forms in the biosphere. Microbial enzymes catalyze 14 different redox reactions involving eight key inorganic N species of different oxidation states (Kuypers et al. 2018). Both eukaryotic and prokaryotic microorganisms degrade organic N compounds releasing ammonia (ammonification). Conversely, the reactive inorganic forms $NH_4^+$, $NO_3^-$ and $NO_2^-$ are assimilated by many microorganisms, which use them for the biosynthesis of the organic N compounds that are part of their cellular components. $NH_4^+$ and $NO_2^-$ are also the sources of energy for chemolithotrophic microorganisms, which can oxidize them to $NO_3^-$ mostly using $O_2$ as an electron acceptor (the process is called 'nitrification') or convert them to $N_2$ under strictly anaerobic conditions (the process is called 'anaerobic ammonia oxidation' or 'anammox'). Nitrogen oxides $NO_3^-$, $NO_2^-$, NO, and $N_2O$ ($NO_x$) can be sequentially reduced to $N_2$ following a dissimilatory pathway by acting as final electron acceptors of respiration, more often under anaerobic conditions (the process is called 'denitrification') (Bitton 2011). Otherwise, $NO_3^-$ and $NO_2^-$ can undergo dissimilatory reduction to ammonia (DNRA) (Canfield et al. 2010). Finally, an alternative denitrifying pathway termed 'oxygenic denitrification' or 'intra-aerobic denitrification' involves the dismutation of NO into $N_2$ and $O_2$ to enable the oxidation of methane under anoxic conditions by some bacteria (Ettwig et al. 2010, Chen and Strouss 2013). In practice, $N_2$-fixation and DNRA drive N retention in the ecosystems, while denitrification and anammox promote N loss (Welsh et al. 2014). A detailed overview of the N-cycle can be found in Chapter 1 in this book; throughout the present chapter, the following sections will focus on the opportunities offered by

N-cycle biological processes for the design of engineered systems achieving efficient and cost-effective N removal from wastewaters.

## 8.3 DIVERSITY OF N-CYCLE MICROORGANISMS IN WASTEWATER TREATMENT PLANTS

Successful biological WWT relies on the metabolic abilities of the microbial consortia inhabiting activated sludge, which allow them to degrade organic matter in order to increase their cellular biomass as well as remove N and P compounds and transform or immobilize organic toxic molecules (Ferrera and Sánchez 2016). The microbial communities of activated sludge are complex, composed of both eukaryotic and prokaryotic microorganisms (metazoa, protozoa, algae, fungi, bacteria, and archaea) and viruses (Seviour and Nielsen 2010). Bacteria, archaea, and protozoa are addressed as the major contributors to the removal of organic matter (Madoni 2011), while the roles of fungi have been only narrowly explored to date (Maza-Márquez et al. 2016). Bacteria are particularly relevant for their involvement in N removal since they are the most abundant microorganisms in activated sludge [$10^{12}$-$10^{13}$ cells/g volatile suspended solids in the sludge (Seviour and Nielsen 2010)], and several steps of the N-cycle rely on enzymatic systems which, according to current knowledge, are exclusive of prokaryotes (Canfield et al. 2010, Bock and Wagner 2013).

### 8.3.1 Assimilation

Assimilation of organic and inorganic N compounds contributes to N removal in WWT plants: for every 100 atoms of carbon assimilated, cells need *ca.* 10 atoms of nitrogen (C/N ratio = 10) (Bitton 2011). Ammonia, typically in its ionic form ($NH_4^+$), is the key compound of the pathways of N assimilation by many cellular organisms, either eukaryotic (plants, algae, fungi) or prokaryotic (bacteria and archaea) since it is directly incorporated into carbon skeletons (Moreno-Vivián and Flores 2007, Daims and Wagner 2010). Many microorganisms, including fungi, algae, and bacteria, can also uptake $NO_3^-$ and/or $NO_2^-$ and perform its assimilatory reduction to $NH_4^+$ (Moreno-Vivián and Flores 2007). However, in the most common scenario, the concentrations of N compounds carried in wastewaters largely exceed the demands of the microbial communities of activated sludge, so only a small portion of the total nitrogen input is removed by assimilation (Seviour et al. 1999).

### 8.3.2 Nitrification

Nitrification is the biological oxidation of ammonia to nitrite and subsequently to nitrate (Bock and Wagner 2013). As stated in Section 8.1.2, ammonia is the major source of N in domestic and municipal wastewaters, hence nitrification is a key step to achieve biological N removal in WWT plants.

Nitrification is mediated by several groups of prokaryotic microorganisms. In nature, this process is mainly driven by chemolithotrophic bacteria and archaea, which are able to fix $CO_2$ and obtain energy by oxidizing ammonia to nitrate in a two-step process (Bitton 2011, Lam and Kuypers 2011):

$$NH_3 + \tfrac{3}{2}\, O_2 \rightarrow NO_2^- + H_2O + H^+ (\Delta G^\circ = -278 \text{ kJ/mol})$$
$$NO_2^- + \tfrac{1}{2}\, O_2 \rightarrow NO_3^- (\Delta G^\circ = -82 \text{ kJ/mol})$$
$$(\Delta G^\circ \text{ values calculated for } 25°C \text{ and at pH} = 7)$$

Until recently, it was believed that either step of chemolithotrophic nitrification was independent and done by different groups of prokaryotic microorganisms, namely ammonia-oxidizing bacteria (AOB) and archaea (AOA) [or ammonia-oxidizing prokaryotes (AOP)] and nitrite-oxidizing bacteria (NOB). However, in 2015, bacteria that were able to perform both steps of nitrification were discovered, and the biological process was accordingly termed 'comammox' (complete ammonia oxidation) (van Kessel et al. 2015, Daims et al. 2016). AOP, NOB, and comammox are difficult to grow and isolate under laboratory conditions, and cultivation-independent molecular methods are crucial in investigating the incidence of these organisms, estimating their diversity, and anticipating their physiological traits.

### 8.3.2.1 Biochemistry of Ammonia Oxidation by Prokaryotes

Ammonia oxidation to nitrite (nitritation) is the first and rate-limiting step of nitrification (Daims and Wagner 2010). In AOB, this step comprises two reactions: 1) oxidation of ammonia to hydroxylamine by the membrane-bound, copper-containing enzyme complex ammonia monooxygenase (AMO) and 2) oxidation of hydroxylamine to nitrite by the periplasmic multi-heme enzyme hydroxylamine oxidoreductase (HAO) (Stahl and de la Torre 2012, Bock and Wagner 2013). Recently, various evidence arose indicating that the product of hydroxylamine oxidation by HAO is nitric oxide rather than nitrite, which is subsequently oxidized to $NO_2^-$ by a yet unidentified enzyme (Caranto and Lancaster 2017, Stein 2019). Figure 8.4 depicts the proposed respiratory pathway for ammonia oxidation in the AOB type species *Nitrosomonas europaea*.

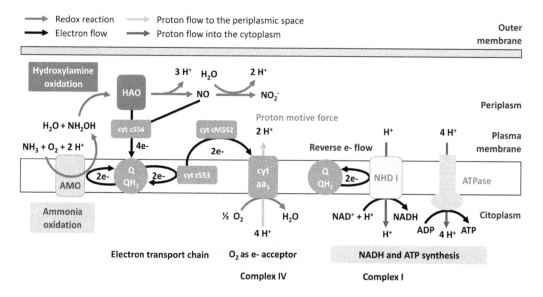

**FIGURE 8.4** Proposed respiratory pathways for ammonia oxidation in *Nitrosomonas europaea*, adapted from Stahl and de la Torre (2012) and Caranto and Lancaster (2017). Ammonia is oxidized to hydroxylamine by the membrane-bound ammonia monooxygenase (AMO) complex, then hydroxylamine is oxidized to NO in the periplasm by hydroxylamine oxidoreductase (HAO). NO is subsequently oxidized to $NO_2^-$, by a mechanism not yet characterized. Four electrons from these oxidations are transferred to the quinone pool by cytochrome c554. Two electrons from the reduced quinone pool return to AMO and are required to initiate ammonia oxidation. The remaining two electrons enter the electron transport chain via cytochrome c553 and cytochrome cM552 to generate the proton motive force necessary for NADH and ATP synthesis.

$O_2$ is required for the nitritation reaction and is also the preferred terminal acceptor of the energy-generating electron transport chain; however, under $O_2$ limitation some species of AOB commit the available $O_2$ for the oxygenation of ammonia to hydroxylamine and use nitrite as an alternative electron acceptor in a process called 'nitrifier denitrification', which generates additional nitric oxide (Wunderlin et al. 2012). Nitrifier denitrification also takes place at lower rates under fully aerobic conditions in most AOB, probably acting as a sink of electrons to speed up the oxidation of hydroxylamine (Stein 2011). The biochemistry and genetics of nitrifier denitrification have not been completely elucidated yet, although *nirK* and *norB* genes (encoding nitrite and nitric oxide reductases, respectively, as it will be described in Section 8.3.3.1) have been identified in many AOB isolates (Stein 2011). AOB more often gain energy by chemolithotrophic oxidation and use $CO_2$ as carbon source through the Calvin-Benson-Bassham cyclic pathway (Koch et al. 2019), but some strains have been described as metabolically versatile and are able to grow mixotrophically on organic compounds, such as pyruvate, amino acids, or fructose, or denitrify using $H_2$ and organic compounds as electron donors (Daims and Wagner 2010).

The mechanism of ammonia oxidation by AOA has not been fully characterized yet, but it is recognized as evolutionarily divergent from that of AOB. Several genes encoding the proteins of the canonical ammonia oxidation pathway in AOB are absent in AOA, most remarkably the HAO enzyme (Stahl and de la Torre 2012). Since no heme-based enzymes are encoded in the known AOA genomes, it is postulated that the oxidation of hydroxylamine should be carried out by copper-based enzymes (Stein 2019). AOA are also autotrophic organisms that fix $CO_2$ through a highly efficient modification of the 3-hydroxypropionate/4-hydroxybutyrate pathway (Könneke et al. 2014), which enables a good adaptation of AOA to nutrient-limited conditions (Alfreider et al. 2018). Unlike AOB, all AOA investigated to date are regarded as physiologically incapable of nitrifier denitrification (Stein 2019), although a study by Jung et al. (2014b) provided contradictory results since $^{15}N$-$NH_4^+$-labeling experiments indicated that AOA strains native from soil generate $N_2O$ by nitrifier denitrification. In addition to this, homologs of *nirK* were found widespread in AOA in environmental samples (Bartossek et al. 2010).

The AMO enzyme complex is conserved through evolution, though its encoding genes are phylogenetically distinct in the AOB, AOA, and comammox clades (Lawson and Lücker 2018). The *amoA* gene encodes the α-subunit of ammonia monooxygenase and is used as the universal phylogenetical and functional molecular marker for all the ammonia-oxidizing organisms (Purkhold et al. 2000, Daims and Wagner 2010, Pester et al. 2012).

### 8.3.2.2  *Diversity of Ammonia-Oxidizing Prokaryotes in Wastewater Treatment Systems*

Excluding comammox bacteria, which will be later described in Section 8.3.2.4, all canonical AOB characterized to date are evolutionarily related to the class *Proteobacteria*, currently comprising 3 genera with 20 species (Table 8.2). Species affiliated to the genus *Nitrosococcus* (*Gammaproteobacteria*) are obligately halophilic (Koops and Pommerening-Röser 2001) and mainly circumscribed to marine and salt-lake waters, while most AOB species form a monophyletic group within the genera *Nitrosomonas* and *Nitrosospira* of the family *Nitrosomonadaceae* (*Betaproteobacteria*). Betaproteobacterial AOB are ubiquitous in terrestrial, freshwater, estuarine, and marine environments and can be grouped in separate clusters according to their phylogeny but also in relation to their ecophysiology and habitats (Soliman and Eldyasti 2018, Sedlacek et al. 2019).

The genus *Nitrosomonas* is subdivided into seven lineages (Clusters 5, 6a, 6b, 7, 8, *N. cryotolerans*, and Cluster 9/Nm143) of which Cluster 5 has no cultivated representatives (Prosser et al. 2014). The clusters comprising the twelve validly or effectively published species names are listed in Table 8.2; however, it must be taken into consideration that eight of these species are at the risk of not being considered validly published any longer since their type strains are currently deposited only in one culture collection, contravening the Rules of the Bacteriological Code (Judicial Commission of the International Committee on Systematics of Prokaryotes, 2008). Species of Cluster 6b (*N. aestuarii/N. marina*), Cluster 9, and *N. cryotolerans* require high salt concentrations and are mostly found in saline or brackish waters (Norton 2011, Sedlacek et al. 2019), while members of Clusters 6a (*N. oligotropha/N. ureae*), 7 (*N. europaea/N. mobilis*) and 8 (*N. communis/N. nitrosa*) are usually found in freshwaters, including wastewaters and their treatment systems (Koops and Pommerening-Röser 2001, Daims and Wagner 2010).

**TABLE 8.2** Known genera and species of ammonia oxidizing bacteria (AOB)

| AOB Species | Characteristics and Main Habitats | References |
|---|---|---|
| **Class *Betaproteobacteria* - Order *Nitrosomonadales* – Family *Nitrosomonadaceae*** | | |
| **Genus *Nitrosomonas*** | | |
| **Cluster 6a**<br>*N. oligotropha*[†]<br>*N. ureae*[†] | Ammonia-sensitive species. Primarily found in oligotrophic freshwaters, also found in wastewater treatment plants and in soils. | Koops and Pommerening - Röser (2005a), Norton (2011) |
| **Cluster 6b**<br>*N. aestuarii*[†]<br>*N. marina*[†] | Require high salt concentration. Mainly found in marine/estuarine waters. | Koops and Pommerening - Röser (2005a), Norton (2011) |
| **Cluster 7**<br><br>*N. europaea*<br>*N. eutropha*[†]<br>*N. halophila*[†]<br>*N. mobilis*** | Tolerate high ammonia concentrations. Widely distributed.<br>Found in eutrophic fresh waters, wastewater treatment plants, soils.<br>Moderately halophilic. Found in brakish waters, soda lakes, also in wastewater treatment plants. | Koops and Pommerening - Röser (2005a), Norton (2011), Almstrand et al. (2014) |
| **Cluster 8**<br>*N. communis*[†]<br>*N. nitrosa*[†] | <br>Mostly inhabits neutral agricultural soils.<br>Tolerate high ammonium levels. Found in wastewater treatment plants, freshwaters (ponds, rivers), soils. | Koops and Pommerening - Röser (2005a), Norton (2011) |
| **Other *Nitrosomonas* species:**<br>*N. cryotolerans**<br><br>*N. stercoris* | <br><br>Deep-branching group, psychrophilic and halophilic. Mostly found in cold, marine waters.<br>Closest lineage is cluster 7. Tolerate higher ammonium levels than any other known species. Isolated from composted cattle manure. | <br><br>Koops and Pommerening - Röser (2005a), Norton (2011)<br>Nakagawa et al. (2019), Oren and Garrity (2017) |
| **Genus *Nitrosospira*** | | |
| <br><br>**Cluster 0**<br><br>*N. lacus* | Often the most abundant AOB in soils, also found in fresh and marine waters, and industrial or municipal wastewater treatment plants.<br>Mostly abundant in sediments, paddy fields, farm soils.<br>Psychrotolerant, grows within a wide pH range. Isolated from a sandy lake sediment. | Siripong and Rittmann (2007), Norton (2011)<br><br>Urakawa et al. (2015) |
| **Cluster 3**<br>*N. briensis*<br>*N. multiformis*<br>*N. tenuis* | Most commonly recovered from terrestrial environments. | Validation list No. 54. (1995), Koops and Pommerening - Röser (2005b), Norton (2011) |
| **Class *Gammaproteobacteria* - Order *Chromatiales* - Family *Chromatiaceae*** | | |
| **Genus *Nitrosococcus*** | | |
| *N. halophilus* | Found in salt lagoons and salt lakes. | Koops and Pommerening - Röser (2005c) |
| *N. nitrosus* | Type culture no longer available. | Koops and Pommerening - Röser (2005c) |
| *N. oceani* | Tolerate or require high salt concentrations. Detected in sea waters globally. | Koops and Pommerening - Röser (2005c), Norton (2011) |
| *N. wardiae** | Isolated from a eutrophic marine sediment. | Wang et al. (2016) |
| *N. watsonii** | Isolated from sea water. | Campbell et al. (2011) |

* not validly published taxonomical names; ** named *Nitrosococcus mobilis*, proposed to be reclassified as *Nitrosomonas mobilis* (Koops and Pommerening-Röser 2005a) [†] deposited in only one culture collection (Judicial Commission of the International Committee on Systematics of Prokaryotes, 2008).

Species of *Nitrosospira* are assembled in five Clusters (0, 1, 2, 3, and 4), and the four species validly published belong to Clusters 0 and 3 (Table 8.2). Sequences of Cluster 0 have been found in soils and freshwater environments, Cluster 1 sequences predominantly belong to ocean waters and sediments, and members of Clusters 2, 3, and 4 have been identified in marine and freshwaters and seem widespread in terrestrial environments (Norton 2011). *Nitrosospira* spp. are most common in soils where they often dominate the ammonia-oxidizing community and are only occasionally found in WWT plants, often in much lower abundance than *Nitrosomonas* (Daims and Wagner 2010). Specific operating conditions, such as long solids retention time (SRT) in combination with low temperature, have been occasionally described to favor the dominance of *Nitrosospira* over *Nitrosomonas* (Siripong and Rittmann 2007, Wells et al. 2009, Yu et al. 2010).

**TABLE 8.3** Salt and ammonia tolerance limits and substrate affinity for ammonia of the different species of *Nitrosomonas* and *Nitrosospira*

| | Limit of NaCl Tolerance (mM) | Limit of NH$_4^+$ Tolerance (mM)* | *K*s for NH$_3$ Range (µM) | References |
|---|---|---|---|---|
| *Nitrosomonas* species | | | | |
| **Cluster 6a** | | | | Koops and Pommerening-Röser (2001, 2005a) |
| *N. oligotropha* | 200 | 50 | 1.9-4.2 | |
| *N. ureae* | 300 | 100 | | |
| **Cluster 6b** | | | | |
| *N. aestuarii* | 700 | 300 | 50-52 | |
| *N. marina* | 800 | 200 | | |
| **Cluster 7** | | | | |
| *N. europaea* | 500 | 400 | | |
| *N. eutropha* | 500 | 600 | 30-61 | |
| *N. halophila* | 1000 | 400 | | |
| *N. mobilis* | 600 | 300 | | |
| **Cluster 8** | | | | |
| *N. communis* | 300 | 200 | 14-46 | |
| *N. nitrosa* | 200 | 100 | | |
| **Other species** | | | | |
| *N. cryotolerans* | 600 | 400 | 42-59 | |
| *N. stercoris* | 400 | 1000 | not available | Nakagawa et al. (2019) |
| *Nitrosospira* species | | | | |
| *N. briensis* | 250 | 200 | | Jian and Bakken (1999), Taylor and Bottomley (2006), Soliman and Eldyasti (2018) |
| *N. multiformis* | 200 | 50 | 0.8-11** | |
| *N. tenuis* | 100 | 100 | | |

\* at pH=7.8
\*\* calculated for isolates not classified at the species level.

Members of the different Clusters of *Nitrosomonas* distribute unevenly in WWT systems, and the activated sludge AOB community may be composed of several populations or dominated by a single *Nitrosomonas* species, depending on the influent wastewater nature, the specific type of technology, and the operational conditions (Daims and Wagner 2010, Soliman and Eldyasti 2018). Activated sludges in facilities treating domestic wastewater usually harbor more diverse AOB assemblages than those receiving mixed influents from both domestic and industrial origin (Wang et al. 2010).

The variables mainly driving the diversity and distribution of *Nitrosomonas* spp. in the environment include ammonia and nitrite concentrations, dissolved oxygen (DO) concentration, pH, temperature, and salinity (Limpiyakorn et al. 2005, Lydmark et al. 2007, Norton 2011, Soliman and Eldyasti 2018). Among the aforementioned factors, salinity and $NH_4^+$ load had been best characterized for their influence on the community structure and population dynamics of nitrifiers in WWT systems. All known species of *Nitrosomonas* classified within Cluster 7 have on average lower substrate affinity than those belonging to Cluster 6a, according to the values of their half-saturation constants (*K*s) (Table 8.3). The same applies to $O_2$ affinity, according to investigations of the population dynamics of AOB through the different layers of a nitrifying biofilm (Gieseke et al. 2001). Members of Cluster 7 have been addressed as r-strategist AOB since they display a moderately higher growth rate and predominate in eutrophic habitats with high N input and turnover, while Cluster 6a strains are considered K-strategists AOB that are able to compete better under oligotrophic conditions where ammonium concentrations are limiting for the growth of the r-strategists AOB (Bollmann et al. 2002). In addition to this, Cluster 7 members are halotolerant and display less sensitivity to high concentrations of ammonia and nitrite than those in Cluster 6a (Koops and Pommerening-Röser 2001, Limpiyakorn et al. 2007, Sedlacek et al. 2019).

The aforementioned features explain the often-reported prevalence of phylotypes related to the *N. europaea*/*N. mobilis* cluster in activated sludge samples, particularly in those systems receiving higher ammonium loads (Otawa et al. 2006, Limpiyakorn et al. 2007, Wells et al. 2009, Daims and Wagner 2010, Cerrone et al. 2013, Wang et al. 2014) or salinity (Limpiyakorn et al. 2005). However, the co-occurrence of *Nitrosomonas* from Cluster 6a, 7 and/or 8 has been widely observed in both laboratory- and full-scale nitrifying bioreactors (Limpiyakorn et al. 2005, Gómez-Villalba et al. 2006, Lydmark et al. 2007, Siripong and Rittmann 2007, Wang et al. 2010), implying that ecological and physiological characteristics of *Nitrosomonas*, such as substrate/oxygen affinity and tolerance to salt and nitrite, very likely differ at the strain level (Wang et al. 2010). Recently, the draft-genome sequences of ten cultivated strains of *Nitrosomonas* spp. belonging to Clusters 6a and 7 were compared in search of conserved and differentiating features among the two groups (Sedlacek et al. 2019). All Cluster 6a genomes lacked genes controlling NO detoxification, which were universal in the Cluster 7 genomes, in accordance with the reported higher tolerance of the latter to nitrite toxicity. Several Cluster 6a strains harbored genes encoding putative hydrogenases and/or ureases, which seem to be strain-specific features that may confer competitive advantages in oligotrophic environments. The comparative analyses also revealed a number of cluster-level conserved differences regarding the strategies to confront the shifts in the concentrations of $CO_2$ and $O_2$, corroborating that a variety of traits is involved in determining niche differentiation and population succession of *Nitrosomonas* spp. in nature and engineered systems.

The initial discovery of aerobic archaea as potential ammonia-oxidizers and the isolation and description of the first cultivated AOA species *Nitrosopumilus maritimus* were published 15 years ago (Treusch et al. 2005, Könneke et al 2005) and led to a complete reexamination of the global N-cycle. Since then, many studies based on molecular methods have assessed the contribution of AOA to the ammonia-oxidizing communities in natural ecosystems, such as soils, geothermal habitats, marine and fresh waters, often concluding that AOA were numerically dominant over AOB. At present, AOA are generally regarded to exert primary control over ammonia oxidation in many habitats (Stahl and de la Torre 2012). In spite of their recent discovery, a significant number of AOA were isolated or enriched in recent years, leading to the description and proposal of several genera and species (Table 8.4) and unveiling a seemingly wider phylogenetical diversity than that of AOB. All known AOA are classified in the Phylum *Thaumarchaeota* within four orders of the Class *Nitrososphaeria* (*Candidatus* Nitrosocaldales, *Nitrosopumilales*, *Candidatus* Nitrosotaleales, and *Nitrososphaerales* [Qin et al. 2017a, 2017b]). Only representatives of the genera *Nitrosarchaeum*, *Nitrosopumilus*, and *Nitrososphaera* have been isolated to date, while most of the ten currently proposed AOA genera are enrichment cultures.

**TABLE 8.4**  Known genera and species of ammonia oxidizing archaea (AOA)

| AOA Species | Characteristics and Main Habitats | References |
|---|---|---|
| **Class *Nitrososphaeria*** | | |
| **1 - Order *Ca.*Nitrosocaldales (group ThAOA)** | | |
| **Family *Ca.* Nitrosocaldaceae** | Thermophilic (> 60°C) and neutrophilic. Obligately aerobic, chemolithoautotrophic growth (ammonia oxidation using $CO_2$ as C source). Globally distributed in various geothermal environments. | Qin et al. (2017b) |
| **Genus *Ca.*Nitrosocaldus** | | |
| *Ca.* N. cavascurensis | Enriched from a hot spring in Ischia, Italy. | Abby et al. (2018) |
| *Ca.* N. islandicus | Enriched from biofilms of an Icelandic hot spring. | Daebeler et al. (2018) |
| *Ca.* N. yellowstonensis | Enriched from the sediment of a terrestrial hot spring (Yellowstone National Park, USA). | De la Torre et al. (2008) |
| **2 - Order *Nitrosopumilales* (group I.1a)** | | |
| **Family*Nitrosopumilaceae*** | Obligatorily aerobic autotrophic ammonia-oxidizing archaea. Neutrophilic and mesophilic. Found in marine waster, freshwater, and soil. Also occurring in wastewater treatment plants. | Qin et al. (2016, 2017b) |
| **Genus *Ca.* Cenarchaeum** | | |
| *Ca.* C. symbiosum | Uncultivated symbiont of marine sponges. | Preston et al. (1996) |
| **Genus *Nitrosarchaeum*** | | |
| N. koreense | Isolated from agricultural soil. | Jung et al. (2018) |
| *Ca.* N. limnae* | Enrichment culture from low-salinity sediments in San Francisco Bay (USA). | Blainey et al. (2011) |
| **Genus *Ca.* Nitrosomarinus** | | |
| *Ca.* N. catalina | Isolated from a water column, near-shore California. | Ahlgren et al. (2017) |
| **Genus *Ca.* Nitrosopelagicus** | | |
| *Ca.* N. brevis | Enriched form open ocean waters. | Santoro et al. (2015) |
| **Genus *Nitrosopumilus*** | Mostly isolated in marine waters, require high salt concentrations to grow (>10%). Some species are photosensitive. | |
| N. adriaticus | Isolated from surface water, Mediterranean Sea. | Bayer et al. (2019) |
| N. cobalaminigenes | Isolated from sea water in Washington, USA. | Qin et al. (2017b) |
| *Ca.* N. koreense | Isolated from marine sediments in the Artic circle. | Park et al. (2014), Qin et al. (2017) |
| N. maritimus | First cultivated AOA, isolated from a tropical marine fish tank at the Seattle Aquarium, Washington, USA. | Könneke et al. (2005), Qin et al. (2017b) |
| N. oxyclinae | Isolated from sea water in Washington, USA. Psychrophilic. | Qin et al. (2017b) |
| N. piranensis | Isolated from surface water, Mediterranean Sea. | Bayer et al. (2016, 2019) |
| *Ca.* N. salarius | Isolated from marine sediments (S. Francisco Bay, USA). | Qin et al. (2017b) |
| *Ca.* N. sediminis | Isolated from marine sediments in the Artic circle. | Park et al. (2014) |
| N. ureiphilus | Isolated form a near-shore surface sediment near Seattle, Washington, USA. | Qin et al. (2017b) |

*(Contd.)*

**TABLE 8.4** Contd.

| | | |
|---|---|---|
| **Family *Ca.* Nitrosotenuaceae** | Aerobic chemolithoautotrophs, gain energy from ammonia oxidation. Widely distributed in soils, freshwater, hot springs, the subsurface, and activated sludge of wastewater treatment plants. Intolerant to high salinity (>0.3%). | Herbold et al. (2016) |
| **Genus *Ca.* Nitrosotenuis** | | |
| *Ca.* N. aquarius | Enriched from a freshwater aquarium biofilter. | Sauder et al. (2018) |
| *Ca.* N. chungbukensis | Isolated from a deep oligotrophic agricultural soil horizon. | Jung et al. (2014a) |
| *Ca.* N. cloacae | Enriched from activated sludge of a municipal wastewater treatment plant in Tianjin, China. | Li et al. (2016) |
| *Ca.* N. uzonensis | Enriched from a thermal spring in Kamchatka. | Lebedeva et al. (2013) |
| **3 - Order *Nitrosospherales* (group I.1b)** | | |
| **Family *Nitrososphaeraceae*** | Obligate aerobes, mesophilic/moderately thermophilic, chemolithoautotrophic (ammonia oxidation and $CO_2$ fixation). Found predominantly in soil and freshwater environments, also associated with sediments, plants, and the human skin. Very often found in wastewater treatment plants. | Limpiyakorn et al. (2013), Kerou et al. (2018) |
| **Genus *Nitrososphaera*** | | |
| *Ca.* N. evergladensis | Enriched from agricultural soil. | Zhalnina et al. (2014) |
| *N. garguensis* | Enriched from a hot-spring under moderately thermophilic conditions (46°C). | Hatzenpichler et al. (2008) |
| *N. viennensis* | Isolated from garden soil in Vienna (Austria). | Stieglmeier et al. (2014) |
| **Genus *Ca.*Nitrosocosmicus** | | |
| *Ca.* N. arcticus | Enriched and isolated from a mineral artic soil (Svalbard, Norway). | Alves et al. (2019) |
| *Ca.* N. hydrocola** | Isolated from a municipal wastewater treatment plant. | Sauder et al. (2017) |
| *Ca.* N. franklandianus* | Isolated from a neutral, arable soil. | Lehtovirta-Morley et al. (2016) |
| *Ca.* N. oleophilus | Isolated from a coal tar-contaminated sediment. | Jung et al. (2016) |
| **4 - Order *Ca.*Nitrosotaleales** | | |
| **Family *Ca.* Nitrosotaleaceae (group I.1a - associated)** | | |
| **Genus *Ca.* Nitrosotalea** | | |
| *Ca.* N. devaniterrae* | Isolated from an acidic agricultural soil. Acidophilic (pH < 5.5). Also detected in wastewater treatment plants. | Lehtovirta-Morley et al. (2011), Merbt et al. (2015) |

\* original species names are displayed as corrected by Oren (2017). \*\* originally proposed as *Ca.* N. exaquare and corrected later by the authors.

Since ammonia is usually the major N compound in municipal and industrial wastewaters and biological treatment technologies under aerated conditions are widespread, it was predicted that AOA, alike AOB, should fulfill an important role in N removal in engineered systems. AOA were first detected in WWT systems by Park et al. (2006) and thereafter many studies surveyed their occurrence in WWT plants based on different types of technologies. Limpiyakorn et al. (2013) generated the first comprehensive phylogeny of AOA in WWT systems from a set of 208 consensus sequences of archaeal *amoA* genes detected in activated sludge samples available in public databases, concluding that a large fraction of sequences (71%) fell within Group 1.1b *Thaumarchaeota*, most closely related to species of the genus *Nitrososphaera* (Table 8.4). The prevalence of *Nitrososphaera* among the AOA populations identified in WWT systems has been corroborated in more recent work (i.e., Pan et al. 2018a, 2018b, Srithep et al. 2018).

Many of the available studies were focused on comparing the abundance of AOA in relation with AOB in WWT systems by molecular methods, such as quantitative PCR (qPCR), reporting either the complete absence/negligible occurrence of AOA, the coexistence of AOA and AOB in similar abundances, or even AOA outcompeting AOB (Calderón et al. 2013, Pan et al. 2018a). AOA seem to dominate the ammonia-oxidizing community only under certain operating conditions as a consequence of their phylogenetical and physiological differences with AOB (Calderón et al. 2013). AOA abundance is favored by low DO concentrations, operation under long hydraulic and solids retention times, and particularly the available concentration of ammonia, which is regarded as the most influential factor driving the abundance of copies of the archaeal *amoA* gene (Calderón et al. 2013). Ammonia acts as the energy substrate for both AOA and AOB, but AOB have much higher values of the kinetic parameter $Ks$ [2-3 orders of magnitude (Lawson and Lücker 2018)], thus displaying less affinity for ammonia than AOA, which are favored under concentrations of the substrate limiting for the growth of AOB but inhibited at higher ammonia concentrations (Stahl and de la Torre 2012). Overall, AOB prevail in treatment systems directly receiving high inputs of inorganic ammonia, whereas systems sustained by the ammonification of organic material select AOA (Stahl and de la Torre 2012).

The numerical dominance of a particular group of AOP does not imply its higher contribution to ammonia-oxidizing activity. AmoA-related monooxygenases of both bacterial and archaeal populations have been described to be involved in the biodegradation of organic substrates (Stahl and de la Torre 2012, Kumwimba and Meng 2019), hence the sole occurrence of *amoA* genes or even their transcripts in the environment is not always indicative of active nitrification. Specifically, the contribution of AOA to N removal in WWT systems has been a matter of controversy among researchers. Mußmann et al. (2011) used fluorescence *in situ* hybridization (FISH) to investigate the incidence of AOB and AOA in 52 municipal and industrial WWT plants across Europe, concluding that AOA were more often absent and were only detected in four industrial facilities, outnumbering AOB (*ca.* 10000-fold) only in a refinery WWT plant. Following the application of a nitrification mathematical model to estimate ammonia-oxidizing biomass in relation to ammonia removal and testing $^{14}CO_2$ fixation in the presence of ammonia through FISH-microautoradiography (FISH-MAR), the authors concluded that the size of the AOA community in the refinery WWT plant could not possibly be sustained by a chemolithoautotrophic lifestyle. The authors postulated that AOA should be living chemoheterotrophically despite harboring and transcribing copies of *amoA* genes, although their putative organic carbon and/or energy source could not be identified. In contrast, more recent studies demonstrated an effective contribution of AOA to ammonia oxidation in full-scale nitrifying biofilms (Roy et al. 2017) and activated sludge-based laboratory-scale systems (Srithep et al. 2018) by measuring ammonia-oxidizing activity in the presence of specific chemical inhibitors of the AOB enzymes.

### 8.3.2.3 Biochemistry of Nitrite Oxidation by Bacteria

The second step of nitrification is nitrite oxidation to nitrate (nitratation). The key enzyme for this process is nitrite oxidoreductase (NXR), which belongs to the type II dimethyl sulfoxide reductase-like family of molybdopterin-binding enzymes (Daims et al. 2016). NXR oxidizes nitrite to nitrate taking oxygen atoms from $H_2O$ and transferring two electrons into the electron transport chain with $O_2$ as terminal electron acceptor (Daims and Wagner 2010):

$$NO_2^- + H_2O \rightarrow NO_3^- + 2H^+ + 2e^-$$
$$2H^+ + 2e^- + \tfrac{1}{2}O_2 \rightarrow H_2O$$

The enzyme is postulated to be associated with the cytoplasmic membrane and consists of three subunits: NxrA, NxrB, and NxrC. The NxrA subunit, which binds to the substrate, is placed either in the periplasmic space or in the cytoplasm, depending on the genus (Figure 8.5). NxrA is periplasmic in *Nitrospina* and *Nitrospira* and putatively in *Candidatus* Nitrotoga and *Ca.* Nitromaritima, while it is located in the cytoplasm in *Nitrobacter*, *Nitrococcus*, and *Nitrolancea* (Daims et al. 2016, Kitzinger et al. 2018). This different positioning significantly influences the optimization of the nitrite oxidation reaction since $H^+$ derived from $NO_2^-$ oxidation contributes to the proton motive force only when NxrA is periplasmic, while the cytoplasmic location of NxrA requires $NO_2^-$ and $NO_3^-$ transport through the plasma membrane. Since nitrite oxidation provides a low energy yield, these differences make the nitrite oxidation pathway better optimized from the energetic point of view in the NOB that bind to their substrate in the periplasmic space (Daims et al. 2016).

As in the case of AOB, NOB use $O_2$ as the preferred electron acceptor and assimilate inorganic C, although they fix $CO_2$ following two different pathways: the Calvin-Benson-Basshan cycle in *Nitrobacter*, *Nitrococcus*, *Nitrotoga*, and *Nitrolancea* and the reductive TCA cycle in *Nitrospina*, *Nitrospira*, and (putatively) *Ca.* Nitromaritima (Ngugi et al. 2016, Alfreider et al. 2018). However, NOB are currently addressed as metabolically versatile and being very often mixotrophs. $O_2$ is not required as a source of oxygen atoms for the oxidation of nitrite to nitrate, and strains of several species are described as oxygen-sensitive and/or microaerophilic. The ability to grow following a full heterotrophic lifestyle, reduce nitrate to nitrite, or aerobically oxidize sulfide has been described in strains of different NOB genera (Daims et al. 2016, Füssel et al. 2017). On the basis of both cultivation-dependent and metagenomic evidence, growth on $H_2$ and formate are predicted to be widespread features in NOB, which may use these additional energy sources concomitantly with aerobic nitrite oxidation, enabling them to colonize a broad range of habitats and withstand environmental changes.

### 8.3.2.4 Diversity of Nitrite-Oxidizing Bacteria in Wastewater Treatment Systems

In contrast to ammonia oxidation, nitrite oxidation is an evolutionarily widespread metabolism in Bacteria, and the seven genera of NOB identified to date fall within a large range of phylogenetic groups, encompassing four different Phyla: *Nitrobacter*, *Ca.* Nitrotoga, and *Nitrococcus*, which belong to the classes *Alpha-*, *Beta-*, and *Gamma-* of the *Proteobacteria*, respectively; *Nitrospina* and *Ca.* Nitromaritima (Phylum *Nitrospinae*), *Nitrospira* (Phylum *Nitrospirae*), and *Nitrolancea* (Phylum *Chloroflexi*), the latter being the only Gram-positive representative (Table 8.5). Among these genera, *Nitrospira* was revealed as the most phylogenetically and functionally diverse (Daims 2014). The genus is currently divided into six sublineages (I-VI), which are widespread in either natural habitats (terrestrial environments, fresh and marine waters, geothermal springs) or engineered systems, such as drinking water facilities and WWT plants (Daims 2014, Daims et al. 2016). In spite of their huge diversity, strains of only seven species could be successfully isolated (Table 8.5) since *Nitrospira* are particularly difficult to grow under laboratory conditions and very

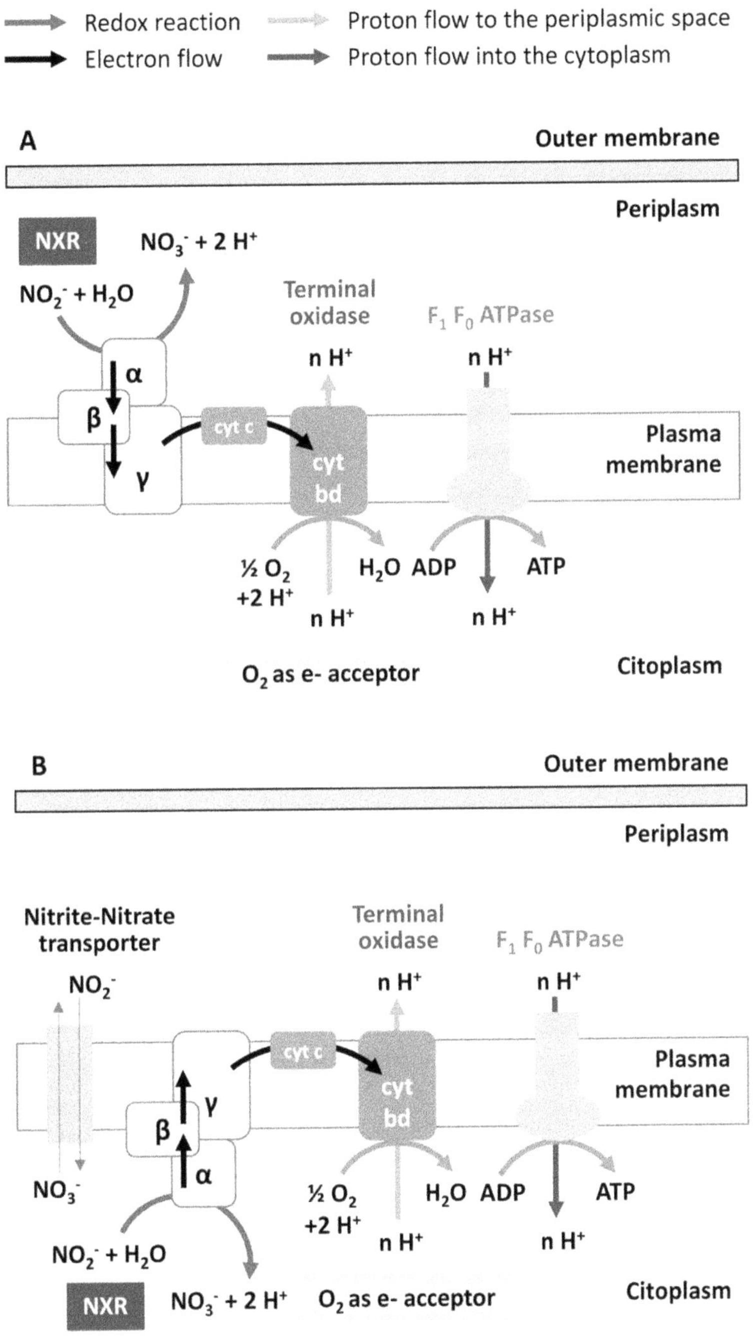

**FIGURE 8.5** Proposed pathways for nitrite oxidation in bacteria: nitrite oxidoreductase (NXR) types and assumed electron flow, adapted from Daims et al. (2016). **A.** Periplasmic location of the nitrite binding subunit NxrA. **B.** Cytoplasmic location of the nitrite binding subunit NxrA.

few can develop colonies on solid media (Daims 2014). The existence of archaea that can oxidize nitrite has been long hypothesized and is sustained by recent metagenomic data which evidence a complex phylogeny of the genes encoding the known nitrite oxidoreductases (Kitzinger et al. 2018), but to date, this capacity has not been demonstrated in any archaeal organism.

**TABLE 8.5** Known species of nitrite oxidizing bacteria (NOB) and comammox bacteria

| NOB Species | Characteristics and Main Habitats | References |
|---|---|---|
| **Phylum *Proteobacteria*** | | |
| **Class *Alphaproteobacteria* - Order *Rhizobiales*** | | |
| **Genus *Nitrobacter*** | Grows lithoautotrophically, all species except *N. alkalicus* also grow chemoorganotrophically. $NO_2^-$ is the preferred energy source and $CO_2$ is the main source of C when grown aerobically. Some species are $O_2$ sensitive. Some strains use $NO_3^-$ as terminal e⁻ acceptor. | Spieck and Bock (2005a) |
| *N. alkalicus* | First isolated from sediments of soda lakes and soda soil in Siberia and Kenya. | Sorokin et al. (1998), Validation list No. 71 (1999) |
| *N. hamburguensis* | Isolated from soils in Germany, Mexico and France. | Bock et al. (1983), Validation list No. 78 (2001) |
| *N. vulgaris* | Isolated from soils, groundwater, fresh and brackish water, sewage sludge. | Bock et al. (1990), Validation list No. 78 (2001) |
| *N. winogradsky* | Isolated from soil. | Skerman et al. (1980) |
| **Class *Betaproteobacteria* - Order *Nitrosomonadales*** | | |
| **Genus *Candidatus* Nitrotoga** | First isolated from permafrost, often identified in cold habitats by molecular methods. Also frequent in sewage treatment plants. | Alawi et al. (2007), Ishii et al. (2017) |
| *Ca.* N. arctica | Isolated from permafrost (tundra polygonal soil in the Siberian Artic). | Alawi et al. (2007) |
| *Ca.* N. fabula | Isolate from activated sludge, optimum temperature higher than previous *Ca.* Nitrotoga enrichments/isolates (24-28°C). Harbors genes of the complete oxidation pathways of $H_2$ and sulfite. | Kitzinger et al. (2018) |
| **Class *Gammaproteobacteria* - Order *Chromatiales*** | | |
| **Genus *Nitrococcus*** | Exclusively detected in marine habitats, requires NaCl. Able to reduce nitrate and oxidize sulfide. | Skerman et al. (1980), Spieck and Bock (2005b), Füssel et al. (2017) |
| *N. mobilis* | Isolated from equatorial Pacific seawater. | |
| **Phylum *Nitrospirae* - Order *Nitrospirales*** | | |
| **Genus *Nitrospira*** | Widespread in nature and engineered systems. Some strains use $H_2$ and formate as substrates, with $O_2$ or $NO_3^-$ as terminal e⁻ acceptors. Microaerophilic. | Daims (2014) |
| *Ca.* N. bockiana | Isolated from internal corrosion deposits from a steel pipeline of the Moscow heating system. | Lebedeva et al. (2008) |
| *N. calida** | Isolated from a microbial mat of the terrestrial geothermal spring Gorjachinsk, Russia. | Lebedeva et al. (2011) |
| *N. defluvii** | Isolated from a municipal wastewater treatment plant in Hamburg, Germany. | Nowka et al. (2015b) |
| *N. japonica** | Enriched and isolated from activated sludge of a municipal wastewater treatment plant in Tokyo, Japan. | Ushiki et al. (2013) |

*(Contd.)*

**TABLE 8.5** Contd.

| | | |
|---|---|---|
| *N. lenta** | Isolated from a municipal wastewater treatment plant in Hamburg, Germany. | Nowka et al. (2015b) |
| *N. marina* | Isolated from ocean water. | Euzeby and Kudo (2001) |
| *N. moscoviensis*[†] | Isolated from an enrichment culture initiated with a sample from a partially corroded area of an iron pipe of a heating system in Moscow. | Ehrich et al. (1995), Validation list No. 78 (2001) |
| *Ca.* N. salsa | Enriched from Dutch coastal North Sea water. | Haaijer et al. (2013) |
| **Phylum *Nitrospinae* - Order *Nitrospinales*** | | Lücker et al. (2013) |
| **Genus *Nitrospina*** | Most frequent NOB in marine habitats. | Lücker et al. (2013) |
| *N. gracilis* | Isolated from surface waters at the South Atlantic ocean. | Skerman et al. (1980) |
| *N. watsonii** | Isolated at 100 m depth in the Black Sea. | Spieck et al. (2014) |
| **Genus *Ca.* Nitromaritima** | Identified by single-cell genomics in the brine-seawater interface layer of the Red Sea. | Ngugi et al. (2016) |
| **Phylum *Chloroflexi* - Order *Sphaerobacterales*** | | |
| **Genus *Nitrolancea*** | | Sorokin et al. (2014) |
| *N. hollandica* | Isolated from a nitrifying bioreactor with high ammonia bicarbonate loading as sole substrate. | Sorokin et al. (2014) |
| **Commamox *Nitrospira* species** | | |
| *Ca.* N. inopinata | Enriched from a microbial biofilm on the walls of a hot water pipe (56°C, pH 7.5), raised from an oil exploration well in Russia. | Daims et al. (2015) |
| *Ca.* N. nitrificans *Ca.* N. nitrosa | Enriched from the anaerobic compartment of a trickling filter connected to a recirculation aquaculture system. | Van Kessel et al. (2015) |

* not validly published taxonomical names
† deposited in only one culture collection (Judicial Commission of the International Committee on Systematics of Prokaryotes, 2008).

The distribution of NOB in nature and engineered systems is uneven (reviewed by Daims et al. 2016, Figure 8.6). *Nitrospina*, *Nitrospinae* clade 2, *Ca.* Nitromaritima, and *Nitrospira* sublineage IV are restricted to marine habitats, while *Nitrospira* sublineages III and VI belong to subsurface/soil and geothermal environments, respectively. *Nitrobacter*, *Nitrococcus*, *Nitrospira* sublineages I, II, and V, and *Nitrotoga* are ubiquitous in freshwater and soils and together with *Nitrolancea* occur in WWT systems, which have been very often selected as models to investigate NOB diversity and biology (Daims et al. 2016).

*Nitrobacter* were long-time believed to be the major nitrite-oxidizing organisms in WWT plants since they are easily isolated from activated sludge samples (Daims and Wagner 2010); however, as soon as cultivation-independent methods became extensive tools in microbial ecology, it was revealed that their contribution to the NOB community is more often minor and that the key role in this N-cycle process regularly relies on members of the *Nitrospira* sublineages I and II (Daims et al. 2016). In this regard, *Nitrospira* was recognized as a core taxon of the activated sludge community in a survey conducted in 269 WWT plants from all over the world (Wu et al. 2019b). After the recent discovery of *Nitrotoga*, it has been established that representatives of this genus frequently coexist with *Nitrospira* in activated sludge (Daims et al. 2016) or even dominate the NOB community under particular conditions, such as low temperature (Lücker et al. 2015, Kitzinger et al. 2018).

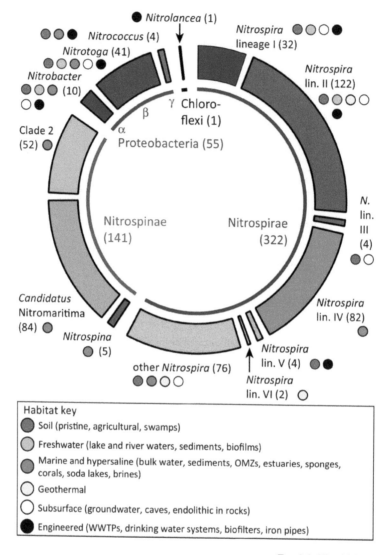

**Trends in Microbiology**

**FIGURE 8.6** Phylogenetic affiliation, species-level diversity, and habitats of nitrite-oxidizing bacteria (NOB). The known nitrite oxidizers belong to the genera *Nitrobacter, Nitrotoga, Nitrococcus, Nitrospira, Nitrospina, Nitrolancea,* and *Candidatus* Nitromaritima. Nitrospinae 'Clade 2' is a monophyletic group without a defined taxonomic status. The inner ring indicates which bacterial phyla contain these genera. To assess the species-level diversity within each genus, the 16S rRNA sequences of known NOB and closely related bacteria were retrieved from the SILVA Ref NR 99 database (release 123, July 2015) and clustered in operational taxonomic units (OTUs) using a sequence identity threshold of 98.7%. The obtained OTU numbers are indicated in parentheses for each NOB group or bacterial phylum as an estimate of the NOB species numbers represented in the sequence database. Wedges in the outer ring are drawn proportionally to the OUT numbers. The six *Nitrospira* lineages are shown separately, whereas 'other *Nitrospira*' does not refer to one coherent lineage but instead comprises all *Nitrospira* sequences outside the already established lineages of this genus. Colored small circles indicate the known major habitats of each NOB group. Most of the collected 16S rRNA sequences stem from uncultured organisms. Reprinted from: Trends in Microbiology vol. 24, Daims H, Lücker S, Wagner M "A New Perspective on Microbes Formerly Known as Nitrite-Oxidizing Bacteria", pp. 699-712 (2016), with permission from Elsevier.

Analogously as described above for AOB, a K/r strategy has been hypothesized for NOB, since the *K*s values for nitrite affinity are on average about one order of magnitude lower for *Nitrospira* and *Ca.* Nitrotoga than for *Nitrobacter* and *Nitrolancea* (Lawson and Lücker 2018). On such a basis, *Nitrospira* and *Ca.* Nitrotoga are regarded as K-strategists, thriving under concentrations of nitrite which are limiting for the r-strategists *Nitrobacter* and *Nitrolancea*. Indeed, it has been experimentally shown that *Nitrobacter* grows faster and outcompete *Nitrospira* only in the presence of high nitrite concentrations (Daims 2014). At the molecular level, the different subcellular localization of NXR (see Section 8.3.2.3) is very likely related to those distinct strategies of adaptation (Daims 2014). However, wide ranges of $K$s values have been measured at the strain level in either *Nitrospira* and *Nitrobacter* (Nowka et al. 2015a, Lawson and Lücker 2018), consistently with the ample phylogenetic diversity of both genera.

As previously stated in Section 8.3.2, in 2015 it was discovered that species of the genus *Nitrospira* were able to perform complete ammonia oxidation. To date, all known commamox *Nitrospira* are classified within sublineage II, phylogenetically interspersed with non-ammonia-oxidizing *Nitrospira* species, and have been identified in a wide array of habitats, including engineered systems, such as drinking water treatment and distribution facilities, biological filters for aquaculture, and conventional WWT plants, where they co-occur with AOB, AOA, and canonical NOB within the nitrifying community. Comammox may become the prevailing nitrifiers under certain operating conditions, particularly low availability of ammonia since these organisms have an extremely high affinity for this substrate, displaying values of $K$s up to 2,500-fold lower than those described for any AOB or the freshwater AOA (Lawson and Lücker 2018). Conversely, they display a lower affinity for nitrite than the canonical NOB *Nitrospira* species and cannot use nitrite as a sole N source since assimilatory nitrite reductases are absent in their genomes (Koch et al. 2019). Complete nitrifiers carry homologs of all the key genes required for ammonia oxidation in AOB, including those encoding HAO, and share the versatile metabolic capacities of the non-ammonia-oxidizing *Nitrospira* (Lawson and Lücker 2018).

Consortia of nitrifying prokaryotes in WWT systems comprise spatially close microcolonies of AOP and NOB co-aggregating in the activated sludge flocs or biofilms (Daims et al. 2016). AOP and NOB were classically regarded as mutualistic symbionts since NOB benefit from AOP in order to access their energy substrate, and NOB aid AOP by detoxifying nitrite (Arp and Bottomley 2006); however, recent evidence point toward a more complex relationship among both metabolic groups, involving the facilitation of iron uptake by AOB through the synthesis of siderophores by NOB, reciprocal feeding with urea or cyanate, and signaling through NO-generation and quorum-sensing mediated by *N*-acyl-homoserine lactones (reviewed by Daims et al. 2016).

### 8.3.2.5 *Other Processes Potentially Contributing to Biological Oxidation of Ammonia and Nitrite in WWT Plants*

#### 8.3.2.5.1 *Heterotrophic Nitrification*

Nitrification by autotrophic chemolithotrophs is regarded as the predominant mechanism of biological ammonia and nitrite oxidation in nature, but nitrification can also be carried out by either bacteria or eukaryotes which use organic compounds instead of $CO_2$ as C source. These organisms are termed 'heterotrophic nitrifiers', falling within a broad definition which includes any biological process through which a reduced form of N (either organic or inorganic) is converted into a more oxidized form (Stein 2011, Zhang et al. 2015). Heterotrophic nitrifiers are able to oxidize ammonia, hydroxylamine, and/or nitrite, including both fungi (i.e., *Aspergillus* and *Absidia*) and species of >20 genera of Gram-negative and Gram-positive bacteria classified within the Phyla *Actinobacteria* (*Arthrobacter, Microbacterium, Rhodococcus*), *Bacteroidetes* (*Chryseobacterium*), *Firmicutes* (*Bacillus*), and *Proteobacteria* (*Achromobacter, Acinetobacter, Aeromonas, Agrobacterium, Alcaligenes, Burkholderia, Comamonas, Diaphorobacter, Enterobacter, Klebsiella, Luteibacter, Marinobacter, Methylococcus, Moraxella, Paracoccus, Providencia, Pseudomonas,* and *Vibrio*) (Wehrfritz et al. 1997, Stein 2011, Zheng et al. 2012, Zhang et al. 2013, Fitzgerald et al. 2015,

Li et al. 2015, Padhi et al. 2017, Holmes et al. 2019). Such phylogenetical diversity makes the development of universal cultivation-based or molecular approaches to investigate the diversity, activity, and contribution to the N-cycle of this metabolic group difficult (Zhang et al. 2018).

More often, oxidation of ammonia through heterotrophic nitrification neither generates energy nor contributes to cell growth and proceeds at a significantly slower rate than autotrophic nitrification (de Boer and Kowalchuk 2001, Stein 2011, Holmes et al. 2019). It has been classically assumed that this process does not make a substantial input to N-turnover either in nature or engineered systems (Bitton 2011). However, populations of heterotrophic nitrifiers may be large in a variety of environments, hence having the potential to make a relevant contribution to the global N-cycle (Holmes et al. 2019), particularly under environmental conditions which inhibit the activity of the autotrophic nitrifiers (Stein 2011, Zhang et al. 2015).

The pathways of heterotrophic ammonia oxidation remain largely understudied and have not been fully clarified, although there is evidence of the involvement of homologs of the AMO and HAO enzyme systems of chemolithotrophic bacteria (de Boer and Kowalchuk 2001, Stein 2011, Padhi et al. 2017, Liu et al. 2018). The possible involvement of other bacterial enzyme systems was recently reviewed elsewhere (Holmes et al. 2019). In some fungi, the oxidation of ammonia or other N compounds is believed to occur through a non-enzymatic mechanism involving reactions with hydroxyl radicals generated to degrade lignin (de Boer and Kowalchuk 2001). Heterotrophic oxidation of $NO_2^-$ to $NO_3^-$ at slow rates has been also described in several genera of Bacteria and Fungi, and it is believed to be mainly mediated by catalase enzymes through a fortuitous mechanism (Sakai et al. 1996, Stein 2011).

*8.3.2.5.2 Phototrophic Nitrite Oxidation*

Besides chemolithotrophic NOB, some purple non-sulfur bacteria of the genera *Rhodopseudomonas* and *Thiocapsa* (*Proteobacteria*) can oxidize nitrite to nitrate anaerobically in the presence of light by using them as electron sources for anoxygenic photosynthesis (Griffin et al. 2007). Strains of both genera frequently occur in aquatic habitats, including activated sludge, but their contribution to nitrite removal remains nearly unexplored (Schott et al. 2010).

### 8.3.3 Denitrification

Denitrification is a biologically driven multi-step process involving the sequential dissimilatory reduction of nitrate to molecular nitrogen via nitrite, nitric oxide, and nitrous oxide intermediates ($NO_3^- \rightarrow NO_2^- \rightarrow NO \rightarrow N_2O \rightarrow N_2$). Denitrification reactions are anaerobic respirations, which occur in the presence of $NO_x$ when the $O_2$ concentration is limiting, provided the availability of a suitable electron donor (Bedmar et al. 2013). The four consecutive reactions are catalyzed by the enzymes 'nitrate reductase', 'nitrite reductase', 'nitric oxide reductase', and 'nitrous oxide reductase', respectively (Lam and Kuypers 2011, Kuypers et al. 2018):

$$2NO_3^- + \tfrac{1}{2}C_2H_3O_2^- \rightarrow 2NO_2^- + HCO_3^- + \tfrac{1}{2}H^+ \ (\Delta G^\circ = -244 \ \text{kJ/mol})$$
$$4NO_2^- + \tfrac{1}{2}C_2H_3O_2^- + \tfrac{7}{2}H^+ \rightarrow 4NO + HCO_3^- + 2H_2O \ (\Delta G^\circ = -371 \ \text{kJ/mol})$$
$$4NO + \tfrac{1}{2}C_2H_3O_2^- \rightarrow 2N_2O + HCO_3^- + \tfrac{1}{2}H^+ \ (\Delta G^\circ = -530 \ \text{kJ/mol})$$
$$2N_2O + \tfrac{1}{2}C_2H_3O_2^- \rightarrow 2N_2 + HCO_3^- + \tfrac{1}{2}H^+ \ (\Delta G^\circ = -600 \ \text{kJ/mol})$$

($\Delta G^\circ$ values per mol organic C, calculated with acetate as $e^-$ donor, at 25°C and pH = 7)

As far as it is known, nearly all denitrifiers are also capable of aerobic respiration and are generally regarded as facultative anaerobes (Shapleigh 2014, Torres et al. 2016). In most denitrifiers, the expression of the genes encoding the $NO_x$ reductases is repressed (both transcriptional and post-transcriptionally) by $O_2$, which is the preferred terminal electron acceptor because much more energy is conserved and fewer enzyme complexes are needed for aerobic respiration compared to denitrification (Chen and Strouss 2013). An exception is aerobic denitrifiers, which will be specifically addressed in section 8.3.3.3.

The ability to denitrify is widespread in the evolution and organisms that can perform one or several of the reactions of the denitrification pathway belong to the three domains of life: Eukarya, Archaea, and Bacteria (Coyne 2018); however, the involved enzymes and the genes encoding them have been mostly investigated among Bacteria. More often, denitrifying bacteria use organic compounds as a source of both energy and carbon (heterotrophic denitrification), although some inorganic molecules including $H_2$, sulfur-reduced forms (i.e., pyrite, $S^0$, sulfite, thiocyanate, thiosulfate) and ferrous iron can be used as electron donors by a limited number of genera of autotrophic bacteria (chemolithoautotrophic denitrification) (Lu et al. 2014, di Capua et al. 2019, Tian and Yu 2020).

Denitrification occurs in all anoxic environments in Earth in which nitrate is present (Lam and Kuypers 2011). Bacteria that can complete the denitrification pathway by fully converting $NO_3^-$ to $N_2$ are scarce since many do not carry (or do not express) the genes for all the required reductases and partial denitrification chains including only a single enzyme or discontinuous chains with two or more enzymes have been found (Shapleigh 2014). Partial denitrification may lead to the accumulation of $N_2O$, which is a very stable and powerful green-house gas displaying nearly 300-fold the warming potential of $CO_2$ over 100 years, being a major contributor to the destruction of the ozone layer (IPCC 2013). A significant fraction of the soil bacterial community (0.5-5%) displays the ability to denitrify, and denitrification is recognized as the primary pathway of N loss and generation of $N_2O$ emissions from soil and waters contaminated with nitrates, often in connection with the abuse of fertilizers (Bedmar et al. 2013). The environmental impacts of denitrification will not be discussed in detail here but are specifically addressed in Chapter 4 in this book.

### *8.3.3.1 Biochemistry of Denitrification*

All the known reductases involved in the denitrification pathways in Bacteria are metalloenzymes using a redox-active metal cofactor: molybdenum for $NO_3^-$ reduction, iron and/or copper for $NO_2^-$ and NO reduction, and copper for $N_2O$ reduction (Tavares et al. 2006). The structure, characteristics and catalytic mechanisms of these enzymes, and their encoding genes and their regulation have been thoroughly reviewed by several authors (Philippot 2002, Shiro 2012, Shapleigh 2014, Coelho and Romao 2015, Torres et al. 2016, Carreira et al. 2017, Horrell et al. 2017, Rinaldo et al. 2017) and will be only briefly addressed here.

The canonical denitrifying respiratory chain in the model denitrifier bacteria *Paracoccus denitrificans* is depicted in Figure 8.7. As previously stated, denitrifiers more often are also capable of aerobic respiration, and both pathways share the core respiratory machinery comprising NADH dehydrogenase (complex I), the quinone pool, the bc1 complex (complex III), and cytochrome c (Chen and Strouss 2013). The different electron transport processes linked to denitrification in prokaryotes have been reviewed in detail elsewhere (Simon and Klotz 2013). Denitrification has been proposed as the ancestor of aerobic respiration, on the basis of the evolutionary relationships among the two pathways (Saraste and Castresana 1994).

The first step of denitrification in Bacteria is catalyzed by the respiratory nitrate reductases, which belong to two types: the membrane-bound Nar complex or the periplasmic Nap (Philippot 2002, Bedmar et al. 2013, Kuypers et al. 2018), both members of the dimethyl sulfoxide reductase family of Mo-containing enzymes (Coelho et al. 2015). Many bacteria carry both Nar and Nap systems (Kuypers et al. 2018), although the periplasmic Nap is found almost exclusively in the Phylum *Proteobacteria* (Simon and Klotz 2013, Bedmar et al. 2013). While Nar reduces nitrate in the cytoplasm and releases protons into the periplasm, Nap does not contribute to the proton motive force (Kuypers et al. 2018).

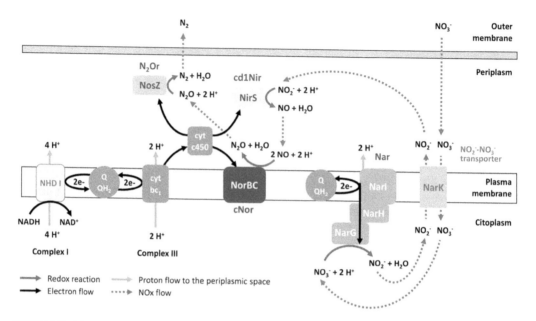

**FIGURE 8.7** Schematic representation of the denitrifying respiratory chain in *Paracoccus denitrificans*, adapted from Chen and Strous (2013), Simon and Klotz (2012) and Korth and Harnisch (2019). The electrons are transferred from NADH to NOx-reductases through NADH dehydrogenase (complex I), the quinone pool and cytochromes bc1 and c. Protons are pumped from the cytoplasm by NADH dehydrogenase, cytochrome bc1 and the cytoplasmic nitrate reductase (NarGHI). Nar: nitrate reductase; Nir: nitrite-reductase; Nor: nitric oxid reductase; Nos: nitrous oxide reductase.

The genes encoding the catalytic subunits of Nar and Nap (*narG* and *napA*, respectively) are frequently used as molecular markers of nitrate reducers (Tavares et al. 2006); however, it is worth noting that nitrate reduction is not unique to the denitrification pathway, and Nar and Nap enzymes may participate in other metabolic processes, such as DNRA (see Section 8.3.5). Consequently, bacteria performing solely nitrate reduction to nitrite and not any other step of the pathway are more often excluded from the canonical definition of 'denitrifier' (Shapleigh 2014).

Nap is a multipurpose enzymatic system which, besides denitrification or DNRA, acts in many bacteria as an electron sink for excess reducing equivalents generated during aerobic growth on reduced C sources, contributing to maintaining the cellular redox balance (Coelho and Romao 2015, Torres et al. 2016). Unlike the rest of the reductases of the denitrification pathway, Nap expression may be high in some bacteria under microaerophilic or fully aerobic conditions (Torres et al. 2016). In this sense, Nap is a key enzyme for aerobic denitrification (see Section 8.3.3.3).

Dissimilatory $NO_2^-$ reduction to NO can be catalyzed by two different and not evolutionarily related enzymes: the copper nitrite reductase (CuNir), and the heme-containing cytochrome cd1 nitrite reductase (cd1Nir) (Horrell et al. 2017, Rinaldo et al. 2017). CuNir is encoded by the *nirK* gene, while *nirS* encodes the structural monomers of cd1Nir, and both genes are used as biomarkers for the detection of $NO_2^-$ reducing bacteria in the environment (van Spanning et al. 2007, Rinaldo et al. 2017). Nitrite reductases do not contribute directly to energy conservation in bacteria since both types are located in the periplasmic space (Kuypers et al. 2018) and may receive electrons from cytochrome *c* or pseudoazurin via the cytochrome bc1 complex (Bedmar et al. 2013). CuNir and cd1Nir fulfill the same functions *in vivo* (Rinaldo and Cutrozzolà 2007) and do not coexist in a single organism with very few exceptions (Kuypers et al. 2018). In a study conducted with > 200 cultivable species of denitrifiers, it was concluded that *nirK*-encoded Nir were prevalent in *Alphaproteobacteria*, *Bacteroidetes*, and *Firmicutes*, while most denitrifying *Betaproteobacteria* carried *nirS*, and both types were found with similar frequencies in *Gammaproteobacteria* (Heylen et al. 2006).

Nitric oxide reductases (Nor) are a diverse group of iron-containing membrane-bound enzymes catalyzing the reduction of NO to $N_2O$, belonging to the heme-copper oxidase super-family (Shiro 2012). These enzymes combine two NO molecules generating $N_2O$ by the input of two electrons (Heylen and Keltjens 2012). Three different types of Nor have been identified in bacteria, cNor, qNor, and $qCu_A$Nor (de Vries et al. 2007, Bedmar et al. 2013), which receive the electrons from different donors, although the structures of their active sites are very similar (Tavares et al. 2006, Simon and Klotz 2013). The cNor consists of the two structural subunits NorB and NorC, containing heme *b* and heme *c* molecules, respectively (van Spanning et al. 2007), while qNor has a single heme *b* containing subunit (NorB). The *norB* genes encoding cNor were found more frequently in cultivable denitrifiers of the *Alphaproteobacteria*, while *Betaproteobacteria* harbored *norB* genes for either cNor or qNor in equal frequencies (Heylen et al. 2007). The *norB* gene is commonly used as a biomarker for the study of NO reducing populations in the environment (Braker and Tiedje 2003). The $qCu_A$ Nor was the first NO reductase described in denitrifying Gram-positive bacteria (van Spanning et al. 2007). It comprises a CuA-type small subunit and a large subunit with two heme *b* molecules and has been mostly characterized in *Bacillus azotoformans* strains (Heylen and Keltjens 2012).

The final step of the denitrification pathway is the reduction of $N_2O$ to $N_2$ (Bedmar et al. 2013). The only known enzyme that catalyzes this step is nitrous oxide reductase ($N_2Or$ or Nos), an homodimeric copper protein designated NosZ (Carreira et al. 2017). With very few exceptions (Torres et al. 2016), NosZ is a periplasmic enzyme which does not directly generate proton motive force as in the case of Nap and NirK/S (Kuypers et al. 2018), and its electron input comes from cytochrome *c* or cupredoxins (Bedmar et al. 2013). For a long time, it was believed that $N_2Or$ was more sensitive to $O_2$ concentration than other nitrogen-oxide reductases, but recent data evidence the ability of some bacteria to carry out $N_2O$ reduction under high $O_2$ partial pressure (Torres et al. 2016).

The *nosZ* gene encoding the catalytic subunit of $N_2Or$ is used as a molecular marker in environmental studies (Jones et al. 2013). $N_2O$ reducers are divided into two different clades (I and II) according to the NosZ protein phylogeny and the organization of the *nos* gene cluster (Hallin et al. 2018). According to molecular surveys of environmental samples, clade I-*nosZ* organisms are more often affiliated to *Alpha*- and *Betaproteobacteria*, while clade II-*nosZ* prokaryotes usually belong to *Bacteroidetes, Deltaproteobacteria, Gammaproteobacteria, Gemmatimonadetes*, and *Archaea* (Hallin et al. 2018). Both clades may vary in distribution and abundance in different habitats (Jones et al. 2013, 2014).

$N_2Or$ is the last enzyme of the denitrification pathway and fulfills a crucial environmental function (Carreira et al. 2017). About 40% of denitrifying bacteria carrying *nir* genes lack the *nosZ* gene, hence potentially contributing to generate $N_2O$ emissions; this feature is more common within *Actinobacteria* and *Acidobacteria* genomes (Hallin et al. 2018). Clade I-*nosZ* organisms are more likely able to complete the denitrifying pathway, while 51% of clade II-*nosZ* prokaryotes lack *nirK/nirS* genes and perform $N_2O$ reduction for purposes other than energy conservation, thus acting as a sink for $N_2O$ in the environment (Hallin et al. 2018).

Denitrification in Archaea is considerably less characterized than in Bacteria, but genes encoding enzymes involved in the four sequential reduction steps of denitrification have been identified in a number of archaeal genomes, and some species such as *Pyrobaculum aerophilum* and *Haloferax denitrificans* are able to complete the full denitrification pathway (reviewed by Philippot 2002, Cabello et al. 2004, Shapleigh 2014). The ability to denitrify is widespread in fungi as well, including *Ascomycota* and *Basidiomycota* (Mothapo et al. 2015) and follows slightly different pathways than in prokaryotes, particularly in the step involving NO reduction to $N_2O$, which is carried out by reductases of the cytochrome P450 superfamily using NAD(P)H as a direct electron donor (Shoun et al. 2012). Fungal denitrifiers characteristically lack $N_2O$ reduction ability (Coyne 2018) and are regarded as significant sources of this greenhouse gas in soils (Mothapo et al. 2015). Denitrification by fungi is further addressed in Chapter 5 in this book.

### 8.3.3.2 Diversity of Denitrifiers in WWT Systems

As stated in Section 8.3.3, the ability to denitrify is widespread within Bacteria, which have been classically addressed as the main organisms responsible for denitrification in WWT plants (Daims and Wagner 2010). The ability to denitrify (partial of completely) has been largely described in members of the *Proteobacteria* but is a trait also known in species of the *Actinobacteria*, *Bacteroidetes, Chloroflexi, Cyanobacteria, Firmicutes, Spirochaetes,* and *Verrucomicrobia* (reviewed by Shapleigh 2014).

The earlier studies attempting the characterization of the denitrifying communities in activated sludge samples relied on the use of plating and isolation methods, allowing for the identification of genera of well-known, easily cultivable denitrifiers, such as *Achromobacter, Alcaligenes, Bacillus, Hyphomicrobium, Methylobacterium, Paracoccus,* and *Pseudomonas* (Daims and Wagner 2010). The introduction of cultivation-independent molecular methods provided a more realistic picture, considerably widening the knowledge of the diversity of denitrifiers in WWT plants and revealing discrepancies among the genera/species most frequently identified by isolation methods and those actually prevailing under real operating conditions. In this sense, Lu et al. (2014) constructed a phylogeny of the potential prokaryotic denitrifiers in wastewater, on the basis of >1,000 partial 16S rRNA gene sequences available in public databases (Figure 8.8). The most frequently detected denitrifiers (59% of sequences) belonged to the *Proteobacteria* (*Alpha-, Beta-, Gamma-* and *Delta-*classes; 13, 28, 9 and 9%, respectively), followed by *Bacteroidetes* (16%), while 18% of the analyzed sequences were related to Archaea. Genera of *Proteobacteria* with recognized denitrifying species, such as *Hyphomicrobium, Paracoccus, Rhodobacter* (*Alphaproteobacteria*), *Acidovorax, Comamonas, Curvibacter, Dechloromonas, Hydrogenophaga, Rhodocyclus, Rhodoferax, Thauera, Thiobacillus, Zoogloea* (*Betaproteobacteria*), *Luteimonas* (*Gammaproteobacteria*), and *Arcobacter* (*Epsilonproteobacteria*), have been more often found in significant relative abundances in the communities of activated sludge samples characterized by metagenomic approaches based on the amplification of the 16S-rRNA coding genes (Wang et al. 2012, Tian et al. 2015, González-Martínez et al. 2016, Wu et al. 2019b). It is worth noting that the most abundant populations identified in these studies widely varied from plant to plant since the diversity of denitrifiers in engineered systems is subjected to the influence of several factors, mainly the nature of the treated wastewater, the type of treatment technology and configuration, and operating conditions, such as SRT, pH, and DO concentration (reviewed by Lu et al. 2014). In particular, the diversity of denitrifiers is very much influenced by the availability and nature of the C sources in the influent wastewater (Lu et al. 2014; Figure 8.8). Denitrification has a high C demand with a required theoretical stoichiometric C/N ratio of 2.86 $COD/NO_3^-$-N (Kumar et al. 2012), although many authors report the need of much higher ratios in practice depending on the nature and level of biodegradability of the organic matter present in wastewater. On average, 4 g biodegradable COD per g $NO_3^-$ reduced are required (Ni et al. 2017). This implies the requisite for the addition of external C sources to wastewaters with low C/N ratios in order to enable denitrification, being methanol, ethanol, or acetate the more often selected substrates (Lu et al. 2014, Tian et al. 2020). Among these, methanol is most frequently used, since it is easily assimilated by denitrifying bacteria and provides the potential for complete denitrification achieving high nitrate removal efficiencies without nitrite accumulation and is easily available at a low operational cost (Li et al. 2018). The addition of C compounds exerts a selective pressure, which influences the diversity and structure of the denitrifying community (Hallin et al. 2006). Bioreactors fed with methanol develop distinctive communities since the ability to use single C compounds to support growth (methylotrophy) is confined to ten families of the *Proteobacteria* and some *Firmicutes* (Chistoserdova and Lidstrom 2014). Denitrifying communities tend to be more diverse when a larger array of C compounds is available in the influent wastewater (Hallin et al. 2006). The impact of the addition of different C sources on the diversity of denitrifiers has been thoroughly reviewed by Lu et al. (2014).

In order to properly characterize the communities in charge of N removal in WWT systems, it must be taken into consideration that denitrifiers are phylogenetically diverse, and the links among the phylogenies of the genes encoding 16S rRNA and the denitrifying enzymes are only partial (Heylen et al. 2006, 2007). Furthermore, this trait can be both vertically and horizontally transferred, and the incidence of the different genes of the denitrification pathway is often species-specific or even strain-specific (Álvarez et al. 2014). Consequently, the application of molecular methods for the detection and/or quantification of denitrifiers in environmental samples should not solely rely on 16S rRNA-probing or 16S rRNA-gene sequencing and several additional strategies are often implemented:

- The use of molecular markers involved in the synthesis of the enzymes of the denitrification pathway has become common (see Section 8.3.3.1). Denitrifying communities in WWT systems are most frequently screened and/or quantified by targeting the genes encoding Nir (*nirS, nirK*) and Nos (*nosZ*). Studies based on these approaches lead to the conclusion that *nirS*-type $NO_2^-$ reducing bacteria often dominate over those carrying the *nirK* gene in WWT plants (Geets et al. 2007, Kim et al. 2011, Song et al. 2014; Wang et al. 2014, Fan et al. 2017, Liao et al. 2018) and corroborated significant differences of the denitrifying community structure and diversity among different treatment facilities, which were already observed using the 16S rRNA gene-based approach (Zhang et al. 2019a).
- Denitrification genes may be present in genomes but not expressed; hence, transcriptional analyzes are required to fully understand the functionality of the activated sludge denitrifying communities and the balance of the activities among the different types of $NO_x$-reducing enzymes (van Doan et al. 2013, Song et al. 2014, Vieira et al. 2019, Wang et al. 2019).
- As previously stated, some of the $NO_x$-reductases are shared among denitrification and other pathways (i.e., DNRA), and most microorganisms are able to perform partial rather than complete denitrification. The combination of molecular methods with *in situ* physiological characterizations, such as DNA stable-isotope probing (DNA-SIP) or FISH-MAR, becomes necessary for the accurate identification of populations carrying out active denitrification in mixed communities (Daims and Wagner 2010). Such studies have been often applied to WWT systems, revealing that the denitrifying communities in activated sludge may be composed of only a few dominant populations, contrasting with the much wider diversity reported in soils (reviewed by Lu et al. 2014).

Most wastewaters frequently carry a significant load of easily biodegradable organic matter and denitrification in WWT plants is assumed to be mostly conducted by chemorganoheterotrophic prokaryotes. However, as stated previously in Section 8.3.3, denitrification can be also performed by chemolithoautotrophic bacteria that are able to fix $CO_2$ and oxidize a variety of inorganic compounds under anoxic conditions, using $NO_3^-$ or $NO_2^-$ as electron acceptors. Both hydrogen- and sulfur-driven autotrophic denitrification may occur in WWT plants, abilities restricted to a few species of the *Proteobacteria*. The current knowledge on denitrification with $H_2$ as an electron donor (hydrogenotrophic denitrification) is limited, although it has been known for long and is well described in the model bacterial species *Paracoccus denitrificans* (*Alphaproteobacteria*) (Häring and Conrad 1991) and has been proposed for species of the genera *Acidovorax, Acinetobacter, Aeromonas, Hydrogenophaga, Ochrobactrum, Pseudomonas, Rhodocyclus, Shewanella,* and *Thauera* (Karanasios et al. 2010, Mao et al. 2013), many of which have been found to occur in WWT in significant abundances by using molecular methods. Sulfur-driven autotrophic denitrification has been described in 13 species of the *Proteobacteria* (*Paracoccus pantotrophus, Thiobacillus denitrificans, Thioalkalivibrio denitrificans, T. nitroreducens, T. thiocyanodenitrificans, Thiohalomonas nitratireducens, T. denitrificans, Thiohalophilus thiocyanoxidans, Thioalkalispira microaerophila, Thiohalorhabdus denitrificans, Thiomicrospira* sp., *Sulfurimonas denitrificans*

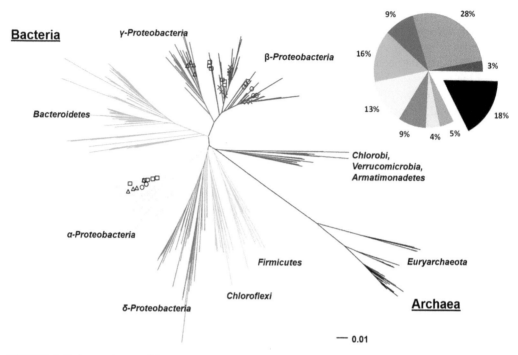

**FIGURE 8.8** Phylogenetic tree of the major phyla of wastewater denitrifying bacteria constructed by the Neighbor-Joining method on the basis of 1003 partial 16S rRNA gene sequences (>500 bp) (Lu et al. 2014, see source for details). Special carbon assimilating populations: □ Methanol; × Acetate; ○ Glycerol; Δ Methane. Reprinted from: Water Research vol. 26, Lu H, Chandran K, Stensel D "Microbial ecology of denitrification in biological wastewater treatment", pp. 237-254 (2014), with permission from Elsevier.

and *S. paralvinellae*) and the currently unclassified species *Thermothrix thiopara;* among these, only *P. pantotrophus* and *Thiobacillus denitrificans* were isolated from WWT plants (Shao et al. 2010). Denitrification by autotrophic bacteria is emerging as a suitable alternative to treat wastewaters with low organic C content or low C/N ratio since it enables completely autotrophic N removal, avoiding the cost of organic C addition (di Capua et al. 2019). The advantages and drawbacks of this novel approach, the most promising inorganic electron donors for denitrification and tested process applications, have been recently reviewed by Tian and Yu (2020).

Furthermore, new biological N removal strategies suitable for wastewaters with a low C/N ratio have been recently proposed, mediated by microorganisms that are able to use methane as an electron donor for denitrification. Methane is generated *in situ* at many WWT facilities through anaerobic digestion of waste sludge (see Section 8.1.3), hence it provides an inexpensive alternative to other C sources commonly used to enhance denitrification (Lu et al. 2014). The aerobic methane oxidation coupled to denitrification (AME-D) process relies on bacteria that oxidize methane through an aerobic pathway and supply electron donors to canonical heterotrophic denitrifiers which achieve the complete conversion of $NO_3^-$ to $N_2$ (Zhu et al. 2016). Nitrite/nitrate-dependent anaerobic methane oxidation (N-DAMO) is accomplished by recently discovered lineages of Archaea and Bacteria (van Kessel et al. 2018). The anaerobic methanotrophic (ANME) archaea *Candidatus* Methanoperedens nitroreducens oxidizes methane via a reverse-methanogenesis pathway while reducing nitrate to ammonia via nitrite (Haroon et al. 2013, van Kessel et al. 2018). Bacteria classified within the Candidate Phylum NC10 complete nitrite-dependent anaerobic methane oxidation through the 'oxygenic denitrification pathway', which will be further addressed in Section 8.3.3.4. Both groups establish syntrophic symbiosis in which NC10 bacteria use the nitrite generated by ANME archaea; such consortia have been enriched in bioreactors inoculated with activated sludge from municipal and industrial WWT plants (van Kessel et al. 2018). The potential of N-DAMO organisms together with anammox bacteria to complete nitrite reduction with methane and ammonia as electron donors has been also explored (van Kessel et al. 2018). For a more detailed examination of the chances offered by the AME-D and N-DAMO processes for the design of more sustainable and environmentally friendly WWT technologies, the readers are directed to recently published reviews by Zhu et al. (2016), Wang et al. (2017), and van Kessel et al. (2018).

### 8.3.3.3 *Aerobic Denitrification*

The ability of microorganisms to denitrify under fully aerobic conditions has been reported in a wide range of bacteria since the end of the XIX century (Robertson and Kuenen 1984), and the first species in which this process was well characterized was *Paracoccus pantotrophus* (formerly named *Thiosphera pantrotropha*) that was able to simultaneously perform aerobic respiration and denitrification (Robertson et al. 1988). To date, aerobic denitrification has been described in isolates of over 30 genera of Gram-negatives (*Alpha-*, *Beta-*, and *Gammaproteobacteria*) and a few strains of Gram-positives of the genera *Bacillus* (*Firmicutes*), *Arthrobacter*, *Gordonia*, *Propionibacterium*, and *Rhodococcus* (*Actinobacteria*), isolated mainly from soils, fresh waters, river biofilms and sediments, and WWT systems (Chen and Strouss 2013, Ji et al. 2015, Lv et al. 2017). In most cases, the aerobic denitrification rates are much lower than anaerobic denitrification by the same strain, highlighting the secondary role of aerobic denitrification versus aerobic respiration (Chen and Strouss 2013).

The enzymology of aerobic denitrification has been primarily investigated in *Proteobacteria* in which this ability has been found connected to the expression and activity of the periplasmic nitrate reductase (Nap) under fully aerobic conditions (Ji et al. 2015). Consequently, the *napA* gene encoding the catalytic unit of Nap is proposed as a molecular signature of aerobic denitrifiers (Huang et al. 2015). As already stated in Section 8.3.3, the expression of the rest of the enzymes of the denitrification pathway is downregulated by $O_2$ concentration in most bacteria, although $N_2O$ is the only enzyme known to be chemically inactivated by $O_2$ (Chen and Strouss 2013). Nonetheless, several denitrifiers are able to express the *nosZ* gene and effectively reduce $N_2O$ to $N_2$ under high $O_2$ partial pressures (Torres et al. 2016).

Aerobic denitrification is understood as a sink for the excess reducing power generated during the oxidation of C compounds in environments where $O_2$ concentrations are low or subjected to sharp gradients. Under such conditions, the generation of NADH becomes faster than the $O_2$ supply if energy substrates are in excess (de Boer and Kowalchuk 2001, Chen and Strouss 2013). In addition to this, facultative anaerobic microorganisms experiencing constant fluctuations of the $O_2$ concentration are unable to regenerate the elements of the corresponding aerobic/anoxic respiratory chains fast enough to adapt to the succession of the available electron acceptors, so they evolved to allow the coexistence of both aerobic respiration and denitrification electron transport chains which are thus simultaneously used (Chen and Strouss 2013). This hypothesis is congruent with the observed enrichment of aerobic denitrifiers in WWT plants working under intermittent aeration conditions (Kong et al. 2006, Wang et al. 2007, Du et al. 2014).

Many heterotrophic nitrifying bacteria (see Section 8.3.2.5.1), particularly those carrying hydroxylamine-oxidizing enzymes, are also aerobic denitrifiers which immediately reduce the $NO_2^-$ they generate to NO, $N_2O$, and/or $N_2$ (Stein 2011). An electron-transfer model, coupling nitrification and denitrification, has been proposed for *Paracoccus pantotrophus* strain LMD82, which provides an explanation for the inability of heterotrophic ammonia oxidation to deliver energy and assumes the role of the whole heterotrophic nitrification-aerobic denitrification (HN-AD) process as a strategy to dissipate reductant (Wehrfritz et al. 1993).

Bacteria that can perform HN-AD have been often isolated from WWT plants, particularly in those systems bearing high organic and/or ammonia loads or high salinity; i.e., coking wastewater, leachate from municipal waste incineration plants, wastes from the chemical/dye industry, slaughterhouse and piggery wastes, or tannery (Khardenavis et al. 2007, Chen and Ni 2011, Zhang et al. 2011, Kundu et al. 2014, Sun et al. 2016, Zhang et al. 2018, Li et al. 2019). Some of the available strains showing HN-AD activities have been tested as inocula for the treatment of real wastewaters, either in batch culture or laboratory-scale reactors (Joo et al. 2006, Li et al. 2015).

Several studies reported effective nitrification in WWT systems where autotrophic nitrifiers were undetectable or present in very low numbers (Cydzik-Kwiatkowska 2015, Fitzgerald et al. 2015, Ma et al. 2015) or in reactors amended with chemical inhibitors of the chemolithotrophic nitrification enzymes (Chai et al. 2019, Pan et al. 2020), suggesting that ammonia oxidation should be mediated by heterotrophic organisms. However, many of the available studies draw conclusions on the identity of the involved organisms solely supported by the detection of genera recognized previously as able to perform HN-AD (often a result of massive parallel sequencing approaches), while far less research has been conducted to demonstrate the actual contribution of such organisms to N removal in engineered systems and elucidate the underlying biological mechanisms. In this respect, evidence of the involvement of *Paracoccus* spp. on ammonia oxidation and denitrification in a sequencing batch reactor (SBR) treating incineration leachate with high free ammonia concentrations was provided by Liu et al. (2018) by the identification and absolute quantification of transcripts of *Paracoccus* spp.-related *amoA* genes. Zhang et al. (2018) observed coexistence of AOB and heterotrophic nitrifiers in the oxic phase of an anoxic/oxic sewage treatment process, estimating autotrophic and heterotrophic ammonia oxidation activities by means of a comprehensive N-transformation activity test, which revealed significant correlations among the changes in the relative abundances of AOB and cultivable heterotrophic nitrifiers and their specific ammonia-oxidizing activities. The diversity of both AOB and heterotrophic nitrifiers significantly shifted depending on the applied COD and ammonia loading rates.

HN-AD provides a promising biotechnological approach to enable the achievement of simultaneous nitrification-denitrification (SND) or concomitant nitrification and denitrification in a single WWT reactor, which will be addressed in Section 8.4.2.

### 8.3.3.4 Oxygenic Denitrification

Oxygenic denitrification was discovered ten years ago after the analysis of the complete genome of the bacterium *Ca.* Methylomirabilis oxyfera (related to Candidate Phylum NC10), enriched in a culture coupling the anaerobic oxidation of methane with the reduction of $NO_2^-$ to $N_2$ (Ettwig et al. 2010). *Ca. M.* oxyfera transcribed and expressed the enzymes of the aerobic methane oxidation pathway but known genes for $N_2$ production were absent in its genome. It was found through isotopic labeling experiments that denitrification by *Ca. M.* oxyfera does not use $N_2O$ as intermediate and instead converts two molecules of NO into $N_2$ and $O_2$, and subsequently consume the $O_2$ intracellularly to oxidize methane. The key step of oxygenic denitrification is the dismutation of NO to generate $N_2$ and $O_2$, which is thought to be mediated in *Ca. M.* oxyfera by a putative NO-dismutase enzyme belonging to the quinol-dependent NO-reductase (qNor) family, although direct biochemical evidence are still missing (Ettwig et al. 2012). Since 2010, another bacterial strain phylogenetically related to the *Gammaproteobacteria* has been identified that is able to perform oxygenic denitrification to oxidize long-chain alkanes, and putative NO dismutase-like sequences have been retrieved from diverse freshwater environments, including WWT systems, pointing toward an ample distribution of this metabolic process (Zhu et al. 2017, 2019). Since oxygenic denitrification does not generate the green-house gas $N_2O$ and can be fueled by the methane generated from anaerobic digestion of sludge, it turns out as an interesting alternative to the methods currently applied for biological N removal from wastewaters. An extensive review of the potential of oxygenic denitrification in the design of novel advanced WWT technologies has been recently published by He et al. (2018).

## 8.3.4    Anaerobic Ammonia Oxidation (Anammox)

Anammox is the coined term for anaerobic ammonia oxidation or the biological oxidation of ammonia to $N_2$ with nitrite as an electron acceptor through the following reaction (Lam and Kuypers 2011):

$$NH_4^+ + NO_2^- \rightarrow N_2 + 2\,H_2O\ (\Delta G^\circ = -357\ kJ/mol)$$
$$(\Delta G^\circ\ \text{values calculated for 25°C and at pH} = 7)$$

Although the major product of the oxidation of $NH_4^+$ is $N_2$, a small amount of $NO_3^-$ is also produced in a ratio 1:1.31:0.22. It is postulated that $NO_3^-$ formation from $NO_2^-$ generates reducing equivalents required for $CO_2$ reduction (Jetten et al. 1998).

The anammox reaction was proposed as a thermodynamically possible chemolithotrophic metabolism by Broda (1977), but evidence of the occurrence of this process was not gathered until nearly 20 years later when it was observed that ammonium became depleted under anoxic conditions in a pilot-scale WWT bioreactor, subsequently leading to the search and discovery of the involved organisms (Op den Camp et al. 2007).

Phylogenetically, all known anammox organisms belong to Bacteria and are classified within the Phylum *Planctomycetes* (Order *Brocadiales*, family *Brocadiaceae*). To date, five candidate genera have been proposed: *Ca.* Anammoxoglobus, *Ca.* Brocadia, *Ca.* Jettenia, *Ca.* Kuenenia, and *Ca.* Scalindua, comprising over 20 species (Mao et al. 2017), none of which has pure culture representatives (Peeters and van Niftrik 2019) in spite of multiple isolating efforts (Op den Camp et al. 2007). Anammox bacteria are regarded as ubiquitous in anoxic ecosystems and have been detected in diverse habitats including soil, freshwaters, WWT systems, and brackish/marine water, being acknowledged for a major contribution to N-cycling in the marine oxygen-minimum zones (Lam and Kuypers 2011, Oshiki et al. 2016, Peeters and van Niftrik 2019). Although the five candidate genera have been detected in WWT plants (Hu et al. 2010), *Ca.* Brocadia, *Ca.* Jettenia, and *Ca.* Kuenenia are the most common in engineered systems (Ma et al. 2016), while *Ca.* Scalindua seems confined to saline waters and are the prevalent anammox bacteria in deep-sea surface sediments (Wu et al. 2019a).

Anammox bacteria convert the substrates $NH_4^+$ and $NO_2^-$ through three consecutive reactions (Figure 8.9) with nitric oxide (NO) and hydrazine ($N_2H_4$) as intermediates (Kartal and Keltjens 2016). Although the free energy associated with the anammox reaction is higher than for aerobic ammonia oxidation, these organisms display very long doubling times and extremely slow growth rates. These facts are attributed to the slow activity of the key enzyme hydrazine synthase (Kartal et al. 2012), which is unique to anammox organisms (Peeters and van Niftrik 2019). Hydrazine synthase is encoded by the *hzsABC* genes of which *hzsA* is used as a molecular marker for anammox in the environment (Harhangi et al. 2012). Anammox bacteria are additionally able to use some organic and inorganic compounds as alternative electron donors (i.e., acetate, formate, propionate, methylamines, ferrous iron) and are autotrophic organisms carrying out $CO_2$ fixation through the Wood-Ljungdahl reductive acetyl CoA pathway (van Niftrik and Jetten 2012).

The *Planctomycetes*, unlike most bacteria, are characterized for the occurrence of internal compartments delimited by membranes (Youssef and Elshahed 2014). Anammox bacteria harbor a large vacuolar cell organelle called the anammoxosome, which is essential for the anammox reaction (Kartal and Keltjens 2016). It is predicted that all the catabolic processes take place coupled to the anammoxosome membrane since all the proteins required for the anammox reaction are located there (van Niftrik and Jetten 2012, Peeters and van Niftrik 2019). A model for a generation of proton motive force and ATP synthesis has been proposed but still requires experimental validation (Peeters and van Niftrik 2019).

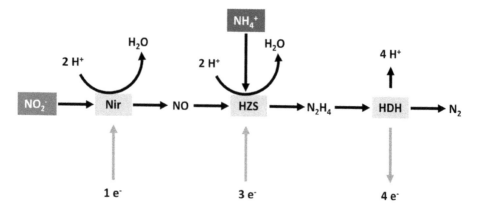

**Nir:** Nitrite reductase; **HZS:** hydrazine synthase; **HDH:** hydrazine dehydrogenase

**FIGURE 8.9** Schematic overview of the main reactions in the anammox pathway and the involved enzymes: 1. nitrite reduction to nitric oxide by a nitrite reductase (Nir), 2. combination among nitric oxide and ammonium to form hydrazine by hydrazine synthase (HZS), and 3. oxidation of hydrazine to dinitrogen gas by hydrazine dehydrogenase (HDH). Adapted from Peeters and Van Niftrik (2019).

The discovery of the anammox process provided an interesting cost-effective and environmentally friendly alternative to conventional methods for N removal from wastewater (Hu et al. 2010). Since these organisms can achieve a completely autotrophic conversion of ammonia to $N_2$ in the absence of $O_2$, external C source and aeration requirements are eliminated, and sludge production is also significantly reduced (Mao et al. 2017, González-Martinez et al. 2018). However, the anammox reaction requires nearly equimolar amounts of ammonia and nitrite, while most domestic or industrial wastewaters are rich in either ammonia or nitrate but carry only trace amounts of nitrite (Table 8.1). Ammonia or nitrate in raw wastewaters need to be partially converted to nitrite prior to the anammox reaction, and the two main strategies implemented for this purpose, partial nitrification-anammox (PN-AMX) and partial denitrification-anammox (PD-AMX) (Ma et al. 2016), will be addressed in Section 8.4.4.

## 8.3.5 Dissimilatory Nitrate Reduction to Ammonia (DNRA)

Dissimilatory nitrate reduction to ammonia (DNRA) is a type of anaerobic respiration that involves the stepwise reduction of nitrate to nitrite and then to ammonium (Figure 8.3). Electron donors for the reaction can be either organic or inorganic molecules, such as $H_2$, $Fe^{2+}$, $CH_4$, $HS^-$, and other reduced S compounds (Kuypers et al. 2018, Kratf et al. 2011, Holmes et al. 2019).

$$\tfrac{1}{2}NO_3^- + \tfrac{1}{2}C_2H_3O_2^- + \tfrac{1}{2}H^+ + \tfrac{1}{2}H_2O \rightarrow \tfrac{1}{2}NH_4^+ + HCO_3^- \; (\Delta G^\circ = -257 \text{ kJ/mol})$$
$$\tfrac{2}{3}NO_2^- + \tfrac{1}{2}C_2H_3O_2^- + \tfrac{5}{6}H_2O \rightarrow \tfrac{2}{3}NH_4^+ + HCO_3^- \; (\Delta G^\circ = -222 \text{ kJ/mol})$$

($\Delta G^\circ$ values per mol organic C, calculated with acetate as $e^-$ donor, at 25°C and pH = 7)

DNRA dominates over denitrification mostly in $O_2$-limited environments where the available reductants are in significant excess over the oxidants ($NO_3^-/NO_2^-$). Since the reduction of $NO_2^-$ to $NH_4^+$ consumes 6 electrons while reduction of $NO_2^-$ to $N_2$ consumes 3 electrons, a general hypothesis is that the DNRA pathway provides a competitive advantage to microorganisms in such habitats in order to maintain the redox balance while sustaining growth (Mohan and Cole 2007). Some fermentative microorganisms also use this reaction as an electron sink to enable ATP synthesis via substrate-level phosphorylation with organic substrates (Lams and Kuypers 2011). DNRA is more often carried out by bacteria, although there is some evidence of this process taking place in eukaryotic microorganisms (Stief et al. 2014). Since nitrate is converted to ammonia bypassing denitrification and $N_2$ fixation, DNRA is considered a short-cut of the biological N-cycle (Mohan and Cole 2007).

In bacteria, the respiratory DNRA occurs in the cytoplasm, periplasm, or both depending on the species and the environmental conditions. The cytoplasmic pathway is restricted to some groups of facultative anaerobic bacteria living under high concentrations of nitrate, while the periplasmic pathway is widespread in Bacteria (Kraft et al. 2011). Both pathways are well described in the model facultative anaerobe *Escherichia coli* (Mohan and Cole 2007). Such as in denitrification, $NO_3^-$ is reduced to $NO_2^-$ by either a cytoplasmic respiratory nitrate reductase (encoded by the *narGHIJ* genes) or the periplasmic NapA. Then, in the cytoplasmic DNRA pathway, $NO_2^-$ reduction to $NH_4^+$ is coupled through NirB, a soluble cytoplasmic NADH-dependent nitrite reductase with a S-Fe center. In parallel, in the periplasmic pathway, the reduction of $NO_2^-$ to $NH_4^+$ is catalyzed without the release of any intermediate by the pentahaem cytochrome *c* nitrite reductase NrfA, which receives electrons from the NrfB donor protein through a membrane integral complex (NrfCD) (Kraft et al. 2011, Lockwood et al. 2011). In other *Proteobacteria*, a membrane-bound tetraheme cytochrome *c* subunit (NrfH) mediates the electron transport to NrfA and anchors the enzyme complex to the membrane (Figure 8.10). NrfA is also able to catalyze the reduction of the potential intermediates NO and $NH_2OH$ (Lockwood et al. 2011) in agreement with evidence of $N_2O$ being also a byproduct of DNRA (Kraft et al. 2011). Most organisms that are able to perform DNRA reduce $NO_3^-$ to $NO_2^-$, but some sulfate-reducing bacteria can only use $NO_2^-$ (Kraft et al. 2011). The *nrfA* gene has been identified in obligate and facultative anaerobic bacteria and is conserved through evolution being used as a molecular marker for the DNRA organisms (Welsh et al. 2014).

Even though DNRA has received little attention compared to other pathways of the biological N-cycle, it is currently addressed as a widely distributed and ecologically relevant process which mitigates the N losses due to conventional denitrification in the terrestrial and aquatic environments (Kraft et al. 2011, van den Berg et al. 2016). In the absence of $O_2$, nitrate is biologically reduced to nitrite, which is not necessarily linked to any specific downstream process and may thus undergo denitrification, anammox, or DNRA, depending on the environmental conditions (Welsh et al. 2014). DNRA activity has been observed in several types of WWT systems under limited $O_2$ availability (Chutivisut et al. 2018), where it is an undesired process since it hampers N removal through denitrification. Consequently, those variables enabling denitrifiers to outcompete microorganisms that can perform DNRA have been investigated in order to control the occurrence of the latter in engineered systems. In this sense, several studies addressed the effect of the organic C to $NO_3^-$ ratio ($OC/NO_3^-$) on the balance of end products of $NO_2^-$ reduction in WWT plants. Overall, high $OC/NO_3^-$ ratios promote the dominance of DNRA over denitrification, while denitrification is favored at low $OC/NO_3^-$ ratios, in agreement with the observations in other habitats. Chutivisut et al. (2018) used stable-isotope labeling of $NO_3^-$ and $NO_2^-$ to follow their fate in an anoxic semi-continuous SBR fed with glucose as the source of C and energy and $NO_3^-$ as the electron acceptor, concluding that DNRA microorganisms were the major contributors to $NO_3^-$ and $NO_2^-$ reduction with a $OC/NO_3^-$ ratio = 8:1, while denitrifiers were the major $NO_3^-$ and $NO_2^-$ reducers with a $OC/NO_3^-$ ratio = 4:1. van den Berg et al. (2016) showed that DNRA and denitrifiers coexisted within a wide range of $OC/NO_3^-$ ratios in a chemostat fed with acetate and nitrate and inoculated with activated sludge from a WWT plant, concluding that both substrates were limiting. DNRA was dominant under nitrate limitation, while denitrification prevailed under acetate limitation. When operating under conditions favoring DNRA over denitrification, several studies reported that the microbial biomass became enriched in bacteria phylogenetically related to *Geobacter* spp. (van den Berg et al. 2015, 2016, Chutivisut et al. 2018).

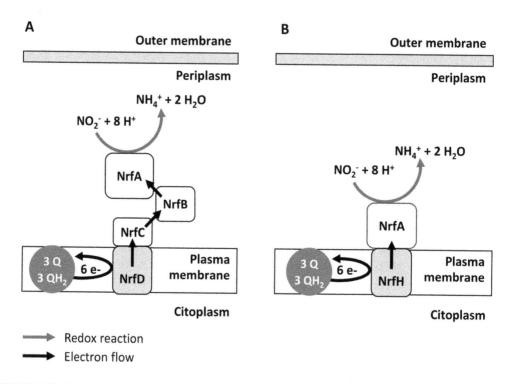

**FIGURE 8.10** Proposed respiratory dissimilatory nitrate reduction to ammonia (DNRA) periplasmic pathways in *Escherichia coli* (*Gammaproteobacteria*) (A) and *Wollinella succinogenes* (*Epsilonproteobacteria*) (B), adapted from Kraft et al. (2011). The electron donors are oxidized by the corresponding enzymes (NADH dehydrogenase, formate dehydrogenase or hydrogenase. not shown), and electrons transferred to the quinone pool.

DNRA metabolism has been described in anammox bacteria, some of which display the ability to reduce nitrite to ammonium only partially, then using both products for anaerobic ammonium oxidation (Lam and Kuypers 2011). Such partial DNRA-anammox processes are promising for the design of new technologies enabling improved biological N removal from wastewaters as an alternative to PN-AMX or PD-AMX strategies (Castro-Barros et al. 2017).

## 8.3.6 N₂-Fixation

$N_2$-fixation is a biological process restricted to prokaryotes through which N returns from the atmospheric to the terrestrial compartment in the form of ammonia by the following reaction (Kuypers et al. 2018):

$$N_2 + 8e^- + 8H^+ + 16ATP \rightarrow 2NH_3 + H_2 + 16ADP + 16P_i$$

The reaction is catalyzed in all known diazotrophic organisms by the enzymatic complex nitrogenase, which breaks the triple bond of $N_2$ adding three H atoms to each N atom with a considerable energy requirement (Newton 2007). All nitrogenases are composed of two metalloenzymes: the catalytic component and the nitrogenase-reductase, also called the iron protein (Fe-protein). Electron donors such as ferredoxin and flavodoxin reduce the Fe-protein, which then transfers electrons to the catalytic unit through the hydrolysis of two ATP molecules for each electron delivered (Dixon and Kahn 2004). There are three distinct types of nitrogenase that carry iron, vanadium, or molybdenum in the active center of its catalytic component of which molybdenum-nitrogenase (Mo-nitrogenase) is widespread among diazotrophs and is the best characterized (Newton 2007). Nitrogenase reductases are encoded by the *nifH* gene, which is used as a molecular marker to detect $N_2$-fixing microorganisms in the environment (Kuypers et al. 2018).

The synthesis of both components of nitrogenase is strongly regulated in prokaryotes at both the transcriptional and post-translational levels in response to environmental factors, mainly $O_2$ concentration and the availability of fixed N forms (Dixon and Kahn 2004). Both components of nitrogenase are extremely sensitive to $O_2$, which irreversibly inactivates the enzyme complex because of the nature of their metallic centers (Oelze 2000, Dixon and Kahn 2004). Diazotrophs have developed different physiological strategies to protect nitrogenase from $O_2$ damage, spanning from the simplest solution of growing diazotrophically only under anaerobic conditions to more sophisticated mechanisms that allow aerobic growth at the expense of biologically-fixed N: compartmentalization or restriction of diazotrophy to $O_2$-protected differentiated cellular structures (i.e., the heterocyst of *Cyanobacteria*), creation of an $O_2$-diffusion barrier by secretion of copious extracellular polymeric substances, conformational protection of the enzyme complex, or respiratory protection, which hypothesizes that $O_2$-scavenging occurs at the cell surfaces due to increased respiratory rates, keeping intracellular $O_2$ low (Oelze 2000, Ureta and Norlund 2002, Dixon and Kahn 2004, Smercina et al. 2019). Since $N_2$-fixation is an energy-demanding process, the synthesis of nitrogenase is strongly repressed when ammonia and other inorganic or organic fixed N sources become available (Dixon and Kahn 2004, Smercina et al. 2019).

$N_2$-fixing organisms are phylogenetically widespread across a range of archaeal and bacterial Phyla, the latter including *Chlorobi, Cyanobacteria, Firmicutes*, and *Proteobacteria* (Smercina et al. 2019). Many organisms with the capacity of fixing molecular N occur in municipal WWT plants (Yu and Zhang 2012, Fan et al. 2017), but $N_2$-fixation is more often negligible in these engineered systems since the amount of available combined N (mostly in the form of ammonia) is usually high enough to repress the expression of nitrogenase. Like DNRA, $N_2$-fixation is undesirable in most WWT systems, which rather aim for the removal of N exceeding the growing requirements of activated sludge microorganism by enhancing the biological processes which release $N_2$ to the atmosphere. However, some industrial wastewaters are characterized for very low N contents (Table 8.1), leading to a high bioavailable carbon to nitrogen ratio (often expressed as the ratio between the biological oxygen demand at five days and total nitrogen, $BOD_5$/TN) which makes the complete oxidation of organic matter by biological processes unfeasible unless the raw wastewater is amended with supplementary nutrients, such as ammonia or urea, thus significantly increasing treatment cost (Gauthier et al. 2000). A $BOD_5$/TN ratio of 100/5 is usually recommended and $N_2$-fixation stops at a $BOD_5$/TN ratio around 100/1.9 (Slade et al. 2011); for reference, the typical $BOD_5$/N ratio of municipal sewage is 100/20 (Awad et al. 2004). In this sense, $N_2$-fixation has been successfully implemented in both laboratory- and full-scale systems as a sustainable and low-cost alternative for the treatment of paper and pulp mill (Gauthier et al 2000, Slade et al. 2004), winery (Welz et al. 2018) and olive oil mill wastewaters (Awad et al. 2004), which are rich in carbohydrates and poor in fixed N, unlike municipal sewage (Table 8.1).

Several studies reported that $N_2$-fixation occurred spontaneously in full-scale WWT plants treating effluents of the papermaking industry. $N_2$-fixation was detected in the primary clarifiers of six pulp and paper mill full-scale WWT facilities (CAS and SBR) by Gauthier et al. (2000) but not in the aeration tanks, which were amended with either ammonia or urea. $N_2$-fixation activity in the clarifiers increased with depth, consequently with the dominance of the facultative anaerobe *Klebsiella* spp. (which only fix $N_2$ under $O_2$-deprivation) among the cultivable populations carrying the *nifH* gene. *Klebsiella* spp. have been isolated from papermaking wastewaters in other previous studies as well, although it is discussed that their prevalence may have been overestimated in earlier work owing to being easier to cultivate than other bacteria (Bowers et al. 2008, Addison et al. 2010). In contrast, Clark et al. (1997) investigated a full-scale facility consisting of four aerated stabilization basins lacking nutrient supplementation, demonstrating that bacterial $N_2$ fixation occurred mostly under aerated conditions, providing enough input of N (ca. 600 kg N/day) to achieve high $BOD_5$ removal rates.

Gapes et al. (1999, 2003) developed and patented a system (N-ViroTech®) aimed for the treatment of nutrient-deficient wastewaters where N limitation was overcome by enhanced $N_2$-fixation rates.

The process relies on the manipulation of growth conditions within the biological system to select and maintain an active $N_2$-fixing community primarily responsible for carbon removal. The N-ViroTech® process was successfully tested at a laboratory, pilot and full-scale on real wastewaters from several papermaking industries, achieving optimal treatment performances whilst providing significant advantages over the conventional operation, such as the self-regulation of N requirements, low N discharges in the effluent, and 25-35% reduced operational cost (Slade et al. 2003, 2004). Optimization of the operating DO concentration is a key component of the technology. Analysis of microbial community markers in pilot-scale N-ViroTech® systems under different DO levels (0.3, 1.0, and 2.2 mg/L) reported a high ratio among the number of copies of *nifH* and those of the 16S rRNA coding genes, in agreement with the high rates of $N_2$-fixation observed, and revealed that the diversity of the $N_2$-fixing community was low but significantly influenced by DO (Bowers et al. 2008, Reid et al. 2008). The lowest DO level tested (0.3 mg/L) provided a competitive advantage to diazotrophic bacteria (Clark et al. 1997, Slade et al. 2003, 2004). Under such conditions, the community was dominated by *Proteobacteria* of the Orders *Rhizobiales* and *Burkholderiales*, which comprise many genera of free-living aerobic diazotrophs (Reid et al. 2008), in agreement with the phylogenetic analysis of partial *nifH* sequences retrieved from the same pilot-scale plant (Bowers et al. 2008). A more recent study (Addison et al. 2010) identified the active diazotrophs in real wastewater from a full-scale pulp and paper treatment plant by $^{15}N_2$ stable-isotope probing and reverse transcription analysis of the $^{15}$N-enriched 16S rRNA. This study revealed a diverse community dominated by members of the *Gammaproteobacteria* (85%) followed by *Firmicutes* (8.2%). *Aeromonas*, *Pseudomonas*, and *Bacillus* were identified as dominant genera, and their ability to fix $N_2$ was assessed by the detection of the *nifH* gene in cultivated representatives.

Welz et al. (2018) successfully tested laboratory-scale $N_2$-fixing sand bioreactors to treat synthetic winery wastewater with a high C/N ratio (193/1). The cultivable $N_2$-fixing bacterial community was dominated by aerobic free-living diazotrophs of the genus *Azotobacter*, which increased their abundance by up to two orders of magnitude at the surface of the filters. A strain of *Azotobacter vinelandii* was also able to degrade organic C in synthetic N-deficient wastewater in batch cultures (Kargi and Özmıhçı 2002, 2004). The potential of strains of *A. vinelandii* and *A. chroococcum* to treat and valorize olive oil mill wastewater has been also tested in several studies (Ehaliotis et al. 1999, Awad et al. 2004). Alongside its $N_2$-fixing ability, *Azotobacter* spp. are able to degrade recalcitrant phenolic compounds commonly present in this kind of wastes (Juárez et al. 2008).

## 8.4 FUNDAMENTALS OF N REMOVAL STRATEGIES IN WASTEWATER TREATMENT PLANTS

As already stated in Section 8.1.3, the early activated sludge configurations have considerably evolved over the years, and a number of further modifications of the CAS process together with other technical solutions have been designed and optimized to take advantage of the abilities of the different groups of microorganisms involved in N-cycling in search of better performance and cost-effectiveness of WWT systems. The description of such configurations, the factors affecting their performance, and the advantages and suitability of each kind of technology for the different types of wastewaters will be only briefly addressed in this section, but the readers are referred to recent comprehensive reviews on the subject (Holmes et al. 2019, Jaramillo et al. 2018, McCarty 2018, Nancharaiah and Reddy 2018, Soliman and Eldyasti 2018, Wilén et al. 2018, Du et al. 2019b, Winkler and Straka 2019, Zhang et al. 2020).

### 8.4.1 Conventional Biological N Removal by Nitrification/Denitrification

Aerobic nitrification coupled to heterotrophic denitrification was the earliest biological process exploited in order to achieve biological N removal from wastewater, and technologies based on this strategy are still the most common in use today, particularly in domestic and municipal WWT plants, where ammonia is the major source of N (McCarty 2018, Winkler and Straka 2019).

In a CAS process (Figure 8.2), the fully-aerated conditions promote the complete oxidation of carbonaceous organic matter to $CO_2$ by chemoorganoheterotrophic microorganisms, releasing additional ammonia through the breakdown of the organic N compounds. Subsequently, under extensive aeration and long enough SRT, ammonia is oxidized to nitrate through nitrification:

$$C_5H_7O_2N + 5O_2 \rightarrow 5CO_2 + 2H_2O + NH_3$$
$$NH_4^+ + 2O_2 \rightarrow NO_3^- + 2H^+ + H_2O$$

This way, CAS processes more often produce treated water effluents bearing undesirable high concentrations of $NO_3^-$ since denitrification (required to dissimilatory reduce $NO_3^-$ to atmospheric $N_2$ gas) mainly occurs under anaerobic or microaerophilic conditions with DO concentrations < 2 mg/L (Lu et al. 2014). Thus, excess N would remain in the effluent of CAS unless further removal strategies are implemented (Daims and Wagner 2010).

The first effective N removal technologies were introduced in the second half of the twentieth century. Wuhrmann (1957) proposed a modified design of CAS consisting of separate aerated and anoxic bioreactors in series, each one providing the distinct redox conditions required by nitrifiers and denitrifiers, respectively (Figure 8.11A). With this configuration (named 'post-denitrification'), most of the organic substrates were consumed in the aerobic zone and denitrification was solely supported by substrates released after death and lysis of the biomass (Ni et al. 2017), implying a very slow denitrification rate unless an external C source was supplemented to the anoxic bioreactor with an additional cost (Seviour et al. 1999). Ludzack and Ettinger (1962) implemented another modification of the CAS process to improve biological N removal by nitrification-denitrification, using the biodegradable organic matter in the influent wastewater as an electron donor for the denitrification process (Figure 8.11B). The system comprised two separate zones that were partially in contact. In this configuration (named 'pre-denitrification'), the influent wastewater entered first an anoxic zone (mixed but not aerated) in which denitrification took place removing most of the organic matter, then the water flowed to an aerated zone in order to enable nitrification. The effluent entered a secondary clarifier from which decanted sludge was recirculated back to the aerated zone (Seviour et al. 1999). Due to the action of aerators, mixing and interchange of the nitrifying and denitrifying biomasses occurred in the interzone, enabling nitrate to enter the anoxic zone to be denitrified to $N_2$; however, variable denitrification performances were achieved due to lack of control over the mixing (Ekama and Wentzel 2008). The pre-denitrification configuration was further modified by Barnard (1973) to improve the performance and control of the process by reintroducing the separation between the anoxic and aerated bioreactors and implementing recirculation of sludge from both the clarifier and the aerated bioreactor to the anoxic bioreactor, the latter providing $NO_3^-$ for denitrification (Modified Ludzack-Ettinger process, MLE, Figure 8.11C) (Ekama and Wentzel 2008). Additional modifications of the pre-denitrification configuration increased the number of reactors and aimed for the simultaneous removal of N and P (Zhu et al. 2008). For further descriptions and historical review of BNR processes, the readers are directed to the recent review of Esfahani et al. (2019).

Conventional nitrification-denitrification strategies are effective, enabling up to 80% N removal from wastewater (Tchobanoglous et al. 2003) but still have some important disadvantages (Zhu et al. 2008, Ni et al. 2017, Winkler and Straka 2019, Zhang et al. 2020):

- High energy demand due to the aeration rates required for nitrification (4.2 g $O_2$ per g $NH_4^+$-N).
- High C demand for denitrification (see Section 8.3.3.2), which implies the addition of external C sources to wastewaters with low C/N ratios, increasing treatment cost.
- Generation of excess waste sludge which must undergo further treatment, increasing the final cost. Reported yields range 0.23-0.65 mg biomass per mg COD removed.

- Generation of green-house gases $CO_2$ and $N_2O$ through the activity of both nitrifiers and denitrifiers.

In the following sections, newer biological N removal strategies developed to overcome these drawbacks will be summarized.

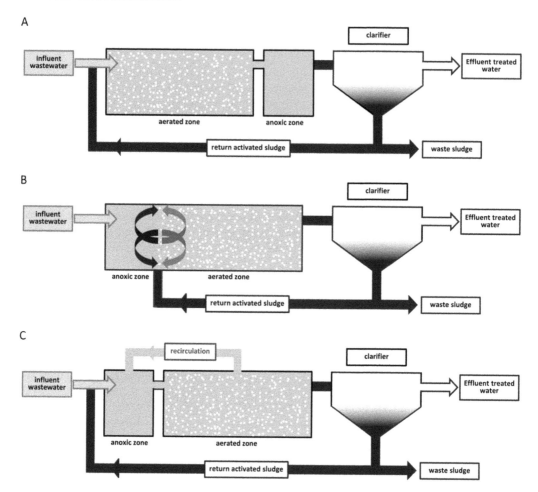

**FIGURE 8.11** Flow diagrams of the conventional biological N removal processes based on nitrification-denitrification: Wuhrmann (A), Ludzack-Ettinger (B) and modified Ludzack-Ettinger. Adapted from Seviour et al. (1999).

## 8.4.2 Simultaneous Nitrification and Denitrification (SND)

SND is broadly defined as the co-occurrence of both nitrification and denitrification in a single WWT reactor and under identical overall operating conditions (Münch et al. 1996, Chai et al. 2019). SND is theoretically limited by the distinct physiological requirements of canonical nitrifiers and denitrifiers, particularly regarding DO concentration: > 1-2 mg/L are required for optimal nitrification, while inhibition of denitrification starts at DO > 0.5 mg/L (Zhu et al. 2008).

It has been extensively reported that SND is achieved in aerated WWT systems due to the existence of anoxic microenvironments, generated when the size and density of cell aggregates enable the creation of a DO concentration gradient descending toward the inner zones. This way, nitrifiers grow in the outer regions with higher DO concentrations, whereas denitrification preferentially occurs at the center of the aggregates where DO concentration is very low (Zhu et al. 2008, Bassin et al. 2012). This physical limitation to $O_2$ diffusion is more easily achieved in biofilm-based technologies (Ma et al. 2017) or suspended growth systems where cells aggregate in macroflocs (flocs with a diameter > 0.15 mm, Judd et al. 2011) or granules (Bassin et al. 2012). Besides the size and structure of cell-aggregates, another key factor is the accurate control of the DO concentration to prevent $O_2$ to completely diffuse through them (Meyer et al. 2005). For instance, it has been reported that $O_2$ penetrates through biofilms to a 2.6 mm depth at DO concentrations > 2.5 mg/L in the bulk liquid (Cao et al. 2017b).

Another explanation for SND is the involvement of bacteria being able to carry out HN-AD (already addressed in Section 8.3.3.3). SND occurs at concentrations up to 5 mg/L (Rajta et al. 2019), which will limit the existence of anoxic microniches and will be inhibitory for canonical denitrifiers, while aerobic denitrifiers are active at DO concentrations up to 8 mg/L (Ji et al. 2015). Also, optimal N removal in SND systems has been observed at DO concentrations <1 mg/L (Ma et al. 2017, Yan et al. 2019), which should inhibit canonical nitrification. The role of HN-AD in SND systems is supported by studies reporting a high relative abundance of genera capable of HN-AD at DO <2.5 mg/L (Ma et al. 2017, Chai et al. 2019). At low DO concentration, $NH_4^+$ oxidation could be as well accomplished by AOB through nitrifier denitrification (as already discussed in Section 8.3.2.1). Accordingly, the contribution of AOB adapted to low DO concentrations to SND was proposed by Park et al. (2006). In conclusion, the N removal pathways involved in SND remain not fully understood to date since the involved microorganisms may be fairly diverse and deeper characterization of this process is still needed.

Besides DO concentrations, other important factors must be carefully controlled to successfully achieve SND, such as adjustment of the C/N and food-to-microorganisms (F/M) ratios, pH, or temperature (He et al. 2009, Yan et al. 2019, Rajta et al. 2019). SND offers an interesting cost-saving choice for N removal in WWT systems by reducing space requirements and simplifying process design. It was estimated that SND improves the denitrification rates by 63%, while at the same time lowers the C source demand of the denitrification process by 40% and significantly reduces the generation of waste sludge (Ma et al. 2017). However, it must be taken into consideration that nitrifier denitrification and aerobic denitrification have been regarded as major sources of $N_2O$ emissions in studies conducted in soils (Wrage et al. 2001, Ji et al. 2015).

### 8.4.3 Short-Cut Biological N Removal

As stated in Section 8.4.1, the conventional nitrification-denitrification processes proceed through the oxidation of $NH_4^+$ to $NO_2^-$ by AOP and oxidation of $NO_2^-$ to $NO_3^-$ by NOB followed by the sequential reduction of $NO_3^-$ to $NO_2^-$, NO, $N_2O$, and $N_2$ by denitrifiers. Since $NO_2^-$ is an intermediary of both nitrification and denitrification, an alternative strategy is to short-cut the nitrification-denitrification pathway by achieving the oxidation of $NH_4^+$ to $NO_2^-$ but avoiding the oxidation of $NO_2^-$ to $NO_3^-$ (partial nitrification or nitritation) (Figure 8.12). This strategy provides the basis for the process, named 'short-cut biological N removal' (Chung et al. 2007), 'nitrification and denitrification via nitrite' (Peng and Zhu 2006) or 'nitritation-denitritation' (Winkler and Straka 2019), and enables an enhanced efficiency of biological N removal while reducing operational cost. Compared with conventional nitrification-denitrification, nitritation-denitritation accomplishes >95% N removal rates while shortening the reaction time and reducing $O_2$ and organic C requirements (25% and 40%, respectively) through skipping $NO_2^-$ oxidation to $NO_3^-$ by NOB and $NO_3^-$ reduction to $NO_2^-$ by denitrifiers, also decreasing excess sludge production (30 and 55% in nitrification and denitrification, respectively) and $CO_2$ emissions (20%) (Peng and Zhu 2006). The major drawback of short-cut biological N removal is the increased generation of $N_2O$ under low DO conditions (Winkler and Straka 2019).

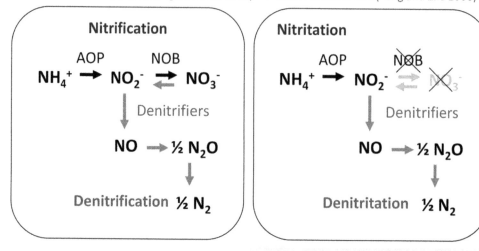

**FIGURE 8.12** Conceptual diagrams explaining the differences between conventional nitrification-denitrification and short-cut biological N removal strategies to achieve biological N removal. AOP: ammonia oxidizing prokaryotes; NOB: nitrite oxidizing bacteria.

Short-cut biological N removal primarily relies on the inhibition of the growth and activity of nitrite oxidizers through different strategies, including control of temperature, SRT, organic carbon loading rates, ammonia concentrations, C/N ratios, pH, and aeration rates, which are thoroughly discussed elsewhere (Ge et al. 2015, Ma et al. 2016, Winkler and Straka 2019). In particular, the application of intermittent aeration as a strategy to enable nitrification-denitrification favors AOB over NOB because the latter had a lower affinity for $O_2$ and recover worse from transient anoxia (Ge et al. 2015, Ma et al. 2016). Achieving long-term stability of the suppression of NOB activity is crucial for the success of the process (Gao et al. 2009), and more often several strategies need to be implemented simultaneously to efficiently inhibit nitrite oxidation to nitrate at full-scale (Ma et al. 2016).

The so-called 'SHARON' process (Stable and High activity Ammonia Removal Over Nitrite) was the first short-cut N removal strategy put into practice (Hellinga et al. 1998), taking advantage of the fastest growth rate of ammonia oxidizers compared to nitrite oxidizers at temperatures >26°C and operating using a SRT longer than required for ammonia oxidizers but too short for nitrite oxidizers, which were washed out the reactor (Schmidt et al. 2003). The effluent with theoretically 100% ammonia converted to nitrite can be fed into a second reactor for denitrification. Alternatively, a single-reactor configuration can be adopted, using a continuously-stirred tank reactor or an SBR with alternating aerated and anoxic phases to enable sequential nitrification and denitrification (Schmidt et al. 2003). Methanol is commonly used as an external C source for denitrification (Schmidt et al. 2003). The SHARON process has proven very efficient to treat high-strength ammonia wastewaters with high temperatures, such as the reject streams from the dewatering of digested sewage sludge, and several full-scale plants with varying configurations were built in The Netherlands for this purpose (Schmidt et al. 2003, van Kempen et al. 2005).

## 8.4.4 Partial Nitrification-Anammox (PN-AMX) and Partial Denitrification-Anammox (PD-AMX)

### 8.4.4.1 Partial Nitrification-Anammox (PN-AMX)

As already stated in Section 8.3.4, most raw wastewaters are rich in either ammonia or nitrate but lack nitrite in significant concentrations; thus, anammox-based biological N removal treatments require the development of strategies providing nitrite roughly in equimolar proportion to ammonia. Approaches aiming to transform half of the ammonia carried in most wastewaters (Table 8.1) into nitrite (partial nitritation) were soon proposed through modifications of partial-nitrification configurations, which were initially aimed for accomplishing short-cut biological N removal (see Section 8.4.3). The earliest PN-AMX systems had a two-stage reactor configuration and were developed by adapting pre-existent SHARON-type reactors to achieve partial nitritation by AOB in order to generate a 1:1 mixture of $NH_4^+$ and $NO_2^-$ (Figure 8.13A). The partially nitritated effluent was fed to a second, anaerobic stage, to enable the anammox reaction and achieve N removal without the need to add external C sources (Schmidt et al. 2003). Newer PN-AMX implementations are mostly based on single-stage configurations due to their lower cost and simpler operation (Mao et al. 2017). Complete autotrophic removal of nitrate over nitrite (CANON® configuration) is based on the development of granular sludge and strict control of DO, which is kept at a concentration suppressing NOB activity but allowing AOB to partially oxidize ammonia to nitrite. Control of the

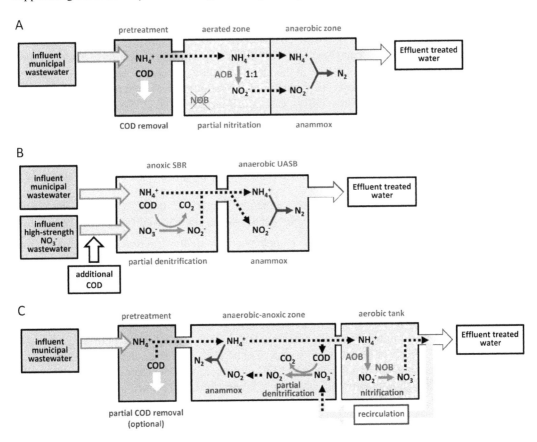

**FIGURE 8.13** Flow diagrams showing examples of biological N removal configurations based on partial nitrification-anammox (PN-AMX) and partial denitrification-anammox (PD-AMX): A. Two-stage PN-AMX for mainstream wastewater treatment (WWT) (Zhang et al. 2019b) B. PD-AMX for mixed influents of municipal and high-strength nitrate wastewater (Du et al. 2019b), and C. PD-AMX for mainstream WWT (Du et al 2019b). COD: chemical oxygen demand; AOB: ammonia oxidizing bacteria; NOB: nitrite oxidizing bacteria.

DO concentration must ensure that all the available $O_2$ is consumed by AOB in the process, allowing the anammox reaction (González-Martínez et al. 2018). The DEMON® (De-ammonification) configuration relies on a patented system to provide intermittent aeration controlled by pH (van Haandel and van der Lubbe 2012). Ammonia oxidation lowers pH until a threshold value which indicates that partial nitrification is achieved and triggers the switching off of aeration, providing anoxia for anammox bacteria; subsequently, the anammox reaction generates alkalinity, and when pH rises to a signal value, aeration is resumed. In addition to this, intermittent aeration is used to keep DO within a low range, inhibiting NOB and preventing the accumulation of nitrite, which is toxic to anammox bacteria at too high concentrations (Wett et al. 2007). Many other configurations taking advantage of the PN-AMX strategy have been designed and patented (reviewed by Lackner et al. 2014, Cao et al. 2017a, Mao et al. 2017) either as two-stage or single-stage processes, based on both suspended growth (floccular or granular sludge) and biofilms (i.e., moving bed biofilm reactors, rotating biological contactors). According to the most recent comprehensive review available (Lackner et al. 2014), over 100 full-scale PN-AMX WWT plants were in operation worldwide by 2014. The most commonly applied technology in full-scale installations (50%) was SBR, 80% of which were operated under the DEMON® single-stage configuration (Lackner et al. 2014).

PN-AMX offers several advantages over conventional nitrification-denitrification treatments, achieving 89% N removal rate while reducing $O_2$ (57%) and C (100%) requirements and sludge production (84%) (Zhang et al. 2019b). However, emissions of $N_2O$ are significantly enhanced, according to measures in full-scale PN-AMX processes, ranging from 0.4-6.6% of the N load (Mao et al. 2017). $N_2O$ is mainly generated in the partial-nitrification process because of the accumulation of high $NO_2^-$ concentrations under DO-limited conditions, which promotes nitrifier denitrification, and are higher in two-stage systems (Mao et al. 2017). The main strategy to minimize $N_2O$ emissions is controlling the accumulation of nitrite in the reactor, keeping it below 5 mg $NO_2^-$-N/L (Mao et al. 2017). PN-AMX processes have other important limitations, which make their application to mainstream WWT difficult (Cao et al. 2017a, González-Martínez et al. 2018):

- Anammox bioreactors require a very long start-up time due to the long doubling times of anammox bacteria (see Section 8.3.4).
- Mesophilic temperatures (≥ 30°C) are required for optimal metabolic activity of anammox bacteria. Nitrogen removal efficiency by anammox drops by tenfold when the temperature is lowered down from 30 to 10°C.
- Heterotrophic denitrifiers outcompete annamox bacteria if the C/N ratio of the source wastewater is too high. A biodegradable COD/N ratio ≤ 2-3 in the influent is usually required.

Most full-scale PN-AMX plants have been implemented to treat warm water reject streams from the anaerobic digestion of sludge, which commonly have low C/N ratios and carry high ammonia contents (500-1,500 mg $NH_4^+$-N/L, Cao et al. 2017a). These characteristics allow anammox bacteria to compete for the substrates $NH_4^+$ and $NO_2^-$ with nitrifying and denitrifying bacteria, despite their much slower growth rates. Besides, high $NH_4^+$ concentrations favor the generation of free ammonia and free nitrous acid at levels inhibitory for NOB (Cao et al. 2017a). In contrast, only two PN-AMX full-scale installations are in operation to treat municipal wastewaters to date, both working stably at temperatures > 25°C (van Haandel and van der Lubbe 2012, Wrinkler and Straka 2019). These systems require pretreatment to reduce ≥ 70% COD of the source wastewater, a step which is often accomplished by a high-rate activated sludge or alternatively by an anaerobic reactor (Cao et al. 2017a, Zhang et al. 2019b).

### 8.4.4.2 *Algal Anaerobic Ammonium Oxidation (Algammox)*

Microalgae are unicellular phototrophic and often mixotrophic organisms comprising a taxonomically miscellaneous group, which includes both eukaryotic algae and prokaryotic cyanobacteria (Maza-Márquez et al. 2018). Microalgae were incorporated to biological WWT technologies in the 1950s since this approach has interesting advantages: the involved organisms are easy to cultivate using inexpensive sunlight and $CO_2$ as energy and C sources, remove N and P nutrients by assimilation and other mechanisms, and contribute to the biodegradation of recalcitrant organic molecules; in addition to this, the microalgal biomass can be valorized in several ways (Maza-Márquez et al. 2018).

Engineered systems enabling the cultivation of microalgae by incorporating a light source (either natural or artificial) are termed photobioreactors (PBRs) (Maza-Márquez et al. 2018). In PBRs, mutualistic relationships are established among microalgae and prokaryotes. Aerobic bacteria respire using the $O_2$ delivered by algae through photosynthesis, and the $CO_2$ generated by the mineralization of organic matter by organotrophs is fixed by algae and autotrophic bacteria (Maza-Márquez et al. 2018). N removal in PBRs also takes place through both assimilation and bacterial nitrification-denitrification (Wang et al. 2018). WWT in PBRs by such microalgae-prokaryotic consortia eliminates aeration cost and reduces $CO_2$ emissions (Maza-Márquez et al. 2018). Several WWT strategies based on phototrophic microalgae-bacteria consortia are being currently developed, although many of them have been only tested at bench scale (reviewed by Wang et al. 2018).

Algal Anaerobic Ammonium Oxidation (Algammox) is a recently proposed adaptation of the PN-AMX strategy for the removal of N from COD-limited wastewaters, which avoids the need of mechanical aeration and addition of electron donors by coupling partial nitrification achieved by algae-nitrifying consortia to the anammox process (Manser et al. 2016). The system was designed as a photosequencing batch reactor alternating light and dark phases. This way, nitrification was supported by $O_2$ generated by photosynthesis during the light phases, while removal of ammonia plus nitrite by anammox bacteria occurred in the dark (anoxic) phases. The activity of NOB was minimized due to low DO and high concentrations of ammonium and nitrite. Controlling the generation of $O_2$ to avoid inhibition of anammox bacteria seems a critical step for the optimization and future development of this novel technology at full-scale (Mukarunyana et al. 2018).

### 8.4.4.3  *Partial Denitrification-Anammox (PD-AMX)*

Partial-denitrification-anammox (PD-AMX) configurations combine nitrification, partial denitrification, and the anammox process (Ma et al. 2016) and have recently emerged as an alternative to PN-AMX, which overcomes some of the limitations of the latter for the application of anammox-based biological N removal to mainstream WWT (Zhang et al. 2020). PD-AMX also offers a cost-saving alternative for treating high-strength nitrate industrial wastewaters (Table 8.1) in combination with municipal sewage carrying ammonia. Nitrate is converted to nitrite through partial denitrification, providing the electron acceptor for the anammox reaction (Figure 8.13) (Du et al. 2019b).

Partial denitrification, or denitratation, is based on blocking the conventional denitrification pathway beyond the $NO_3^-$ to $NO_2^-$ reduction step and can be achieved through two different approaches (Zhang et al. 2019b). The first one relies on the enrichment of bacteria which reduce $NO_3^-$ to $NO_2^-$ but lack the rest of reductases of the denitrification pathway (Zhang et al. 2020). Such nitrate-reducing bacteria can use either inorganic or organic electron donors and are widespread in natural habitats and engineered systems, including CAS and other WWT processes; however, concerns arise that their prevalence could not be long-term kept in a full-scale WWT facility under real operating conditions (Zhang et al. 2019b). The second approach is based on adjusting the operating conditions to force an imbalance of the denitrification pathway by preventing the activity of nitrite reductases while keeping high nitrate reduction rates. In this respect, low COD/N ratios (1.5-3.0) and alkaline pH (9.0) limit the C oxidation rate favoring nitrate reductase versus nitrite reductase in the competition for electrons. These and other strategies applied to successfully truncate the denitrification pathway after nitrate reduction have been recently reviewed by Zhang et al. (2020).

PD-AMX is less advantageous than PN-AMX but still offers benefits over conventional nitrification-denitrification, which include achieving > 90% N removal rate reducing $O_2$ and C source consumption (48% and 54%, respectively) with 66% less sludge production (Zhang et al. 2019b) and decreased $N_2O$ emissions (Du et al. 2019a). Compared to PN-AMX, in PD-AMX there is no need to inhibit NOB, which facilitates plant design and process control and provides higher flexibility to manage low/high strength wastewaters by either two- or single-stage configurations (Figure 8.13) (reviewed by Du et al. 2019b).

## 8.5 CONCLUDING REMARKS

A significant amount of effort and resources have been devoted to improve and evolve the design and performance of WTT systems since they were first implemented in the early twentieth century. Most recently developed advanced biological N removal technologies move away from the effective but energetically demanding nitrification-denitrification strategy in search of feasible approaches providing economic and environmental advantages. Since wastewaters from different sources are characterized for important differences in their composition and load of C and N compounds, no single biological N removal process can be considered of choice, and a careful evaluation must be made in order to select, among the wide array of technologies available, those better at meeting the demands for each specific application. The forefront of current research focuses to achieve the stable generation of high-quality treated water effluents by enhancing the efficiency of N removal but at the same time searching for the elimination of cost derived from aeration and C demand and minimizing the generation of greenhouse gases. Operability of plants is also an important issue, and in this respect the feasibility of plant start-up, its management, and control need also to be taken into consideration. In this regard, fully autotrophic N removal technologies, such as those based on autotrophic denitrification, anammox and N-DAMO processes, or their combinations, provide promising opportunities to successfully achieve the aforementioned goals. Overcoming the limitations currently constraining the application of such novel processes to mainstream WWT and implementing the available configurations to reach a zero-waste model for wastewater management is the ambitious challenge to be faced in the near future.

## 8.6 REFERENCES

Abby, S., Melcher, M., Kerou, M., Krupovich, M., Stieglmeier, M., Rossel, C., et al. 2018. *Candidatus* Nitrosocaldus cavascurensis, an ammonia-oxidizing, extremely thermophilic archaeon with a highly mobile genome. Frontiers in Microbiology 9: 28.

Addison, S.L., McDonald, I.R. and Lloyd-Jones, G. 2010. Identifying diazotrophs by incorporation of nitrogen from $^{15}N_2$ into RNA. Applied Microbiology and Biotechnology 87: 2313-2322.

Ahlgren, N.A., Chen, Y., Needham, D.M., Parada, A.E., Sachdeva, R., Trinh, V., et al. 2017. Genome and epigenome of a novel marine *Thaumarchaeota* strain suggest viral infection, phosphorothioation DNA modification, and multiple restriction systems. Environmental Microbiology 19: 2434-2452.

Akgul, D., Yuzer, B., Yapsakli, K. and Mertoglu, B. 2018. Nitrogen converters in various landfill leachates. Polish Journal of Environmental Studies 27: 1941-1948.

Alawi, M., Lipski, A., Sanders, T., Pfeiffer, E.M. and Spieck, E. 2007. Cultivation of a novel cold-adapted nitrite-oxidizing betaproteobacterium from the Siberian Arctic. ISME Journal 1: 256-264.

Alfreider, A., Grimus, V., Luger, M., Ekblad, A., Salcher, M.M. and Summerer, M. 2018. Autotrophic carbon fixation strategies used by nitrifying prokaryotes in freshwater lakes. FEMS Microbiology Ecology 94: fiy163.

Almstrand, R., Persson, F. and Hermansson, H. 2014. Biofilms in nitrogen removal: population dynamics and spatial distribution of nitrifying and anammox bacteria. pp. 227-260. *In*: Marco, D. (ed.). Metagenomics of the Microbial Nitrogen Cycle. Caister Academic Press, Poole, UK.

Álvarez, L., Bricio, C., Blesa, A., Hidalgo, A. and Berenguer, J. 2014. Transferable denitrification capability of *Thermus thermophilus*. Applied and Environmental Microbiology 80: 19-28.

Alves, R.J.E., Kerou, M., Zappe, A., Bittner, R., Abby, S.S., Schmidt, H.A., et al. 2019. Ammonia oxidation by the arctic terrestrial Thaumarchaeote *Candidatus* Nitrosocosmicus arcticus is stimulated by increasing temperatures. Frontiers in Microbiology 10: 1571.

Arantes, M.K., Alves, H.J., Sequinel, R. and da Silva, E.A. 2017. Treatment of brewery wastewater and its use for biological production of methane and hydrogen. International Journal of Hydrogen Energy 42: 26243-26256.

Arden, E. and Locket, W.T. 1914. Experiments on the oxidation of sewage without the aid of filters. Journal of the Society of Chemical Industry 35: 523-539.

Arp, D.J. and Bottomley, P.J. 2006. Nitrifiers: more than 100 years from isolation to genome sequences. Microbe 1: 229-234.

Awad, A., Salman, H. and Hung, Y.T. 2004. Olive oil waste treatment. pp. 737-810. *In*: Wang, L.K., Hung, Y.T., Lo, H.H. and Yapijakis, C. (eds.). Handbook of Industrial and Hazardous Wastes Treatment. CRC Press, Boca Raton, Florida, USA.

Barnard, J.L. 1973. Biological denitrification. Journal of the Water Pollution Control Federation 72: 705-720.

Bartossek, R., Nicol, G.W., Lanzen, A., Klenk, H.P. and Schleper, C. 2010. Homologues of nitrite reductases in ammonia-oxidizing archaea: diversity and genomic context. Environmental Microbiology 12: 1075-1088.

Bassin, J.P., Kleerebezem, R., Dezotti, M. and van Loosdrecht, M.C.M. 2012. Simultaneous nitrogen and phosphate removal in aerobic granular sludge reactors operated at different temperatures. Water Research 46: 3805-3816.

Bayer, B., Vojvoda, J., Offre, P., Alves, R.J.E., Elisabeth, N.H., Garcia, J.A., et al. 2016. Physiological and genomic characterization of two novel marine thaumarchaeal strains indicates niche differentiation. ISME Journal 10: 1051-1063.

Bayer, B., Vojvoda, J., Reinthaler, T., Reyes, C., Pinto, M. and Herndl, G.J. 2019. *Nitrosopumilus adriaticus* sp. nov. and *Nitrosopumilus piranensis* sp. nov., two ammonia-oxidizing archaea from the Adriatic Sea and members of the class *Nitrososphaeria*. International Journal of Systematic and Evolutionary Microbiology 69: 1892-1902.

Bedmar, E.J., Bueno, E., Correa, D., Torres, M.J., Delgado, M.J. and Mesa, S. 2013. pp. 164-182. Ecology of denitrification in soils and plant-associated bacteria. *In*: Rodelas, B. and González-López, J. (eds.). Beneficial Plant-Microbe Interactions. Ecology and Applications. CRC Press, Boca Raton, Florida, USA.

Bhandari, V.M., Sorokhaibam, L.G. and Ranade, V.V. 2016. Industrial wastewater treatment for fertilizer industry—a case study. Desalination and Water Treatment 57: 1-11.

Bitton, G. 2011. Wastewater Microbiology. Wiley-Blackwell.

Blainey, P.C., Mosier, A.C., Potanina, A., Francis, C.A. and Quake, S.R. 2011. Genome of a low-salinity ammonia-oxidizing archaeon determined by single-cell and metagenomic analysis. PLoS One 6: E16626.

Bock, E., Sundermeyer-Klinger, H. and Stackebrandt, E. 1983. New facultative lithoautotrophic nitrite-oxidizing bacteria. Archives of Microbiology 136: 281-284.

Bock, E., Koops, H.-P., Moller, U.C. and Rudert, M. 1990. A new facultatively nitrite-oxidizing bacterium, *Nitrobacter vulgaris* sp. nov. Archives of Microbiology 153: 105-110.

Bock, E. and Wagner, M. 2013. Oxidation of inorganic nitrogen compounds as an energy source. pp. 83-118. *In*: Rosemberg, E., de Long, E.F., Lory, S., Stackebrandt, E. and Thompson, F. (eds.). The Prokaryotes 4th Ed. – Prokaryotic Physiology and Biochemistry. Springer-Verlag.

Bollmann, A., Bär-Gilissen, M.J. and Laanbroek, H.J. 2002. Growth at low ammonium concentrations and starvation response as potential factors involved in niche differentiation among ammonia-oxidizing bacteria. Applied and Environmental Microbiology 68: 4751-4757.

Bowers, T.H., Reid, N.M. and Lloyd-Jones, G. 2008. Composition of *nifH* in a wastewater treatment system reliant on $N_2$ fixation. Applied Microbiology and Biotechnology 79: 811-818.

Braker, G. and Tiedje, J.M. 2003. Nitric oxide reductase (*norB*) genes from pure cultures and environmental samples. Applied and Environmental Microbiology 69: 3476-3483.

Broda, E. 1977. Two kinds of lithotrophs missing in nature. Zeitschrift Für Allgemeine Mikrobiologie 17: 491-493.

Bustamante, M.A., Paredes, C., Moral, R., Moreno-Caselles, J., Pérez-Espinosa, A. and Pérez-Murcia, M.D. 2005. Uses of winery and distillery effluents in agriculture: characterisation of nutrient and hazardous components. Water Science and Technology 51: 145-151.

Cabello, P., Roldán, M.D. and Moreno-Vivián, C. 2004. Nitrate reduction and the nitrogen cycle in archaea. Microbiology 150: 3527-3546.

Calderón, K., González-Martínez, A., Gómez-Silván, C., Osorio, F., Rodelas, B. and González-López, J. 2013. Archaeal diversity in biofilm technologies applied to treat urban and industrial wastewater: recent advances and future prospects. International Journal of Molecular Sciences 14: 18572-18598.

Campbell, M.A., Chain, P.S., Dang, H., El Sheikh, A.F., Norton, J.M., Ward, N.L., et al. 2011. *Nitrosococcus watsonii* sp. nov., a new species of marine obligate ammonia-oxidizing bacteria that is not omnipresent in the world's oceans: calls to validate the names 'Nitrosococcus halophilus' and 'Nitrosomonas mobilis'. FEMS Microbiology Ecology 76: 39-48.

Canfield, D.E., Glazer, A.N. and Falkowski, P.G. 2010. The evolution and future of earth's nitrogen cycle. Science 330: 192-196.

Cao, Y., van Loosdrecht, M.C.M. and Daigger, G. 2017. Mainstream partial nitritation-anammox in municipal wastewater treatments: status, bottlenecks and further studies. Applied and Environmental Microbiology 101: 1365-1383.

Cao, Y., Zhang, C., Rong, H., Zheng, G. and Zhao, L. 2017. The effect of dissolved oxygen concentration (DO) on oxygen diffusion and bacterial community structure in moving bed sequencing batch reactor (MBSBR). Water Research 108: 86-94.

Caranto, J.D. and Lancaster, K.M. 2017. Nitric oxide is an obligate bacterial nitrification intermediate produced by hydroxylamine oxidoreductase. Proceedings of the National Academy of Sciences of the USA 114: 8217-8222.

Carreira, C., Pauleta, S.R. and Moura, I. 2017. The catalytic cycle of nitrous oxide reductase—the enzyme that catalyzes the last step of denitrification. Journal of Inorganic Biochemistry 177: 423-434.

Carrera, J.M., Baeza, J.A., Vicent, T. and Lafuente, J. 2003. Biological nitrogen removal of high-strength ammonium industrial wastewater with two-sludge system. Water Research 37: 4211-4221.

Castro-Barros, C.M., Jia, M., van Loosdrecht, M.C.M., Volcke, E.I.P. and Winkler, M.K.H. 2017. Evaluating the potential for dissimilatory nitrate reduction by anammox bacteria for municipal wastewater treatment. Bioresource Technology 233: 363-372.

Cerrone, F., Poyatos, J.M., Molina-Muñoz, M., Cortés-Lorenzo, C., González-López, J. and Rodelas, B. 2013. Prevalence of *Nitrosomonas* cluster 7 populations in the ammonia-oxidizing community of a submerged membrane bioreactor treating urban wastewater under different operation conditions. Bioprocess and Biosystems Engineering 36: 901-910.

Chai, H., Xiang, Y., Chen, R., Shao, Z., Gu, L. and He, Q. 2019. Enhanced simultaneous nitrification and denitrification in treating low carbon-to-nitrogen wastewater: treatment performance and nitrogen removal pathway. Bioresource Technology 280: 51-58.

Chen, J. and Strouss, M. 2013. Denitrification and aerobic respiration, hybrid electron transport chains and co-evolution. Biochimica. et Biophysica. Acta 1827: 136-144.

Chen, Q. and Ni, J. 2011. Heterotrophic nitrification-aerobic denitrification by novel isolated bacteria. Journal of Industrial Microbiology and Biotechnology 38: 1305-1310.

Cheng, D.L., Ngo, H.H., Guo, W.S., Chang, S.W., Nguyen, D.D. and Kumar, S.M. 2019. Microalgae biomass from swine wastewater and its conversion to bioenergy. Bioresource Technology 275: 109-122.

Chistoserdova, L. and Lidstrom, M.E. 2014. Aerobic methylotrophic prokaryotes. pp. 267-285. *In*: Rosemberg, E., de Long, E.F., Lory, S., Stackebrandt, E. and Thompson, F. (eds.). The Prokaryotes–Prokaryotic Physiology and Biochemistry. Springer-Verlag.

Chung, J., Bae, W., Lee, Y.W. and Rittmann, B.E. 2007. Shortcut biological nitrogen removal in hybrid biofilm/suspended growth reactors. Process Biochemistry 42: 320-328.

Chutivisut, P., Isobe, K., Powtongsook, S., Pungrasmi, W. and Kurisu, F. 2018. Distinct microbial community performing dissimilatory nitrate reduction to ammonium (DNRA) in a high $C/NO_3^-$ reactor. Microbes and Environment 33: 264-271.

Clark, T.A., Dare, P.H. and Bruce, M.E. 1997. Nitrogen fixation in an aerated stabilization basin treating bleached kraft mill wastewater. Water Environment Research 69: 1039-1046.

Coelho, C. and Romao, M.J. 2015. Structural and mechanistic insights on nitrate reductases. Protein Science 24: 1901-1911.

Cooper, P.F. 2001. Historical aspects of wastewater treatment. pp. 11-38. *In*: Lens, P., Zeeman, G. and Lettinga, G. (eds.). Decentralised Sanitation and Reuse: Concepts, Systems and Implementation. IWA Publishing, London, UK.

Council Directive 91/271/EEC of 21 May 1991 concerning urban waste-water treatment, OJL 135, 30.5.1991, pp. 40-52.

Coyne, M.S. 2018. Denitrification in soil. pp. 96-140. *In*: Lal, R. and Stewart, B.A. (eds.). Soil Nitrogen Uses and Environmental Impacts. CRC Press, Boca Raton, Florida, USA.

Cydzik-Kwiatkowska, A. 2015. Bacterial structure of aerobic granules is determined by aeration mode and nitrogen load in the reactor cycle. Bioresource Technology 181: 312-320.

Daebeler, A., Herbold, C., Vierheilig, J., Sedlacek, C.J., Pjevac, P., Albertsen, M., et al. 2018. Cultivation and genomic analysis of a *Candidatus* Nitrosocaldus islandicus, an obligately thermophilic, ammonia-oxidizing thaumarchaeon from a hot spring biofilm in Graendalur valley, Iceland. Frontiers in Microbiology 9: 193.

Daims, H. and Wagner, M. 2010. The microbiology of nitrogen removal. pp. 259-280. *In*: Seviour, R.J. and Nielsen, P.H. (eds.). Microbial Ecology of Activated Sludge. IWA Publishing, London, UK.

Daims, H. 2014. The family *Nitrospiraceae*. pp. 733-750. *In*: Rosemberg, E., de Long, E.F., Lory, S., Stackebrandt, E. and Thompson, F. (eds.). The Prokaryotes 4th Ed. – Prokaryotic Physiology and Biochemistry. Springer-Verlag.

Daims, H., Lebedeva, E.V., Pjevac, P., Han, P., Herbold, C., Albertsen, M., et al. 2015. Complete nitrification by *Nitrospira* bacteria. Nature 528: 504-509.

Daims, H., Lücker, S. and Wagner, M. 2016. A new perspective on microbes formerly known as nitrite-oxidizing bacteria. Trends in Microbiology 24: 699-712.

de Boer, W. and Kowalchuk, G.A. 2001. Nitrification in acid soils: micro-organisms and mechanisms. Soil Biology and Biochemistry 33: 853-866.

de la Torre, J.R., Walker, C.B., Ingalls, A.E., Knneke, M. and Stahl, D.A. 2008. Cultivation of a thermophilic ammonia-oxidizing archaeon synthesizing crenarchaeol. Environmental Microbiology 10: 810-818.

de Vries S., Suharti and Pouvreau LAM. 2007. Nitric oxide reductase: structural variations and catalytic mechanism. pp. 57-66. *In*: Bothe, H., Ferguson, S.J. and Newton, W.E. (eds.). Biology of the Nitrogen Cycle. Elsevier, Amsterdam, The Netherlands.

di Capua, F., Pirozzia, F., Lens, P.N.L. and Esposito, G. 2019. Electron donors for autotrophic denitrification. Chemical Engineering Journal 362: 922-937.

Dixon, R.A. and Kahn, D. 2004. Genetic regulation of biological nitrogen fixation. Nature Reviews Microbiology 2: 621-631.

Du, C., Cui, C., Shi, S. and Ma, F. 2014. Identification of a highly efficient aerobic denitrifying bacterium in SBR and denitrification optimization. Advanced Materials Research 955-959: 376-382.

Du, R., Cao, S., Peng, Y., Zhang, H. and Wang, S. 2019a. Combined partial denitrification (PD)-anammox: a method for high nitrate wastewater treatment. Environment International 126: 707-716.

Du, R., Peng, Y., Ji, J., Shi, L., Gao, R. and Li, X. 2019b. Partial denitrification providing nitrite: opportunities of extending application for anammox. Environment International 131: 105001.

Durai, G. and Rajasimman, M. 2011. Biological treatment of tannery wasterwater - a review. Journal of Environmental Science and Technology 4: 1-17.

Ehaliotis, C., Papadopoulou, K., Kotsou, M., Mari, I. and Balis, C. 1999. Adaptation and population dynamics of *Azobacter vinelandii* during aerobic biological treatment of olive-mill wastewater. FEMS Microbiology Ecology 30: 301-311.

Ehrich, S., Behrens, D., Lebedeva, E., Ludwig, W. and Bock, E. 1995. A new obligately chemolithoautotrophic, nitrite-oxidizing bacterium, *Nitrospira moscoviensis* sp. nov. and its phylogenetic relationship. Archives of Microbiology 164: 16-23.

Ekama, G.A. and Wentzel, M.C. 2008. Biological nitrogen removal. pp. 97-156. *In*: Henze, M., van Loosdrecht, M.C.M., Ekama, G.A. and Brdjanovic, D. (eds.). Biological Wastewater Treatment: Principles, Modelling and Design. IWA Publishing, London, UK.

Emerson, K., Russo, R.C., Lund, R.E. and Thurston, R.V. 1975. Aqueous ammonia equilibrium calculations: effect of pH and temperature. Journal of the Fisheries Research Board of Canada 32: 2379-2383.

Esfahani, E.B., Zeidabadi, F.A., Bazargan, A. and McKay, G. 2019. The modified Bardenpho process. pp. 1551-1592. *In*: Hussain, C.M. (ed.). Handbook of Environmental Materials Management. Springer Nature Switzerland AG.

Ettwig, K.F., Butler, M.K., Le Paslier, D., Pelletier, E., Mangenot, S., Kuypers, M.M., et al. 2010. Nitrite-driven anaerobic methane oxidation by oxygenic bacteria. Nature 464: 543-548.

Ettwig, K.F., Speth, D.R., Reimann, J., Wu, M.L., Jetten, M.S. and Keltjens, J.T. 2012. Bacterial oxygen production in the dark. Frontiers in Microbiology 3: 273.

Euzeby, J.P. and Kudo, T. 2001. Corrigenda to the validation lists. International Journal of Systematic and Evolutionary Microbiology 51: 1933-1938.

Fan, X.Y., Gao, J.F., Pan, K.L., Li, D.C. and Dai, H.H. 2017. Temporal dynamics of bacterial communities and predicted nitrogen metabolism genes in a full-scale wastewater treatment plant. RSC Advances 7: 56317-56327.

Ferrera, I. and Sánchez, O. 2016. Insights into microbial diversity in wastewater treatment systems: how far have we come? Biotechnology Advances 34: 790-802.

Finnegan, W., Clifford, E., Goggins, J., O'Leary, N., Dobson, A., Rowan, N., et al. 2018. Dairy Water: striving for sustainability within the dairy processing industry in the Republic of Ireland. Journal of Dairy Research 85: 366-374.

Fitzgerald, C.M., Camejo, O., Zachary Oshlag, J. and Noguera, D.R. 2015. Ammonia-oxidizing microbial communities with efficient nitrification at low-dissolved oxygen. Water Research 70: 38-51.

Friedler, E., Butler, D. and Alfiya, Y. 2013. Wastewater composition. In: Larsen, T.A., Udert, K.M. and Lienert, J. (eds). Source Separation and Decentralization for Wastewater Management. IWA Publishing, London, UK.

Füssel, J., Lücker, S., Yilmaz, P., Nowka, B., van Kessel, M.A.H.J., Bourceau, P., et al. 2017. Adaptability as the key to success for the ubiquitous marine nitrite oxidizer Nitrococcus. Scientific Advances 3: e1700807.

Gao, D., Peng, Y., Li, B. and Liang, H. 2009. Shortcut nitrification–denitrification by real-time control strategies. Bioresource Technology 100: 2298-2300.

Gapes, D.J., Frost, N.M., Clark, T.A., Dare, P.H., Hunter, R.G. and Slade, A.H. 1999. Nitrogen-fixation in the treatment of pulp and paper wastewaters. Water Science and Technology 40: 85-92.

Gapes, D.J., Clark, T., Frost, N.M. and Slade, A.H. 2003. Wastewater treatment process for nitrogen-deficient feed in controlled environment. US Patent 6623641B1.

Gauthier, F., Neufeld, J.D., Driscoll, B.T. and Archibald, F.S. 2000. Coliform bacteria and nitrogen fixation in pulp and paper mill effluent treatment systems. Applied and Environmental Microbiology 66: 5155-5160.

Ge, S., Wang, S., Yang, X., Qiu, S., Li, B. and Peng, Y. 2015. Detection of nitrifiers and evaluation of partial nitrification for wastewater treatment: a review. Chemosphere 140: 85-98.

Geets, J., de Cooman, M., Wittebolle, L., Heylen, K., Vanparys, B., De Vos, P., et al. 2007. Real-time PCR assay for the simultaneous quantification of nitrifying and denitrifying bacteria in activated sludge. Applied Microbiology and Biotechnology 75: 211-221.

Gerardi, MH. 2006. Wastewater Bacteria. Wiley-Interscience.

Ghafari, S., Hasan, M. and Aroua, M.K. 2008. Bio-electrochemical removal of nitrate from water and wastewater—a review. Bioresource Technology 99: 3965-3974.

Gieseke, A., Purkhold, U., Wagner, M., Amann, R. and Schramm, A. 2001. Community structure and activity dynamics of nitrifying bacteria in a phosphate-removing biofilm. Applied and Environmental Microbiology 67: 1351-1362.

Glass, C. and Silverstein, J. 1999. Denitrification of high-nitrate, high-salinity wastewater. Water Research 33: 223-229.

Gómez-Villalba, B., Calvo, C., Vílchez, R., González-López, J. and Rodelas, B. 2006. TGGE analysis of the diversity of ammonia-oxidizing and denitrifying bacteria in submerged filter biofilms for the treatment of urban wastewater. Applied Microbiology and Biotechnology 72: 393-400.

González-Martínez, A., Rodriguez-Sanchez, A., Lotti, T., Garcia-Ruiz, M.J., Osorio, F., Gonzalez-Lopez, J., et al. 2016. Comparison of bacterial communities of conventional and A-stage activated sludge systems. Scientific Reports 5: 18786.

González-Martínez, A., Muñoz-Palazón, B., Rodríguez-Sánchez, A. and González-López, J. 2018. New concepts in anammox processes for wastewater nitrogen removal: recent advances and future prospects. FEMS Microbiology Letters 365: fny031.

Griffin, B.M., Schott, J. and Schink, B. 2007. Nitrite, an electron donor for anoxygenic photosynthesis. Science 316: 1870.

Haaijer, S.C., Ji, K., van Niftrik, L., Hoischen, A., Speth, D., Jetten, M.S., et al. 2013. A novel marine nitrite-oxidizing Nitrospira species from Dutch coastal North Sea water. Frontiers in Microbiology 4: 60.

Hallin, S., Throbäck, I.N., Dicksved, J. and Pell, M. 2006. Metabolic profiles and genetic diversity of denitrifying communities in activated sludge after addition of methanol or ethanol. Applied and Environmental Microbiology 72: 5445-5452.

Hallin, S., Philippot, L., Löffler, F.E., Sandford, R.A. and Jones, C.M. 2018. Genomic and ecology of novel $N_2O$-reducing microorganisms. Trends in Microbiology 26: 43-55.

Harhangi, H.R., Le Roy, M., van Alen, T., Hu, B.L., Groen, J., Kartal, B., et al. 2012. Hydrazine synthase, a unique phylomarker with which to study the presence and biodiversity of anammox bacteria. Applied and Environmental Microbiology 78: 752-758.

Häring, V. and Conrad, R. 1991. Kinetics of $H_2$ oxidation in respiring and denitrifying Paracoccus denitrificans. FEMS Microbiology Letters 78: 259-264.

Haroon, M.F., Hu, S., Shi, Y., Imelfort, M., Keller, J., Hugenholtz, P., et al. 2013. Anaerobic oxidation of methane coupled to nitrate reduction in a novel archaeal lineage. Nature 500: 567-570.

Hatzenpichler, R., Lebedeva, E.V., Spieck, E., Stoecker, K., Richter, A., Daims, H., et al. 2008. A moderately thermophilic ammonia-oxidizing crenarchaeote from a hot spring. Proceedings of the National Academy of Sciences of the USA 105: 2134-2139.

He, S.B., Xue, G. and Wang, B.Z. 2009. Factors affecting simultaneous nitrification and de-nitrification (SND) and its kinetics model in membrane bioreactor. Journal of Hazardous Materials 168: 704-710.

He, Z., Feng, Y., Zhang, S., Wang, X., Wu, S. and Pan, X. 2018. Oxygenic denitrification for nitrogen removal with less greenhouse gas emissions: microbiology and potential applications. Science of the Total Environment 621: 453-464.

Hellinga, C., Schellen, A.A.J.C., Mulder, J.W., van Loosdrecht, M.C.M. and Heijnen, J.J. 1998. The SHARON process: an innovative method for nitrogen removal from ammonium-rich wastewater. Water Science and Technology 37: 135-142.

Herbold, C.W., Lebedeva, E., Palatinszky, M. and Wagner, M. 2016. *Candidatus* Nitrosotenuaceae. *Bergey's Manual of Systematics of Bacteria and Archaea* (online version, https://doi.org/10.1002/9781118960608. fbm00263).

Heylen, K., Gevers, D., Vanparys, B., Wittebolle, L., Geets, J., Boon, N., et al. 2006. The incidence of *nirS* and *nirK* and their genetic heterogeneity in cultivated denitrifiers. Environmental Microbiology 8: 2012-2021.

Heylen, K., Vanparys, B., Gevers, D., Wittebolle, L., Boon, N. and De Vos, P. 2007. Nitric oxide reductase (*norB*) gene sequence analysis reveals discrepancies with nitrite reductase (*nir*) gene phylogeny in cultivated denitrifiers. Environmental Microbiology 9: 1072-1077.

Heylen, K. and Keltjens, J. 2012. Redundancy and modularity in membrane-associated dissimilatory nitrate reduction in *Bacillus*. Frontiers in Microbiology 3: 371.

Holmes, D.E., Dang, Y. and Smith, J. 2019. Nitrogen cycling during wastewater treatment. Advances in Applied Microbiology 106: 113-192.

Horrell, S., Kekilli, D., Strange, R.W. and Hough, M.A. 2017. Recent structural insights into the function of copper nitrite reductases. Metallomics 9: 1470-1482.

Hu, B.L., Zheng, P., Tang, C.J., Chen, J.W., van der Biezen, E., Zhang, L., et al. 2010. Identification and quantification of anammox bacteria in eight nitrogen removal reactors. Water Research 44: 5014-5020.

Huang, T.L., Guo, L., Zhang, H.H., Su, J.F., Wen, G. and Zhang, K. 2015. Nitrogen-removal efficiency of a novel aerobic denitrifying bacterium, *Pseudomonas stutzeri* strain ZF31, isolated from a drinking-water reservoir. Bioresource Technology 196: 209-216.

IPCC. 2013. Climate Change 2013: The Physical Science Basis. Working Group I Contribution to the IPCC 5th Assessment Report. IPCC, Cambridge, United Kingdom and New York, NY, USA.

Ishii, K., Fujitani, H., Soh, K., Nakagawa, T., Takahashi, R. and Tsuneda, S. 2017. Enrichment and physiological characterization of a cold-adapted nitrite-oxidizing *Nitrotoga* sp. from an Eelgrass sediment. Applied and Environmental Microbiology 83: e00549-17.

Iskander, S.M., Brazil, B., Novak, J.T. and He, Z. 2016. Resource recovery from landfill leachate using bioelectrochemical systems: opportunities, challenges, and perspectives. Bioresource Technology 201: 347-354.

Jaramillo, F., Orchard, M., Muñoz, C., Zamorano, M. and Antileo, C. 2018. Advanced strategies to improve nitrification process in sequencing batch reactors - a review. Journal of Environmental Management 218: 154-164.

Jetten, M.S.M., Strous, M., Van de Pas-Schoonen, K.T., Schalk, J., van Dongen, U.G.J.M., van de Graaf, A.A., et al. 1998. The anaerobic oxidation of ammonium. FEMS Microbiology Reviews 22: 421-437.

Ji, B., Yang, K., Zhu, L., Jiang, Y., Wang, H., Zhou, J., et al. 2015. Aerobic denitrification: a review of important advances of the last 30 years. Biotechnology Bioprocesses and Engineering 20: 643-651.

Jiang, Q.Q. and Bakken, LR. 1999. Comparison of *Nitrosospira* strains isolated from terrestrial environments. FEMS Microbiology Ecology 30: 171-186.

Jones, C.M., Graf, D.R.H., Bru, D., Philippot, L. and Hallin, S. 2013. The unaccounted yet abundant nitrous oxide-reducing microbial community: a potential nitrous oxide sink. ISME Journal 7: 417-426.

Jones, C.M., Spor, A., Brennan, F.P., Breuil, M.C., Bru, D., Lemanceau, P., et al. 2014. Recently identified microbial guild mediates soil $N_2O$ sink capacity. Nature Climate Change 4: 801-805.

Joo, H.S., Hirai, M. and Shoda, M. 2006. Piggery wastewater treatment using *Alcaligenes faecalis* strain No. 4 with heterotrophic nitrification and aerobic denitrification. Water Research 40: 3029-3036.

Juárez, M.J.B., Zafra-Gómez, A., Luzón-Toro, B., Ballesteros-García, O.A., Navalón, A., González, J., et al. 2008. Gas chromatographic–mass spectrometric study of the degradation of phenolic compounds in wastewater olive oil by *Azotobacter chroococcum*. Bioresource Technology 99: 2392-2398.

Judd, S. 2011. The MBR Book. Principles and Applications of Membrane Bioreactors in Water and Wastewater Treatment, 2$^{nd}$ Ed. Elsevier Science.

Judicial Commission of the International Committee on Systematics of Prokaryotes. 2008. Status of strains that contravene rules 27(3) and 30 of the international code of nomenclature of bacteria. Opinion 81. International Journal of Systematic and Evolutionary Microbiology 58: 1755-1763.

Jung, M.Y., Park, S.J., Kim, S.J., Kim, J.G., Sinninghe Damste, J.S., Jeon, C.O., et al. 2014a. A mesophilic, autotrophic, ammonia-oxidizing archaeon of thaumarchaeal group I.1a cultivated from a deep oligotrophic soil horizon. Applied and Environmental Microbiology 80: 3645-3655.

Jung, M.Y., Well, R., Min, D., Giesemann, A., Park, S.J., Kim, J.G., et al. 2014b. Isotopic signatures of N$_2$O produced by ammonia-oxidizing archaea from soils. ISME Journal 8: 1115-1125.

Jung, M.Y., Kim, J.G., Sinninghe Damsté, J.S., Rijpstra, W.I., Madsen, E.L., Kim, S.J., et al. 2016. A hydrophobic ammonia-oxidizing archaeon of the Nitrosocosmicus clade isolated from coal tar-contaminated sediment. Environmental Microbiology Reports 8: 983-992.

Jung, M.Y., Islam, M.A., Gwak, J.H., Kim, J.G. and Rhee, S.K. 2018. *Nitrosarchaeum koreense* gen. nov., sp. nov., an aerobic and mesophilic, ammonia-oxidizing archaeon member of the phylum *Thaumarchaeota* isolated from agricultural soil. International Journal of Systematic and Evolutionary Microbiology 68: 3084-3095.

Karanasios, K.A., Vasiliadou, I.A., Pavlou, S. and Vayenas, D.V. 2010. Hydrogenotrophic denitrification of potable water: a review. Journal of Hazardous Materials 180: 20-37.

Kargi, F. and Özmıhçı, S. 2002. Improved biological treatment of nitrogen deficient wastewater by incorporation of N$_2$-fixing bacteria. Biotechnology Letters 24: 1281-1284.

Kargi, F. and Özmıhçı, S. 2004. Batch biological treatment of nitrogen deficient synthetic wastewater using *Azotobacter* supplemented activated sludge. Bioresource Technology 94: 113-117.

Kartal, B., van Niftrik, L., Keltjens, J.T., Op den Camp, H.J. and Jetten, M.S. 2012. Anammox-growth physiology, cell biology, and metabolism. Advances in Microbial Physiology 60: 211-262.

Kartal, B. and Keltjens, J.T. 2016. Anammox biochemistry: a tale of heme c proteins. Trends in Biochemical Sciences 41: 998-1011.

Kerou, M., Alves, R.J.E. and Schleper, C. 2018. Nitrososphaerales. Bergey's Manual of Systematics of Bacteria and Archaea (online version, https://doi.org/10.1002/9781118960608.obm00124).

Khardenavis, A.A., Kapley, A. and Purohit, H.J. 2007. Simultaneous nitrification and denitrification by diverse *Diaphorobacter* sp. Applied Microbiology and Biotechnology 77: 403-409.

Kim, Y.M., Cho, H.U., Lee, D.S., Park, D. and Park, J.M. 2011. Influence of operational parameters on nitrogen removal efficiency and microbial communities in a full scale activated sludge process. Water Research 45: 5785-5795.

Kitzinger, K., Koch, H., Lucker, S., Sedlacek, C.J., Herbold, C., Schwarz, J., et al. 2018. Characterization of the first *Candidatus* Nitrotoga isolate reveals metabolic versatility and separate evolution of widespread nitrite-oxidizing bacteria. mBio 9: e01186-18.

Koch, H., van Kessel, M.A.H.J. and Lücker, S. 2019. Complete nitrification: insights into the ecophysiology of comammox *Nitrospira*. Applied Microbiology and Biotechnology 103: 177-189.

Kong, Q.X., Wang, X.W., Jin, M., Shen, Z.Q. and Li, J.W. 2006. Development and application of a novel and effective screening method for aerobic denitrifying bacteria. FEMS Microbiology Letters 260: 150-155.

Könneke, M., Bernhard, A.E., de la Torre, J.R., Walker, C.B., Waterbury, J.B. and Stahl, D.A. 2005. Isolation of an autotrophic ammonia-oxidizing marine archaeon. Nature 437: 543-546.

Könneke, M., Schubert, D.M., Brown, P.C., Hugler, M., Standfest, S., Schwander, T., et al. 2014. Ammonia-oxidizing archaea use the most energy-efficient aerobic pathway for CO$_2$ fixation. Proceedings of the National Academy of Sciences of the USA 111: 8239-8244.

Koops, H.P. and Pommerening-Röser, A. 2001. Distribution and ecophysiology of the nitrifying bacteria emphasizing cultured species. FEMS Microbiology Ecology 37: 1-9.

Koops, H.P. and Pommerening-Röser, A. 2005a. Genus I. *Nitrosomonas* Winogradsky 1892, 127AL (Nom. Cons. Opin. 23 Jud. Comm. 1958, 169). pp. 865-867. *In*: Brenner, D.J., Krieg, N.R., Staley, J.T. and Garrity, G.M. (eds.). Bergey's Manual of Systematic Bacteriology, 2nd Ed., Vol. 2. The Proteobacteria-Part C- The Alpha-, Beta-, Delta-, and Epsilonproteobacteria, Springer, Berlin, Germany.

Koops, H.P. and Pommerening-Röser, A. 2005b. Genus III. *Nitrosospira* Winogradsky and Winogradsky 1933, 406AL. pp. 868-869. *In*: Brenner, D.J., Krieg, N.R., Staley, J.T. and Garrity, G.M. (eds.). Bergey's Manual of Systematic Bacteriology, 2nd Ed., Vol. 2. The Proteobacteria-Part C- The Alpha-, Beta-, Delta-, and Epsilonproteobacteria, Springer, Berlin, Germany.

Koops, H.P. and Pommerening-Röser, A. 2005c. Genus VIII. *Nitrosococcus* Winogradsky 1892, 127AL (Nom. Cons. Opin. 23 Jud. Comm. 1958b, 169). pp. 21-22. *In*: Brenner, D.J, Krieg, N.R., Staley, J.T. and Garrity, G.M. (eds.). Bergey's Manual of Systematic Bacteriology, 2nd Ed., Vol. 2. The Proteobacteria-Part B- The Gammaproteobacteria, Springer, Berlin, Germany.

Korth, B. and Harnisch, F. 2019. Spotlight on the energy harvest of electroactive microorganisms: the impact of the applied anode potential. Frontiers in Microbiology 10: 1352.

Kraft, B., Strous, M. and Tegetmeyer, H.E. 2011. Microbial nitrate respiration – genes, enzymes and environmental distribution. Journal of Biotechnology 155: 104-117.

Kumar, M., Lee, P.Y., Fukusihma, T., Whang, L.M. and Lin, J.G. 2012. Effect of supplementary carbon addition in the treatment of low C/N high-technology industrial wastewater by MBR. Bioresource Technology 113: 148-153.

Kumwimba, M. and Meng, F. 2019. Roles of ammonia-oxidizing bacteria in improving metabolism and cometabolism of trace organic chemicals in biological wastewater treatment processes: a review. Science of the Total Environment 659: 419-441.

Kundu, P., Pramanik, A., Dasgupta, A., Mukheriee, S. and Mukheriee, J. 2014. Simultaneous heterotrophic nitrification and aerobic denitrification by *Chryseobacterium* sp. R31 isolated from abattoir wastewater. BioMed Research International 436056: 12.

Kuypers, M.M.M., Marchant, H.K. and Kartal, B. 2018. The microbial nitrogen-cycling network. Nature Reviews Microbiology 16: 263-276.

Lackner, S., Gilbert, E.M., Vlaeminck, S.E., Joss, A., Horn, H. and van Loosdrecht, M.C.M. 2014. Full-scale partial nitritation/anammox experiences-an application survey. Water Research 55: 292-303.

Lam, P. and Kuypers, M.M.M. 2011. Microbial nitrogen cycling processes in oxygen minimum zones. Annual Reviews of Marine Science 3: 317-345.

Lawson, C.E. and Lücker, S. 2018. Complete ammonia oxidation: an important control on nitrification in engineered systems? Current Opinion in Biotechnology 50: 158-165.

Lebedeva, E.V., Alawi, M., Maixner, F., Jozsa, P.-G., Daims, H. and Spieck, E. 2008. Physiological and phylogenetic characterization of a novel lithoautotrophic nitrite-oxidizing bacterium, *Candidatus* Nitrospira bockiana. International Journal of Systematic and Evolutionary Microbiology 58: 242-250.

Lebedeva, E.V., Off, S., Zumbragel, S., Kruse, M., Shagzhina, A., Lucker, S., et al. 2011. Isolation and characterization of a moderately thermophilic nitrite-oxidizing bacterium from a geothermal spring. FEMS Microbiology Ecology 75: 195-204.

Lebedeva, E.V., Hatzenpichler, R., Pelletier, E., Schuster, N., Hauzmayer, S., Bulaev, A., et al. 2013. Enrichment and genome sequence of the group I.1a ammonia-oxidizing Archaeon Ca. Nitrosotenuis uzonensis representing a clade globally distributed in thermal habitats. PLoS One 8(11): e80835.

Lehtovirta-Morley, L.E., Stoecker, K., Vilcinskas, A., Prosser, J.I. and Nicol, G.W. 2011. Cultivation of an obligate acidophilic ammonia oxidizer from a nitrifying acid soil. Proceedings of the National Academy of Sciences of the USA 108: 15892-15897.

Lehtovirta-Morley, L.E., Ross, J., Hink, L., Weber, E.B., Gubry-Rangin, C., Thion, C., et al. 2016. Isolation of *Candidatus* Nitrosocosmicus franklandus', a novel ureolytic soil archaeal ammonia oxidiser with tolerance to high ammonia concentration. FEMS Microbiology Ecology 92: fiw057.

Li, C.N., Yang, J.H., Wang, X., Wang, E.T., Li, B.Z., He, R.X., et al. 2015. Removal of nitrogen by heterotrophic nitrification-aerobic denitrification of a phosphate accumulating bacterium *Pseudomonas stutzeri* YG-24. Bioresource Technology 182: 18-25.

Li, D., Liang, X., Jin, Y., Wu, C. and Zhou, R. 2019. Isolation and nitrogen removal characteristics of an aerobic heterotrophic nitrifying-denitrifying bacterium, *Klebsiella* sp. TN-10. Applied Biochemistry and Biotechnology 188: 540-554.

Li, E., Jin, X. and Lu, S. 2018. Microbial communities in biological denitrification system using methanol as carbon source for treatment of reverse osmosis concentrate from coking wastewater. Journal of Water Reuse and Desalination 8: 360-371.

Li, Y., Ding, K., Wen, X., Zhang, B., Shen, B. and Yang, Y. 2016. A novel ammonia-oxidizing archaeon from wastewater treatment plant: its enrichment, physiological and genomic characteristics. Scientific Reports 6: 23747.

Liao, R., Miao, Y., Li, J., Li, Y., Wang, Z., Du, J., et al. 2018. Temperature dependence of denitrification microbial communities and functional genes in an expanded granular sludge bed reactor treating nitrate-rich wastewater. RSC Advances 8: 42087-42094.

Limpiyakorn, T., Shinohara, Y., Kurisu, F. and Yagi, O. 2005. Communities of ammonia-oxidizing bacteria in activated sludge of various sewage treatment plants in Tokyo. FEMS Microbiology Ecology 54: 205-217.

Limpiyakorn, T., Kurisu, F., Sakamoto, Y. and Yagi, O. 2007. Effects of ammonium and nitrite on communities and populations of ammonia-oxidizing bacteria in laboratory-scale continuous flow reactors. FEMS Microbiology Ecology 60: 501-512.

Limpiyakorn, T., Fürhacker, M., Haberl, R., Chodanon, T., Srithep, P. and Sonthiphand, P. 2013. *amoA*-encoding archaea in wastewater treatment plants: a review. Applied Microbiology and Biotechnology 97: 1425-1439.

Liu, X., Shu, Z., SUn, D., Dang, Y. and Holmes, D.E. 2018. Heterotrophic nitrifiers dominate reactors treating incineration leachate with high free ammonia concentrations. ACS Sustainable Chemistry and Engineering 6: 1504-15049.

Liu, Y., Ngo, H.H., Guo, W., Peng, L., Wang, D. and Ni, B. 2019. The roles of free ammonia (FA) in biological wastewater treatment processes: a review. Environment International 123: 10-19.

Lockwood, C., Butt, J.N., Clarke, T.A. and Richardson, D.J. 2011. Molecular interactions between multihaem cytochromes: probing the protein-protein interactions between pentahaem cytochromes of a nitrite reductase complex. Biochemical Society Transactions 39: 263-268.

Lu, H., Chandran, K. and Stensel, D. 2014. Microbial ecology of denitrification in biological wastewater treatment. Water Research 64: 237-254.

Lücker, S., Nowka, B., Rattei, T., Spieck, E. and Daims, H. 2013. The genome of *Nitrospina gracilis* illuminates the metabolism and evolution of the major marine nitrite oxidizer. Frontiers in Microbiology 4: 27.

Lücker, S., Schwarz, J., Gruber-Dorninger, C., Spieck, E., Wagner, M. and Daims, H. 2015. *Nitrotoga*-like bacteria are previously unrecognized key nitrite oxidizers in full-scale wastewater treatment plants. ISME Journal 9: 708-720.

Ludzack, F.J. and Ettinger, M.B. 1962. Controlling operation to minimize activated sludge effluent nitrogen. Journal of the Water Pollution Control Federation 34: 920-931.

Lv, P., Luo, J., Zhuang, X., Zhang, D., Huang, Z. and Bai, Z. 2017. Diversity of culturable aerobic denitrifying bacteria in the sediment, water and biofilms in Liangshui River of Beijing, China. Scientific Reports 7: 10032.

Lydmark, P., Almstrand, R., Samuelsson, K., Mattsson, A., Sorensson, F., Lindgren, P.E., et al. 2007. Effects of environmental conditions on the nitrifying population dynamics in a pilot wastewater treatment plant. Environmental Microbiology 9: 2220-2233.

Ma, B., Wang, S., Cao, S., Miao, Y., Jia, F., Du, R., et al. 2016. Biological nitrogen removal from sewage via anammox: recent advances. Bioresource Technology 200: 981-990.

Ma, Q., Qu, Y., Shen, W., Zhang, Z., Wang, J., Liu, Z., et al. 2015. Bacterial community compositions of coking wastewater treatment plants in steel industry revealed by Illumina high-throughput sequencing. Bioresource Technology 179: 436-443.

Ma, W.W., Han, Y.S., Ma, W.C., Han, H.J., Zhu, H., Xu, C.Y., et al. 2017. Enhanced nitrogen removal from coal gasification wastewater by simultaneous nitrification and denitrification (SND) in an oxygen-limited aeration sequencing batch biofilm reactor. Bioresource Technology 244: 84-91.

Madoni, P. 2011. Protozoa in wastewater treatment processes: a minireview. Italian Journal of Zoology 78: 3-11.

Manser, N.D., Wang, M., Ergas, S.J., Mihelcic, J.R., Mulder, A., van de Vossenberg, J., et al. 2016. Biological nitrogen removal in a photo-sequencing batch reactor with an algal-nitrifying bacterial consortium and anammox granules. Environmental Science and Technology Letters 3: 175-179.

Mao, N., Ren, H., Geng, J., Ding, L. and Xu, K. 2017. Engineering application of anaerobic ammonium oxidation process in wastewater treatment. World Journal of Microbiology and Biotechnology 3: 153.

Mao, Y., Xia, Y. and Zhang, T. 2013. Characterization of *Thauera*-dominated hydrogen-oxidizing autotrophic denitrifying microbial communities by using high-throughput sequencing. Bioresource Technology 128: 703-710.

Maza-Márquez, P., Vílchez-Vargas, R., Kerckhof, F.M., Aranda, E., González-López, J. and Rodelas, B. 2016. Community structure, population dynamics and diversity of fungi in a full-scale membrane bioreactor (MBR) for urban wastewater treatment. Water Research 105: 507-519.

Maza-Márquez, P., González-Martínez, A., Juárez-Jiménez, B., Rodelas, B. and González-López, J. 2018. Microalgae-bacteria consortia for the removal of phenolic compounds from industrial wastewaters. pp. 135-184. *In*: Prasad, R. and Aranda, E. (eds.). Approaches in Bioremediation: The New Era of Environmental Microbiology and Nanobiotechnology. Springer International Publishing.

McCarty, P.L. 2018. What is the best biological process for nitrogen removal: when and why? Environmental Science and Technology 52: 3835-3841.

Merbt, S.N., Auguet, J.C., Blesa, A., Martí, E. and Casamayor, E.O. 2015. Wastewater treatment plant effluents change abundance and composition of ammonia-oxidizing microorganisms in mediterranean urban stream biofilms. Microbial Ecology 69: 66-74.

Meyer, R.L., Zeng, R.J., Giugliano, V. and Blackall, L.L. 2005. Challenges for simultaneous nitrification, denitrification, and phosphorus removal in microbial aggregates: mass transfer limitation and nitrous oxide production. FEMS Microbiology Ecology 52: 329-338.

Mohan, S.B. and Cole, J.A. 2007. The dissimilatory reduction of nitrate to ammonia by anaerobic bacteria. pp. 93-106. *In*: Bothe, H., Ferguson, S.J. and Newton, W.E. (eds.). Biology of the Nitrogen Cycle. Elsevier, Amsterdam, The Netherlands.

Moreno-Vivián, C. and Flores, E. 2007. Nitrate assimilation in bacteria. pp. 263-282. *In*: Bothe, H., Ferguson, S.J. and Newton, W.E. (eds.). Biology of the Nitrogen Cycle. Elsevier, Amsterdam, The Netherlands.

Mothapo, N., Chen, H., Cubeta, M.A., Grossman, J.M., Fuller, F. and Shi, W. 2015. Phylogenetic, taxonomic and functional diversity of fungal denitrifiers and associated $N_2O$ production efficacy. Soil Biology and Biochemistry 83: 160-175.

Mukarunyana, B., van de Vossenberg, J., van Lier, J.B. and van der Steen, P. 2018. Photo-oxygenation for nitritation and the effect of dissolved oxygen concentrations on anaerobic ammonium oxidation. Science of the Total Environment 634: 868-874.

Münch, E.V., Lant, P. and Keller, J. 1996. Simultaneous nitrification and denitrification in bench-scale sequencing batch reactors. Water Research 30: 277-284.

Mußmann, M., Brito, I., Pitcher, A., Sinninghe Damsté, J.S., Hatzenpichler, R., Richter, A., et al. 2011. *Thaumarchaeotes* abundant in refinery nitrifying sludges express *amoA* but are not obligate autotrophic ammonia oxidizers. Proceedings of the National Academy of Sciences of the USA 108: 16771-16776.

Nakagawa, T., Tsuchiya, Y. and Takahashi, R. 2019. Whole-genome sequence of the ammonia-oxidizing bacterium *Nitrosomonas stercoris* type strain KYUHI-S, isolated from composted cattle manure. Microbiology Resources Announcements 8: e00742-19.

Nancharaiah, Y.V. and Reddy, G.K.K. 2018. Aerobic granular sludge technology: mechanisms of granulation and biotechnological applications. Bioresource Technology 247: 1128-1143.

Newton, W.E. 2007. Physiology, biochemistry, and molecular biology of nitrogen fixation. pp. 109-129. *In*: Bothe, H., Ferguson, S.J. and Newton, W.E. (eds.). Biology of the Nitrogen Cycle. Elsevier, Amsterdam, The Netherlands.

Ngugi, D.K., Blom, J., Stepanauskas, R. and Stingl, U. 2016. Diversification and niche adaptations of *Nitrospina*-like bacteria in the polyextreme interfaces of Red Sea brines. ISME Journal 10: 1383-1399.

Ni, B.J., Pan, Y., Guo, J., Virdis, B., Hu, S., Chen, X., et al. 2017. Denitrification processes for wastewater treatment. pp. 368-418. *In*: Moura, I., Moura, J.J.G., Pauleta, S.R. and Maia, L.B. (eds.). Metalloenzymes in Denitrification: Applications and Environmental Impacts. RSC Publishing, Cambridge, UK.

Niaounakis, M. and Halvadakis, C. 2006. Olive Processing Waste Management. Waste Management Series, Vol. 5. London: Elsevier.

Nielsen, P.H., Saunders, A.M., Hansen, A.A., Larsen, P. and Nielsen, J.L. 2012. Microbial communities involved in enhanced biological phosphorus removal from wastewater—a model system in environmental biotechnology. Current Opinion in Biotechnology 23: 452-459.

Nielsen, P.H., McIlroy, S.J., Albertsen, M. and Nierychlo, M. 2019. Re-evaluating the microbiology of the enhanced biological phosphorus removal process. Current Opinion in Biotechnology 57: 111-118.

Norton, J.M. 2011. Diversity and environmental distribution of ammonia-oxidizing bacteria. pp. 39-55. *In*: Ward, B., Arp, D.J. and Klotz, M.G. (eds.). Nitrification. ASM Press, Washington.

Nowka, B., Daims, H. and Spieck, E. 2015a. Comparison of oxidation kinetics of nitrite-oxidizing bacteria: nitrite availability as a key factor in niche differentiation. Applied and Environmental Microbiology 81: 745-753.

Nowka, B., Off, S., Daims, H. and Spieck, E. 2015b. Improved isolation strategies allowed the phenotypic differentiation of two *Nitrospira* strains from widespread phylogenetic lineages. FEMS Microbiology Ecology 91: fiu031.

Oehmen, A., Lemos, P.C., Carvalho, G., Yuan, Z., Keller, J., Blackall, L.L., et al. 2007. Advances in enhanced biological phosphorus removal: from micro to macro scale. Water Research 41: 2271-2300.

Oelze, J. 2000. Respiratory protection of nitrogenase in *Azotobacter* species: is a widely held hypothesis unequivocally supported by experimental evidence? FEMS Microbiology Reviews 24: 321-333.

Op den Camp, H.J.M., Jetten, M.S.M. and Strous, M. 2007. Anammox. *In*: Bothe, H., Ferguson, S.J. and Newton, W.E. (eds.). Biology of the Nitrogen Cycle. pp. 245-262. Elsevier, Amsterdam, The Netherlands.

Oren, A. 2017. A plea for linguistic accuracy – also for *Candidatus taxa*. International Journal of Systematic and Evolutionary Microbiology 67: 1085-1094.

Oren, A. and Garrity, G.M. 2017. Validation List No. 173. List of new names and new combinations previously effectively, but not validly, published. International Journal of Systematic and Evolutionary Microbiology 67: 1-3.

Oshiki, M., Satoh, H. and Okabe, S. 2016. Ecology and physiology of anaerobic ammonium oxidizing bacteria. Environmental Microbiology 18: 2784-2796.

Otawa, K., Asano, R., Ohba, Y., Sasaki, T., Kawamura, E., Koyama, F., et al. 2006. Molecular analysis of ammonia-oxidizing bacteria community in intermittent aeration sequencing batch reactors used for animal wastewater treatment. Environmental Microbiology 8: 1985-1996.

Padhi, S.K., Tripathy, S., Mohanty, S. and Maiti, N.K. 2017. Aerobic and heterotrophic nitrogen removal by *Enterobacter cloacae* CF-S27 with efficient utilization of hydroxylamine. Bioresource Technology 232: 285-296.

Pan, K.L., Gao, J.F., Fan, X.Y., Li, D.C. and Dai, H.H. 2018a. The more important role of archaea than bacteria in nitrification of wastewater treatment plants in cold season despite their numerical relationships. Water Research 145: 552-561.

Pan, K.L., Gao, J.F., Li, H.Y., Fan, X.Y., Li, D.C. and Jiang, H. 2018b. Ammonia-oxidizing bacteria dominate ammonia oxidation in a full-scale wastewater treatment plant revealed by DNA-based stable isotope probing. Bioresource Technology 256: 152-159.

Pan, Z., Zhou, J., Lin, Z., Wang, Y., Zhao, P., Zhou, J., et al. 2020. Effects of COD/TN ratio on nitrogen removal efficiency, microbial community for high saline wastewater treatment based on heterotrophic nitrification-aerobic denitrification process. Bioresource Technology 301: 122726.

Park, H.D., Wells, G.F., Bae, H., Criddle, C.S. and Francis, C.A. 2006. Occurrence of ammonia-oxidizing archaea in wastewater treatment plant bioreactors. Applied and Environmental Microbiology 72: 5643-5647.

Park, J.B., Craggs, R.J. and Sukias, J.P. 2009. Removal of nitrate and phosphorus from hydroponic wastewater using a hybrid denitrification filter (HDF). Bioresource Technology 100: 3175-3179.

Park, S.J., Ghai, R., Martin-Cuadrado, A.B., Rodriguez-Valera, F., Chung, W.H., Kwon, K., et al. 2014. Genomes of two new ammonia-oxidizing archaea enriched from deep marine sediments. PLoS One 9: 1-10.

Peeters, S.H. and van Niftrik, L. 2019. Trending topics and open questions in anaerobic ammonium oxidation. Current Opinion in Chemical Biology 49: 45-52.

Peng, Y. and Zhu, G. 2006. Biological nitrogen removal with nitrification and denitrification via nitrite pathway. Applied Microbiology and Biotechnology 73: 15-26.

Pester, M., Rattei, T., Flechl, S., Gröngröft, A., Richter, A., Overmann, J., et al. 2012. *amoA*-based consensus phylogeny of ammonia-oxidizing archaea and deep sequencing of *amoA* genes from soils of four different geographic regions. Environmental Microbiology 14: 525-539.

Philippot, L. 2002. Denitrifying genes in bacterial and Archaeal genomes. Biochimica. et Biophysica. Acta 1577: 355-376.

Preston, C.M., Wu, K.Y., Molinski, T.F. and De Long, E.F. 1996. A psychrophilic crenarchaeon inhabits a marine sponge: *Cenarchaeum symbiosum* gen. nov., sp. nov. Proceedings of the National Academy of Sciences of the USA 93: 6241-6246.

Prosser, J.I., Head, I. and Stein, L.Y. 2014. The family *Nitrosomonadaceae*. pp. 901-918. *In*: Rosemberg, E., de Long, E.F., Lory, S., Stackebrandt, E. and Thompson, F. (eds.). The Prokaryotes – Alphaproteobacteria and Betaproteobacteria. Springer-Verlag.

Purkhold, U., Pommerening-Roser, A., Juretschko, S., Schmid, M., Koops, H. and Wagner, M. 2000. Phylogeny of all recognized species of ammonia oxidizers based on comparative 16S rRNA and *amoA* sequence analysis: implications for molecular diversity surveys. Applied and Environmental Microbiology 66: 5368-5382.

Qin, W., Martens-Habbena, W., Kobelt, J.N. and Stahl, D.A. 2016. *Candidatus* Nitrosopumilaceae. Bergey's Manual of Systematics of Bacteria and Archaea (online version, https://doi.org/10.1002/9781118960608. fbm00262).

Qin, W., Heal, K.R., Ramdasi, R., Kobelt, J.N., Martens-Habbena, W., Bertagnolli, A.D., et al. 2017a. *Nitrosopumilus maritimus* gen. nov., sp. nov., *Nitrosopumilus cobalaminigenes* sp. nov., *Nitrosopumilus oxyclinae* sp. nov., and *Nitrosopumilus ureiphilus* sp. nov., four marine ammonia-oxidizing archaea of the phylum *Thaumarchaeota*. International Journal of Systematic and Evolutionary Microbiology 67: 5067-5079.

Qin, W., Jewell, T.N.M., Russell, V.V., Hedlund, B.P., de la Torre, J.R. and Stahl, D.A. 2017b. *Candidatus* Nitrosocaldales. Bergey's Manual of Systematics of Bacteria and Archaea (online version, https://doi. org/10.1002/9781118960608.obm00120).

Rajta, A., Bhatia, R., Setia, H. and Pathania, P. 2019. Role of heterotrophic aerobic denitrifying bacteria in nitrate removal from wastewater. Journal of Applied Microbiology (In press) doi: 10.1111/jam.14476.

Reid, N.M., Bowers, T.H. and Lloyd-Jones, G. 2008. Bacterial community composition of a wastewater treatment system reliant on $N_2$ fixation. Applied Microbiology and Biotechnology 79: 285-292.

Rinaldo, S. and Cutruzzolà, F. 2007. Introduction to the biochemistry and molecular biology of denitrification. pp. 37-55. *In*: Bothe, H., Ferguson, S.J. and Newton, W.E. (eds.). Biology of the Nitrogen Cycle. Elsevier.

Rinaldo, S., Giardina, G. and Cutruzzolà, F. 2017. Nitrite Reductase – Cytochrome cd1. pp. 59-90. *In*: Moura, I., Moura, J.J.G., Pauleta, S.R. and Maia, L.B. (eds.). Metalloenzymes in Denitrification: Applications and Environmental Impacts. RSC Publishing, Cambridge, UK.

Robertson, L.A. and Kuenen, J.G. 1984. Aerobic denitrification-old wine in new bottles? Antonie Van Leeuwenhoek 50: 525-544.

Robertson, L.A., van Niel, E.W., Torremans, R.A. and Kuenen, J.G. 1988. Simultaneous nitrification and denitrification in aerobic chemostat cultures of *Thiosphaera pantotropha*. Applied and Environmental Microbiology 54: 2812-2818.

Roy, D., McEvoy, J., Blonigen, M., Amundson, M. and Khan, E. 2017. Seasonal variation and *ex situ* nitrification activity of ammonia-oxidizing archaea in biofilm-based wastewater treatment processes. Bioresource Technology 244: 850-859.

Sakai, K., Ikehata, Y., Ikenaga, Y., Wakayama, M. and Moriguchi, M. 1996. Nitrite oxidation by heterotrophic bacteria under various nutritional and aerobic conditions. Journal of Fermentation and Bioengineering 82: 613-617.

Santoro, A.E., Dupont, C.L., Richter, R.A., Craig, M.T., Carini, P., McIlvin, M.R., et al. 2015. Genomic and proteomic characterization of *Candidatus* Nitrosopelagicus brevis: an ammonia-oxidizing archaeon from the open ocean. Proceedings of the National Academy of Sciences of the USA 112: 1173-1178.

Saraste, M. and Castresana, J. 1994. Cytochrome oxidase evolved by tinkering with denitrification enzymes. FEBS Letters 341: 1-4.

Sauder, L.A., Albertsen, M., Engel, K., Schwarz, J., Nielsen, P.H., Wagner, M., et al. 2017. Cultivation and characterization of *Candidatus* Nitrosocosmicus exaquare, an ammonia-oxidizing archaeon from a municipal wastewater treatment system. ISME Journal 11: 1142-1157.

Sauder, L.A., Engel, K., Lo, C.C., Chain, P. and Neufeld, J.D. 2018. *Candidatus* Nitrosotenuis aquarius, an ammonia-oxidizing archaeon from a freshwater aquarium biofilter. Applied and Environmental Microbiology 84: e01430-18.

Schmidt, I., Sliekers, O., Schmid, M., Bock, E., Fuerst, J., Kuenen, J.G., et al. 2003. New concepts of microbial treatment processes for the nitrogen removal in wastewater. FEMS Microbiol. Reviews 27: 481-492.

Schott, J., Griffin, B.M. and Schink, B. 2010. Anaerobic phototrophic nitrite oxidation by *Thiocapsa* sp. strain KS1 and *Rhodopseudomonas* sp. strain LQ17. Microbiology 156: 2428-2437.

Sedlacek, C.J., McGowan, B., Suwa, Y., Sayavedra-Soto, L., Laanbroek, H.J., Stein, L.Y., et al. 2019. A physiological and genomic comparison of *Nitrosomonas* cluster 6a and 7 ammonia-oxidizing bacteria. Microbial. Ecology 78: 985-994.

Seviour, R.J., Lindrea, K.C., Griffiths, P.C. and Blackall, L.L. 1999. The activated sludge process. pp. 44-75. *In*: Seviour, R.J. and Blackall, L.L. (eds.). The Microbiology of Activated Sludge. Chapman and Hall, London, UK.

Seviour, R.J. 2010. An overview of the microbes in activated sludge. pp. 1-56. *In*: Seviour, R.J. and Nielsen, P.H. (eds.). Microbial Ecology of Activated Sludge. IWA Publishing, London, UK.

Seviour, R.J. and Nielsen, P.H. 2010. Microbial communities in activated sludge plants. pp. 95-138. *In*: Seviour, R.J. and Nielsen, P.H. (eds.). Microbial Ecology of Activated Sludge. IWA Publishing, London, UK.

Shao, M.F., Zhang, T. and Fang, H.H.P. 2010. Sulfur-driven autotrophic denitrification: diversity, biochemistry, and engineering applications. Applied Microbiology and Biotechnology 88: 1027-1042.

Shapleigh, J.P. 2014. Denitrifying prokaryotes. pp. 405-426. *In*: Rosemberg, E., de Long, E.F., Lory, S., Stackebrandt, E. and Thompson, F. (eds.). The Prokaryotes – Prokaryotic Physiology and Biochemistry. Springer-Verlag.

Shen, J., He, R., Han, W., Sun, X., Li, J. and Wang, L. 2009. Biological denitrification of high-nitrate wastewater in a modified anoxic/oxic-membrane bioreactor (A/O-MBR). Journal of Hazardous Materials 172: 595-600.

Shiro, Y. 2012. Structure and function of bacterial nitric oxide reductases. Nitric oxide reductase, anaerobic enzymes. Biochimica. et Biophysica. Acta 1817: 1907-1913.

Shoun, H., Fushinobu, S., Jiang, L., Kim, S. and Wakagi, T. 2012. Fungal denitrification and nitric oxide reductase cytochrome P450nor. Philosophical Transactions of the Royal Society B. 367: 1186-1194.

Simon, J. and Klotz, M.G. 2013. Diversity and evolution of bioenergetic systems involved in microbial nitrogen compound transformations. Biochimica. et Biophysica. Acta 1827: 114-135.

Siripong, S. and Rittmann, B.E. 2007. Diversity study of nitrifying bacteria in full-scale municipal wastewater treatment plants. Water Research 41: 1110-1120.

Skerman, V.B.D., McGowan, V. and Sneath, P.H.A. (eds). 1980. Approved lists of bacterial names. International Journal of Systematic Bacteriology 30: 225-420.

Slade, A.H., Anderson, S.M. and Evans, B.G. 2003. Nitrogen fixation in the activated sludge treatment of thermomechanical pulping wastewater: effect of dissolved oxygen. Water Science and Technology 48: 1-8.

Slade, A.H., Gapes, D.J., Stuthridge, T.R., Anderson, S.M., Dare, P.H., Pearson, H.G.W., et al. 2004. N-ViroTech® – a novel process for the treatment of nutrient limited wastewaters. Water Science and Technology 50: 131-139.

Slade, A.H., Thorn, G.J.S. and Dennis, M.A. 2011. The relationship between BOD: N ratio and wastewater treatability in a nitrogen-fixing wastewater treatment system. Water Science and Technology 63: 627-632.

Smercina, D.N., Evans, S.E., Friesen, M.L. and Tiemann, L.K. 2019. To fix or not to fix: controls on free-living nitrogen fixation in the rhizosphere. Applied and Environmental Microbiology: 85: e02546-18.

Soliman, M. and Eldyasti, A. 2018. Ammonia-oxidizing bacteria (AOB): opportunities and applications-a review. Reviews in Environmental Science and Biotechnology 17: 285-321.

Song, K., Suenaga, T., Hamamoto, A., Satou, K., Riya, S., Hosomi, M., et al. 2014. Abundance, transcription levels and phylogeny of bacteria capable of nitrous oxide reduction in a municipal wastewater treatment plant. Journal of Bioscience and Bioengineering 118: 289-297.

Sonune, A. and Gathe, R. 2004. Developments in wastewater treatment methods. Desalination 167: 55-63.

Sorokin, D.Y., Muyzer, G., Brinkhoff, T., Kuenen, J.G. and Jetten, M.S. 1998. Isolation and characterization of a novel facultatively alkaliphilic *Nitrobacter* species, *N. alkalicus* sp. nov. Archives of Microbiology 170: 345-352.

Sorokin, D.Y., Vejmelkova, D., Lucker, S., Streshinskaya, G.M., Rijpstra, W.I., Sinninghe Damste, J.S., et al. 2014. *Nitrolancea hollandica* gen. nov., sp. nov., a chemolithoautotrophic nitrite-oxidizing bacterium isolated from a bioreactor belonging to the phylum *Chloroflexi*. International Journal of Systematic and Evolutionary Microbiology 64: 1859-1865.

Spieck, E., Keuter, S., Wenzel, T., Bock, E. and Ludwig, W. 2014. Characterization of a new marine nitrite-oxidizing bacterium, *Nitrospina watsonii* sp. nov., a member of the newly proposed phylum *Nitrospinae*. Systematic and Applied Microbiology 37: 170-176.

Spieck, E. and Bock, E. 2005a. Genus VI. Nitrobacter Winogradsky 1892, 127AL (Nom. Cons. Opin. 23 Jud. Comm. 1958, 169). pp. 461-468. *In*: Brenner, D.J., Krieg, N.R., Staley, J.T. and Garrity, G.M. (eds.). Bergey's Manual of Systematic Bacteriology, 2nd Ed., Vol. 2. The Proteobacteria-Part C- The Alpha-, Beta-, Delta-, and Epsilonproteobacteria, Springer, Berlin, Germany.

Spieck, E. and Bock, E. 2005b. Genus IV. Nitrococcus Watson and Waterbury 1971, 224AL. pp. 52-55. *In*: Brenner, D.J., Krieg, N.R., Staley, J.T. and Garrity, G.M. (eds.). Bergey's Manual of Systematic Bacteriology, 2nd Ed., Vol. 2. The Proteobacteria-Part B-The Gammaproteobacteria. Springer, Berlin, Germany.

Srithep, P., Pornkulwat, P. and Limpiyakorn, T. 2018. Contribution of ammonia-oxidizing archaea and ammonia-oxidizing bacteria to ammonia oxidation in two nitrifying reactors. Environmental Science and Pollution Research 25: 8676-8687.

Stahl, D.A. and de la Torre, J.R. 2012. Physiology and diversity of ammonia-oxidizing Archaea. Annual Reviews of Microbiology 66: 83-101.

Stein, L.Y. 2011. Heterotrophic nitrification and nitrifier denitrification. pp. 95-114. *In*: Ward, B., Arp, D.J. and Klotz, M.G. (eds.). Nitrification. ASM Press, Washington.

Stein, L.Y. and Klotz, M.G. 2016. The nitrogen cycle. Current Biology 26: R83-R101.

Stein, L.Y. 2019. Insights into the physiology of ammonia-oxidizing microorganisms. Current Opinion in Chemical Biology 49: 9-15.

Stief, P., Fuchs-Ocklenburg, S., Kamp, A., Manohar, C.S., Houbraken, J., Boekhout, T., et al. 2014. Dissimilatory nitrate reduction by *Aspergillus terreus* isolated from the seasonal oxygen minimum zone in the Arabian Sea. BMC Microbiology 11: 14-35.

Stieglmeier, M., Klingl, A., Alves, R.J., Rittmann, S.K., Melcher, M., Leisch, B., et al. 2013. *Nitrososphaera viennensis* gen. nov., sp. nov., an aerobic and mesophilic, ammonia-oxidizing archaeon from soil and a member of the archaeal phylum *Thaumarchaeota*. International Journal of Systematic and Evolutionary Microbiology 64: 2738-2752.

Sun, Z., Lv, Y., Liu, Y. and Ren, R. 2016. Removal of nitrogen by heterotrophic nitrification-aerobic denitrification of a novel metal resistant bacterium *Cupriavidus* sp. S1. Bioresource Technology 220: 142-150.

Tavares, P., Pereira, A.S., Moura, J.J.G. and Moura, I. 2006. Metalloenzymes of the denitrification pathway. Journal of Inorganic Biochemistry 100: 2087-2100.

Taylor, A.E. and Bottomley, P.J. 2006. Nitrite production by *Nitrosomonas europaea* and *Nitrosospira* sp. AV in soils at different solution concentrations of ammonium. Soil Biology and Biochemistry 38: 828-836.

Tchobanoglous, G., Burton, F.L. and Stensel, H.D. (eds.) 2003. Metcalf and Eddy: Wastewater Engineering: Treatment and Reuse, 4th Ed. New York: McGraw-Hill.

Tian, M., Zhao, F., Shen, X., Chu, K., Wang, J., Chen, S., et al. 2015. The first metagenome of activated sludge from full-scale anaerobic/anoxic/oxic ($A^2O$) nitrogen and phosphorus removal reactor using Illumina sequencing. Journal of Environmental Sciences 35: 181-190.

Tian, T. and Yu, H.Q. 2020. Denitrification with non-organic electron donor for treating low C/N ratio wastewaters. Bioresource Technology 299: 122686.

Torres, M.J., Simon, J., Rowley, G., Bedmar, E.J., Richardson, D.J., Gates, A.J., et al. 2016. Nitrous oxide metabolism in nitrate-reducing bacteria: physiology and regulatory mechanisms. Advances in Microbial Physiology 68: 343-432.

Treusch, A.H., Leininger, S., Kletzin, A., Schuster, S.C., Klenk, H.P. and Schleper, C. 2005. Novel genes for nitrite reductase and Amo-related proteins indicate a role of uncultivated mesophilic *Crenarchaeota* in nitrogen cycling. Environmental Microbiology 7: 1985-1995.

Urakawa, H., Garcia, J.C., Nielsen, J.L., Le, V.Q., Kozlowski, J.A., Stein, L.Y., et al. 2015. *Nitrosospira lacus* sp. nov., a psychrotolerant, ammonia-oxidizing bacterium from sandy lake sediment. International Journal of Systematic and Evolutionary Microbiology 65: 242-250.

Ureta, A. and Nordlund, S. 2002. Evidence for conformational protection of nitrogenase against oxygen in *Gluconacetobacter diazotrophicus* by a putative FeSII protein. Journal of Bacteriology 184: 5805-5809.

Ushiki, N., Fujitani, H., Aoi, Y. and Tsuneda, S. 2013. Isolation of *Nitrospira* belonging to sublineage II from a wastewater treatment plant. Microbes and Environments 28: 346-353.

Validation list no. 54. 1995. Validation of the publication of new names and new combinations previously effectively published outside the IJSB. International Journal of Systematic Bacteriology 45: 619-620.

Validation List No. 71. 1999. Validation of publication of new names and new combinations previously effectively published outside the IJSEM. International Journal of Systematic Bacteriology 49: 1325-1326.

Validation List No. 78. 2001. Validation of publication of new names and new combinations previously effectively published outside the IJSEM. International Journal of Systematic Bacteriology 51: 1-2.

van den Berg, E.M., van Dongen, U., Abbas, B. and van Loosdrecht, M.C. 2015. Enrichment of DNRA bacteria in a continuous culture. ISME Journal 9: 2153-2161.

van den Berg, E.M., Boleij, M., Kuenen, J.G., Kleerebezem, R. and van Loosdrecht, M.C. 2016. DNRA and denitrification coexist over a broad range of acetate/N-$NO_3^-$ ratios, in a chemostat enrichment culture. Frontiers in Microbiology 7: 1842.

van Doan, T., Lee, T.K., Shukla, S.K., Tiedje, J.M. and Park, J. 2013. Increased nitrous oxide accumulation by bioelectrochemical denitrification under autotrophic conditions: kinetics and expression of denitrification pathway genes. Water Research 47: 7087-7097.

van Haandel, A. and van der Lubbe, J. 2012. Handbook of Biological Wastewater Treatment, 2nd Ed. IWA Publishing, London, UK.

van Kempen, R., Meijer, S.C.F., ten Have, C.C.R., Mulder, J.W., Duin, J.O.J., Uijterlinde, C.A., et al. 2005. SHARON for improved WWTP nitrogen effluent quality evaluated. Water Science and Technology 52: 55-62.

van Kessel, M.A.H.J., Speth, D.R., Albertsen, M., Nielsen, P.H., Op den Camp, H.J.M., Kartal, B., et al. 2015. Complete nitrification by a single microorganism. Nature 528: 555-559.

van Kessel, M.A.H.J., Stultiens, K., Slegers, M.F., Guerrero Cruz, S., Jetten, M.S., Kartal, B., et al. 2018. Current perspectives on the application of N-damo and anammox in wastewater treatment. Current Opinion in Biotechnology 50: 222-227.

van Niftrik, L. and Jetten, M.S.M. 2012. Anaerobic ammonium-oxidizing bacteria: unique microorganisms with exceptional properties. Microbiology and Molecular Biology Reviews 7: 585-596.

van Spanning, R.J.M., Richardson, D.J. and Ferguson, S.J. 2007. Introduction to the biochemistry and molecular biology of denitrification. pp. 3-20. In: Bothe, H., Ferguson, S.J. and Newton, W.E. (eds.). Biology of the Nitrogen Cycle. Elsevier.

Vieira, A., Galinha, C., Oehmen, A. and Carvalho, G. 2019. The link between nitrous oxide emissions, microbial community profile and function from three full-scale WWTPs. Science of the Total Environment 651: 2460-2472.

von Sperling, M. 2007. Biological Wastewater Treatment in Warm Climate Regions, Vol I-Wastewater characteristics. IWA publishing.

Vymazal, J. 2014. Constructed wetlands for treatment of industrial wastewaters: a review. Ecological Engineering 7: 724-751.

Wang, D., Wang, Y., Liu, Y., Ngo, H.H., Lian, Y., Zhao, J., et al. 2017. Is denitrifying anaerobic methane oxidation-centered technologies a solution for the sustainable operation of wastewater treatment plants? Bioresource Technology 234: 456-465.

Wang, D., He, Y. and Zhang, XX. 2019. A comprehensive insight into the functional bacteria and genes and their roles in simultaneous denitrification and anammox system at varying substrate loadings. Applied Microbiology and Biotechnology 103: 1523-1533.

Wang, L., Lim, C.K., Dang, H., Hanson, T.E. and Klotz, M.G. 2016. D1FHS, the type strain of the ammonia-oxidizing bacterium *Nitrosococcus wardiae* spec. nov.: enrichment, isolation, phylogenetic, and growth physiological characterization. Frontiers in Microbiology 7: 512.

Wang, M., Keeley, R., Zalivina, N., Halfhide, T., Scott, K., Zhang, Q., et al. 2018. Advances in algal-prokaryotic wastewater treatment: a review of nitrogen transformations, reactor configurations and molecular tools. Journal of Environmental Management 217: 845-857.

Wang, P., Li, X., Xiang, M. and Zhai, Q. 2007. Characterization of efficient aerobic denitrifiers isolated from two different sequencing batch reactors by 16S-rRNA analysis. Journal of Bioscience and Bioengineering 103: 563-567.

Wang, X., Wen, X., Criddle, C., Wells, G., Zhang, J. and Zhao, Y. 2010. Community analysis of ammonia-oxidizing bacteria in activated sludge of eight wastewater treatment systems. Journal of Environmental Sciences 22: 627-634.

Wang, X., Hu, M., Xia, Y., Wen, X. and Ding, K. 2012. Pyrosequencing analysis of bacterial diversity in 14 wastewater treatment systems in China. Applied and Environmental Microbiology 78: 7042-7047.

Wang, Z., Zhang, X.X., Lu, X., Liu, B., Li, Y., Long, C., et al. 2014. Abundance and diversity of bacterial nitrifiers and denitrifiers and their functional genes in tannery wastewater treatment plants revealed by high-throughput sequencing. PLoS One 9: e113603.

Wehrfritz, J.M., Reilly, A., Spiro, S. and Richardson, D.J. 1993. Purification of hydroxylamine oxidase from *Thiosphaera pantotropha*: identification of electron acceptors that couple heterotrophic nitrification to aerobic denitrification. FEBS Letters 335: 246-250.

Wehrfritz. J.M., Carter, J.P., Spiro, S. and Richardson, DJ. 1997. Hydroxylamine oxidation in heterotrophic nitrate-reducing soil bacteria and purification of a hydroxylamine-cytochrome c oxidoreductase from a *Pseudomonas* species. Archives of Microbiology 166: 421-424.

Wells, G., Park, H.D., Yeung, C.H., Eggleston, B., Francis, C.A. and Criddle, C.S. 2009. Ammonia-oxidizing communities in a highly aerated full-scale activated sludge bioreactor: betaproteobacterial dynamics and low relative abundance of *Crenarchaea*. Environmental Microbiology 11: 2310-2328.

Welsh, A., Chee-Sanford, J.C., Connor, L.M., Löffler, F.E. and Sanford, R.A. 2014. Refined NrfA phylogeny improves PCR-based *nrfA* gene detection. Applied and Environmental Microbiology 80: 2110-2119.

Welz, P.J., Holtman, G., Haldenwang, R. and le Roes-Hill, M. 2016. Characterisation of winery wastewater from continuous flow settling basins and waste stabilisation ponds over the course of 1 year: implications for biological wastewater treatment and land application. Water Science and Technology 74: 2036-2050.

Welz, P.J., Ramond, J.B., Braun, L., Vikram, S. and Le Roes-Hill, M. 2018. Bacterial nitrogen fixation in sand bioreactors treating winery wastewater with a high carbon to nitrogen ratio. Journal of Environmental Management 207: 192-202.

Wett, B., Murthy, S., Takács, I., Hell, M., Bowden, G., Deur, A., et al. 2007. Key parameters for control of DEMON deammonification process. Water Practice 1(5): 1-11.

Wiessman, U., Choi, I.S. and Dombrowski, E.M. 2007. Fundamentals of Biological Wastewater Treatment. Wiley VCH Verlag GmbH and Co. KGaA, Weinheim.

Wilén, B.M., Liébana, R., Persson, F., Modin, O. and Hermansson, M. 2018. The mechanisms of granulation of activated sludge in wastewater treatment, its optimization, and impact on effluent quality. Applied Microbiology and Biotechnology 102: 5005-5020.

Wrage, N., Velthof, G.L., van Beusichem, M.L. and Oenema, O. 2001. Role of nitrifier denitrification in the production of nitrous oxide. Soil Biology and Biochemistry 33: 1723-1732.

Winkler, M.K.H. and Straka, L. 2019. New directions in biological nitrogen removal and recovery from wastewater. Current Opinion in Biotechnology 57: 50-55.

Wu, J., Hong, Y., Chang, X., Jiao, L., Li, Y., Liu, X., et al. 2019a. Unexpectedly high diversity of anammox bacteria detected in deep-sea surface sediments of the South China Sea. FEMS Microbiology Ecology 95: fiz013.

Wu, L., Ning, D., Zhang, B., Li, Y., Zhang, P., Shan, X., et al. 2019b. Global diversity and biogeography of bacterial communities in wastewater treatment plants. Nature Microbiology 7: 1183-1195.

Wuhrmann, K. 1957. The tertiary treatment: ways and achievements in the elimination of elements causing eutrophication. Schweizerische Zeitschrift Für Hydrologie 19: 409-427.

Wunderlin, P., Mohn, J., Joss, A., Emmenegger, L. and Siegrist, H. 2012. Mechanisms of $N_2O$ production in biological wastewater treatment under nitrifying and denitrifying conditions. Water Research 46: 1027-1037.

Yan, L., Liu, S., Liu, Q., Zhang, M., Liu, Y., Wen, Y., et al. 2019. Improved performance of simultaneous nitrification and denitrification via nitrite in an oxygen-limited SBR by alternating the DO. Bioresource Technology 275: 153-162.

Youssef, N.H. and Elshahed, M.S. 2014. The Phylum *Planctomycetes*. pp. 759-810. *In*: Rosemberg, E., de Long, E.F., Lory, S., Stackebrandt, E. and Thompson, F. (eds.). The Prokaryotes – Other major lineages of Bacteria and the Archaea. Springer-Verlag.

Yu, K. and Zhang, T. 2012. Metagenomic and metatranscriptomic analysis of microbial community structure and gene expression of activated sludge. Plos One 7: e38183.

Yu, T., Qi, R., Li, D., Zhang, Y. and Yang, M. 2010. Nitrifier characteristics in submerged membrane bioreactors under different sludge retention times. Water Research 44: 2823-2830.

Zhalnina, K.V., Dias, R., Leonard, M.T., Dorr de Quadros, P., Camargo, F.A., Drew, J.C., et al. 2014. Genome sequence of *Candidatus* Nitrososphaera evergladensis from group I.1b enriched from Everglades soil reveals novel genomic features of the ammonia-oxidizing archaea. PLoS One 9: e101648.

Zhang, D., Li, W., Huang, X., Qin, W. and Liu, M. 2013. Removal of ammonium in surface water at low temperature by a newly isolated *Microbacterium* sp. strain SFA13. Bioresource Technology 137: 147-152.

Zhang, H., Feng, J., Chen, S., Zhao, Z., Li, B., Wang, Y., et al. 2019a. Geographical patterns of *nirS* gene abundance and *nirS*-type denitrifying bacterial community associated with activated sludge from different wastewater treatment plants. Microbial. Ecology 77: 304-316.

Zhang, J., Müller, C. and Cai, Z. 2015. Heterotrophic nitrification of organic N and its contribution to nitrous oxide emissions in soils. Soil Biology and Biochemistry 84: 199-209.

Zhang, J., Wu, P., Hao, B. and Yu, Z. 2011. Heterotrophic nitrification and aerobic denitrification by the bacterium *Pseudomonas stutzeri* YZN-001. Bioresource Technology 102: 9866-9869.

Zhang, M., Wang, S., Ji, B. and Liu, Y. 2019b. Towards mainstream deammonification of municipal wastewater: partial nitrification-anammox versus partial denitrification-anammox. Science of the Total Environment 692: 393-401.

Zhang, X., Zheng, S., Zhang, S. and Duan, S. 2018. Autotrophic and heterotrophic nitrification-anoxic denitrification dominated the anoxic/oxic sewage treatment process during optimization for higher loading rate and energy savings. Bioresource Technology 263: 84-93.

Zhang, Z., Zhang, Y. and Chen, Y. 2020. Recent advances in partial denitrification in biological nitrogen removal: from enrichment to application. Bioresource Technology 298: 122444.

Zheng, H.Y., Liu, Y., Gao, X.Y., Ai, G.M., Miao, L.L. and Liu, Z.P. 2012. Characterization of a marine origin aerobic nitrifying-denitrifying bacterium. Journal of Bioscience and Bioengineering 114: 33-37.

Zheng, T., Li, P., Ma, X., Sun, X., Wu, C., Wang, Q., et al. 2019. Pilot-scale experiments on multilevel contact oxidation treatment of poultry farm wastewater using saran lock carriers under different operation model. Journal of Environmental Sciences 77: 336-345.

Zhu, B., Bradford, L., Huang, S., Szalay, A., Leix, C., Weissbach, M., et al. 2017. Unexpected diversity and high abundance of putative nitric oxide dismutase (*nod*) genes in contaminated aquifers and wastewater treatment systems. Applied and Environmental Microbiology 83: e02750-16.

Zhu, B., Wang, J., Bradford, L.M., Ettwig, K., Hu, B. and Lueders, T. 2019. Nitric oxide dismutase (*nod*) genes as a functional marker for the diversity and phylogeny of methane-driven oxygenic denitrifiers. Frontiers in Microbiology 10: 1577.

Zhu, G., Peng, Y., Li, B., Guo, J., Yang, Q. and Wang, S. 2008. Biological removal of nitrogen from wastewater. Reviews of Environmental Contamination and Toxicology 192: 159-195.

Zhu, J., Wang, Q., Yuan, M., Tan, G.A., Sun, F., Wang, C., et al. 2016. Microbiology and potential applications of aerobic methane oxidation coupled to denitrification (AME-D) process: a review. Water Research 90: 203-215.

# Nitrogenized and Chlorinated Compounds Pollutants From Industrial Wastewater: Their Environmental Impacts and Bioremediation Strategies

Jonathan Parades-Aguilar[1], F. Javier Almendariz-Tapia[2],
Roberto Vázquez-Euán[3], Marco A. López-Torres[1],
Luis R. Martínez-Córdova[1] and Kadiya Calderón[1,*]

## 9.1 INTRODUCTION

In recent decades, overpopulation and human activities have considerably raised pollution worldwide. The contaminants released by these activities have a major impact on the environment (i.e., soil and water) and human health. Therefore, several solutions have been recently explored in order to remove these pollutants and prevent contamination. One of the fast, less expensive, and eco-friendly way to treat contaminants, in comparison to other remediation strategies, is bioremediation by bacteria, which could be efficiently applied before wastewater discharges to the environment to generate less or non-harmful products. On the other hand, strict regulation for industrial releases and wastewater treatments are still necessary to reduce or avoid environmental pollution.

The aim of the present chapter is to discuss the environmental and human health effects caused by nitrogenized and chlorinated compounds usually released by human activities, such as industries, mining, and aquaculture, with a special interest in nitrifying bacteria, the use of their metabolic processes to oxidize nitrogenized and chlorinated compounds, and their possible applications in bioremediation to degrade these compounds. Furthermore, some bioremediation strategies are presented based on the use of bioreactors and treatment systems that can be implemented to remove contaminants from wastewater, such as biofilm and granular systems, and improve the future perspectives in this field.

---

[1] DICTUS. Blvd. Luis Donaldo Colosio S/N. CP. 83000, Hermosillo, Sonora, Mexico.
[2] Departamento de Ingeniería Química y Metalurgia, Universidad de Sonora. Blvd. Luis Donaldo Colosio S/N. CP.83000, Hermosillo, Sonora, Mexico.
[3] CONACYT-Departamento de Investigaciones Científicas y Tecnológicas, Universidad de Sonora. Blvd. Luis Donaldo Colosio S/N. CP. 83000, Hermosillo, Sonora, México.
* Corresponding author: kadiya.calderon@unison.mx.

## 9.2 INDUSTRIAL WASTEWATER COMPOSITION AND DEPOSITION

The increased development of human population and its activity in agriculture and industrialization have directly impacted land degradation. Lower crop production and water and soil contamination derived from industrial residues have become two of the main factors where those activities have been impacted. Even if several methods were designed and developed in order to reduce these effects, the uncontrolled and unregulated activities made those processes generate secondary contamination and impact the environment more (Pakshirajan et al. 2015). Industries, such as textiles, tanneries, batteries, pesticides, ore refineries, petrochemicals, paper manufacturing, metal plating, and mining and agro-industries like aquaculture, are the main contributors to pollution and environmental degradation (Pakshirajan et al. 2015, EPA 2009, Maulin 2017, O'Connor et al. 2018).

Mining is one of the most remunerated activities around the world, but its environmental and human health impacts are irreversible and related to hazardous residual compounds. Heavy metals are common pollutants derived from mining activity, either because they are wastes of the interest minerals or because they are part of its extraction process. Their ecological impact is mainly due to their high toxicity and their resistance to environmental degradation (León-García et al. 2018). Also, heavy metals have been demonstrated to cause some negative effects on human health, such as cancer, liver pathologies, and neurologic and reproductive disorders (Jaishankar et al. 2014). Among the main metals derived from mining activity are iron (Fe), copper (Cu), manganese (Mn), zinc (Zn), lead (Pb), nickel (Ni), cadmium (Cd), mercury (Hg), selenium (Se), and arsenic (As) (León-García et al. 2018).

Particularly, on the Arizona-Mexico border, there exists high mining activity; however, its toxic release affected around 22,000 people from localities placed along the Sonora River, who raised concerns of possible dermatologic, gastrointestinal, ophthalmologic, neurologic, and cardiovascular effects that were potentially associated with the disaster (Ibarra and Moreno 2017, Rivera Carvajal et al. 2019).

Another case of mining contamination is clearly exemplified for gold mining extraction where lots of arsenic sources are wasted into the biosphere, including waste soil and rocks, residual water from ore concentrations, the roasting of some types of gold-containing ores to remove sulfur and sulfur oxides, and bacterially-enhanced leaching (Eisler 2004). Thus, arsenic concentrations near gold mining operations are frequently elevated in abiotic materials and are released in toxic doses for humans and animals.

However, the impasse reached by environmental authorities around the world on enforcing regulation compliance by the mining group has also raised social complaints (Ibarra and Moreno 2017), and even if several metals criteria to protect human health and natural resources have been studied and discussed, rigorous measures from governments and countries are required in order to reduce this irreversible damage.

On the other hand, aquaculture is another industrial activity that can generate great amounts of nitrogen compounds. The contamination of this activity is mostly related due to the ammonia nitrogen accumulation in pond systems, which is generated by the excessive food used and animal excretion. Although the main damage of aquaculture is to the environment, the nitrogen compounds formed, such as ammonia and nitrite, can result in pathological disturbances and mortality of aquatic species, such as shrimp larvae, as well as affecting the activity itself and causing an unbalance of these ecosystems (Rurangwa and Verdegem 2015, Hoang et al. 2016, Lu et al. 2018). Furthermore, in least developed and developed countries, rapid industrial growth has led to elevated discharges of toxic chemicals and nutrients into water bodies.

In a particular case of study in Mexico, different nitrogenized, chlorinated, and metal compounds have been detected in groundwater and in drinking water sources at the Mexican-US border area due to the improper disposal of wastes from electronic industries that infiltrate through the soil and groundwater. Unfortunately, some of these chemical compounds are not biodegradable and tend to accumulate in the environmental systems (Beamer 2012, López-Galvez 2014, Berrelleza-Valdez et al. 2019).

Even if many countries around the world follow standards for nitrogen, phosphorous, organic matter, heavy metals, and organic compound discharges from wastewater, industries often face problems in meeting these requirements (Pakshirajan et al. 2015).

## 9.2.1  Nitrogenized Residual Compounds

Many different sources of pollution are responsible for contributing to the environment with nitrogen residual compounds, particularly nitrogen oxides (NOx)—a group designated as gaseous chemical compounds formed by the combination of oxygen and nitrogen—of which the most important compounds are nitric oxide (NO) and nitrogen dioxide ($NO_2$). Both oxides are considered as air pollutants, and they are responsible for acid rain which damages vegetation and the structure of a building. Exposure to oxides of nitrogen represents a risk for public health, especially NOx has been associated with adverse health effects to humans and animals (EPA 2010).

Even though NOx are produced naturally during the bacterial decomposition of organic nitrates, natural fires of vegetation, electrical storms, volcanic eruptions, etc., in general, the NOx air pollution is related to the large population cities because their principal source is related to human activities, which contribute through motorized vehicles, especially those with diesel motors, to the combustion of coal or natural gas among others sources (Cros et al. 2015). However, in rural areas with mining and agricultural practices, the amount of NOx can be even higher (Hendryx et al. 2019). In this sense, explosives are so commonly used in open-pit mining to extract rocks from the ground by blasting through the primary rock fragmentation method (Mulenga 2020). Currently, most of the commercial explosives used in mining and civil blasting applications are based on mixtures of ammonium nitrate with fuel (ANFO explosives); when they explode, they release a high amount of NOx toxic gases as by-products, principally when the explosives used do not explode perfectly (Araos and Onederra 2017).

Although the NOx from blasting constitutes only a small proportion of the total NOx emissions from human activities, the blasting produces a sudden localized release of gases with potentially high toxic concentrations of NOx (500 ppm), exceeding up to 3,000 times the international standards (Oluwoye et al. 2017), which represents a health risk not only for miners but also for the entire population living in the surrounding communities if they are exposed to the NOx gases before the plumes can dissipate (Attalla et al. 2008, Fugiel et al. 2017, Hendryx et al. 2019).

With respect to the impact on ecosystems, the acid rain and dry deposition of NOx increase the nitrogen contents in soil and terrestrial water bodies, causing eutrophication and depletion of aquatic oxygen and other negative consequences that can affect marine life, cause leaf and root damage, and contribute to soil and water acidity. NOx also can impact stratospheric ozone depletion (Oluwoye et al. 2017).

The NOx contamination is difficult to eliminate or reduce since existing technologies employed in a post-combustion plant do not apply to the blasting activities because the post-explosion atmospheric mixture does not yield itself to capturing, scrubbing, treatment, or reprocessing. However, new technology has emerged for NOx gas abatements, such as alkalimetric neutralization, application of stabilizing and scavenging additives, chemical trapping, reburning-like techniques, and the use of alternative oxidizing agents (Oluwoye et al. 2017). The last option has been the most promising because the ammonium nitrate can be replaced by hydrogen peroxide as an oxidizer in the explosive formulation with similar heat release and velocity of detonation but without NOx hazards (Araos and Onederra 2017).

Aquaculture is one of the most booming productive food systems; it has grown enormously in recent years and its potential is still very great due to the imminent growth of the worldwide population and the reduction of spaces to produce food (Avnimelech et al. 2008). The growth of aquaculture has been achieved through both expansion in area and intensity of production; however, an intensive aquaculture system needs a high quantity of artificial feed and that brings water quality deterioration because cultivated animals, like fishes or crustaceans, only assimilate 20-25% of the protein in feed and the remainder is excreted as ammonia and organic nitrogen in dregs and unconsumed feed, which eventually ends in the sediment (Ekasari et al. 2012, Little et al. 2018) and at the same time contaminates the water ponds and other natural effluents where this contaminated water is poured.

The accumulation of nitrogenous compounds, such as ammonium, nitrites, and nitrates, are toxic to biota, especially ammonium which is the most dangerous of all. Most nitrate results from nitrification—the process by which the ammonium produced during bacterial decomposition and animal excretion is first converted to nitrites and subsequently to nitrates by aerobic autotrophic bacteria (Páez-Osuna and Frías-Espericueta 2001).

## 9.2.2 Chlorinated Residual Compounds

The presence of chlorinated volatile organic compounds (Cl-VOCs) in the environment has been increasing since the beginning of the twentieth century due to human activities. Industries commonly use Cl-VOCs as solvents for metal degreasing, dry cleaning, and as a constituent of different chemical products such as refrigerants. Chlorination of water for its disinfection is also a common process that generates Cl-VOCs when chlorine reacts with organic matter present in water. Most of Cl-VOCs are sweet smell colorless liquids at room temperature, and they are characterized by their high volatility and strong recalcitrance to degradation. The improper disposal of industries could cause contamination of waters, soil, and air, and there is evidence that exposure to them can cause damage to human health and develop different kinds of cancer (EPA 2009, Beamer et al. 2012, Huang et al. 2014).

Due to their effects on human health, some Cl-VOCs have been listed as priority pollutants by international organizations and are strictly regulated (European Commission 2008, EPA 2009). Table 9.1 shows some of the most used Cl-VOCs by industries and their maximum contaminant level (MCL) in drinking water as reported by the EPA (2009).

**TABLE 9.1** Common chlorinated volatile organic compounds used by industries and their maximum contaminant level in drinking water. Modified from EPA (2009)

| Contaminant | Maximum Contaminant Level (mg/L) | Potential Health Effects | Use |
|---|---|---|---|
| Chlorobenzene (CB) | 0.1 | Liver and kidney problems | Solvent used in chemical and agrochemical industries |
| Dichloromethane (DM) | 0.005 | Liver and kidney problems, increased risk of cancer | Solvent used in chemical industries |
| Hexachlorobenzene (HCB) | 0.001 | Liver and kidney problems, reproductive difficulties, increased risk of cancer | Agrochemical, fungicide |
| Trichloroethylene (TCE) | 0.005 | Liver and kidney problems, increased risk of cancer | Solvent, product of tetrachloroethylene degradation |
| Vinyl chloride (VC) | 0.002 | Increased risk of cancer | Manufacture of Polyvinyl chloride (PVC) products, product of TCE degradation |

Diverse strategies have been studied for the removal of recalcitrant xenobiotics, such as Cl-VOCs, including bioremediation by bacteria (Huang et al. 2014). Through different studies, it has become known that the degradation of chlorinated compounds can be performed by three main mechanisms: dehalogenation, direct oxidation, and co-metabolism. Dehalogenation or reductive dechlorination consists of the chemical reduction of a halogenated compound by a dehalogenase enzyme and the release of inorganic chloride ions ($Cl^-$). This process is generally performed by anaerobic bacteria since it uses the halogenated compound as the electron acceptor by halorespiration. *Dehalococcoides ethenogenes, Dehalospirillum multivorans,* and *Enterobacter agglomerans* are some of the bacteria that use this process (García-Solares et al. 2013, Matteucci et al. 2015, Yoshikawa et al. 2017). Otherwise, direct oxidation is an aerobic or anaerobic oxidation process in which bacteria obtain energy and organic carbon from the chlorinated compound—generally, chlorinated ethenes. Finally, cometabolism is a form of oxidation that involves oxygenase enzymes that directly oxidize the halogenated compound but without a benefit for the bacteria. Some of the advantages of cometabolism over other mechanisms remain in the fact that aerobic degradation processes generally are faster than anaerobic processes and that in cometabolism even other products generated by the degradation could be oxidized (Pant and Pant 2010).

Cometabolism is also the mechanism by which Ammonia Oxidizing Bacteria (AOB) *Nitrosomonas europaea* use to degrade chlorinated compounds, such as chlorinated methane and ethenes, including the carcinogenic compound trichloroethylene, via its ammonia monooxygenase enzyme (Vannelli et al. 1990, Wahman et al. 2005).

On the other hand, trichloroethylene (TCE) is one of the most dangerous chlorinated compounds which AOB *Nitrosomonas europaea* has been shown as capable of degrading. It is a colorless, non-flammable liquid commonly used in electronic industries as a detergent and degreaser of electronic parts during computer assembly and can easily pollute sources of water and soil when it is released to the environment by these industries, making it harmful to ecosystems as well as to human health. It can also pollute air since it evaporates easily (Shukla et al. 2014, ATSDR. 2015). Its chemical structure is shown in Figure 9.1.

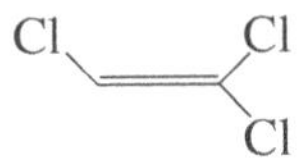

**FIGURE 9.1** Chemical structure of trichloroethylene.

Prolonged exposure to TCE may cause injuries to the nervous system, liver, and kidneys and promote inflammatory diseases. It has even been cataloged as carcinogenic (EPA 2009, Selmin et al. 2014, Shukla et al. 2014, Khare et al. 2018). Considering these factors, different agencies have established contaminant levels for TCE; the Environmental Protection Agency of United States (EPA) established a maximum contaminant level (MCL) of 0.005 mg/L of TCE in potable water, while for workplace standards, the Occupational Safety and Health Administration (OSHA) established a permissible exposure limit (PEL) of 100 ppm over an 8-hour workday (EPA 2009, ATSDR 2015). By 2000s, TCE had been detected in at least 852 of the 1,430 sites of the National Priority List (NPL) of the EPA in the United States with concentrations of up to 900 µg/L in some sites (Shukla et al. 2014).

Due to the concerns over the serious effects of TCE as a pollutant and its property to persist in the environment, bioremediation has become one of the eco-friendlier, less expensive, and faster ways to treat TCE contamination. Several studies related to TCE biodegradation have been conducted in the last decades. Most of them lightly vary on their experimental designs and systems used to prove the biodegradation, but TCE determination is usually carried out by gas chromatography due to the volatility of the compound (Shukla et al. 2014).

The first study that demonstrated the capacity of a nitrifying bacteria to co-metabolize TCE was performed by Arciero et al. (1989) using suspensions of *N. europaea* cells in batch systems. These experiments reached the complete disappearance of 10 µM of TCE from the headspace of bottles containing *N. europaea* at rates of 1 µM /mg protein. Later, some studies, such as those performed by Vannelli et al. (1990), Wahman et al. (2005), and Kocamemi and Çeçen (2010) confirmed this biodegradation capacity. More recently, Berrelleza-Valdez et al. (2019) demonstrated that a nitrifying consortium of AOB and nitrite oxidizing bacteria (NOB) isolated from a wastewater treatment plant by a novel process using a continuous stirred-tank bioreactor had the capacity to biodegrade high levels of TCE from synthetic contaminated water (500 mg/L of TCE dilution) in batch and two different natural laboratory-scale packed-bed bioreactors. Thus, nitrifying bacteria, which could be present in so many ecosystems, can be applied for the bioremediation of Cl-VOCs, such as TCE, in environmental and engineering systems.

## 9.3 NITRIFYING SYSTEMS: HARNESSING THE NITROGEN CYCLE OF BACTERIA

The use of biological systems can be successfully used to take advantage of the metabolic processes of microorganisms. As we know, microorganisms have been used over time to elaborate on food products, like bread, beer, or wine, in the chemical industry to produce solvents and in medicine to generate antibiotics. They are also very important biological and ecological regulators for ecosystems' maintenance due to their variety of roles. Specifically, bacteria are involved in the biogeochemical cycles of some of the most important elements in life, such as carbon, sulfur, phosphorus, and nitrogen (Bertrand et al. 2015, Calderón et al. 2018). They can be present in any habitat, and their enzymes can catalyze a large number of reactions; for these reasons, they can be easily exploited for biotechnological purposes.

In this sense, nitrifying bacteria are a group present in different ecosystems; their functions are fundamental for all living beings and can be used for human benefit. They are chemolithoautotrophic microorganisms extremely important in the nitrogen cycle since they are involved in the nitrification processes (Kumar et al. 2014). Nitrifying bacteria may be present in natural or industrial nitrogen species-rich environments, such as surface waters with an influent of nitrogen-species or in wastewater treatment plants where they play a fundamental role in oxidizing inorganic nitrogen compounds by specific enzymes to get energy. Based on their source of energy, nitrifying bacteria are classified as AOB and NOB while both groups predominantly fix carbon dioxide as a carbon source (Farges et al. 2012, Pérez et al. 2014, Feng et al. 2016).

AOB perform the oxidation of ammonium to nitrite via two enzymes in the presence of oxygen: the transmembrane protein ammonia monooxygenase (AMO) that oxidizes ammonium ($NH_4^+$) to hydroxylamine and the periplasmic protein hydroxylamine oxidoreductase (HAO) that oxidizes hydroxylamine to nitrite ($NO_2^-$). This mechanism is shown in Figure 9.2 and its general reaction is as follows:

$$2H^+ + 2e^- + NH_3 + O_2 \longrightarrow NH_2OH + H_2O \longrightarrow NO_2^- + 5H^+ + 4e^-$$

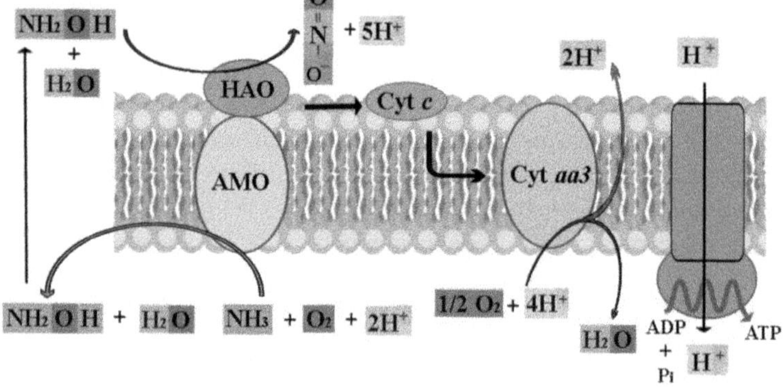

**FIGURE 9.2** Ammonium oxidation mechanism by ammonia oxidizing bacteria. Adapted from Madigan et al. (2017).

The most representative bacteria of the AOB group are part of the *Nitrosomonas* genus (Papp et al. 2016). On the other hand, NOB perform the oxidation of nitrite to nitrate ($NO_3$) by the transmembrane protein nitrite oxidoreductase (NOR) (Kumar et al. 2014, Ralebitso-Senior and Orr 2016). The molecular mechanism of these bacteria is shown in Figure 9.3, and its general reaction is also depicted as follows:

$$NO_2^- + H_2O \longrightarrow NO_3^- + 2H^+ + 2e^-$$

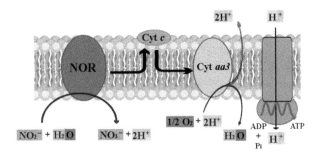

**FIGURE 9.3** Nitrite oxidation mechanism by nitrite oxidizing bacteria. Adapted from Madigan et al. (2017).

For this group, the most representative bacteria are part of the *Nitrobacter* and *Nitrospira* genus (Farges et al. 2012). In nature, both groups of bacteria work together simultaneously; while AOB provides the source of energy for NOB, they consume the nitrite, which even though is a product of the metabolism of AOB, is harmful to them at high concentrations and can affect the first step of the nitrification process (Pérez et al. 2014). The genus, phylogenetic group, and habitat of the principal AOB and NOB are listed in Table 9.2.

**TABLE 9.2** Genus, phylogenetic group and habitat of ammonia oxidizing bacteria and nitrite oxidizing bacteria

| Group | Genus | Phylogenetic Group | Habitat |
|---|---|---|---|
| Ammonia oxidizing bacteria (AOB) | *Nitrosomonas* | Beta | Sewage (Itoh et al. 2013, Rezaee et al. 2013), soil (Dubey et al. 2014, Wang et al. 2014a), surface waters (French et al. 2012, Zhu et al. 2018). |
| | *Nitrosococcus* | Gamma | Sewage (Kumar et al. 2012, Pal et al. 2012), surface waters (Simister et al. 2012, Vipindas et al. 2015, Zhu et al. 2018). |
| | *Nitrosospira* | Beta | Sewage (Kumar et al. 2012, Pal et al. 2012), soil (Wang et al. 2014a), surface waters (Feng et al. 2016). |
| Nitrite oxidizing bacteria (NOB) | *Nitrobacter* | Alpha | Sewage (Kumar et al. 2012, Pal et al. 2012, Itoh et al. 2013, Rezaee et al. 2013), soil (Poly et al. 2008, Wang et al. 2014a), surface waters (Vipindas et al. 2015). |
| | *Nitrococcus* | Gamma | Surface waters (Füssel et al. 2017). |
| | *Nitrospira* | Nitrospirae | Sewage (Kumar et al. 2012, Pal et al. 2012), soil (Wang et al. 2014b), surface waters (Simister et al. 2012, Vipindas et al. 2015). |

The process of nitrification carried out by nitrifying bacteria has been harnessed in different biological systems, such as bioreactors with suspended or immobilized biomass. In suspended biomass systems, stirred-tank bioreactors are generally used, while for immobilized biomass systems, nitrifying bacteria have been fixed in different kinds of carriers (Farges et al. 2012, Pungrasmi et al. 2016). Usually, these systems are used in wastewater treatment plants where AOB and NOB work together for the removal of inorganic nitrogen compounds, such as those described previously, but this removal can also be applied for other purposes. Moreover, AOB, like *Nitrosomonas europaea*, have been demonstrated since the 1990s to degrade organic compounds, such as chlorine, into carbon dioxide ($CO_2$) (Arciero et al. 1989, Vannelli et al. 1990, Wahman et al. 2005), and this capacity can be also used to degrade some toxic organic compounds by bioremediation into less harmful products in biological systems (Papp et al. 2016).

As in nitrification, nitrate is generally accumulated rather than nitrite, while levels of nitrate can be regulated by denitrifying microorganisms, a great accumulation of nitrogen compounds caused by the release of them to the environment by industries and other human activities could have some negative effects, such as acid rain and eutrophication, which could lead to perturbation in ecosystems or even cause human diseases, like methemoglobinemia in babies (Kumar et al. 2014). On the other hand, nitrogen compound removal could also be applied in systems in which generated products are moved between a series of compartments to be transformed and used in a specific way. Thus, various biological systems have been developed to stimulate and carry on the nitrification process of AOB and NOB.

In the aquaculture industry, the contamination of this activity is mostly related to the ammonia nitrogen accumulation in the systems, generated by the excessive food used and animal excretion. Although the main damage of aquaculture is to the environment, the nitrogen compounds formed, such as ammonia and nitrite, can result in pathological disturbances and mortality of aquatic species, such as shrimp larvae, affecting also the activity itself. For more sustainable, efficient, and environmentally friendly aquaculture, RAS, or recirculating aquaculture system, for the removal of nitrogen compounds of the brackish wastes has been developed and used recently. These are closed systems in which the water circulates, being filtered by biological and mechanical filtration in order to be reused for the culture of aquatic organisms. While mechanical filtration mainly removes suspended solids, biological filtration uses nitrifying bacteria adhered to specific surfaces to convert the ammonia and nitrite, highly toxic for aquatic organisms, to a nontoxic form such as nitrate (Rurangwa and Verdegem 2015, Hoang et al. 2016).

One of the problems in harnessing the process of nitrification in biological systems is the difficulty of obtaining high concentrations of suspended nitrifying bacteria because of their low growth rate and the inhibition caused by the intermediate nitrification products like nitrite (Papp et al. 2016). These limitations could be overcome with immobilized biomass systems since in this way bacteria can proliferate when attached to a carrier, such as plastic beads, foams, or stones, and also improve the nitrification process and prevent the washout of the bacteria. For that reason, fixed-film systems, such as biofilters, could be considered as the most used systems for nitrogen compound removal in aquaculture nowadays. The most common systems are the fluidized bed reactor, trickling filter, and floating packed-bed reactor (Hoang et al. 2016).

For an efficient process and system management of RAS, it is essential to understand the microbial community composition and its interactions with the farmed organisms. The start-up of these systems often includes inoculation with nitrifying bacteria. In RAS, the distribution of the nitrifying bacteria depends on oxygen and nutrient availability. In the biofilm, the aerobic part is usually dominated by AOB, like *Nitrosomonas* sp., and NOB, such as *Nitrobacter* sp., due to the high ammonium and nitrite concentrations. Another AOB genus present in lower proportions is *Nitrosospira*. The NOB *Nitrospira* is most abundant in the anoxic interface because it uses a small amount of oxygen and nitrite (Rurangwa and Verdegem 2015).

In aquaculture, AOB and NOB consortia have been immobilized on synthetic carriers, such as polystyrene and low-density polyethylene beads in packed-bed bioreactors by Kumar et al. (2008), reaching undetectable levels of ammonia and a higher rate of survival of shrimp larvae. These authors reported high removal percentages for ammonia (78%) and nitrite (79%) in the spent water. Finally, by fluorescence *in situ* hybridization, they demonstrated the presence of AOB belonging to *Nitrosococcus mobilis* and NOB belonging to *Nitrobacter* sp. in the system. Natural inorganic carriers have been also used in RAS. Pungrasmi et al. (2016) used pumice stone in a 100 L biofilter tank and found that ammonia and nitrite concentrations were below 1 mg-N/L throughout a 121-day of culture of tilapia.

Nitrifying bacteria have been tested even in autonomous life support systems (LSS) developed for human missions in space, such as the Micro-Ecological Life Support System Alternative (MELiSSA) project of the European Space Agency that initiated in 1989 and is recognized as the most advanced effort to develop an LSS. These artificial systems are used to regenerate the atmosphere and provide water and food for a crew by reprocessing their wastes. The function of nitrifying bacteria in these systems is to convert the nitrogen present in wastes, like urea, and proteins into suitable sources of food. All the process of reprocessing is carried out through different compartments where diverse organisms are transforming the wastes, such as in a treatment plant (Imhof et al. 2017).

Specifically, MeLiSSA is a project in development in which pilot plant was inaugurated in 2009 and is located at Universitat Autònoma de Barcelona. In the MeLiSSA project, the wastes are recollected in Compartment I (liquefying compartment) where they are degraded into more simple units, such as ammonium, peptides and amino acids, and $CO_2$ and organic acids by a consortium of anaerobic bacteria (acidogenic, acetogenic, and methanogenic bacteria). Then, these products enter Compartment II (photoheterotrophic compartment), which is responsible for the elimination of the terminal products of the liquefying compartment by phototrophic bacteria. The nitrification process of AOB and NOB is performed in Compartment III that leads to the formation of nitrate ($NO_3^-$). The nitrate produced is used then by plants as a source of nitrogen in Compartment IV. Finally, Compartments IVa and IVb take care of the $CO_2$ exhaustion and the productions of oxygen and food by microalgae and plants (Farges et al. 2012, Imhof et al. 2017).

The nitrifying bacteria used in Compartment III comprise a consortium of AOB, such as *Nitrosomonas europaea* and *Nitrosospira,* that oxidize ammonium to nitrite to obtain energy and NOB, like *Nitrobacter* and *Nitrospira*, that oxidize nitrite and produce nitrate. Due to the low growth of nitrifying bacteria, the type of system used for Compartment III was a bioreactor packed with polystyrene where the bacteria could fix and proliferate with less generation of degradation products, like nitrite (Farges et al. 2012).

From industrial pollutants, such as BTEX, that include Benzene, toluene, ethylbenzene, and xylene, and in order to bioremediate diverse organic compounds, such as volatile organic compounds (VOCs), single strands of nitrifying bacteria as *N. europaea* have been applied with high success (Yoshikawa et al. 2017); however, this method is not feasible for industries every time. In this sense, another way to take advantage of bacteria is through the use of the specific enzymes that they use in their metabolism but for the catalysis of other substrates (co-substrates), which they transform without getting a benefit of them by cometabolism. In the case of nitrifying bacteria, the AOB *Nitrosomonas europaea* have been shown to be capable of degrading organic compounds, such as methane, methanol, ethylene, and propylene since the 1980s. This oxidation capacity of organic compounds has been attributed since then to the ammonia monooxygenase enzyme (AMO) of *N. europaea* due to the following main reasons proposed by Hyman et al. (1988):

1. The reactions products are compatible with a monooxygenase-catalyzed reaction.
2. The oxidations are inhibited by specific inhibitors of AMO (thiourea and acetylene).
3. The amount of hydrocarbon oxidized depends on the ammonium ion concentrations.

Therefore, it was determined that AMO could catalyze the oxidation of several compounds and not just ammonia. This is similar to the methane monooxygenase of methanotrophs, such as *Methylococcus capsulatus* and *Methylosinus trichosporium* OB3b (1-3), which share certain features with AMO because both can oxidize ammonia and hydrocarbons other than methane (Hyman et al.

1988; Fitch et al. 1993). Thus, some bacterial strains or their specific enzymes can be used for varied purposes and can be implemented in industries to prevent environmental contamination after the necessary feasibility studies and research.

## 9.4 BIOLOGICAL WASTEWATER TREATMENT SYSTEMS FOR INDUSTRIAL EFFLUENTS

Biological processes are well-perceived as a cost-effective and environmentally friendly method for wastewater treatment because they exploit the microbial metabolism and permit colloidal organic pollutants dissolved in water to be transformed into harmless substances (Show and Lee 2017, Yu et al. 2017). The biotechnologies available nowadays represent both well-established and novel technologies, although several aspects of their performance remain to be tested; the use of novel biocatalysts and reactor designs a fundamental understanding of microbial community dynamics, the mechanisms occurring within a bioreactor, the assessment of the performance of bioreactors during long-term operation, and its modeling which are some of these missing points (Pakshirajan et al. 2015).

In terms of water treatments to eliminate organic compounds, activated sludge processes are the most used technique by far (Semrany et al. 2012), although alternative processes such as biofilm systems or granules systems also exist (see the next sections). Activated sludge systems contain flocs, and sludge properties are mainly determined by the size, shape, density, and strength of the sludge flocs, which consist of microorganisms, extracellular polymeric substances, organic debris, and inorganic particles (Christensen et al. 2015). Therefore, the floc and sludge properties determine the type of pollutant removal from industrial wastewater and in turn, the composition of the incoming wastewater affects the properties of the sludge produced, especially the organic compounds, all of which are important factors that must be used for the selection of the appropriate type of biological process employed in the treatment system.

Microbial metabolism in wastewater treatment can be carried out under both aerobic and anaerobic conditions. The major advantage of the aerobic processes is their capacity to handle a wide variety of low-strength soluble wastewaters, while in anaerobic microbial communities are superior specifically at high temperatures and high concentrations of soluble and insoluble wastes. It is worthwhile to evaluate to what extent both aerobic and anaerobic technologies are currently evolving, either as direct competitors or as complementary treatments to one another. The best choice must be the most environmentally desirable and cost-effective choice (Show and Lee 2017). In some cases, like the pulp and paper wastewater, an integrated treatment combined biological processes operating under different environmental conditions (aerobic and anaerobic) or physicochemical and biological processes have been used (Ashrafi et al. 2015).

The anaerobic sludge treatment technology has been successfully applied for decades to a wide spectrum of industrial wastewater at full scale as a consolidated technology. The key mechanism is the immobilization of microorganisms, although the modern anaerobic high-rate reactors employ flocculent biomass, which is retained in the system by advanced gas-liquid sludge separation devices (Van Lier et al. 2015). Activated sludge in the aeration tanks uses the current state vector as purifying microorganisms by adsorption and concentrated on the surface of the activated sludge microorganisms to decompose organic matter (Yu et al. 2017).

Unfortunately, several compounds such as petroleum and some industrial wastes, pesticides, pharmaceuticals, and hormonal substances appear to be refractory to traditional (aerobic or anaerobic) activated sludge treatments and can join the different water surfaces from other natural environments, especially contaminated soils. Amending activated sludge with additional microorganisms or an efficient consortium that can produce versatile enzymes to enhance biodegradability or 'bioaugmentation' is considered as a possible solution for the recalcitrance of bio-refractory organic compounds when indigenous microorganisms are not able to physiologically perform the purification process. In this way, the bio-purification of wastewaters with chlorinated and nitrogenated compounds and other industrial pollutants is now possible that removes anything which is harmful to the aquatic life through the addition of specific microorganisms as well as decide what type of pollutant needs to be eliminated. For example, the nitrifying bacteria grow more slowly than the general heterotrophic community and are less resistant to toxicity. Supplementation through bioaugmentation of these bacteria may be beneficial to a treatment system characterized by a high toxic nitrogen compound loading, like ammonia, in aquaculture ponds, industries of palm oil, coke, and petrochemical wastewaters because they improve ammonia removal (Raper et al. 2018).

On the other hand, membrane bioreactors (MBR) are an alternative to activated sludge where a membrane is used instead of the clarifier. The outcome of the process is treated wastewater (effluent), return sludge, and excess sludge (Christensen et al. 2015). Though MBR have made enormous progress and have become a promising approach in wastewater treatment, there are still several limitations on conventional MBR and thus there emerged integrated MBR for wastewater treatment technology. Nevertheless, more and more new configurations of MBR have been proposed for practical application regarding revolution in environmental engineering. MBR have shown good performance in terms of high organic removal and they may become an attractive alternative for water reuse and recycling in the near future (Neoh et al. 2016).

For saline wastewater originating from agriculture, aquaculture, and many industrial sectors that usually contain high levels of salts and other contaminants, wetlands are another alternative to the activated sludge process and have been successfully used for a treating a wide variety of wastewaters which adversely affect both aquatic and terrestrial ecosystems and are eco-friendly and cost-effective compared to bioreactors or physicochemical technologies. In these systems, like other types of wastewater treating processes mentioned, the microorganisms play the most important role in the removal process of many contaminants (Liang et al. 2017).

Different methods for treating industrial wastewater containing heavy metals often involve technologies for the reduction of toxicity in order to meet technologies based on treatment standards. The physicochemical removal processes are more commonly used; however, these are expensive methods and create toxic sludges. For this reason, an alternative approach is the bio-absorption, which is also eco-friendly and more effective at removing heavy metals from wastewater. Biological removal of heavy metals in wastewater involves the use of biological techniques to eliminate pollutants from wastewater. In this process, microorganisms play a key role in settling solids in the solution (Gunatilake 2015).

Moreover, wastewater is now being considered as one of the resources for water, plant fertilizing nutrients, and a potential source of electricity generation using microbial fuel cells (MFCs) (He et al. 2017), which are biochemical devices that use bacteria as a biocatalyst to convert chemical energy present in organic matter (e.g., glucose) into electricity.

An MFC consists of an anaerobic anode chamber, a cathode chamber, and a proton exchange membrane or salt bridge that separates both chambers and only permits the transfer of proton ($H^+$) from the anode chamber to the cathode chamber (Seow et al. 2016), which has allowed the use of some different types of wastewater for simultaneous treatment and the generation of electricity but also for the production and recovery of value-added products, like bioflocculants, bioplastics, biosurfactants, chemicals, electricity, hydrogen, methane, and many other compounds from wastewater.

This is very important, especially because the current wastewater treatment technologies are commonly characterized by energy-intensive, large quantities of residuals generation, and incapability of recovering the potential resources available in wastewater (He et al. 2017). Instead of removing these contaminants from wastewater through a physical or chemical method, MFC provide an alternative method for wastewater treatment by harnessing the chemical energy within the biodegradable compounds using bacteria and subsequently generating sustainable and clean electricity. The substrates that were successfully used to remove pollutants and produce energy using MFC included paper recycling wastewater, domestic wastewater, food processing wastewater, starch processing wastewater, chocolate industry wastewater, mustard tuber wastewater, textile wastewater, mining and allied industry, and any other whose wastewater contain substrates rich in a mixture of carbon, nitrogen, and/or organic sources (Pandey et al. 2016, Seow et al. 2016).

## 9.5  BIOFILM SYSTEMS

Biological wastewater treatment (WWT) frequently harnesses biofilms to remove anthropogenic contaminants. Biofilms are complex biological systems that encompass polymicrobial aggregates as organic matter and in some cases also inorganic deposits. In most biofilms, the microorganisms account for less than 10% of the dry mass, whereas the matrix can account for over 90%. The matrix is composed of extracellular material denominated extracellular polymeric substances (EPS) of a different nature, such as lipids, polysaccharides, extracellular nucleic acids, or proteins, which are mostly produced by the microorganisms themselves and in where the biofilm cells are embedded. This EPS allows that biofilms to form the scaffold for the three-dimensional architecture and is responsible for adhesion to surfaces and cohesion in the biofilm (Flemming and Wingender 2010, Calderón et al. 2013). The nature of biofilm structure is dynamic as the cells anchored to the surface eventually disperse and revert into the planktonic mode of living, which then enables the colonization of new niches (Hall-Stoodley et al. 2004, Calderón et al. 2013).

The steps that lead to the formation of microbial biofilms have been extensively described, but it is generally accepted that the process starts when microbes associated with a surface change from a reversible to an irreversible mode of attachment to it followed by the aggregation of cells and their subsequent proliferation. The extraordinary tolerance of biofilms to antimicrobial compounds, heavy metals, and other damaging agents derive from a complex mixture of physical, chemical, and physiological factors—the metabolic heterogeneity of the community, the particular physiological state of the microorganisms in the different biofilm layers, the support of syntrophic and other mutualistic interactions, and the development of specialized subpopulations of resistant phenotypes and persisted cells (Calderón et al. 2013, Miao et al. 2019). The relative contribution of each of these mechanisms (and possibly others) varies with the type of biofilm and the nature of the environment where they develop.

The use of inert carrier materials to support biofilm development is often required, although under certain operating conditions microorganisms yield structures called granules-dense aggregates of self-immobilized cells with the characteristics of biofilms maintained in suspension.

## 9.6  GRANULAR SYSTEMS

Biogranulation can generate two types of granular sludge, which are aerobic granular sludge and anaerobic granular sludge and both of these can be developed in a fixed sequencing cycle of feeding, reacting, settling, and decanting under a single sequencing batch reactor system (Seow et al. 2016). Aerobic granulation technology is superior to a conventional system, such as the activated sludge process, and due to this it, is drawing increasing global interest in a quest for an efficient and innovative technology for wastewater treatment in the industrial sector. Besides, it is cost-effective, and removing organic matter has shown promising results (Yu et al. 2014, Baeten et al. 2018, Sengar et al. 2018). Pollutant removal relies on microorganisms that grow in approximately spherical aggregates that can freely move within a reactor and therefore the requirement of space and energy is highly reduced. Compared to flocs, granules have a higher biomass density and a more regular shape; they are mechanically stronger, more highly resistant to organic load and toxic substances, and they can become larger. Other advantages offered by the granulation system are good settleability and ability to withstand a high organic loading rate (Baeten et al. 2018).

The aerobic granulation technology has been successfully applied for nitrogen removal from wastewater since nitrification and denitrification processes can take place simultaneously in a single reactor through the different layers inside the granules that have both aerobic and anoxic zones

(Kishida et al. 2008). The aerobic zone is dominated by microorganisms involved in nitrification, whereas the heterotrophic microorganisms present in the anoxic zone are mostly responsible for denitrification (Sarma and Tay 2018), which allows for the achievement of >80% $NH_4^+$ removal from wastewater (Tsuneda et al. 2006). Because the biomass in a wastewater treatment system is created by multi-species microbial communities, a variety of pollutants can be removed (Cydzik-Kwiatkowska and Wojnowska-Baryła 2015), including for excessive high-strength ammonium ($NH_4^+$) wastewaters from petroleum refineries, landfills, livestock farms, and other industrial sources (Yu et al. 2014) or hypersaline wastewater from fish procession industries (Corsino et al. 2016).

Nitrogenized toxic compounds, especially ammonium, can be removed from industrial wastewater directly in a single step without having to take the two-step nitrification/denitrification using the anammox granular sludge or granular partial-nitritation-anammox technology (Abma et al. 2007, Li et al. 2016).

The anammox bacterium was a revolution in the biological nitrogen cycle; they are able to create a short cut in the nitrogen cycle by oxidizing ammonium directly with nitrite as an electron acceptor to dinitrogen gas under anoxic conditions. This conversion is called anaerobic oxidation or anammox (Speth et al. 2016). For the purpose of wastewater treatment, anammox is extremely beneficial, energy-efficient, and more sustainable over the conventional nitrification/denitrification process because it significantly decreases oxygen and organic consumption, resulting in cost saving in chemical additions and sludge disposal (Li et al. 2016). In granular reactors, 60% of power consumption can be reduced by the less or null aeration performance (Abma et al. 2007).

A full-scale granular sludge anammox process has been a proven success in full scale (Abma et al. 2007), although there are still problems to be solved like for the industrial effluents, such as coke ovens, fertilizer, antibiotics, and digested black water that contain high concentrations of organic matter and nitrogen which cause the slow growth rate of anammox bacteria and constrain some industrial applications (Tomar and Gupta 2017).

Many new technologies and processes were originally developed for domestic or urban wastewater systems and failed to properly represent specific industrial processes because industrial wastewater is not the same, even between the diverse types of industries, as a result of different production schemes within the factories (Feldman et al. 2017). Therefore, it is necessary to implement specific studies for each different industrial wastewater. It is possible to implement a combination of processes with a multi-scale concept, which may rely on temporal or spatial aspects of microbial metabolism for successful results. Microbial diversity and abundance in the granules depend on operational conditions, such as the type of organic substrate, the loading rate, hydraulic retention time, or aeration intensity; so for successful nitrogen removal from industrial wastewater, operational conditions of the granular system must be carefully selected (Cydzik-Kwiatkowska and Wojnowska-Baryła 2015).

## 9.7 CONCLUSIONS AND FUTURE PERSPECTIVES

Nowadays different treatment systems are currently used in order to remove different contaminants from wastewater. The most developed systems are referred to reduce organic matter; however, industrial wastewater contains contaminants other than only organic matter. Different substances and organic compounds, such as nitrogenated, chlorinated, or heavy metals, are some pollutants that are not only difficult to eliminate from water but also could strongly impact negatively in the health of environmental systems and the human population. Industrial waste is a major source of environmental pollution in the world, especially in the least developed countries where the limits of the established norms in cases that exist are higher than in other countries or where there are absolutely no laws in order to regulate pollutants present in environmental systems. For example, for the United States, the EPA has established a maximum contaminant level of 0.005 mg/L of Trichlorethylene (TCE) in potable water, one of the more toxic chlorinated compounds, used and released by industries to the environmental water systems and soil. However, on the Southern side of the border country, in Mexico, TCE has been detected in different areas, but there is currently no law that establishes its maximum contaminant level. Although later on, in the mid-2000s, a Mexican Official Norm (NOM) was formulated which proposed a maximum concentration of 0.070 mg/L in water, clearly higher concentration than those established internationally (Alfán-Guzmán, 2010).

Therefore, efforts by governments and authorities are really necessary to strictly regulate industrial discharges in environmental systems in order to prevent contamination and pollution. Moreover, the most appropriate means of controlling nitrogenized compounds (i.e., ammonium and nitrite) and chlorinated compounds is the prevention of contamination, that is to have strict waste management for human activities.

In general, treatment systems are well used to remove domestic sewages and biodegradable industrial wastes (such as suspended solids, grease, and oils) but not the industries' discharge toxic pollutants as their industrial treatment systems are not designed to deal with it (EPA 2011). Thus, pretreatments in industrial wastes to obtain non- or less toxic compounds, such as nitrification and dichlorination processes, should be preferably applied before they are discharged in order to prevent environmental damages to human health, as established in EPA objectives. Bioremediation is an eco-friendly, less expensive, and faster way to treat contaminants in comparison to traditional remediation strategies, although further research is still needed to find the potential of microorganisms to apply in treatment systems and the more effective ways to use them, especially based on their mechanisms, dynamics, performance, and consortiums.

On the other hand, efficient biocatalysts bioreactor systems are required to be properly tested to improve bioremediation technologies. However, the utilization of microorganisms highly present in nature, such as nitrifying microorganisms, is a promising choice due to the diverse capacities of these bacteria.

## 9.8 REFERENCES

Abma, W.R., Schultz, C.E., Mulder, J.W., Van der Star, W.R.L., Strous, M., Tokutomi, T., et al. 2007. Full-scale granular sludge Anammox process. Water Science and Technology 55(8-9): 27-33.

Agency for Toxic Substances and Disease Registry (ATSDR). 2015. ToxFAQs™ for Trichloroethylene (TCE). ATSDR.cdc.gov. Report Update May 12, 2015. http://www.atsdr.cdc.gov/toxfaqs/TF.asp?id=172&tid=30.

Alfán-Guzmán, R. 2010. Efecto de compuestos organoclorados sobre el crecimiento de una población microbiana aerobia. Bachelor´s thesis. Instituto Politécnico Nacional. Unidad Profesional Interdisciplinaria de Biotecnología. Ciudad de México, México.

Araos, M. and Onederra, I. 2017. Detonation characteristics of a NOx-free mining explosive based on sensitised mixtures of low concentration hydrogen peroxide and fuel. Central European Journal of Energetic Materials 14(4): 759-774.

Arciero, D., Vannelli, T., Logan, M. and Hopper, A.B. 1989. Degradation of trichloroethylene by the ammonia-oxidizing bacterium *Nitrosomonas europaea*. Biochemical and Biophysical Research Communications 159(2): 640-643. doi: 10.1016/0006-291X(89)90042-9.

Ashrafi, O., Yerushalmi, L. and Haghighat, F. 2015. Wastewater treatment in the pulp-and-paper industry: a review of treatment processes and the associated greenhouse gas emission. Journal of Environmental Management 158: 146-157.

Attalla, M.I., Day, S.J., Lange, T., Lilley, W. and Morgan, S. 2008. NO$_x$ emissions from blasting operations in open-cut coal mining. Atmospheric Environment 42(34): 7,874-7,883.

Avnimelech, Y., Verdegem, M.C.J., Kurup, M. and Keshavanath, P. 2008. Sustainable land-based aquaculture: rational utilization of water, land and feed resources. Mediterranean Aquaculture Journal 1(1): 45-55.

Baeten, J.E., Batstone, D.J., Schraa, O.J., Van Loosdrecht, M.C. and Volcke, E.I. 2018. Modelling anaerobic, aerobic and partial nitritation-anammox granular sludge reactors: a review. Water Research 1(149): 322-341.

Beamer, P.I., Luik, C.E., Abrell, L., Campos, S., Martínez, M.E. and Sáez, A.E. 2012. Concentration of trichloroethylene in breast milk and household water from Nogales, Arizona. Environmental Science and Technology 46(16): 9,055-9,061. doi: 10.1021/es301380.

Berrelleza-Valdez, F., Parades-Aguilar, J., Peña-Limón, C.E., Certucha-Barragán, M.T., Gámez-Meza, N., Serrano-Palacios, D., et al. 2019. A novel process of the isolation of nitrifying bacteria and their development in two different natural lab-scale packed-bed bioreactors for trichloroethylene bioremediation. Journal of Environmental Management 241: 211-218. doi: 10.1016/j.jenvman.2019.04.037.

Bertrand, J.C., Bonin, P., Caumette, P., Gatusso, J.P., Grégory, G., Guyoneaud, G., et al. 2015. Biogeochemical cycles. pp. 511-617. *In*: Bertrand, J.C., Caumette, P., Lebaron, P., Matheron, R. and Normand, P. (eds.). Environmental Microbiology: Fundamentals and Applications. Springer, Dordrecht.

Calderón, K., González-Martínez, A., Gómez-Silván, C., Osorio, F., Rodelas, B. and González-López, J. 2013. Archaeal diversity in biofilm technologies applied to treat urban and industrial wastewater: recent advances and future prospects. International Journal of Molecular Sciences 9(14): 18,572-18,598.

Calderón, K., Philippot, L., Bizouard, F., Breuil, M.C., Bru, D. and Spor, A. 2018. Compounded disturbance chronology modulates the resilience of soil microbial communities and N-Cycle related functions. Frontiers in Microbiology 6(9): 2,721.

Christensen, M.L., Keiding, K., Nielsen, P.H. and Jørgensen, M.K. 2015. Dewatering in biological wastewater treatment: a review. Water Research 82: 14-24.

Corsino, S.F., Capodici, M., Morici, C., Torregrossa, M. and Viviani, G. 2016. Simultaneous nitritation–denitritation for the treatment of high-strength nitrogen in hypersaline wastewater by aerobic granular sludge. Water Research 88: 329-336.

Cros, C.J., Terpeluk, A.L., Crain, N.E., Juenger, M.C. and Corsi, R.L. 2015. Influence of environmental factors on removal of oxides of nitrogen by a photocatalytic coating. Journal of the Air and Waste Management Association 65(8): 937-947.

Cydzik-Kwiatkowska, A. and Wojnowska-Baryła, I. 2015. Nitrogen-converting communities in aerobic granules at different hydraulic retention times (HRTs) and operational modes. World Journal of Microbiology and Biotechnology 31(1): 75-83.

Dubey, M., Yadav, G., Kapuria, A., Ghosh, A., Muralidharan, M., Lal, D. et al. 2014. Exploring bacterial diversity from contaminated soil samples from river Yamuna. Microbiology 83(5): 585-588. doi: 10.1134/S002626171405009.

Eisler, R. 2004. Mercury hazards from gold mining to humans, plants, and animals. Reviews of Environmental Contamination and Toxicology 181: 139-198.

European Commission. 2008. Priority substances and certain other pollutants according to Annex II of directive 2008/105/EC. Available online at: http://ec.europa.eu/environment/water/water-framework/priority_substances.htm.

Ekasari, J. and Maryam, S. 2012. Evaluation of biofloc technology application on water quality and production performance of red tilapia *Oreochromis sp.* cultured at different stocking densities. Hayati Journal of Biosciences 19(2): 73-80.

Farges, B., Poughon, L.R., Creuly, C., Dussap, C.G. and Lasseur, C. 2012. Axenic cultures of *Nitrosomonas europaea* and *Nitrobacter winogradskyi* in autotrophic conditions: a new protocol for kinetic studies. Applied Biochemistry and Biotechnology 167(5): 1,076-1,091. doi: 10.1007/s12010-012-9651-6.

Feldman, H., Flores-Alsina, X., Ramin, P., Kjellberg, K., Jeppsson, U., Batstone, D.J., et al. 2017. Modelling an industrial anaerobic granular reactor using a multi-scale approach. Water Research 126: 488-500.

Feng, G., Sun, W., Zhang, F., Karthik. L. and Li, Z. 2016. Inhabitancy of active *Nitrosopumilus*-like ammonia-oxidizing archaea and *Nitrospira* nitrite-oxidizing bacteria in the sponge *Theonella swinhoei*. Scientific Reports 6: 24,966. doi: 10.1038/srep24966.

Fitch, M.W., Graham, K.P., Arnold, R.G., Agarwal, S.K., Phelps, P., Speitel, G.E. Jr., et al. 1993. Phenotypic characterization of cooper-resistant mutants of *Methylosinus trichosporium* OB3b. Applied and Environmental Microbiology 59(9): 2771-2776.

Flemming, H.C. and Wingender, J. 2010. The biofilm matrix. Nature Reviews Microbiology 8: 623-633. doi: 10.1038/nrmicro2415.

French, E., Kozlowski, A., Mukherjee, M., Bullerjahn, G. and Bollman, A. 2012. Ecophysiological characterization of ammonia-oxidizing archaea and bacteria from freshwater. Applied and Environmental Microbiology 78(16): 5,773-5,780.

Fugiel, A., Burchart-Korol, D., Czaplicka-Kolarz, K. and Smoliński, A. 2017. Environmental impact and damage categories caused by air pollution emissions from mining and quarrying sectors of European countries. Journal of Cleaner Production 143: 159-168.

Füssel, J., Lücker, J., Yilmaz, P., Nowka, B., van Kessel, M.A.H.J., Bourceau P., et al. 2017. Adaptability as the key to success for the ubiquitous marine nitrite oxidizer *Nitrococcus*. Sciences Advances 3(11). doi: 10.1126/sciadv.1700807.

García-Solares, S.M., Ordaz, A., Monroy-Hermosillo, O. and Guerrero-Barajas, C. 2013. Trichloroethylene (TCE) biodegradation and its effect on sulfate reducing activity in enriched sulfidogenic cultures prevenient from UASB maintained at 20°C. International Biodeterioration and Biodegradation 83: 92-96. doi: 10.1016/j.ibiod.2013.04.011.

Gunatilake, S.K. 2015. Methods of removing heavy metals from industrial wastewater. Journal of Multidisciplinary Engineering Science Studies (JMESS) 1(1): 12-18.

Hall-Stoodley, L., Costerton, J.W. and Stoodley, P. 2004. Bacterial biofilms: from the natural environment to infectious diseases. Nature Reviews Microbiology 2: 95-108.

He, L., Du, P., Chen, Y., Lu, H., Cheng, X., Chang, B., et al. 2017. Advances in microbial fuel cells for wastewater treatment. Renewable and Sustainable Energy Reviews 71: 388-403.

Hendryx, M., Higginbotham, N., Ewald, B. and Connor, L.H. 2019. Air Quality in Association with Rural Coal Mining and Combustion in New South Wales Australia. The Journal of Rural Health 35(4): 518-527.

Hoang, P.H., Nguyen, H.T., Trung, T.T., Tran, T.T., Do, L.P. and Le, T.H. 2016. Isolation and selection of nitrifying bacteria with high biofilm formation for treatment of ammonium polluted aquaculture water. Journal of Vietnamese Environment 8: 33-40.

Huang, B., Lei, C., Wei, C. and Zeng, G. 2014. Chlorinated volatile organic compounds (Cl-VOCs) in environment – sources, potential human health impacts, and current remediation technologies. Environment International 71: 118-138.

Hyman, M.R., Murton, I.B. and Arp, D.J. 1988. Interaction of ammonia monooxygenase from *Nitrosomonas europaea* with alkanes, alkenes, and alkynes. Applied and Environmental Microbiology 54(12): 3,187-3,190.

Ibarra, M. and Moreno, J. 2017. La justicia ambiental en el Rio Sonora. RevIISE - Revista de Ciencias Sociales y Humanas 10: 135-155.

Imhof, B., Weiss, P., Vermeulen, A., Flynn, E., Hyams, R., Kerrigan, C., et al. 2017. MeLiSSA: the agency's perspective. pp. 288-290. *In*: Armstrong, R. (ed.). Star Ark, a Living, Self-Sustaining Spaceship. Springer, Chinchester, United Kingdom.

Itoh, Y., Sakagami K., Uchino, Y., Boonmak, C., Oriyama, T., Tojo, F., et al. 2013. Isolation and characterization of a thermotolerant ammonia-oxidizing bacterium *Nitrosomonas sp.* JPCCT2 from a thermal power station. Microbes and Environments 28(4): 432-435. doi: 10.1264/jsme2.ME13058.

Jaishankar, M., Tseten, T., Anbalagan, N., Mathew, B.B. and Beeregowda, K.N. 2014. Toxicity, mechanism and health effects of some heavy metals. Interdisciplinary Toxicology 7(2): 60-72. doi: 10.2478/intox2014-0009.

Khare, S., Gokulan, K., Williams, K., Bai, S., Gilbert, K.M. and Blossom, S.J. 2018. Irreversible effects of trichloroethylene on the gut microbial community and gut-associated immune responses in autoimmune-prone mice. Journal of Applied Toxicology 39(2): 209-220. doi: 10.1002/jat.37.

Kishida, N., Tsuneda, S., Sakakibara, Y., Kim, J.H., and Sudo, R. 2008. Real-time control strategy for simultaneous nitrogen and phosphorus removal using aerobic granular sludge. Water Science and Technology 58(2): 445-450.

Kocamemi, B.A. and Çeçen, F. 2010. Biological removal of the xenobiotic trichloroethylene (TCE) through cometabolism in nitrifying systems. Bioresource Technology 101: 430-433. doi: 10.1016/j.biortech.2009.07.079.

Kumar, R., Achuthan, C., Manju, N.J., Philip, R.I. and Singh, S.B. 2008. Activated packed bed bioreactor for rapid nitrification in brackish water hatchery systems. Journal of Industrial Microbiological Biotechnology 36(3): 355-365. doi: 10.1007/s10295-008-0504-9.

Kumar, R., Parmar, B.S., Walia, S. and Saha, S. 2014. Nitrification inhibitors: classes and its use in nitrification management. pp. 103-124. *In*: Rakshit, A., Singh, H.B. and Sen, A. (eds.). Nutrient Use Efficiency: From Basics to Advances. Springer, New Delhi, India.

León-García, G.L., Meza-Figueroa, D.M., Valenzuela-García, et al. 2018. Study of heavy metal pollution in arid and semi-arid regions due to mining activity: Sonora and Bacanuchi rivers. International Journal of Environmental Sciences and Natural Resources 11: 1-11.

Li, X., Sun, S., Badgley, B.D., Sung, S., Zhang, H. and He, Z. 2016. Nitrogen removal by granular nitritation–anammox in an upflow membrane-aerated biofilm reactor. Water Research 94: 23-31.

Liang, Y., Zhu, H., Bañuelos, G., Yan, B., Zhou, Q., Yu, X., et al. 2017. Constructed wetlands for saline wastewater treatment: a review. Ecological Engineering 98: 275-285.

Little, D.C., Young, J.A., Zhang, W., Newton, R.W., Al Mamun, A. and Murray, F.J. 2018. Sustainable intensification of aquaculture value chains between Asia and Europe: a framework for understanding impacts and challenges. Aquaculture 493: 338-354.

Lu, S., Liu, X., Liu, C., Wang, X. and Cheng, G. 2018. Review of ammonia-oxidizing bacteria and archaea in freshwater ponds. Reviews in Environmental Science and Biotechnology 18: 1-10. doi: 10.1007/s11157-018-9486-x.

López-Galvez, N. 2014. Soil Analysis of Organic and Inorganic Contaminants in Goat Canyon (Cañón de Los Laureles), at the U.S-Mexico Border. Master's Thesis. San Diego State University.

Madigan, M.T., Bender, K.S., Buckley, D.H., Sattley, W.M. and Stahl, D.A. 2017. Brock Biology of Microorganisms. Upper Saddle River, Pearson, 2017.

Matteucci, F., Ercole, C. and Del gallo, M. 2015. A study of chlorinated solvent contamination of the aquifers of an industrial area in central Italy: a possibility of bioremediation. Frontier in Microbiology 6: 1-9.

Maulin, P.S. 2017. Environmental bioremediation of industrial effluent. Journal of Molecular Biology and Biotechnology 2: 1.

Miao, L., Yang, G., Tao, T. and Peng, Y. 2019. Recent advances in nitrogen removal from landfill leachate using biological treatments - a review. Journal of Environmental Management 1(235): 178-185.

Mulenga, S. 2020. Evaluation of factors influencig rocks fragmentation by blasting using interrelations diagrams methods. Journal of Physical Sciences 2(1): 1-16.

Neoh, C.H., Noor, Z.Z., Mutamim, N.S.A. and Lim, C.K. 2016. Green technology in wastewater treatment technologies: integration of membrane bioreactor with various wastewater treatment systems. Chemical Engineering Journal 283: 582-594.

O'Connor, M.P., Coulthard, R.M. and Plata, D.L. 2018. Electrochemical deposition for the separation and recovery of metals using carbon nanotube-enabled filters. Environ. Sci. Water Res. Technol. 4: 58.

Oluwoye, I., Dlugogorski, B.Z., Gore, J., Oskierski, H.C. and Altarawneh, M. 2017. Atmospheric emission of $NO_x$ from mining explosives: a critical review. Atmospheric Environment 167: 81-96.

Pakshirajan, K., Rene, E.R. and Ramesh, A. 2015. Biotechnology in environmental monitoring and pollution abatement 2015. BioMed Research International 963803: 1-3. doi: 10.1155/2015/963803.

Pal, L., Kraigher, B., Brajer-Humar, B., Levstek, M. and Mandic-Mulec, I. 2012. Total bacterial and ammonia oxidizer community structure in moving bed biofilm reactors treating municipal wastewater and inorganic synthetic wastewater. Bioresource Technology 110: 135-143.

Pandey, P., Shinde, V.N., Deopurkar, R.L., Kale, S.P., Patil, S.A. and Pant, D. 2016. Recent advances in the use of different substrates in microbial fuel cells toward wastewater treatment and simultaneous energy recovery. Applied Energy 168: 706-723.

Pant, P. and Pant, S. 2010. A review: advances in microbial remediation of trichloroethylene (TCE). Journal of Environmental Sciences 22(1): 116-126. doi: 10.1016/S10010742(09)60082-6.

Papp, B., Török, T., Sándor, E., Fekete, E., Flipphi, M. and Karaffa, L. 2016. High cell density cultivation of the chemolithoautotrophic bacterium *Nitrosomonas europaea*. Folia. Microbiologica. 61(3): 191-198. doi: 10.1007/s12223-015-0425-8.

Páez-Osuna, F. and Frías-Espericueta, M.G. 2001. Toxicidad de los compuestos de nitrógeno en camarones. pp. 253-276. *In*: Páez-Osuna, F. (ed.). Camaronicultura y medio ambiente. Colegio de Sinaloa, Mazatlán, Sinaloa, México.

Pérez, J., Buchanan, A. and Mellbye, B. 2014. Interactions of *Nitrosomonas europaea* and *Nitrobacter winogradskyi* grown in co-culture. Achieves of Microbiology 197: 79-89. doi: 10.1007/s00203-014-1056-1.

Poly, F., Wertz, S., Brothier, E. and Degrange, V. 2008. First exploration of Nitrobacter diversity in soils by a PCR cloning-sequencing approach targeting functional gene nxrA. FEMS Microbiology Ecology 63: 132-140. doi: 10.1111/j.1574-6941.2007.00404.x.

Pungrasmi, W., Phinitthanaphak, P. and Powtongsook, S. 2016. Nitrogen removal from a recirculating aquaculture system using a pumice bottom substrate nitrification-denitrification tank. Ecological Engineering 95: 357-363.

Ralebitso-Senior, T.K. and Orr, C.H. 2016. Elucidating the impacts of biochar applications on nitrogen cycling microbial communities. pp. 163-187. *In*: Ralebitso-Senior, T.K. and Orr, C.H. (eds.). Biochar Application. Essential Soil Microbial Ecology. Elsevier, Amsterdam, Netherlands.

Raper, E., Stephenson, T., Anderson, D.R., Fisher, R. and Soares, A. 2018. Industrial wastewater treatment through bioaugmentation. Process Safety and Environmental Protection 118: 178-187.

Rezaee, A., Naimi, N., Hashemi, S.E. and Hosseini, H. 2013. Molecular identification of nitrifying bacteria in activated sludge. Journal of Materials and Environmental Science 4(5): 601-604.

Rivera Carvajal, R., Duarte-Tagles, H. and Idrovo, Á.J. 2019. Mining leachate contamination and subfecundity among women living near the USA-Mexico border. Environmental Geochemistry and Health 14. doi: 10.1007/s10653-019-00275-w.

Rurangwa, E. and Verdegem, M. 2015. Microorganisms in recirculating aquaculture systems and their management. Reviews in Aquaculture 7: 117-130.

Sarma, S.J. and Tay, J.H. 2018. Carbon, nitrogen and phosphorus removal mechanisms of aerobic granules. Critical Reviews in Biotechnology 38(7): 1,077-1,088.

Selmin, O.I., Makwana, O. and Runyan, R.B. 2014. Environmental sensitivity to trichloroethylene (TCE) in the developing heart. pp. 153-169. *In*: Gilbert, K.M. and Blossom, S.J. (eds.). Trichloroethylene: Toxicity and Health Risks. Molecular and Integrative Toxicology. Springer, London.

Semrany, S., Favier, L., Djelal, H., Taha, S. and Amrane, A. 2012. Bioaugmentation: possible solution in the treatment of bio-refractory organic compounds (Bio-ROCs). Biochemical Engineering Journal 69: 75-86.

Sengar, A., Basheer, F., Aziz, A. and Farooqi, I.H. 2018. Aerobic granulation technology: laboratory studies to full scale practices. Journal of Cleaner Production 197(1): 616-632.

Seow, T.W., Lim, C.K., Nor, M.H.M., Mubarak, M.F.M., Lam, C.Y., Yahya, A., et al. 2016. Review on wastewater treatment technologies. International Journal of Applied Environmental Sciences 11: 111-126.

Show, K.Y. and Lee, D.J. 2017. Anaerobic treatment versus aerobic treatment. pp. 205-230. *In*: Lee, D.J., Jegatheesan, V., Ngo, H.H., Hallenbeck, P.C. and Pandey, A. (eds.). Current Developments in Biotechnology and Bioengineering. Elsevier.

Shukla, A.K., Upadhyay, S.N. and Dubey, S.K. 2014. Current trends in trichloroethylene biodegradation: a review. Critical Reviews in Biotechnology 8,551(2): 1-15. doi: 10.3109/07388551.2012.727080.

Simister, R., Taylor, M.W., Tsai, P., Fan, L., Bruxner, T.J., Crowe, M.L. and Webster, N. 2012. Thermal stress responses in the bacterial biosphere of the Great Barrier Reef sponge, *Rhopaloeides odorabile*. Environmental Microbiology 14: 3,232-3,246. doi: 10.1111/1462-2920.12010.

Speth, D.R., Guerrero-Cruz, S., Dutilh, B.E. and Jetten, M.S. 2016. Genome-based microbial ecology of anammox granules in a full-scale wastewater treatment system. Nature Communications 7: 11,172.

Tomar, S. and Gupta, S.K. 2017. Symbiosis of denitrification, anammox and anaerobic pathways–an innovative approach for confiscating the major bottlenecks of anammox process. Chemical Engineering Journal 313: 355-363.

Tsuneda, S., Oglwara, M., Ejiri, Y. and Hirata, A. 2006. High-rate nitrification using aerobic granular sludge. Water Science and Technology 53: 147-154.

U.S. Environmental Protection Agency (EPA). 2009. National primary drinking water regulations. EPA. gov. Report update may 2009. Available online at: https://www.epa.gov/sites/production/files/2016-06/documents/npwdr_complete_table.pdf.

U.S. Environmental Protection Agency (EPA). 2010. Final Revisions to the Primary National Ambient Air Quality Standard for Nitrogen Dioxide ($NO_2$). Washington, DC: U.S. Environmental Protection Agency, Office of Air Quality Planning and Standards.

U.S. Environmental Protection Agency (EPA). 2011. Introduction to the National Pretreatment Program. Report Update June 2011. Available online at: https://www3.epa.gov/npdes/pubs/pretreatment_program_intro_2011.pdf.

Van Lier, J.B., Van der Zee, F.P., Frijters, C.T.M.J. and Ersahin, M.E. 2015. Celebrating 40 years anaerobic sludge bed reactors for industrial wastewater treatment. Reviews in Environmental Science and Bio/Technology 14(4): 681-702.

Vannelli, T., Logan, M., Arciero, D.M. and Hooper, A.B. 1990. Degradation of halogenated aliphatic compounds by the ammonia-oxidizing bacterium *Nitrosomonas europaea*. Applied and Environmental Microbiology 56(4): 1169-1171.

Vipindas, P.V., Abdulaziz, A., Chekidhenkuzhiyil, J., Lallu, K.R., Fausia, K.H., Balachandran, K.K., et al. 2015. Bacterial domination over archaea in ammonia oxidation in a monsoon-driven tropical estuary. Microbiology Ecology 69: 544-553. doi: 10.1007/s00248-014-0519-x.

Wahman, D.G., Katz, L.E. and Speitel, G.E. 2005. Cometabolism of trihalomethanes by *Nitrosomonas europaea*. Applied and Environmental Microbiology 71(12): 7980-7986. doi: 10.1128/AEM.71.12.7980.

Wang, B., Zhao, J., Guo, Z., Ma, J., Xu, H. and Jia, Z. 2014a. Differential contributions of ammonia oxidizers and nitrite oxidizers to nitrification in four paddy soils. International Society for Microbial Ecology Journal 9: 1062-1075.

Wang, X., Wang, C., Bao, L. and Xie, S. 2014b. Abundance and community structure of ammonia-oxidizing microorganisms in reservoir sediment and adjacent soils. Applied Microbiology and Biotechnology 98: 1,883-1,892. doi: 10.1007/s00253-013-5174.

Yoshikawa, M., Zhang, M. and Toyota, K. 2017. Biodegradation of volatile organic compounds and their effects on biodegradability under co-existing conditions. Microbes and Environments 32(3): 188-200.

Yu, L., Han, M. and He, F. 2017. A review of treating oily wastewater. Arabian Journal of Chemistry 10: S1913-S1922.

Yu, X., Wan, C., Lei, Z., Liu, X., Zhang, Y., Lee, D.J., et al. 2014. Adsorption of ammonium by aerobic granules under high ammonium levels. Journal of the Taiwan Institute of Chemical Engineers 45(1): 202-206.

Zhu, W., Wang, C., Hill, J., He, Y., Tao, B., Mao, Z., et al. 2018. A missing link in the estuarine nitrogen cycle: coupled nitrification-denitrification mediated by suspended particulate matter. Scientific Reports 8: 2,282. doi: 10.1038/s41598-018-20688-4.

# 10

# Simultaneous Nitrification and Denitrification Processes in Granular Sludge Technology

Barbara Muñoz-Palazón[1,*], Miguel Hurtado-Martinez[2] and Jesus Gonzalez-Lopez[1]

## 10.1  INTRODUCTION

Human activities have had a large effect on biogeochemical cycles. Since the development of agriculture, industrialization, and urbanization, among others, humans have altered the normal balance of nutrients in cycles. Of special interest is the nitrogen cycle, so nitrogen enrichment could produce many environmental issues, such as eutrophication of aquatic environment, depletion of oxygen, loss of ecosystem function, or death of aquatic life. For these reasons, nitrogen removal is now one of the main goals for scientific research.

One of the sources of this imbalance of nitrogen comes from the discharge of wastewater treatment plants (WWTPs) from urban centers to natural water bodies. The conventional treatment of wastewater is mainly based on biological processes that are able to remove organic matter, carried out with activated sludge, which are well-established systems with predictable operations, though they require larger spaces with bioreactors and secondary clarifiers. However, the fact that conventional activated sludge systems (CAS) have been constructed for ammonia and organic matter oxidation has made it necessary to enlarge the surface of the wastewater treatment plant for nitrogen and phosphorous removal, which entails a high cost to build new bioreactors, new pumps, and consequently a high operational cost. After the publication of the EU Water Framework Directive (WFD), the level of organic matter, suspended solids nitrogen, and phosphorous have been well-established to discharge in water body receivers. For that matter, the search for a novel technology that saves cost and energy and is environmentally friendly and compact could meet the requirement demanded.

Several novel nitrogen removal processes have been developed to enhance nitrification and denitrification. The research has focused on novel processes that include different technologies, such as simultaneous nitrification and denitrification (SND), shortcut nitrification and denitrification, anaerobic ammonium oxidation (ANAMMOX), aerobic deammonitrification, completely autotrophic nitrogen removal over nitrite (CANON), and oxygen-limited autotrophic nitrification-denitrification (OLAND) (Muller et al. 1995, Strous et al. 1999, Fux et al. 2002, Third et al. 2001, Schmidt et al. 2003, Nielsen et al. 2002, Peng and Zhu 2006). These technologies were implemented

[1] Department of Microbiology, Water Research Institute and Faculty of Pharmacy of University of Granada.
[2] Water Research Institute, University of Granada.
* Corresponding author: bmp@ugr.es

as another step to the biological process with the aim of treating rejected water from anaerobic digestion to exclusively remove nitrogen from industrial processes. In this context, an innovative and recent technology, mostly operates on a sequencing batch reactor, called aerobic granular sludge has been incorporated into the catalog of effective and efficient systems for wastewater treatment. This technology is able to remove organic matter, nitrogen, and phosphorous in the same chamber due to granular conformation allowing a tridimensional structure to support aerobic, anoxic, and anaerobic niches. Also, processes based on aerobic granular sludge are expected to be an energy-efficient alternative to activated sludge.

## 10.1.1 Aerobic Granular Sludge Technology

Granular sludge was originally described in anaerobic sewage treatment systems, being the first observation made in a bubble column (Morgenroth et al. 1997); 20 years later, the aerobic granular sludge (AGS) is focused on biological treatment of urban wastewater, high organic loading, and industrial wastewater. Recently, aerobic granular sludge technology has gained high interest in the scientific community, mainly for two reasons; the first reason is that the technology is based on a compact design, and the second reason is that the organic matter and nutrients removal occurs in the same chamber, which supports a promising alternative to conventional wastewater treatment, becoming a mature option to implement in municipal wastewater treatment plants. Furthermore, acknowledgment of granular structure, formation process, system performance, and microbial community had generated attention in the background of this technology.

Aerobic granular sludge (AGS) exhibits innumerable advantages in comparison with CAS, such as excellent sludge settleability, regular, smooth, and nearly round in shape, dense and strong microbial structure, a higher concentration of MLSS, high biomass retention, compact design, ability to withstand at high organic loading, nutrients recovery, high resistant to toxicity, and shorter hydraulic retention time (Adav et al. 2008). When AGS is compared with conventional activated sludge (CAS), the novel technology presents 23-40% lesser electricity requirements, 20-25% reduction in operation cost, and 50-75% in surface requirements (Adav et al. 2008, Bengtsson et al. 2018, Sarma et al. 2018), increasing the daily wastewater handling capacity; also, since AGS is applied, secondary clarifiers are not needed.

Granulation of activated sludge comprises a complex process that occurs under specific environmental conditions. For the granulation, high hydrodynamic shear forces, a large ratio of height to diameter of a reactor, feast-famine feeding, dissolved oxygen concentration, feeding strategy, and selective wash out of microbial organisms are needed. All these parameters allow the formation of high-density granules along with soft and compact granules; however, the only factor that seems to contribute to the granular formation is hydraulic pressure (Liu et al. 2005).

The granulation is the result of several abiotic and biotic interactions, resulting in compact spherical aggregates, which is self-immobilized in a matrix of extracellular polymeric substances (EPS) (Figure 10.1). The granulation process was described as occurring in several steps, comprising (1) cell to cell contact, (2) attractive forces among cells to aggregate, (3) maturation of microbial aggregates producing EPS, which allow the attachment of cells to aggregation, and (4) formation of tridimensional matrix. Granulation is described as a consequence of dynamic aggregation and breakage of flocs (Zhou et al. 2014).

Cell-to-cell contact and aggregation are dependent on the mechanisms and physicochemical properties; one of the most important properties is hydrophobicity for the initial step of granulation (Liu et al. 2003). The increase in the hydrophobicity property of cells could be due to changes in the protein/polysaccharide ratio caused by changes in EPS and microbial populations. Also, a relevant mechanism involved in the granulation is quorum sensing (QS), which regulates several functions, such as EPS production (Zhang and Li 2016). So, higher quorum sensing activity seems to be linked to the high production of gel-forming EPS and higher microbial attachment potential (Lv et al. 2014).

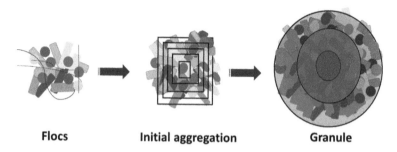

**Flocs**          **Initial aggregation**          **Granule**

**FIGURE 10.1**   Processes of granular formation following the description of Sarma et al. 2017.

Due to the oxygen gradient that originates in the granular structure, different layers in the granule can be detected. Thus, it is possible to differentiate aerobic and anoxic zones in one granule or include anaerobic zones in the core of the granule. Following the models, in general terms, nitrifiers are in the external outer layers, while the phosphate-accumulating bacteria and denitrifying bacteria are found in the inner layers.

The large aggregate size of AGS makes it possible for several metabolic activities to obtain energy and a carbon source, taken place simultaneous nitrification, denitrification, and phosphorus removal in one single reactor due to large diffusion gradients of electron donors and acceptors, creating different redox conditions (Figure 10.2). Moreover, the size of the granules seems to reach a stable mean diameter comprising growth ratio, predation, hydrodynamic shear force, attrition, and breakage. In addition to this, the granular size is a crucial parameter for efficiency since it is linked with microbial activity.

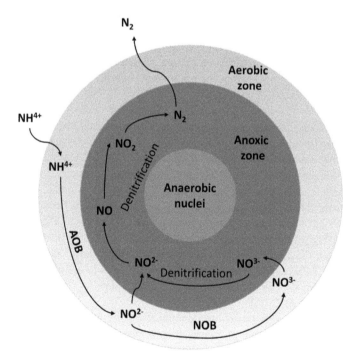

**FIGURE 10.2** Representation of nitrogen removal pathway by nitrification-denitrification in layers of aerobic granular sludge, following the schematic nitrogen pathway described by Nancharaiah et al. 2018.

## 10.1.2  Configuration and Operational Conditions Affecting Granulation

For the development of aerobic granular sludge, several operational and environmental conditions are compromised for the granular formation and granular stability. Some of these are hydrodynamic shear force, cycle time, volume exchange ratio, feeding strategy, feast-famine regime, settling velocity, reactor configuration, carbon source, and organic loading ratio among others.

### 10.1.2.1  Bioreactor Configuration

The design of the bioreactor is crucial for the right development of the granules due to it being necessary to cultivate them, almost exclusively, in a cylindrical-column sequencing batch reactor with fine bubbles aeration. The sequential batch reactor steps consist of aeration time, to allow the consumption of substrates, then the sludge settling stage, typically between 2-10 minutes, effluent discard, and finally filling the bioreactor with raw wastewater. On the contrary, aerobic granular sludge has recently been reported in a continuous-flow system, with a flat geometry, composed by two mixed chambers in a series and a settler to promote the granular formation (Cofré et al. 2018).

On the other hand, the column design allows for microbial aggregates to move continuously in a homogeneous and circular motion (Figure 10.3), which promotes regular granules with minimum surface free energy (Liu et al. 2005). Also, the main factors in the design of the reactor are the height/diameter ratio and volume/exchange ratio. The height/diameter ratio determines the size of the granules due to a lower height/diameter ratio, allowing larger growth in comparison with a high ratio with the same airlift velocity (Muñoz-Palazón et al. 2018, Gonzalez-Martinez et al. 2018).

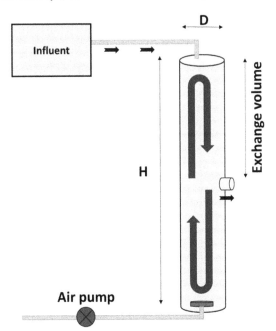

**FIGURE 10.3** Column airlift sequential batch reactor.

The bioreactor column with a high height/diameter ratio (H/D) seems beneficial to granules formation due to an improvement in selection pressure through wash-out of biomass and granules with better settling ability (Kong et al. 2009). H/D values between 20 and 30 are mostly used in research for granular formation, while that H/D value of 64 is mostly used to get nitrifying granules (Tsuneda et al. 2003). High H/D ratios increase the settling ability, but this high ratio is not optimal for the design of a full-scale plant. Reactor height/diameter ratios have been shown to be very flexible in the practice, which is important for the application of aerobic granule technology. Moreover, higher H/D improves the circular flow, creating more efficient hydraulic attrition for microbial aggregates and higher oxygen transference (Liu et al. 2005). In this same sense, it has been reported that reactors with a high H/D ratio have a shorter start-up process and are more active in comparison with those with a low H/D ratio (Kong et al. 2009, Awang et al. 2016). Related to the volume/exchange ratio, it has been described that a high exchange ratio with short settling time promotes the selection of denser and more compact granules with better settling capability, and the granulation process is getting faster (Nancharaiah and Kiran Kumar Reddy 2018).

### 10.1.2.2 Cycle and Settling Time

The cycle time is represented by the reaction time, which consists of aeration, settling, effluent discard, and filling (Figure 10.4). The cycle time is directly correlated with hydraulic retention time and volume exchange. Long cycles could result in hydrolysis of biomass causing a negative effect in aggregation and growth of filamentous bacteria (Liu et al. 2005). In contrast, a shorter cycle time could block the growth of microorganisms due to insufficient reaction time to metabolize substrates, and it could generate a breakage of biomass and then wash-out of biomass (Liu et al. 2005).

On the other hand, settling time is an essential parameter to avoid excessive wash-out of biomass by effluent, although during the start-up it is necessary to increase microorganisms to be able to form granules. Thus, a short settling time has been used for aerobic granular sludge; in other ways, some authors had reported the poor settling abilities and fluffy characteristics of this operation. Accordingly, denser and compact granules are formed under high pressure exerted by the selection of fast-decanting microorganisms by settling velocity in SBR (Kong et al. 2009, Liu et al. 2005).

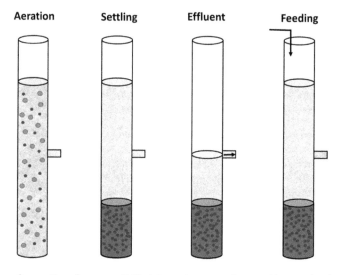

**FIGURE 10.4** Stage of operation of a sequential batch reactor operating aerobic granular sludge technology.

### 10.1.2.3 Hydrodynamic Shear Force

Mostly, reactors are designed as an airlift column (Figure 10.3), where the shear force is created by fine bubbles from the bottom to the top, expressed as up-flow superficial velocity, usually used 1-2 cm s$^{-1}$. Many researchers have found that a high aeration rate accelerates the granulation process, forming smaller, denser, and more compact granules than with low shear force pressure (Wilén et al. 2018). Also, a high shear force promotes the production of an extracellular polymeric substance that favors the adhesion cell-to-cell. Thus, the up-flow air velocities $\leq 0.8$ cm s$^{-1}$ can generate unstable granules with filamentous organisms that are more porous (Lochmatter and Holliger 2014).

### 10.1.2.4 Temperature

Aerobic granular sludge technology has been operated successfully at different temperatures, typically 7-30°C but even temperatures as low as 5°C; while at 3°C, the granules become irregularly shaped and the breakdown of granules has been reported (Gonzalez-Martinez et al. 2018). However, it has been described that at 5°C, the granules have structural stability and their nutrients removal capacity is decreased, unless that AGS bioreactor inoculated with cold-adapted biomass showed high efficiency of organic carbon and nutrient removal at 7°C (Gonzalez-Martinez et al. 2017, Gonzalez-Martinez et al. 2018). Higher efficiency was found in the operation of AGS technology at higher temperatures (20-30°C) in terms of organic matter and nitrogen; in contrast, the phosphorous removal was detected due to unfavorable growth conditions for phosphate accumulating-organisms (Lopez-Vazquez et al. 2009).

### 10.1.2.5 Dissolved Oxygen

Granular sludge technology could operate under a wide range of dissolved oxygen (DO) concentration, which is an important parameter in biological wastewater treatment. Aerobic granular sludge had been performance positive at a low concentration of DO ranging from 2-6 mg $O_2$ $L^{-1}$ (Liu et al. 2005) and even at a high DO up to 11 mg $O_2$ $L^{-1}$ (Gonzalez-Martinez et al. 2017). Mosquera-Corral et al. (2005) showed that dissolved oxygen and nitrogen removal are operational parameters highly related to AGS systems. These authors suggested that OD concentrations of approximately 40% determine the best nitrogen removal ratio at a temperature of 25°C as denitrification process could be carried out.

### 10.1.2.6 Feeding Strategy

The aeration stage in sequential batch reactors consists of two phases: the degradation phase (when the substrates are depleted) and the aerobic phase (where there is not a substrate). This period is the denominated feast-famine period. Short periods of famine promote denser and compact granules and higher hydrophobicity surface, accelerate the microbial aggregation, and create appropriate substrate and oxygen gradients in the granule (Gao et al. 2011). Also, feast-famine periods promote the storage of organic matter; thus, the proliferation of slow-growth microorganisms, such as nitrifying bacteria, is possible (Pronk et al. 2015). Moreover, some authors have reported that the extended famine conditions encouraged granule stability probably because long starvation periods favor microorganisms' ability to store compounds (Corsino et al. 2017).

### 10.1.2.7 Inoculum

The origin of sludge, environmental parameters, and operational conditions of seed sludge have a crucial role in the development of granules due to the microbial community of sludge being affected by sequencing batch conditions. Also, the temperature-adapted activated sludge generated a larger difference in the startup period and consequently in the performance of granules. The heterogeneity of microorganisms residing in activated seed sludge is important to granulation as hydrophilic bacteria would be less likely to attach to sludge flocs compared with the hydrophobic counterpart (Adav et al. 2008, Zita and Hermansson 1997), whose presence has been reported in the effluent of a wastewater treatment plant. The presence of high amounts of hydrophobic cells in the inoculum favors a faster granulation process, regular shape, and high settleability (Wilén et al. 2008). It was further found that the shell of the aerobic granule exhibited a higher hydrophobicity than the core of granule (Wang et al. 2005, Zhi-Wu et al. 2005).

### 10.1.2.8 Organic Source and Organic Loading Rate

Aerobic granular sludge has been cultivated with several organic carbon sources such as phenol, acetate, ethanol, methanol, or glucose, among others, as described by Liu et al. (2005); hence, the granular formation seems independent of carbon source, but the microbial community conformed the granules have a close relationship with the organic source used, although both nitrifying and denitrifying bacteria are found in communities regardless of substrate.

Aerobic granular sludge can support a wide range of organic loading rates of organic matter and nitrogen, directly affecting the granular structure and microbial composition as well as the removal performance in the system. Some research has shown that high organic loading rate (OLR) (4.5 kg $m^{-3}$ $day^{-1}$) results in higher granulation and more porous and bigger granules in comparison with lower OLR (between 1.5-3 kg $m^{-3}$ $day^{-1}$) ( Li et al. 2008).

### 10.1.3   Role of EPS

The extracellular polymeric substances (EPS) play an essential role in the development, formation, and structural stability of granules (Figure 10.5). The EPS allows the formation of a dense tridimensional matrix composed by aggregated of bacteria, archaea, fungi, and protozoans. Some authors hypothesize that very high hydraulic shear force and consequently an increase in aeration flow results in an increase of polysaccharides content; after that, the granules breakdown and disintegrate when polysaccharides are lost (Tay et al. 2001). The hydrodynamic shear force increased the production of EPS with the goal of carrying out the structural stability; conversely, several hypotheses suppose that shear force is not necessary for granules formation. The latest research identified two gel-forming exopolysaccharides; one of them was alginate with high levels of glucuronic acid (69%) (Lin et al. 2010) and a novel heteropolysaccharide called Granulan (Seviour et al. 2010). Despite big scientific efforts, the role of different EPS has not been identified, but the distribution and composition of EPS have been notoriously different in activated sludge than in granular sludge, for instance having an adhesive property and being negative charged (Seviour et al. 2012, Seviour et al. 2010). Mainly, differences are found in the content of protein, registering 50% more proteins in AGS than in activated sludge (McSwain et al. 2005). Studies conducted using a confocal laser scanning microscope and specific fluorophores showed that microbial cells and EPS were located in external layers of granules, while the core is devoid of cells and is mostly composed of proteins (McSwain et al. 2005). The presence of metals such as calcium, magnesium, and iron in influents could reduce the time to form granules in AGS bioreactors (Sarma et al. 2017). The metals encourage granulation due to promoting EPS production, facilitate the auto- and co-aggregation of microorganisms, and act as nuclei to microorganism attachment (Nancharaiah et al. 2018).

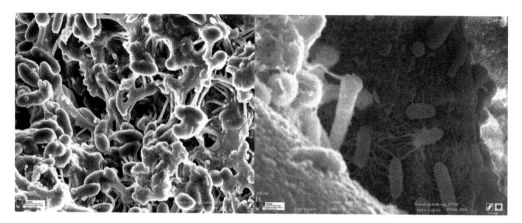

**FIGURE 10.5** Aggregation of microorganisms embedded in extracellular polymeric substances.

### 10.1.4   Microbial Community in Aerobic Granular Sludge

The microbial community showed a diversity and richness of species in a certain biological treatment wastewater. The microbial community is driven by the design of technology, the composition of the influent, the different process focus, and the stability of the system among other conditions. Recent studies have shown that wastewater treatment processes are more stable when higher diversity in microbial communities is found (Rowan et al. 2003, Ma et al. 2013). Aerobic granular sludge is a biofilm containing a multi-species of microbial domains, which interact with each other positively, neutrally, or negatively. The most important negative interaction in granules is predation with the major cause of bacterial mortality being protists, predatory bacteria, and virus bacteriophages (Jousset 2012). Also, Dolinšek et al. (2013) reported that the predation of *Nitrosospira sp.* by *Micavibrio*-like bacteria could directly cause a decrease in nitrogen removal. On the other hand, the protist organisms have been scarcely studied, which can have a role in biofilm production and aggregation of the structure as well as the high consumption of particulate matter (Muñoz-Palazon et al. 2019, Liébana et al. 2016).

## A. Bacteria

Several phylotypes have been described as dominant in the bacterial community in aerobic granular sludge. The changes in the community could be induced by the seed sludge and operational conditions, such as temperature, dissolved oxygen, or organic loading rates. Mainly, they have been described relative to metabolisms or role that could play on ecology. *Thauera* and *Meganema* have been reported in recent studies as nuclei of granulation due to filamentous properties that they possess and the production of EPS (Muñoz-Palazón et al. 2018). The genus *Zooglea* is widely reported in granules and is hypothesized to be of great importance to granulation due to it being known for floc-forming and charge to produce glue-like extracellular polymers (Gonzalez-Martinez et al. 2018, Li et al. 2008). *Acinetobacter* has been found several times, which is a genus described as heterotrophic nitrifying-aerobic denitrifying bacteria, with tolerance to low temperatures (Gonzalez-Martinez et al. 2018; Yao et al. 2013).

The bacterial community in the granule suffers deep changes when the system is subject to micropollutants, such as pharmaceutical compounds or products of personal care, promoting the ability of phylotypes to metabolize or co-metabolize the pharmaceutical products (Kang et al. 2018).

## B. Fungi

Fungi organisms have been reported with a relevant role in granular structure conformation because they increase the surface available for bacterial attachment (Weber et al. 2007). Also, fungal hyphae cause no bulking events when they are involved in a sewage biofilm. Thus, filamentous fungi can act as a precursor to biofilm to immobilize the tridimensional matrix around them that are highly resistant to antifungal and pollution (Figure 10.6). Fungal organisms improve the settling ability by a formation of filaments that trap flocs and by a generation of an extracellular polymeric substance that allows for joining flocs (Avella et al. 2014, Weber et al. 2007). Certainly, the outgrowth of the filamentous fungal community has been reported as an operational problem in aerobic granular sludge systems at low temperatures (De Kreuk et al. 2005), causing irregular structures, bad settling characteristics, and biomass washout; however, if the seed sludge is cold-adapted as Gonzalez-Martinez tested (2017, 2018), the granulation process is successful. The most-reported fungi in aerobic granular sludge were *Trichosporon* (Muñoz-Palazón et al. 2018, Gonzalez-Martinez et al. 2018). *Trichosporonaceae* members could be ubiquitous in wastewater treatment, so it has been reported as appearing in activated sludge, but the high relative abundance of the phylotype in granular sludge leads to the hypothesis that it has a crucial role in granular sludge formation and can extend from the area available for colonization of other microorganisms (Figure 10.6). Other studies point to *Scopularipsis, Tremellomycetes*, and *Pleosporaceae* as phylotypes dominant in both granules and flocs (Liu et al. 2017, Weber et al. 2007). However, all studies have pointed out that high ignorance relative to fungal taxonomy does not allow for a deeper knowledge (Weber et al. 2007).

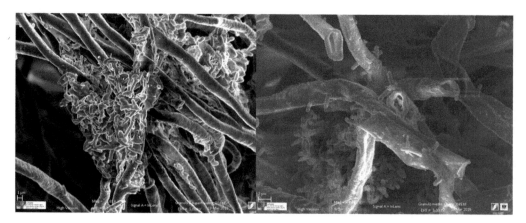

**FIGURE 10.6** Colonization of hyphae of fungi by bacteria in aerobic granular sludge systems.

## C. Archaea

The population of archaea in aerobic granular sludge in comparison with anaerobic processes has been minimally studied. Muñoz-Palazon et al. (2018) and Liu Jun et al. (2017) pointed out that genera such as *Methanosaeta, Methanobacterium, Methanobrevibacter, Methanosphaera, Methanolinea, Methanospirillum, Thermoplasmata,* and *Methanosaeta* were highly represented in the archaea community. The phylum *SM1K20* was found in aerobic granular sludge under low temperature, increasing their presence with the drop in temperature until 3°C (Gonzalez-Martinez et al. 2018). Despite a lack of knowledge about archaea in aerobic granular sludge, it has been found in small amounts, which can yet play an unknown role (Winkler et al. 2012). In this sense, qPCR analysis, reported by Muñoz-Palazón et al. (2018), showed that the archaea population was lower in aerobic granular sludge technologies than in activated sludge systems.

## D. Eukarya Domain

Some authors report that eukaryote organisms such as metazoan or protozoan organisms are essential to controlling the microorganism populations (Muñoz-Palazon et al. 2019, Winkler et al. 2012). In aerobic granular sludge *Vorticella*-like ciliated protozoa have been identified as actively grazing on bacteria by the fluorescence *in situ* hybridization (FISH) technique (Winkler et al. 2012). On the other hand, Muñoz-Palazón et al. (2019) studied the eukaryotic community comprising granules, which are divided into two big groups, both fungi and ciliates; the authors showed that under unstable conditions, operating with high toxic phenolic compounds, the granules were disintegrated and the relative abundance of ciliate growth during these stages. These results suggest that ciliates proliferate when the disintegration of biomass occurs, which can consume a high amount of particulate matter.

## 10.2 NITROGEN REMOVAL

Nitrification (oxidation of ammonia to nitrate via nitrite/nitrate) and denitrification (reduction of nitrate to molecular nitrogen via nitrite, nitric oxide, and nitrous oxides) are crucial processes in the nitrogen cycle involved in water treatment. The nitrogen cycle is naturally balanced; however, the strong pressure exerted by anthropic activities has determined its alteration on our planet.

Conventional nitrogen removal requires aerobic and anaerobic/anoxic conditions for ammonia oxidation and after denitrification, respectively. Thus, conventional treatments for wastewater have a different chamber for controlling the oxygen dissolved in the aerobic chamber, and a near absence of dissolved oxygen is mandatory for denitrification with the presence of organic matter.

However, aerobic granular sludge disposes of both conditions in a millimeter of granular sludge due to several layers of compound granules. In the external layer, aerobic conditions are found, thus autotrophic ammonia oxidizing bacteria (AOB) and nitrite oxidizing bacteria (NOB) are located there with high DO, while the diffusion limitation of oxygen along the granule allows for the nitrate reduction by anoxic/anaerobic heterotrophic bacteria. The diversity of redox conditions lengthwise allows the simultaneous nitrification-denitrification (SND), even under aerobic conditions. The parameters that can affect the SND process are granular size, porosity, dissolved oxygen, and microbial community. Also, the size of the granule is limited because if the granule is too small, the oxygen penetrates to the core, and if the oxygen dissolved is too high, the diffusion limitation can be a detriment. If it occurs, a short anoxic phase after reaction time promotes the low dissolved oxygen conditions and consequently the denitrification process.

The nitrogen removal ratio during the startup of AGS technology is normally low because strong wash-out of biomass is mainly made for granular formation, which leads to a decrease in sludge retention time and leakage of organic carbon from the anaerobic feeding phase into the aerobic phase (Wilén et al. 2018). Usually, slow-growing bacteria such as ammonia oxidizing-bacteria and nitrite oxidizing-bacteria change strongly during startup; however, the nitrifying microorganisms are recovering relatively fast (Muñoz-Palazon et al. 2018). The COD/N ratio is really significant for promoting nitrifying bacteria due to high COD/N ratio proliferate heterotrophic bacteria, while AOB and nitrite oxidizing bacteria (NOB) decrease remarkably. The simultaneous nitrification-denitrification (SND) mechanisms depend on the concentration of organic matter, thus explorations of N removal mechanisms at high COD/N loading in aerobic granular sludge are important.

## 10.2.1   Nitrification in Aerobic Granular Sludge

Nitrification is defined as the chemical transformation of $NH_4^+$ oxidizer via $NO_2^-$ to $NO_3^-$. The nitrification in water treatment is carried out by autotrophic and heterotrophic nitrification, depending on influent composition, although the dominant process in wastewater is chemoautotrophic nitrification and heterotrophic nitrification is a small part of the nitrification process.

The most usual strategy to remove ammonium from wastewater is nitrification over nitrate. Nitrification concerns the oxidation of ammonium to nitrite carried out by ammonium oxidizing-bacteria (AOB) and oxidation of nitrite to nitrate, called nitritation, under aerobic conditions carried out by nitrite oxidizing bacteria (NOB).

$$NH_4^+ + 1.5\ O_2 \rightarrow NO_2^- + H_2O + 2H^+$$
$$NO_2^- + 0.5\ O_2 \rightarrow NO_3^-$$

In contrast, ammonium removal over nitrite requires much less $O_2$ and electron donors than over nitrate. In this way, ammonia oxidizing bacteria oxidize ammonia to nitrite and next, this nitrite would be consumed by heterotrophic bacteria for denitrification. The metabolic pathway to remove nitrogen expends 40% and 25% of less electron donors and oxygen, respectively, in comparison with nitrogen removal over nitrate (Lochmatter and Holliger, 2014). Oxidation of nitrite to nitrate can be avoided by decreasing the dissolved oxygen as AOB have more affinity with DO than NOB as well as operating at a temperature above 25°C (Nancharaiah et al. 2018).

Ammonium and nitrite oxidizing bacteria are chemolitoautotrophic microorganisms, which use inorganic molecules as an electron donor, and their carbon source is carbon dioxide or in practice, bicarbonate. Heterotrophic nitrification provides an alternative pathway to the removal of nitrogen in aerobic granular sludge; also, some heterotrophic nitrifying microorganisms show the capability of aerobic denitrification (Roberson et al. 1989), which is based on the use of organic carbon to convert ammonium into molecular nitrogen gas and thus the organic matter and nitrogen are removed in the same step. Yao et al. (2013) reported that heterotrophic nitrifying bacteria could co-exist in aerobic granular sludge with autotrophic nitrifiers.

## 10.2.2 Ammonium Oxidation Pathways of Bacteria and Archaea in Aerobic Granular Sludge

Nitrifying bacteria are ubiquitous in all soil and water reservoirs and have is important been found even in Antarctic ice, hot springs, and alkaline biotopes (Arrigo et al. 1993). AOB have been classified into five genera: *Nitrosomonas, Nitrosococcus, Nitrosospira, Nitrosovibrio,* and *Nitrosolobus* (Koops et al. 2006); all these genera are related to β-proteobacteria except *Nitrosolobus*.

Recently, it was discovered that AOB do not need a high dissolved oxygen concentration to oxidize ammonia, and it was affirmed that these bacteria can act better with low oxygen concentrations than with low ammonium concentrations. Under anoxic conditions, *Nitrosomonas eutropha* and *Nitrosomonas europeae* as well as possibly other AOB can oxidize ammonia in the presence of pyruvate and with nitrite as electron acceptor or nitrous oxide gas (Abeliovich et al. 1992, Schmidt and Bock 1997). Thus, it is possible that under anoxic conditions, the AOB population suffers physiological adaptations such as alternative electron acceptors or affinity for low DO rather than to a change in the composition of nitrifying microorganisms (Bothe et al. 2000).

Autotrophic ammonia oxidizer microorganisms possess two key enzymes that are essential for energy conservation during the oxidation process—ammonia monooxygenase (AMO) and hydroxylamine oxidoreductase (HAO). Both enzymes are dependent on each other as they generate substrates and electrons of each other. The AMO enzyme has three subunits, AMO-A, AMO-B, and AMO-C, which are composed of different structures and sizes. The subunits are encoded by the genes *amoA*, *amoB,* and *amoC* located in the amo operon. Therefore, the AMO enzyme catalyzes the reaction of ammonia to hydroxylamine:

$$NH_3 + O_2 + 2H^+ + 2e^- \rightarrow NH_2OH + H_2O$$

Both electrons required for this process are produced by the oxidation of hydroxylamine to nitrite (HAO).

$$NH_2OH + H_2O \rightarrow NO_2^- + 5H^+ + 4e^-$$

The oxygen atoms in $NO_2^-$ proceed from the water and form $O_2$. Two of the four electrons produced by the HAO enzyme are transferred via cytochrome *C554* to AMO or deviant to electron transport chain (Arciero et al. 1991). The HAO enzyme is an uncommon enzyme with a complex structure, located in periplasmic space.

In addition to this, the ammonia oxidizing bacteria are not the only organism responsible for ammonium oxidation, and *Archaea* is a very important domain with high activity, which is a well-known metabolic pathway for ammonia oxidation. Like *Bacteria*, ammonia oxidizing archaea (AOA) are ubiquitous in soil, freshwater, marine sediment, and wastewater (Erguder et al. 2009). Moreover, Erguder et al. (2009) reported that AOA are microorganisms that are very important for the global nitrogen cycle, even in some reservoirs where AOA are less represented than AOB.

Many studies point out that the AMO enzyme in the nitrification process plays a relevant role in the degradation of pharmaceutical compounds, synthetic estrogens, bisphenol A (BPA), and triclosan (Batt et al. 2006, Kim et al. 2007, Roh et al. 2009, Maeng et al. 2013, Sathyamoorthy et al. 2013). Thus, it has been proposed that the AMO enzyme also plays an important role in the degradation of micropollutants in biological wastewater technology. Despite most of them are involved with the removal of inhibition of AOB/AOA the ratio of removal without the presence of nitrifying organisms is slower, showing that the other co-metabolic pathway can be playing in presence of AOB/AOA-enriched sludge (Margot et al. 2016).

## 10.2.3 Nitrifying Microbial Community in Aerobic Granular Sludge

The nitrifying microorganisms found are relative to the structure of granule and operational conditions. *Nitrosomonas sp.* was detected in an airlift sequential batch reactor operated under a high concentration of dissolved oxygen at 20°C (Bassin et al. 2011), and *Nitrobacter sp.* has been identified as dominant in acetate-fed aerobic granules (Winkler et al. 2012). Usually, total numbers of NOB compared to AOB is expected to be even lower in systems where simultaneous nitrification-denitrification takes place, such as aerobic granular sludge. Both nitrite and nitrate can be used as an electron acceptor by denitrifying bacteria to produce nitrogen gas. For this reason, NOB would compete against denitrifying bacteria for nitrite when ratios NOB/AOB are even lower than 0.5. This trend could be changed if it is promoting the optimal conditions for the growth of NOB, it could make if NOB is not only based on nitrite produce by AOB if not even other substrates, such as organic compounds, suggesting a mixotrophic metabolism of NOB (Winkler et al. 2012). Recent studies show that the NOB/AOB ratio is higher in aerobic granular sludge than in activated sludge (Winkler et al. 2012).

In contrast, some studies on aerobic granular sludge-based on acetate-feeding did not show any conventional AOB with amplification of the 16SrRNA hypervariable region, such as *Nitrosomonas* (Muñoz-Palazon et al. 2018, Gonzalez-Martinez et al. 2018, Gonzalez-Martinez et al. 2017), although the ammonia oxidation and total nitrogen removal had high performance. On the contrary, the *Xanthomonadaceae*, *Rhodococcus*, and *Sphingomonas* genera were presented, which had been reported by Fitzgerald et al. (2015) as aerobic bacteria involved in ammonia oxidation were at low DO.

Studies based on absolute quantification carried out through Real Time PCR during the increase of the AOB population showed that the biodiversity of *Nitrospira* was reduced (Feng, Wang et al. 2007). Also, Szabó et al. (2016) showed that in aerobic granular sludge systems, the number of AOB microorganisms was strongly decreased during the first seven days of operation, while during the remaining 21 days, the number of AOB increased notably as did, consequently, the oxidizing activity. Val del Río et al. (2012) and Szabó et al. (2016) reported the great relevance of sludge retention time in promoting slow-growing nitrifying microorganisms in the start-up period is made trought wash-out of biomass during settling time for optimizing the AOB/NOB ratio and also the removal ratio of the process. Following the trend, the number of copies of the bacterial 16S rRNA gene, as well the MLSS concentration, usually increased during the first month of operation in AGS systems, but this fact could be attributed (data confirmed by qPCR analysis) to heterotrophic growth due to AOB microorganisms that were kept stable (Muñoz-Palazon et al. 2018, Szabó et al. 2016).

## 10.3 DENITRIFICATION IN AEROBIC GRANULAR SLUDGE

Denitrification is an important process within the global nitrogen cycle, which is getting the attention of the scientific community as well as the industrial and agricultural sectors. The understanding of this process within the biogeochemical cycle and the microorganisms are of fundamental interest to keep the balance of nitrogen and consequently the industrial procedures. In the biosphere, this process is based on a microbial dissimilatory process where nitrate and nitrite are reduced to nitrogen gaseous compounds, such as nitric oxide (NO), nitrous oxide ($N_2O$), and dinitrogen ($N_2$). This biological process is the main contributor to the emission of nitrous oxide, which is a major greenhouse gas and the single most dominant ozone-depleting substance. The denitrification process is based on the reduction in the following reduction chain:

$$NO_3^- \rightarrow NO_2^- \rightarrow NO \rightarrow N_2O \rightarrow N_2$$

Denitrification causes nitrogen loss from the natural ecosystem, such as from soils, as well as in wastewater treatment. Biological denitrification is also important for nutrient removal in water due to high nitrate concentration in water resources, such as drinking water, which has been identified as one of the hazardous contaminants that may reduce to nitrosamines in humans and is suspected of causing gastric cancer (Ghafari et al. 2008). Usually, the nitrate in groundwater originates through the increasing addition of fertilizers to the soil. Other sources of nitrate are the discharges of treated wastewater from WWTPs and are usually caused by the complete ammonia oxidation process that converts ammonia to nitrate.

The nitrate removal from water resources could be treated by physicochemical or biological methods. The most used physicochemical technologies for nitrogen removal are reverse osmosis, electro-dialysis, carbon adsorption, metallic iron-aided, or ion exchange, but their expensive cost and high energy consumption make the implementation of these methods difficult. On the other hand, the biological denitrification methods provide a promising chance for nitrate removal, saving cost, and being environmental-friendly in comparison with physicochemical treatment. Mostly, the biological denitrification process is carried out by facultative anaerobic microorganisms in the absence of oxygen. Microbial removal of nitrate may be the most economical strategy for the reclamation of nitrate polluted waters and wastewaters (Soares 2000). The biological denitrification process could be carried out in heterotrophic or autotrophic ways. The most common way is heterotrophic denitrifiers, which require carbon source for growth and development, while nitrate acts as an electron donor and autotrophs use inorganic compounds as an energy source and carbon dioxide as a carbon source.

## 10.3.1   Denitrification in Aerobic Granular Sludge

The biological denitrification process is carried out with the aim of removing nitrate with an external contribution of carbon. Commonly, the denitrification process is maintained by heterotrophic microorganisms that need organic matter and anaerobic conditions to perform the process. Methanol is normally used as an external carbon source due to its low cost and high carbon content (Costa et al. 2017), although other carbon sources can be used for this process, such as sodium acetate ($CH_3$ COONa). However, methanol has been suggested by several researchers as the best candidate for a carbon source in denitrification processes (Tam et al. 1992). The nitrate removal ratio is deeply related to the carbon concentration that is present in the system. An excess of organic matter induces the absence of the feast-famine regime, which promotes the destabilization of aerobic granular sludge and, on the other hand, causes toxicity in the receiver water bodies (Mohseni-Bandpi et al. 2013).

Granular aerobic systems are very efficient technologies to develop denitrification processes due to the peculiar characteristics of the system. As mentioned above, a granular aerobic system consists of biomass formed by a great diversity of microorganisms, among which are *Bacteria*, *Archaea*, *Fungi*, and *Protozoa*, within a polymeric matrix that make them suitable for the denitrification process in the core of the granule as a consequence of microbial stratification in the granule according to the gradient of oxygen available.

The AGS technology can be considered a good alternative for nitrate removal in industrial and urban wastewaters. Moreover, this technology allows for the simultaneous removal of organic matter, nitrogen, and phosphorus (Pronk et al. 2015). In conclusion, AGS systems can be considered an excellent alternative for wastewater treatment for the joint removal of nitrogen via denitrification.

Nowadays, besides being a technology used to treatment of urban and industrial wastewater, AGS systems have also been studied for the treatment of groundwater polluted with nitrate, phosphorus, and other chemical compounds (Guiot et al. 2007). We cannot forget that groundwater is used massively as a resource in the supply of drinking water to a large part of the world's population and that, with a certain frequency, it can be polluted with nitrates coming from the agricultural fertilizers. High nitrate concentrations (greater than 50 mg $L^{-1}$) are considered toxic by the WHO as these occasions not suitable for human consumption.

Although granular denitrification systems for groundwater treatment require the addition of small amounts of organic matter, the results obtained ensure low operating costs and operational advantages in relation to other technologies which, like reverse osmosis, consume high amounts of energy and generate highly polluting brines.

## 10.3.2 Nitrate Reduction Pathways in Aerobic Granular Sludge

The denitrification process takes place in several stages and begins with the activation of the nitrate reductase enzyme of denitrifying microorganisms (NAR), which is encoded by the nas, nar, and nap operons. The genes most studied in nitrate reduction are narG and napA (Bru et al. 2011). These enzymes are able to reduce the molecule of nitrate up to nitrite, generating energy that is used for its growth; this stage is given in conditions of low pressure of oxygen. Then nitrite is converted to nitric oxide (NO) by nitrite reductase (NIR) with nirK and nirS being the most common markers, which are convergent evolution and usually do not appear in the same denitrifying organisms. As in the previous case, the process requires low oxygen pressures for its operation. After, nitric oxide is transformed into nitrous oxide by cytochrome bc nitric oxide reductase (NOR). Finally, the reduction of $N_2O$ to $N_2$ is carried out by the nitrous oxide reductase enzyme (NOS), achieving a complete nitrogen removal in the form of a gas that is discharged into the atmosphere (Ge et al. 2012). Mostly, the nosZ gene is widely used as a marker (Figure 10.7).

The energy produced during the denitrification process decreased with the successive reduction of the substrate to their oxidation number (Levy-Booth et al. 2014). Also, Bru et al. (2011) reported that functional genes in each step of denitrification decrease their absolute abundance.

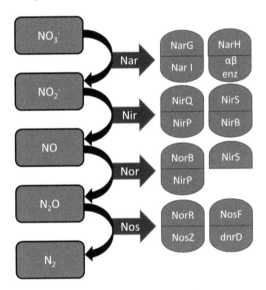

**FIGURE 10.7** The denitrification pathway.

## 10.3.3 Denitrifying Microbial Community in Aerobic Granular Sludge

Several denitrifying microorganisms had been reported in granular sludge as nitrate reducers. The diversity found in numerous studies points to different genera of bacteria isolated from sludge, wastewater, and within the wastewater treatment plants. Among them, the genera *Alcaligenes*, *Zooglea*, *Agrobacterium*, *Acinetobacter*, *Pseudomonas*, and *Comamonas* have been reported in some studies related to the operation of aerobic granular sludge systems as dominant denitrifiers (Joo et al. 2005, Chen et al. 2011, Yao et al. 2013, Ji et al. 2014).

Also, some of these genera were described as being involved in the stability of the granule and include the genus *Zooglea*, which was defined as essential for the main structure of the granule core (Muñoz-Palazon et al. 2018; Gonzalez-Martinez et al. 2018).

On the other hand, genera *Pseudomonas* and *Bacillus* have been described in several studies on aerobic granular sludge, suggesting their potential activity as phosphate accumulating-organisms (PAO) (Gunther et al. 2009).

The type of carbon source is known to have a strong impact on the structure of denitrifying microbial communities and thus on denitrification efficiency in biological treatment. Thus, many studies suggest that the members of the *Comamodacaeae* family play a major role in the denitrification process in the presence of acetate in aerobic granular sludge (Adav et al. 2010), and the genus *Methylophilus* has been widely reported as the dominant genus feeding on methanol as a carbon source in biological reactors (Torresi et al. 2017, Lu et al 2014).

## 10.4 OPERATIONAL PARAMETERS IN AEROBIC GRANULAR SLUDGE FOR SUCCESS NITRIFICATION-DENITRIFICATION

One of the greatest advantages of aerobic granular sludge is its obtaining simultaneous nitrification and denitrification (SND) in a single reactor under aerobic conditions. Thus, the control of dissolved oxygen in the bioreactor is important since it should be high enough to sustain nitrification at the outer layer and low enough to prevent oxygen from penetrating the deeper anoxic parts of the granule where denitrification can take place. Another relevant parameter is the carbon supply, which is directly correlated with the COD/N ratio, which is important for efficient SND since it acts as a strong selection pressure parameter for either enrichment of heterotroph-denitrifying or autotrophic-nitrifying communities. Liu et al. (2004) investigated the selection for nitrifying bacteria at different COD/N ratios, ranging from 20 to 3. The lower ratios produced smaller and more compact granules enriched in nitrifiers. Wei et al. (2012) investigated different COD/N ratios for ammonium-rich synthetic wastewater and obtained the highest nitrogen removal at COD/N ratio of 9. Also, the last relevant property for nitrification-denitrification is the granular size and particularly the granular density (Wan et al. 2009). Therefore, it could be concluded that simultaneous nitrification-denitrification processes (SND) in aerobic granular sludge systems can be successful if the operational conditions are performed correctly and the granular kept the denser and the gradients into layers.

## 10.5 TECHNICAL VERSATILITY OF GRANULAR AEROBIC SLUDGE SYSTEMS

Currently, aerobic granular sludge technology is being used on a laboratory-scale, pilot-scale, and full-scale for the treatment of a wide variety of urban and industrial effluents. Thus, regardless of the use of AGS systems for the treatment of urban wastewater and organic matter and nitrogen removal, other possible applications are dazzled.

### 10.5.1 Phosphorus and Nitrogen Removal

Simultaneous removal of phosphorus and nitrogen by aerobic granules has been extensively studied. The coexistence of heterotrophic, nitrifying, denitrifying, phosphate, and glycogen accumulating-microorganisms allows for the removal of nitrogen and phosphate following the nitrification and denitrification pathway. The dissolved oxygen in these processes has an essential role because of low dissolved oxygen results in a low nitrification rate and high denitrification rate. This parameter has to be controlled related to the conformation and size of the granules.

Phosphorous accumulating-organisms in granules have been studied with concomitant uptake of soluble organic carbon and release of phosphate in the anaerobic stage followed by rapid phosphate uptake in the aerobic granular stage. The phosphorous accumulated showed an increasing trend with a decrease in the substrate P/COD ratio.

### 10.5.2 Treatment of Phenolic Compounds

Phenolic compounds are chemicals that determine the polluted processes of water bodies, causing environmental effects that are certainly harmful to life in these ecosystems. For instance, many wastewaters such as those generated in agri-food, pharmaceutical, and pesticide production industries contain high phenolic compounds concentration (Nancharaiah et al. 2018), and consequently they can produce serious environmental problems.

The nature of phenolic compounds is very diverse, thus there have been many studies focused on the treatment of different phenolic compounds. In this context, aerobic granular sludge systems reached a high removal ratio for higher concentrations of phenolic compounds proceed of the olive oil industry, such as caffeic acid, hydroxybenzoic acid, and protocatechuic acid kept the stability under high organic loading (Muñoz-Palazon et al. 2019). Also, other authors reported that granules degraded phenol at 1.18 g phenol $g^{-1}$ VSS $d^{-1}$. The granular conformation allows the presence of a toxic gradient from the outer layers until the nuclei and consequently this pressure does not exert than in flocs (Adav et al. 2008).

### 10.5.3 Pharmaceutical Compounds

Nowadays, pharmaceutical compounds and personal care products (PPCPs) are widely used by the population. Thus, the scientific community has paid attention to them due to there being detected in sewage, groundwater, surface water, soil, and other environmental reservoirs in recent years (Daughton and Temes 1999). Usually, pharmaceutical compounds are recalcitrant and can possibly affect non-target organisms (Petrie et al. 2014). Between them, there are many different substances, such as antiepileptic, anti-inflammatory, antidepressant, analgesic, antibiotics, or personal care products, which are generating problems derived from keeping in touch with them in water, soils, and food, causing multi-resistance organisms. Moreover, it has been studied how the adsorption of pharmaceutical compounds by aerobic granular sludge shows a high ability to be retained on the surface due to the surface layer charge (Mihciokur et al. 2016). Many studies have shown high bioadsorption and biodegradation for antibiotics, and antibiotics-resistant bacteria (ARB) kept the nitrogen removal capacity stable, defined extracellular polymer substance as essential to contribute to the stability of granules and consequently the nitrogen removal cycle within granules (Xia et al. 2015). Jun Hu et al. (2012) demonstrated that the paracetamol could be effectively biodegraded and mineralized by granular sludge until a concentration of 1,000 mg/L, exhibiting a maximum specific paracetamol biodegradation ratio of 0.315 g paracetamol/g-VSS·h.

### 10.5.4 Heavy Metals

The presence of heavy metals in water bodies from industrial activities represents a serious environmental problem, even at a low concentration, especially for aquatic life. Many industrial effluents contain nickel, cadmium, zinc, chromium, or copper in higher concentrations, which are carcinogenic and constitute a huge problem for human health and wildlife. Biological technologies have been applied for metal removal and recovery (Liu et al. 2003); among them, aerobic granular sludge has shown an excellent granules property with high-porosity structure and heterogeneity of the composition treating high metal concentration influent. The EPS contributes to biosorption and binds metals from the aqueous solution due to the surface charge of granules (Nancharaiah et al. 2018). Moreover, the way precipitated metals can promote the nuclei of granular formation, providing a surface for microbial attachment, has been studied (Xu et al. 2014).

### 10.5.5 Textile, Landfill and Piggery Wastewaters

Aerobic granular sludge technology had been tested for the treatment of a wide diversity of industrial effluents, such as textile, piggery, olive oil, or landfill, among others. These wastewaters contain high amounts of recalcitrant compounds, being frequently discharged to natural water bodies. For instance, the textile industry generates effluents with more than 6,900 organic and inorganic different additives and 8,000 different organic dyes (Brik, et al. 2006). AGS technology had been tested with real textile wastewater, showing up as an innovative technology with high efficiency. Particularly, AGS systems can be achieved even at high organic loading (2.4-2.6 Kg COD m$^{-3}$ d$^{-1}$) and with relatively low hydraulic retention time (HRT) (Lotito et al. 2012). Also, some authors have reported denitrification processes that are carried out by the AGS system in textile effluents (Franca et al. 2014, Lotito et al. 2012), making this biological technology more efficient for this industry. Also, an AGS system used for landfill effluent gave results of high efficiency in COD and nitrogen removal, involving nitrification and denitrification under a high ammonia concentration (Wei et al. 2012), suggesting that AGS technology can be more efficient than other applied technologies for the landfill leachate treatment (Ren et al. 2017). Among others, piggery wastewater is a challenge to treat by aerobic granular sludge due to its high organic matter and ammonia concentrations. Moreover, piggery wastewater contains a wide diversity of pharmaceutical compounds, such as antibiotics. However, although these effluents have a certain difficulty in their treatment, an AGS system can treat these wastes efficiently for nitrogen removal, including denitrification as some authors corroborate (Wang et al. 2019). Despite the high presence of antibiotics, AGS granular sludge exhibited minimal affection in terms of granular stability for long-term operation (Chen et al. 2018).

## 10.6   FUTURE TRENDS

The challenge in aerobic granular sludge technology is the stability of granular biomass regardless of the characteristics of influent and changes in operational conditions. The suspended solids in the effluent are crucial for full-scale application due to avoiding the secondary settlement makes of this parameter essential. Because it is certain that there are some unstable parameters of an AGS system operating with real wastewater.

The microbial community is already well studied; however, the interactions between different groups such as nitrifying-denitrifying are largely unknown as well the role that fungal and ciliate organisms play in the granules.

Future studies could be based on the metabolically active genes in the different steps of the sequencing batch process to understand how the nitrification, denitrification, and phosphorous removal occur in the system as well as studies based on interesting genes belonging to archaea and fungi.

Finally, more information about the emission of greenhouse gases by the denitrification process is a challenge with the aim to improve this technology to avoid the emission of greenhouse gases to the atmosphere, which contributes to climate change. Widely unknown is the production of these gases by biological wastewater treatment plants; for this reason, optimal conditions for reducing the production of greenhouse gases that promote climate change should be developed.

## 10.7   REFERENCES

Abeliovich, A., Vonshak, A., Unit, E.M. and Boker, S. 1992. Anaerobic metabolism of Nitrosomonas europaea Microbiology 9: 267-270.

Adav, S.S., Lee, D.J., Show, K.Y. and Tay, J.H. 2008. Aerobic granular sludge: recent advances. Biotechnology Advances 26: 411-423. https://doi.org/10.1016/j.biotechadv.2008.05.002.

Adav, S.S., Lee, D.J. and Lai, J.Y. 2010. Microbial community of acetate utilizing denitrifiers in aerobic granules. Applied Microbiology and Biotechnology 85(3): 753-762.

Arciero, D.M., Hooper, A.B. and Balny, C. 1991. Spectroscopic and rapid kinetic studies of reduction of cytochrome c554 by hydroxylamine oxidoreductase from nitrosomonas europaea. Biochemistry 30: 11466-11472. https://doi.org/10.1021/bi00112a014.

Arrigo, K.R., Robinson, D.H. and Sullivan, C.W. 1993. A high resolution study of the platelet ice ecosystem in McMurdo sound, Antarctica: photosynthetic and bio-optical characteristics of a dense microalgal bloom. Marine Ecology Progress Series 98: 173-185. https://doi.org/10.3354/meps098173.

Avella, A.C., Chappe, P., Guinot-Thomas, P., de Donato, P. and Gorner, T. 2014. Fungal treatment for wastewater settleability. Environmental Engineering Science 31(1): 18-23. https://doi.org/10.1089/ees.2013.0118.

Awang, N.A. and Shaaban, M.G. 2016. Effect of reactor height/diameter ratio and organic loading rate on formation of aerobic granular sludge in sewage treatment. International Biodeterioration and Biodegradation 112: 1-11. https://doi.org/10.1016/j.ibiod.2016.04.028.

Bassin, J.P., Pronk, M., Muyzer, G., Kleerebezem, R., Dezotti, M. and van Loosdrecht, M.C.M. 2011. Effect of elevated salt concentrations on the aerobic granular sludge process: linking microbial activity with microbial community structure. Applied and Environmental Microbiology 77: 7942-7953. https://doi.org/10.1128/AEM.05016-11.

Batt, A.L., Kim, S. and Aga, D.S. 2006. Enhanced biodegradation of lopromide and trimethoprim in nitrifying activated sludge. Environmental Science and Technology 40: 7367-7373. https://doi.org/10.1021/es060835v.

Bengtsson, S., de Blois, M., Wilén, B.M. and Gustavsson, D. 2018. A comparison of aerobic granular sludge with conventional and compact biological treatment technologies. Environmental Technology (United Kingdom) 3330: 1-10. https://doi.org/10.1080/09593330.2018.1452985.

Bothe, H., Jost, G., Schloter, M., Ward, B.B. and Witzel, K.P. 2000. Molecular analysis of ammonia oxidation and denitrification in natural environments. FEMS Microbiology Reviews 24: 673-690. https://doi.org/10.1016/S0168-6445(00)00053-X.

Brik, M., Schoeberl, P., Chamam, B., Braun, R. and Fuchs, W. 2006. Advanced treatment of textile wastewater towards reuse using a membrane bioreactor. Process Biochemistry 41: 1751-1757. https://doi.org/10.1016/j.procbio.2006.03.019.

Bru, D., Ramette, A., Saby, N.P.A., Dequiedt, S., Ranjard, L., Jolivet, C., et al. 2011. Determinants of the distribution of nitrogen-cycling microbial communities at the landscape scale. The ISME Journal 5(3): 532.

Chen, G., Huang, J., Tian, X., Chu, Q., Zhao, Y. and Zhao, H. 2018. Effects of influent loads on performance and microbial community dynamics of aerobic granular sludge treating piggery wastewater. Journal of Chemical Technology and Biotechnology 93: 1443-1452. https://doi.org/10.1002/jctb.5512.

Chen, Q. and Ni, J. 2011. Heterotrophic nitrification-aerobic denitrification by novel isolated bacteria. Journal of Industrial Microbiology and Biotechnology 38: 1305-1310.

Chen, X., Sin, G. and Ni, B.J. 2019. Impact of granule size distribution on nitrous oxide production in autotrophic nitrogen removal granular reactor. Science of Total Environment 689: 700-708. https://doi.org/10.1016/j.scitotenv.2019.06.490.

Cofré, C., Campos, J.L., Valenzuela-Heredia, D., Pavissich, J.P., Camus, N., Belmonte, M., et al. 2018. Novel system configuration with activated sludge like-geometry to develop aerobic granular biomass under continuous flow. Bioresource Technology 267: 778-781. https://doi.org/10.1016/j.biortech.2018.07.146.

Corsino, S.F., di Biase, A., Devlin, T.R., Munz, G., Torregrossa, M. and Oleszkiewicz, J.A. 2017. Effect of extended famine conditions on aerobic granular sludge stability in the treatment of brewery wastewater. Bioresource Technology 226: 150-157. https://doi.org/10.1016/j.biortech.2016.12.026.

Costa, D.D., Kempka, A.P. and Skoronski, E. 2017. A contaminação de mananciais de abastecimento pelo nitrato: O panorama do problema no Brasil, suas consequências E as soluções potenciais. The contamination of fresh water by nitrate: the background of the problem in Brazil, the consequences and th. REDE-Revista Eletrônica do PRODEMA 10(2): 49-61.

de Sousa Rollemberg, S.L., Mendes Barros, A.R., Milen Firmino, P.I. and Bezerra dos Santos, A. 2018. Aerobic granular sludge: cultivation parameters and removal mechanisms. Bioresource Technology 270: 678-688. https://doi.org/10.1016/j.biortech.2018.08.130.

Dolinšek, J., Lagkouvardos, I., Wanek, W., Wagner, M. and Daims, H. 2013. Interactions of nitrifying bacteria and heterotrophs: identification of a Micavibrio-like putative predator of Nitrospira spp. Applied and Environmental Microbiology 79: 2027-2037. https://doi.org/10.1128/AEM.03408-12.

Daughton, C.G. and Ternes, T.A. 1999. Pharmaceutical and personal care products in the environment: agents of subtle change? Environmental Health Perspectives 107: 907-938. https://doi.org/10.1289/ehp.99107s6907.

Erguder, T.H., Boon, N., Wittebolle, L., Marzorati, M. and Verstraete, W. 2009. Environmental factors shaping the ecological niches of ammonia-oxidizing archaea. FEMS Microbiology Reviews 33: 855-869. https://doi.org/10.1111/j.1574-6976.2009.00179.x.

European Comission, 1991. Guideline 91/676/CEE.

Fitzgerald, C.M., Camejo, P., Oshlag, J.Z. and Noguera, D.R. 2015. Ammonia-oxidizing microbial communities in reactors with efficient nitrification at low-dissolved oxygen. Water Research 70: 38-51. https://doi. org/10.1016/j.watres.2014.11.041.

Franca, R.D.G., Vieira, A., Mata, A.M.T., Carvalho, G.S., Pinheiro, H.M. and Lourenço, N.D. 2015. Effect of an azo dye on the performance of an aerobic granular sludge sequencing batch reactor treating a simulated textile wastewater. Water Research 85: 327-336. https://doi.org/10.1016/j.watres.2015.08.043.

Fux, C., Boehler, M., Huber, P., Brunner, I. and Siegrist, H. 2002. Biological treatment of ammonium-rich wastewater by partial nitritation and subsequent anaerobic ammonium oxidation (anammox) in a pilot plant. Journal of Biotechnology 99(3): 295-306. https://doi.org/10.1016/S0168-1656(02)00220-1.

Gao, D., Liu, L., Liang, H. and Wu, W.M. 2011. Comparison of four enhancement strategies for aerobic granulation in sequencing batch reactors. Journal of Hazardous Materials 186: 320-327. https://doi. org/10.1016/j.jhazmat.2010.11.006.

Ge, S., Peng, Y., Wang, S., Lu, C., Cao, X. and Zhu, Y. 2012. Nitrite accumulation under constant temperature in anoxic denitrification process: the effects of carbon sources and COD/NO 3-N. Bioresource Technology 114: 137-143. https://doi.org/10.1016/j.biortech.2012.03.016.

Ghafari, S., Hasan, M. and Aroua, M.K. 2008. Bio-electrochemical removal of nitrate from water and wastewater—a review. Bioresource Technology 99(10): 3965-3974. https://doi.org/10.1016/j. biortech.2007.05.026.

Gonzalez-Martinez, A., Muñoz-Palazon, B., Rodriguez-Sanchez, A., Maza-Márquez, P., Mikola, A., Gonzalez-Lopez, J., et al. 2017. Start-up and operation of an aerobic granular sludge system under low working temperature inoculated with cold-adapted activated sludge from Finland. Bioresource Technology 239: 180-189. https://doi.org/10.1016/j.biortech.2017.05.037.

Gonzalez-Martinez, A., Muñoz-Palazon, B., Maza-Márquez, P., Rodriguez-Sanchez, A., Gonzalez-Lopez, J. and Vahala, R. 2018. Performance and microbial community structure of a polar Arctic Circle aerobic granular sludge system operating at low temperature. Bioresource Technology 256: 22-29. https://doi. org/10.1016/j.biortech.2018.01.147.

Guiot, S.R., Kuhn, R., Lévesque, M.J. and Cimpoia, R. 2007. Ultrastructure of a bio-electrolytic methanogenic/ methanotrophic granular biofilm for the complete degradation of tetrachloroethylene in contaminated groundwater. Water Science and Technology 55(8-9): 465-471. https://doi.org/10.2166/wst.2007.292.

Gunther, S., Trutnau, M., Kleinsteuber, S., Hause, G., Bley, T., Roske, I., et al. 2009. Dynamics of polyphosphate-accumulating bacteria in wastewater treatment plant microbial communities detected via DAPI (4',6'-diamidino-2-phenylindole) and tetracycline labeling. Applied Microbiology and Biotechnology 75: 2111-2121. https://doi.org/10.1128/AEM.01540-08.

Guo, F., Zhang, S.H., Yu, X. and Wei, B. 2011. Variations of both bacterial community and extracellular polymers: the inducements of increase of cell hydrophobicity from biofloc to aerobic granule sludge. Bioresource Technology 102: 6421-6428. https://doi.org/10.1016/j.biortech.2011.03.046.

Hu, J., Zhou, L., Zhou, Q., Wei, F., Zhang, L. and Chen, J. 2012. Biodegradation of paracetamol by aerobic granules in a sequencing batch reactor (SBR). Advanced Materials Research 441: 531-535. https://doi. org/10.4028/www.scientific.net/AMR.441.531.

Ji, B., Wang, H. and Yang, K. 2014. Tolerance of an aerobic denitrifier (Pseudomonas stutzeri) to high $O_2$ concentrations. Biotechnology Letters 36: 719-722. https://doi.org/10.1007/s10529-013-1417-x.

Joo, H., Hirai, M. and Shoda, M. 2005. Characteristics of ammonium removal by heterotrophic nitrification-aerobic denitrification by Alcaligenes faecalis No. 4. Journal Bioscience and Bioengineering 100: 184-191. https://doi.org/10.1263/jbb.100.184.

Jousset, A. 2012. Ecological and evolutive implications of bacterial defences against predators. Environmental Microbiology 14: 1830-1843. https://doi.org/10.1111/j.1462-2920.2011.02627.x.

Kang, A.J., Brown, A.K., Wong, C.S., Huang, Z. and Yuan, Q. 2018. Variation in bacterial community structure of aerobic granular and suspended activated sludge in the presence of the antibiotic sulfamethoxazole. Bioresource Technology 261: 322-328. https://doi.org/10.1016/j.biortech.2018.04.054.

Kim, J.Y., Ryu, K., Kim, E.J., Choe, W.S., Cha, G.C. and Yoo, I.K. 2007. Degradation of bisphenol A and nonylphenol by nitrifying activated sludge. Process Biochemistry 42: 1470-1474. https://doi.org/10.1016/j. procbio.2007.06.010.

Kong, Y., Liu, Y.Q., Tay, J.H., Wong, F.S. and Zhu, J. 2009. Aerobic granulation in sequencing batch reactors with different reactor height/diameter ratios. Enzyme and Microbial Technology 45: 379-383. https://doi. org/10.1016/j.enzmictec.2009.06.014.

Koops, H.P., Purkhold, U., Pommerening-Röser, A., Timmermann, G. and Wagner, M. 2006. The Lithoautotrophic Ammonia-Oxidizing Bacteria. *In*: Dworkin, M., Falkow, S., Rosenberg, E., Schleifer, KH., Stackebrandt, E. (eds). The Prokaryotes. Springer, New York, NY. https://doi.org/10.1007/0-387-30745-1_36.

Levy-Booth, D.J., Prescott, C.E. and Grayston, S.J. 2014. Microbial functional genes involved in nitrogen fixation, nitrification and denitrification in forest ecosystems. Soil Biology and Biochemistry 75: 11-25.

Li, A.J., Yang, S.F., Li, X.Y. and Gu, J.D. 2008. Microbial population dynamics during aerobic sludge granulation at different organic loading rates. Water Research 42: 3552-3560. https://doi.org/10.1016/j.watres.2008.05.005.

Liébana, R., Arregui, L., Santos, A., Murciano, A., Marquina, D. and Serrano, S. 2016. Unravelling the interactions among microbial populations found in activated sludge during biofilm formation. FEMS Microbiology Ecology 92: 1-13. https://doi.org/10.1093/femsec/fiw134.

Lin, Y.M., Nierop, K.G.J., Girbal-Neuhauser, E., Adriaanse, M. and van Loosdrecht, M.C.M. 2015. Sustainable polysaccharide-based biomaterial recovered from waste aerobic granular sludge as a surface coating material. Sustainable Materials and Technologies 4: 24-29. https://doi.org/10.1016/j.susmat.2015.06.002.

Liu, J., Li, J., Tao, Y., Sellamuthu, B. and Walsh, R. 2017. Analysis of bacterial, fungal and archaeal populations from a municipal wastewater treatment plant developing an innovative aerobic granular sludge process. World Journal of Microbiology and Biotechnology 33: 1-8. https://doi.org/10.1007/s11274-016-2179-0.

Liu, Y., Yang, S.F., Liu, Q.S. and Tay, J.H. 2003. The role of cell hydrophobicity in the formation of aerobic granules. Current Microbiology 46: 270-274. https://doi.org/10.1007/s00284-002-3878-3.

Liu, Y., Wang, Z.W., Qin, L., Liu, Y.Q. and Tay, J.H. 2005. Selection pressure-driven aerobic granulation in a sequencing batch reactor. Applied Microbiology and Biotechnology 67: 26-32. https://doi.org/10.1007/s00253-004-1820-2.

Lochmatter, S. and Holliger, C. 2014. Optimization of operation conditions for the startup of aerobic granular sludge reactors biologically removing carbon, nitrogen, and phosphorous. Water Research 59: 58-70. https://doi.org/10.1016/j.watres.2014.04.011.

Lopez-Vazquez, C.M., Oehmen, A., Hooijmans, C.M., Brdjanovic, D., Gijzen, H.J., Yuan, Z., et al. 2009. Modeling the PAO-GAO competition: effects of carbon source, pH and temperature. Water Research 43: 450-462. https://doi.org/10.1016/j.watres.2008.10.032.

Lotito, A.M., Fratino, U., Mancini, A., Bergna, G. and Di Iaconi, C. 2012. Effective aerobic granular sludge treatment of a real dyeing textile wastewater. International Biodeterioration and Biodegradation 69: 62-68. https://doi.org/10.1016/j.ibiod.2012.01.004.

Lu, H., Chandran, K. and Stensel, D. 2014. Microbial ecology of denitrification in biological wastewater treatment. Water Research 64: 237-254. https://doi.org/10.1016/j.watres.2014.06.042.

Lv, Y., Wan, C., Lee, D.J., Liu, X. and Tay, J.H. 2014. Microbial communities of aerobic granules: granulation mechanisms. Bioresource Technology 169: 344-351. https://doi.org/10.1016/j.biortech.2014.07.005.

Ma, J., Wang, Z., Yang, Y., Mei, X. and Wu, Z. 2013. Correlating microbial community structure and composition with aeration intensity in submerged membrane bioreactors by 454 high-throughput pyrosequencing. Water Research 47: 859-869. https://doi.org/10.1016/j.watres.2012.11.013.

Maeng, S.K., Choi, B.G., Lee, K.T. and Song, K.G. 2013. Influences of solid retention time, nitrification and microbial activity on the attenuation of pharmaceuticals and estrogens in membrane bioreactors. Water Research 47: 3151-3162. https://doi.org/10.1016/j.watres.2013.03.014.

Margot, J., Lochmatter, S., Barry, D.A. and Holliger, C. 2016. Role of ammonia-oxidizing bacteria in micropollutant removal from wastewater with aerobic granular sludge. Water Science and Technology 73: 564-575. https://doi.org/10.2166/wst.2015.514.

McSwain, B.S., Irvine, R.L., Hausner, M. and Wilderer, P.A. 2005. Composition and distribution of extracellular polymeric substances in aerobic flocs and granular sludge. Applied and Environmental Microbiology 71: 1051-1057. https://doi.org/10.1128/AEM.71.2.1051-1057.2005.

Mihciokur, H. and Oguz, M. 2016. Removal of oxytetracycline and determining its biosorption properties on aerobic granular sludge. Environmental Toxicology and Pharmacology 46: 174-182. https://doi.org/10.1016/j.etap.2016.07.017.

Mohseni-Bandpi, A., Elliott, D.J. and Zazouli, M.A. 2013. Biological nitrate removal processes from drinking water supply - a review. Journal of Environmental Health Science and Engineering 11: 35. https://doi.org/10.1186/2052-336X-11-35.

Morgenroth, E., Sherden, T., van Loosdrecht, M.C.M., Heijnen, J.J. and Wilderer, P.A. 1997. Aerobic granular sludge in a sequencing batch reactor. Water Research 31: 3191-3194. https://doi.org/10.1016/S0043-1354(97)00216-9.

Mosquera-Corral, A., De Kreuk, M.K., Heijnen, J.J. and van Loosdrecht, M.C.M. 2005. Effects of oxygen concentration on N-removal in an aerobic granular sludge reactor. Water Research 39: 2676-2686. https://doi.org/10.1016/j.watres.2005.04.065.

Muller, E.B., Stouthamer, A.H. and Van Verseveld, H.W. 1995. Simultaneous $NH_3$ oxidation and $N_2$ production at reduced $O_2$ tensions by sewage sludge subcultured with chemolithotrophic medium. Biodegradation 6(4): 339-349.

Muñoz-Palazon, B., Pesciaroli, C., Rodriguez-Sanchez, A., Gonzalez-Lopez, J. and Gonzalez-Martinez, A. 2018. Pollutants degradation performance and microbial community structure of aerobic granular sludge systems using inoculums adapted at mild and low temperature. Chemosphere 204: 431-441. https://doi.org/10.1016/j.chemosphere.2018.04.062.

Muñoz-Palazon, B., Rodriguez-Sanchez, A., Hurtado-Martinez, M., de Castro, I.M., Juarez-Jimenez, B., Gonzalez-Martinez, A., et al. 2019. Performance and microbial community structure of an aerobic granular sludge system at different phenolic acid concentrations. Journal of Hazardous Materials 376: 58-67. https://doi.org/10.1016/j.jhazmat.2019.05.015.

Nancharaiah, Y.V. and Kiran Kumar Reddy, G. 2018. Aerobic granular sludge technology: mechanisms of granulation and biotechnological applications. Bioresource Technology 247: 1128-1143. https://doi.org/10.1016/j.biortech.2017.09.131.

Nielsen, J.L. and Nielsen, P.H. 2002. Enumeration of acetate-consuming bacteria by microautoradiography under oxygen and nitrate respiring conditions in activated sludge. Water Research 36: 421-428. https://doi.org/10.1016/S0043-1354(01)00224-X.

Petrie, B., Barden, R. and Kasprzyk-Hordern, B. 2014. A review on emerging contaminants in wastewaters and the environment: current knowledge, understudied areas and recommendations for future monitoring. Water Research 72: 3-27. https://doi.org/10.1016/j.watres.2014.08.053.

Pronk, M., de Kreuk, M.K., de Bruin, B., Kamminga, P., Kleerebezem, R. and van Loosdrecht, M.C.M. 2015. Full scale performance of the aerobic granular sludge process for sewage treatment. Water Research 84: 207-217. https://doi.org/10.1016/j.watres.2015.07.011.

Ren, Y., Ferraz, F., Lashkarizadeh, M. and Yuan, Q. 2017. Comparing young landfill leachate treatment efficiency and process stability using aerobic granular sludge and suspended growth activated sludge. Journal of Water Process Engineering 17: 161-167. https://doi.org/10.1016/j.jwpe.2017.04.006.

Robertson, L.A., Cornelisse, R., De Vos, P., Hadioetomo, R. and Kuenen, J.G. 1989. Aerobic denitrification in various heterotrophic nitrifiers. Antonie van Leeuwenhoek 56(4): 289-299. https://doi.org/10.1007/BF00443743.

Roh, H., Subramanya, N., Zhao, F., Yu, C.P., Sandt, J., Chu, K.H., et al. 2009. Biodegradation potential of wastewater micropollutants by ammonia-oxidizing bacteria. Chemosphere 77: 1084-1089. https://doi.org/10.1016/j.chemosphere.2009.08.049.

Rowan, A.K., Snape, J.R., Fearnside, D., Barer, M.R., Curtis, T.P. and Head, I.M. 2003. Composition and diversity of ammonia-oxidising bacterial communities in wastewater treatment reactors of different design treating identical wastewater. FEMS Microbiology Ecology 43: 195-206. https://doi.org/10.1016/S0168-6496(02)00395-1.

Sarma, S.J., Tay, J.H. and Chu, A. 2017. Finding knowledge gaps in aerobic granulation technology. Trends in Biotechnology 35: 66-78. https://doi.org/10.1016/j.tibtech.2016.07.003.

Sarma, S.J. and Tay, J.H. 2018. Aerobic granulation for future wastewater treatment technology: challenges ahead. Environmental Science: Water Research and Technology 4: 9-15. https://doi.org/10.1039/c7ew00148g.

Sathyamoorthy, S., Chandran, K. and Ramsburg, C.A. 2013. Biodegradation and cometabolic modeling of selected beta blockers during ammonia oxidation. Environmental Science and Technology 47: 12835-12843. https://doi.org/10.1021/es402878e.

Schmidt, I. and Bock, E. 1997. Anaerobic ammonia oxidation with nitrogen dioxide by Nitrosomonas eutropha. Archives of Microbiology 167: 106-111. https://doi.org/10.1007/s002030050422.

Schmid, M., Walsh, K., Webb, R., Rijpstra, W. I., van de Pas-Schoonen, K., Verbruggen, M. J. and Damsté, J.S.S. 2003. Candidatus "Scalindua brodae", sp. nov., Candidatus "Scalindua wagneri", sp. nov., two new species of anaerobic ammonium oxidizing bacteria. Systematic and Applied Microbiology 26(4): 529-538.

Seviour, T., Lambert, L.K., Pijuan, M. and Yuan, Z. 2010. Structural determination of a key exopolysaccharide in mixed culture aerobic sludge granules using NMR spectroscopy. Environmental Science Technology 44: 8964-8970. https://doi.org/10.1021/es102658s.

Seviour, T., Yuan, Z., van Loosdrecht, M.C.M. and Lin, Y. 2012. Aerobic sludge granulation: a tale of two polysaccharides? Water Research 46: 4803-4813. https://doi.org/10.1016/j.watres.2012.06.018.

Soares, M.I.M. 2000. Biological denitrification of groundwater. Water, Air, and Soil Pollution 123(1-4): 183-193. https://doi.org/10.1023/A:1005242600186.

Strous, M., Kuenen, J.G. and Jetten, M.S. 1999. Key physiology of anaerobic ammonium oxidation. Applied and Environmental Microbiology 65(7): 3248-3250. 10.1128/AEM.65.7.3248-3250.1999.

Szabó, E., Hermansson, M., Modin, O., Persson, F. and Wilén, B.M. 2016. Effects of wash-out dynamics on nitrifying bacteria in aerobic granular sludge during start-up at gradually decreased settling time. Water (Switzerland) 8(5): 172. https://doi.org/10.3390/w8050172.

Tam, N.F.Y., Wong, Y.S. and Leung, G. 1992. Effect of exogenous carbon sources on removal of inorganic nutrient by the nitrification-denitrification process. Water Research 26: 1229-1236. https://doi.org/10.1016/0043-1354(92)90183-5.

Tay, J.H., Liu, Q.S. and Liu, Y. 2001. The role of cellular polysaccharides in the formation and stability of aerobic granules. Letters in Applied Microbiology 33: 222-226. https://doi.org/10.1046/j.1472-765X.2001.00986.x.

Third, K.A., Sliekers, A.O., Kuenen, J.G. and Jetten, M.S.M. 2001. The CANON system (completely autotrophic nitrogen-removal over nitrite) under ammonium limitation: interaction and competition between three groups of bacteria. Systematic and Applied Microbiology 24(4): 588-596. https://doi.org/10.1078/0723-2020-00077.

Torresi, E., Casas, M.E., Polesel, F., Plósz, B.G., Christensson, M. and Bester, K. 2017. Impact of external carbon dose on the removal of micropollutants using methanol and ethanol in post-denitrifying moving bed biofilm reactors. Water Research 108: 95-105. https://doi.org/10.1016/j.watres.2016.10.068.

Tsuneda, S., Nagano, T., Hoshino, T., Ejiri, Y., Noda, N. and Hirata, A. 2003. Characterization of nitrifying granules produced in an aerobic upflow fluidized bed reactor. Water Research 37: 4965-4973. https://doi.org/10.1016/j.watres.2003.08.017.

Val del Río, Á., Morales, N., Figueroa, M., Mosquera-Corral, A., Campos, J.L. and Méndez, R. 2012. Effect of coagulant-flocculant reagents on aerobic granular biomass. Journal of Chemical Technology and Biotechnology 87: 908-913. https://doi.org/10.1002/jctb.3698.

Wan, J. and Sperandio, M. 2009. Possible role of denitrification on aerobic granular sludge formation in sequencing batch reactor. Chemosphere 75: 220-227. https://doi.org/10.1016/j.chemosphere.2008.11.069.

Wang, F., Xia, S., Liu, Y., Chen, X. and Song, Z.J. 2007. Community analysis of ammonia and nitrite oxidizers in start-up of aerobic granular sludge reactor. Journal of Environmental Sciences 19: 996-1002. https://doi.org/10.1016/S1001-0742(07)60162-4.

Wang, S., Ma, X., Wang, Y., Du, G., Tay, J.H. and Li, J. 2019. Piggery wastewater treatment by aerobic granular sludge: granulation process and antibiotics and antibiotic-resistant bacteria removal and transport. Bioresource Technology 273: 350-357. https://doi.org/10.1016/j.biortech.2018.11.023.

Wang, Z.W., Liu, Y. and Tay, J.H. 2005. Distribution of EPS and cell surface hydrophobicity in aerobic granules. Applied Microbiology and Biotechnology 69: 469-473. https://doi.org/10.1007/s00253-005-1991-5.

Weber, S.D., Ludwig, W., Schleifer, K.H. and Fried, J. 2007. Microbial composition and structure of aerobic granular sewage biofilms. Applied and Environmental Microbiology 73: 6233-6240. https://doi.org/10.1128/AEM.01002-07.

Weber, S.D., Hofmann, A., Pilhofer, M., Wanner, G., Agerer, R., Ludwig, W., et al. 2009. The diversity of fungi in aerobic sewage granules assessed by 18S rRNA gene and ITS sequence analyses. FEMS Microbiology Ecology 68: 246-254. https://doi.org/10.1111/j.1574-6941.2009.00660.x.

Wei, Y., Ji, M., Li, R. and Qin, F. 2012. Organic and nitrogen removal from landfill leachate in aerobic granular sludge sequencing batch reactors. Waste Management 32: 448-455. https://doi.org/10.1016/j.wasman.2011.10.008.

Wilén, B.M., Onuki, M., Hermansson, M., Lumley, D. and Mino, T. 2008. Microbial community structure in activated sludge floc analysed by fluorescence in situ hybridization and its relation to floc stability. Water Research 42: 2300-2308. https://doi.org/10.1016/j.watres.2007.12.013.

Wilén, B.M., Liébana, R., Persson, F., Modin, O. and Hermansson, M. 2018. The mechanisms of granulation of activated sludge in wastewater treatment, its optimization, and impact on effluent quality. Applied Microbiology and Biotechnology 102: 5005-5020. https://doi.org/10.1007/s00253-018-8990-9.

Winkler, M.K.H., Bassin, J.P., Kleerebezem, R., Sorokin, D.Y. and van Loosdrecht, M.C.M. 2012. Unravelling the reasons for disproportion in the ratio of AOB and NOB in aerobic granular sludge. Applied Microbiology and Biotechnology 94: 1657-1666. https://doi.org/10.1007/s00253-012-4126-9.

Xu, P., Liu, L., Zeng, G., Huang, D., Lai, C., Zhao, M. and Zhang, C. 2014. Heavy metal-induced glutathione accumulation and its role in heavy metal detoxification in Phanerochaete chrysosporium. Applied Microbiology and Biotechnology 98(14): 6409-6418. https://doi.org/10.1007/s00253-014-5667-x.

Yan, L., Zhang, S., Hao, G., Zhang, X., Ren, Y., Wen, Y., et al. 2016. Simultaneous nitrification and denitrification by EPSs in aerobic granular sludge enhanced nitrogen removal of ammonium-nitrogen-rich wastewater. Bioresource Technology 202: 101-106. https://doi.org/10.1016/j.biortech.2015.11.088.

Yan, X., Li, Q., Chai, L., Yang, B. and Wang, Q. 2014. Formation of abiological granular sludge - a facile and bioinspired proposal for improving sludge settling performance during heavy metal wastewater treatment. Chemosphere 113: 36-41. https://doi.org/10.1016/j.chemosphere.2014.04.038.

Yao, S., Ni, J., Ma, T. and Li, C. 2013. Bioresource technology heterotrophic nitrification and aerobic denitrification at low temperature by a newly isolated bacterium, Acinetobacter sp. HA2. Bioresource Technology 139: 80-86. https://doi.org/10.1016/j.biortech.2013.03.189.

Yao, S., Ni, J., Ma, T. and Li, C. 2013. Heterotrophic nitrification and aerobic denitrification at low temperature by a newly isolated bacterium *Acinetobacter* sp. HA2. Bioresource Technology 139: 80-86. https://doi.org/10.1016/j.biortech.2013.03.189.

Zhang, H., Zhao, Z., Li, S., Chen, S., Huang, T., Li, N., et al. 2019. Nitrogen removal by mix-cultured aerobic denitrifying bacteria isolated by ultrasound: performance, co-occurrence pattern and wastewater treatment. Chemical Engineering Journal 372: 26-36. https://doi.org/10.1016/j.cej.2019.04.114.

Zhang, W. and Li, C. 2016. Exploiting quorum sensing interfering strategies in gram-negative bacteria for the enhancement of environmental applications. Frontiers in Microbiology 6: 1-15. https://doi.org/10.3389/fmicb.2015.01535.

Zhao, X., Chen, Z., Wang, X., Li, J., Shen, J. and Xu, H. 2015. Remediation of pharmaceuticals and personal care products using an aerobic granular sludge sequencing bioreactor and microbial community profiling using Solexa sequencing technology analysis. Bioresource Technology 179: 104-112. https://doi.org/10.1016/j.biortech.2014.12.002.

Zhou, D., Niu, S., Xiong, Y., Yang, Y. and Dong, S. 2014. Microbial selection pressure is not a prerequisite for granulation: dynamic granulation and microbial community study in a complete mixing bioreactor. Bioresource Technology 161: 102-108. https://doi.org/10.1016/j.biortech.2014.03.001.

Zhou, M., Ye, H. and Zhao, X. 2014. Isolation and characterization of a novel heterotrophic nitrifying and aerobic denitrifying bacterium Pseudomonas stutzeri KTB for bioremediation of wastewater. Biotechnology and Bioprocess Engineering 19: 231-238. https://doi.org/10.1007/s12257-013-0580-1.

Zita, A. and Hermansson, M. 1997. Determination of bacterial cell surface hydrophobicity of single cells in cultures and in wastewater *in situ*. FEMS Microbiology Letters 152: 299-306. https://doi.org/10.1016/S0378-1097(97)00214-0.

# Anaerobic Removal of Nitrogen: Nitrate-Dependent Methane Oxidation and Bioelectrochemical Processes

Alejandro Rodriguez-Sanchez[1,*], Beatriz Gil-Pulido[2],
Alan Dobson[2] and Niall O'Leary[2]

## 11.1 INTRODUCTION

Removal of nitrogen from wastewater is a primary concern for the water treatment industry worldwide due to the significant ecosystem impacts associated with uncontrolled nitrogen discharge to water bodies (Muñoz-Palazon et al. 2018). In response to this risk, regulatory frameworks within developing countries impose a legal onus on wastewater treatment facilities to achieve stringent limits for permissible nitrogen loads in discharge waters. To date, nitrogen removal from wastewater has largely been achieved via biological nitrification and denitrification mechanisms. The two-stage process involves oxidation of nitrogenous compounds to nitrate under aerobic conditions followed by nitrate reduction to molecular nitrogen under anaerobic conditions (Gonzalez-Martinez et al. 2018a). However, despite this approach being widely implemented in full-scale treatment systems, several disadvantages have been acknowledged, such as high aeration costs, infrastructural and capital demands associated with the physical separation of both stages, the need to provide a carbon source for nitrate reduction and the release of greenhouse gas to the atmosphere (Gonzalez-Martinez et al. 2018a). As a result, there are significant commercial opportunities for the development of more efficient and sustainable technologies in this field. One such example is the anammox process that is capable of direct ammonia oxidation to nitrogen in the absence of oxygen and exogenous carbon, thus offsetting several disadvantages of nitrification/denitrification systems (Joss et al. 2009). However, anammox technology applications have yet to fully deliver on their potential for the treatment of urban wastewater effluents as nitrogen removal efficiencies of less than 80% have been reported (Liu et al. 2019). As a result, the scientific community has continued to seek alternative technologies, and over the last decade, two novel and promising approaches have been developed: the nitrate/nitrite-dependent anaerobic methane oxidation (n-DAMO) process and the bioelectrochemical processes.

## 11.2 NITRITE- AND NITRATE-DEPENDENT OXIDATION OF METHANE

Coupled oxidation of methane and reduction of nitrate and/or nitrite was widely hypothesized to occur in natural environments, but the mechanism was first discovered in 2006 (Raghoebarsing et

[1] Department of Horticulture and Landscape Architecture, Purdue University, 625 Agricultural Mall Drive, West Lafayette, 47907, Indiana, United States of America.
[2] Department of Microbiology, University College Cork, Food Science and Technology Building, Cork, T12 K8AF, County Cork, Ireland.
[*] Corresponding author: rodri719@purdue.edu

al. 2006). A microbial consortium enriched from anoxic canal sediment comprising 20% archaeal and 80% bacterial members, respectively, was found to mediate the co-removal of nitrate/nitrite and methane when grown in minimal medium (Raghoebarsing et al. 2006). However, it was subsequently determined that the bacterial species were solely responsible for the process.

### 11.2.1 Candidatus *Methylomirabilis Oxyfera*: The n-DAMO Bacteria

Candidatus *Methylomirabilis oxyfera* (belonging to the candidate division NC10) was identified as the bacterial phylotype associated with this metabolism. The bacterium was found to harbor a range of essential nitrogen metabolism genes with the following polypeptides identified in the proteome; NapA (reduction of nitrate to nitrite), NirSJF (nitrite to nitric oxide) and NorZ (nitric oxide to nitrous oxide), respectively (Ettwig et al. 2010). Further analyzes identified nitrite as the sole electron acceptor capable of facilitating methane oxidation. However, no genes were identified in the assembled genome for nitrous oxide reductase (*nosDFYLZ*), typically associated with full denitrification to dinitrogen (Ettwig et al. 2010). In addition to this, *M. oxyfera* was found to harbor the aerobic methane oxidation pathway, initiated by molecular oxygen and the *pmoCAB* encoded monooxygenase. The proposed pathway for nitrite reduction would thus require the dismutation of nitrous oxide into both dinitrogen and molecular oxygen (Ettwig et al. 2010). *M. oxyfera* was, therefore, deemed to present the first evidence of a novel 'intra-aerobic' pathway of nitrite reduction and coupled methane oxidation, although it may not be limited to methane-oxidizing bacteria. Luesken and co-workers report that oxygenesis and consumption are likely to be tightly controlled in *M. oxyfera* as exogenous oxygen exposure (2-8%) resulted in significant pathway downregulation (Luesken et al. 2012).

Archaeal members with the capacity for coupled anaerobic nitrate reduction and methane oxidation have also been identified, namely Candidatus *Methanoperedens nitroreducens* (Haroon et al. 2013). This archaeon contained genes for the complete reverse methanogenesis pathway (*mcrABCDEG* and *mer*) in addition to nitrate reduction genes (*narGH*). Interestingly, the *narGH* genes present in the genome of *M. nitroreducens* were found to be acquired horizontally from *M. oxyfera*, although they do not appear to be functionally expressed in *M. oxyfera* in the intra-aerobic pathway which relies on nitrite as the functional electron acceptor (Ettwig et al. 2010). Furthermore, Cand. *M. nitroreducens* was not found to harbor any known genes related to nitrite reduction. Thus, it has been hypothesized that in the DAMO process, Cand. *M. nitroreducens* would contribute to nitrate-AMO, while Cand. *M. oxyfera* would perform nitrite-AMO in a syntrophic co-existence (Wang et al. 2017a). The coupled nitrate reduction and methane oxidation yield nitrite, which is thought to be the cause of associations between Cand. *M. nitroreducens* and bacteria with anammox metabolism (Haroon et al. 2013). A summary of the main characteristics of Cand. *M. oxyfera* and Cand. *M. nitroreducens* are offered in Table 11.1.

**TABLE 11.1** Characteristics of Candidatus *Methanomyrabilis oxyfera* and Candidatus *Methanoperedens nitroreducens* in n-DAMO processes (adapted from Wang et al. 2017a)

| Organism | Candidatus *Methanomyrabilis Oxyfera* | Candidatus *Methanoperedens Nitroreducens* |
|---|---|---|
| Domain | *Bacteria* | *Archaea* |
| Phylum | CN10 | ANME-2 |
| Electron acceptor | $NO_2^-$ | $NO_3^-$ |
| Proposed reaction | $3\,CH_4 + 8\,NO_2^{-2} + 8\,H^+ \rightarrow 3\,CO_2 + 4\,N_2 + 10\,H_2O$ | $CH_4 + 4\,NO_3^- \rightarrow CO_2 + 4\,NO_2^- + 2\,H_2O$ |
| Energy | $\Delta G = -765$ kJ mol$^{-1}$ CH$_4$ | $\Delta G = -574$ kJ mol$^{-1}$ CH$_4$ |
| Related gene | *pmo* gene cluster | *mcr* gene cluster |

Cand. *M. oxyfera* has been isolated from sediments contaminated by agricultural runoff or wastewater (Shen et al. 2015a). In addition to this, molecular analyzes have identified the presence of this bacterium in freshwater, marine and wetland ecosystems where it is thought to play an important role in nitrogen geocycling (Shen et al. 2015, Chen et al. 2016). Cand. *M. oxyfera* has also been identified in full-scale wastewater treatment plants (Luesken et al. 2011a), albeit at lower relative abundance than in natural environs (e.g., paddy soils), which may be an operational consequence of short solids retention times (Hu and Ma 2016). Nevertheless, the unique metabolic capability presence of Cand. *M. oxyfera* in wastewater treatment systems, coupled with its unique metabolic capability, has prompted recent research into the development of dedicated bioreactor systems for n-DAMO.

## 11.2.2  n-DAMO Bioprocesses for Wastewater Treatment

The metabolism of n-DAMO bacteria offers two main advantages over typical heterotrophic denitrifiers. One is the absence of $N_2O$ production caused by the intracellular nitrous oxide dismutation by Cand. *M. oxyfera*. $N_2O$ is a significant greenhouse gas with a warming capacity ~300 times higher than $CO_2$ in addition to being the most impactful ozone-depleting substance in the twenty-first century. It has been proposed that wastewater treatment plants may contribute up to 10% of anthropogenic $N_2O$ release associated with conventional heterotrophic denitrification technologies (Kampschreur et al. 2009). Hence, the potential for high-efficiency n-DAMO systems to offset $N_2O$ production/emission has significant environmental importance. The other advantage is the higher affinity for nitrite of n-DAMO bacteria allowing them to potentially outcompete heterotrophic denitrifiers under certain operational conditions (Ren et al. 2019). However, a limiting aspect of the n-DAMO process is that it is dependent on a constant supply of methane for the survival of n-DAMO microorganisms (Hu et al. 2019). As a consequence, the n-DAMO bioprocess could only be applied to wastewaters with sufficient methane content, such as anaerobic digester reject water.

The first attempt to apply an n-DAMO approach to decontaminate urban wastewater under anaerobic conditions and the room temperature was accomplished less than a decade ago (Kampman et al. 2012). An enriched culture of Cand. *M. oxyfera* was shown to treat synthetic and urban wastewater with consumption rates of 33.5 mg-$NO_2^-$-N $L^{-1}$ day$^{-1}$ and 37.8 mg-$NO_2^-$-N $L^{-1}$ day$^{-1}$, respectively. However, the system was reportedly prone to biomass washout. The subsequent incorporation of membrane bioreactor technology was found to stabilize the biomass but did not enhance nitrogen consumption rates (Kampman et al. 2014). Shortened hydraulic retention times and copper dosing were identified as beneficial to nitrogen removal in the n-DAMO membrane bioreactor.

He and his co-workers employed batch experiments to identify optimal temperature, pH and salinity conditions for n-DAMO metabolism in enriched cultures of Cand. *M. oxyfera*. They reported the highest activity at 35°C of temperature, pH of 7.5 and 0 g $L^{-1}$ NaCl (He et al. 2015a). Among these three parameters, salinity had the most detrimental effect on n-DAMO activity, although the adaptation of n-DAMO bacteria was observed at 20 g $L^{-1}$ NaCl which suggested that acclimatization to saline effluents might be accomplished.

In a follow-up study, high nitrogen removal efficiency (up to 98.5%) was achieved when operating n-DAMO processes under SBR conditions (He et al. 2015b). A notable feature of this study was the reported presence of heterotrophic denitrifiers in the system. These microorganisms contributed to approximately 17% of the nitrogen removal, despite being outcompeted by n-DAMO bacteria. Thus, the possibility of n-DAMO/heterotrophic nitrification bioprocess cannot be ruled out. Similar SBR systems have also demonstrated high ammonium removal efficiencies (up to 3 mmol-N day$^{-1}$) (Bhattacharjee et al. 2016). Both SBR studies, above noted, highlighted a strong influence of bioreactor operational time. n-DAMO bacteria are slow-growing and both bioreactor systems require substantial start-up periods (between 300-400 days), which would have obvious implications for n-DAMO implementation at full scale (He et al. 2015b, Bhattacharjee et al. 2016).

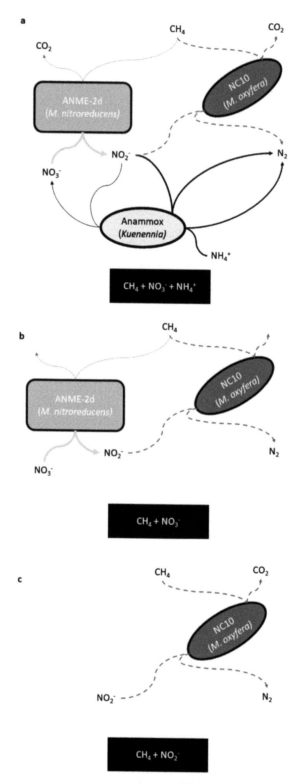

**FIGURE 11.1** Interactions between populations in bioreactors fed with methane and different nitrogen sources (adapted from Haroon et al. 2013). a) Interaction between Candidatus *Methanoperedens nitroreducens* and anammox bacteria; b) interaction between Candidatus *Methanoperedens nitroreducens* and Candidatus *Methylomirabilis oxyfera*; c) Candidatus *Methylomirabilis oxyfera*.

An alternative configuration that has been investigated is the upflow anaerobic sludge blanket (UASB) technology (Ma et al. 2017). The carbon source was found to have a marked effect in this setting as n-DAMO represented 57.9% of total bacteria when feeding sodium acetate in comparison with the 69.4% and 77.7% when feeding methanol and $H_2/CO_2$, respectively. A correlation was also reported between the relative abundance of n-DAMO bacteria and the reduction of $N_2O$ emissions. At 77.7%, n-DAMO bacterial community abundance $N_2O$ emissions were between 3.5 to 4.2-fold lower than for the other two systems demonstrating 69.4% and 57.9% n-DAMO relative abundance (Ma et al. 2017).

In the development of n-DAMO technologies, it is also important to consider the synergy and competition that occurs in co-cultures of n-DAMO archaea and bacteria. The synergy between the two involves n-DAMO bacteria feeding on the nitrite generated from the metabolism of nitrate by n-DAMO archaea (Figure 11.1). The competition arises from the requirement that both microorganisms have for methane. Experimental evaluation of the ability of n-DAMO archaea/bacteria co-culture to remove nitrogen and methane in SBR fed with nitrate and methane demonstrated that synergy between the two can be achieved (Li et al. 2018). With respect to competition, n-DAMO archaea were found to have a methane affinity one order of magnitude higher than their bacterial counterparts. Indeed, this may explain the temporal increase in n-DAMO archaea observed in the test system during prolonged operation (Li et al. 2018). The results suggest that the activity of n-DAMO archaea could be of importance when nitrate is present in n-DAMO systems.

More recent experiments have sought to stimulate the growth of n-DAMO bacteria by introducing different growth factors in a magnetically stirred gas lift reactor (MSGLR) (Wang et al. 2019a). It was found that the addition of small amounts of nucleobase (5 µg $L^{-1}$), betaine (200 µg $L^{-1}$) and nucleobase + betaine increased the conversion rate of nitrite within the MSGLR by 6-, 8- and 10-fold, respectively. MSGLR dosing with nucleobase + betaine achieved nitrite removal efficiencies of up to 99% under a nitrite load rate of 70 mg-N $L^{-1}$. The results provide a promising demonstration of the potential for n-DAMO technologies to deliver high-efficiency treatment of nitrogen-polluted effluents but much work remains to deliver robust and optimized systems for industrial-scale deployment.

### 11.2.3 Future Perspectives for n-DAMO: The n-DAMO/Anammox Systems

The two most problematic issues in wastewater treatment processes today are the shortage of exogenous carbon sources and the emission of greenhouse gases, which can potentially be circumvented via n-DAMO technologies (Wang et al. 2017b). The current vision for the potential implementation of n-DAMO technology involves the combination of n-DAMO with anammox technology for the removal of nitrogen from mainstream urban wastewater (Wang et al. 2017b).

Several reports have identified both groups of microorganisms co-habiting diverse natural environments, such as forest soils (Meng et al. 2016), agricultural soil (Hui et al. 2017), mangrove sediments (Chen and Gu, 2017) and riverbeds (Shen et al. 2019). While competitive relationships have been proposed between anammox and n-DAMO bacteria for nitrite, their frequent cohabitation also suggests the possibility of synergistic interactions. In this sense, n-DAMO bacteria may utilize small quantities of nitrate generated by anammox bacterial metabolism. The identification of their coexistence in bioreactors has been reported from anammox granular technologies, promoting further technology development (Figure 11.2) (Zhu et al. 2011).

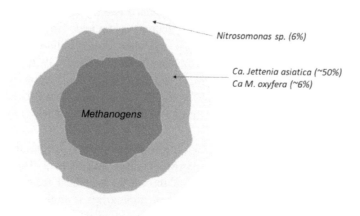

*Nitrosomonas sp. (6%)*

*Ca. Jettenia asiatica (~50%)*
*Ca M. oxyfera (~6%)*

*Methanogens*

**FIGURE 11.2** Spatial ordination of nitrifiers, anammox bacteria, n-DAMO bacteria and methanogens in anammox sludge (adapted from Zhu et al. 2012).

The introduction of n-DAMO bacteria can provide a few advantages over conventional, stand-alone anammox systems. The most important feature is that n-DAMO provide a mechanism for removing excess nitrite when anammox bioprocesses operate out of balance. Anammox systems need to produce a nitrite-to-ammonium ratio of 1.32 for anammox removal of nitrogen to take place. In an unbalanced system, excess nitrite would not be removed from the effluent, but the presence of n-DAMO could resolve this (Liu et al. 2019). Another improvement would be the capacity for the removal of nitrate in the anammox system, which can be inadvertently produced (but not removed) via anammox metabolism. The presence of n-DAMO archaea could provide an efficient way to reduce nitrate to nitrite, providing higher efficiencies in nitrogen removal (Cogert et al. 2019). In addition to this, it has been proposed that n-DAMO archaea could synergize with anammox bacteria for the anaerobic treatment of effluents containing nitrate and methane as observed in constructed and natural wetlands (Wang et al. 2019b). Moreover, prospective modeling studies have reported that coupled n-DAMO/anammox systems will be more attractive than anammox systems due to lower operational costs and greenhouse gas generation (Cogert et al. 2019).

Key operational factors for the successful implementation of this coupled technology for the treatment of anaerobic digester reject water must also be taken into consideration One of the most important factors is the provision of oxygen for ammonium oxidation to nitrite, which can inhibit n-DAMO and anammox bacteria (van Kessel et al. 2018). In the case of n-DAMO bacteria, oxygen inhibition can begin at exposures as low as 2% (Luesken et al. 2012). Moreover, increasing oxygen concentrations select for aerobic methane oxidizers, which outcompetes n-DAMO bacteria and results in their washout from the systems (Luesken et al. 2012). Thus, oxygen concentrations need to be limited in coupled n-DAMO/anammox systems to prevent oxygen inhibition of important bacterial groups, oxidation of ammonium to nitrate and growth of heterotrophs (van Kessel et al. 2018).

Another important factor is the slow growth rates of both n-DAMO and anammox bacteria. In order to avoid the washout of these microorganisms from the system, efficient mechanisms for biomass retention are needed. Among the technologies studied for the development of n-DAMO/anammox bioprocesses, those that proved to be most efficient in terms of biomass retention were membrane bioreactors (MBR), membrane biofilm reactors (MBfR) and sequencing batch reactors (SBR) (van Kessel et al. 2018).

With respect to the dynamics and performance of coupled technologies, Luesken and colleagues investigated a 50/50 mixture of anammox bacteria cultured from an enrichment of n-DAMO bacteria (Luesken et al. 2011b). The co-culture consumed up to 20 mM day$^{-1}$ of ammonium and converted up to 100 mg-N m$^3$ day$^{-1}$. It was also found, however, that anammox bacteria contributed more than n-DAMO bacteria to the removal of nitrite (77% versus 33%), suggesting that anammox bacteria may outcompete n-DAMO bacteria for nitrite over time (Luesken et al. 2011b). A complementary experiment in which n-DAMO bacteria were cultivated from an anammox bacteria inoculum also showed this higher affinity of anammox bacteria for nitrite (Zhu et al. 2011). In this regard, n-DAMO bacteria would need to compete with regular methane-oxidizing bacteria for methane and with anammox bacteria for nitrite, which would put them at a disadvantage in full-scale systems. Investigation of n-DAMO/anammox bacteria co-cultures for the treatment of ammonium and methane wastewater have also reported a similar, dominant contribution of anammox bacteria to nitrogen removal (~70%) when compared to n-DAMO bacteria (Stultiens et al. 2019). Despite this inherent competition in co-culture systems, Cand. *M. oxyfera* has been identified as an important community member in the inner layers of anammox granules. Therefore, it is possible that a synergy between anammox and n-DAMO bacteria can be established in bioreactors wherein spatial distribution and co-location play influential roles (Hu et al. 2012).

Membrane biofilm reactors (MBfR) have also been evaluated for the treatment of synthetic anaerobic reject water. The presence of biofilm over the membrane fibers was important for the distribution of microorganisms involved in nitrogen oxidation, reduction and removal, such as ammonium oxidizing bacteria (AOB), nitrite oxidizing bacteria (NOB), anammox and n-DAMO bacteria. AOB were found to be positioned near the surface of the biofilm where oxygen was more readily available. Below these, anammox and n-DAMO bacteria began to emerge with anammox bacteria demonstrating much higher abundance (Chen et al. 2015). The findings suggest that the competition and/or synergy of different bacterial groups, such as AOB, NOB, anammox and n-DAMO bacteria, are crucial for the development of anaerobic removal of nitrogen in the presence of methane. It was also found that aeration plays a very important role in nitrogen removal within the system. Low oxygen loadings (1.56 g m$^{-3}$ day$^{-1}$) resulted in high activities for AOB and anammox, linked to high nitrogen removal (91.3%) but negligible methane removal due to low n-DAMO bacterial activity. Increased oxygen loading (1.83 g m$^{-3}$ day$^{-1}$) promoted the growth of AOB and n-DAMO bacteria with a concomitant increase in methane removal (97.1%) but resulted in decreased numbers of anammox and nitrogen removal efficiency (93%). At higher loading (1.88 g m$^{-3}$ day$^{-1}$), both nitrogen removal and methane removal dropped to negligible levels (Chen et al. 2015). Thus, the design and optimization of n-DAMO/anammox bioreactor systems would require careful control over oxygen loading within the system.

An approach that has been reported to prevent oxygen toxicity in n-DAMO/anammox technologies is the development of granular sludge under aerobic conditions. Investigation of the functioning of n-DAMO/anammox system under hollow fiber membrane bioreactor (HfMBR) configuration demonstrated the development of both attached biofilm over the hollow fibers and granular biomass (Fu et al. 2017). Both types of biomass accounted for n-DAMO microorganisms and anammox bacteria, but their relative abundances differed significantly. n-DAMO archaea were more abundant in granular configuration than in attached biofilm, while n-DAMO and anammox bacteria had much smaller relative abundances. In addition to this, more biomass developed under the granular configuration (approximately 10/1 ratio). Since n-DAMO archaea are capable of reducing nitrate to nitrite to oxidize ammonium, it is possible that they can develop an important role in granular n-DAMO/anammox systems.

Lee and Rittmann previously demonstrated that increasing partial pressures of hydrogen (the electron donor for contaminant oxidation) had a significant impact on nitrate removal in a hydrogen-based MBfR (Lee and Rittmann 2002).

On this basis, it was deemed possible that methane partial pressure can play an important performance role in methane-based MBfR systems, particularly as the high-efficiency nitrate removal in such systems (684 ± 10 mg-N L$^{-1}$ d$^{-1}$) is dependent on DAMO archaea, DAMO bacteria and annamox bacteria (Chen et al. 2015). Methane partial pressure impacts were investigated in

MBfR and revealed increases of 58% in the nitrate reduction rate by DAMO archaea and 303% in the nitrite reduction rate by DAMO bacteria in short term tests with increased methane gas pressure (Cai et al. 2018). The authors further suggested that methane partial pressure could offer flexibility in the system operation as nitrate/nitrite removal could be completely removed by adjusting methane partial pressure in response to fluctuating influent nitrogen loading (Cai et al. 2018).

The impact of influent COD on membrane aerated biofilm reactors (MABRs) treating anaerobic digestion process reject water (rich in ammonium and dissolved methane) has recently been investigated (Wu and Zhang 2017). Such waters can contain up to 800 mg COD $L^{-1}$ creating the opportunity for heterotrophic bacteria to compete for available electron acceptors (nitrate, oxygen) potentially impacting anammox and DAMO microbial metabolism. The study revealed the at low C/N ratios of 0.1 (100 mg COD/1000 mg N $L^{-1}$) did not impact negatively on either N or $CH_4$ removal. Furthermore, heterotrophs were found to be confined to outer layers of biofilm with nitrate production by anammox bacteria providing the necessary electron acceptor. The findings suggest that coupled n-DAMO/anammox systems are efficient for the treatment of waters with low C/N ratios with limited competition spatially or metabolically from heterotrophic bacteria (Wu and Zhang 2017).

## 11.3 BIOELECTROCHEMICAL PROCESSES

A bioelectrochemical system (BES) is a process in which the interaction between microorganisms and electron donors/acceptors are used to simultaneously decontaminate wastewater and produce electricity (Jain and He 2018). Microorganisms tend to form biofilms over electrodes present in the technological configuration of the BES, named as anode (negative electrode) and cathode (positive electrode). One well-established BES technology is the microbial fuel cell (MFC), which can be used to remediate wastewater and generate electricity by creating an environment for thermodynamically favorable redox reactions (Mook et al. 2013). In a conventional MFC, the anode and cathode are

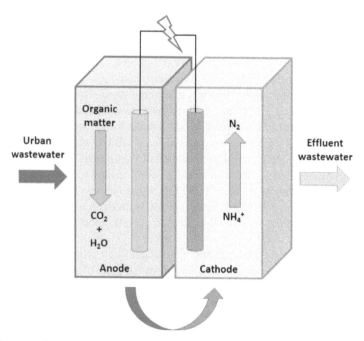

**FIGURE 11.3** Schematic of bioelectrochemical reactor treating industrial wastewater (adapted from Jain and He 2018).

located in either two separate chambers or a single chamber with membrane separation. In the anodic chamber, microorganisms oxidize compounds, typically organic matter, under anaerobic conditions. The oxidation yields electrons, which are then transferred to the anode via redox mediator compounds, respiratory enzymes or a conductive pilus known as a nanowire. The electrons pass to the cathode in the cathodic chamber in which the reduction of terminal electron acceptors occurs (Gude 2016). An example is illustrated in Figure 11.3. In contrast, a microbial electrolysis cell is the reverse of MFC in which an electric current is applied to the bioreactor in order to generate methane or hydrogen (Katuri et al. 2019).

## 11.3.1 Bioelectrochemical Removal of Nitrogen: Principles

In the anodic chamber, microbial anaerobic oxidation of organic matter occurs to produce ATP and oxidize the organic matter through glycolysis to carbon dioxide. The process also transforms NAD+ and FAD to NADH or FADH, respectively. These compounds transfer the oxidation derived electrons to the cell membrane cell from which they travel through direct or mediated electron transfer mechanisms to the anode (Mook et al. 2013).

A direct electron transfer mechanism is rare among microorganisms, and only a few genera with this capacity have been identified, such as *Geobacter*, *Oligotropha*, *Rhodoferax* or *Shewanella*. These specific microorganisms possess membranes with high electrical conductivity which permits the transfer of electrons through them (Du et al. 2007, Watanabe et al. 2008, Huang et al. 2011, Puig et al. 2011). Moreover, for direct electron transfer to take place, a physical contact between the microorganism or the electrically-conductive biofilm and the anode/cathode must take place. However, most microorganisms do not possess this characteristic and thus require a mediation process to transfer the electrons to the anode/cathode and represent the most common mode of electron transfer in MFCs (Rosenbaum et al. 2011). A schematic of exoelectrogenic transport is presented in Figure 11.4.

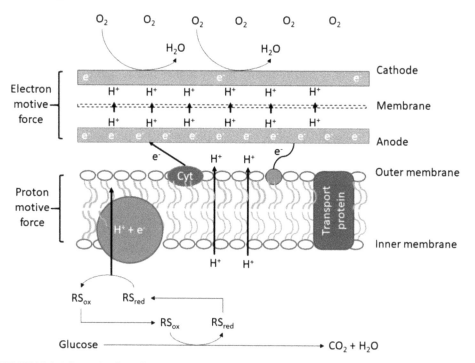

FIGURE 11.4 Schematic of exoelectrogenic transport of electrons (adapted from Palanasamy et al. 2019).

The usual electron acceptor used in cathodic chambers of MFC is oxygen. However, it has been shown that nitrate has a similar redox potential and can, therefore, compete as an electron acceptor (+0.74 V at pH 7 for $NO_3^-/0.5N_2$ versus +0.82 V at pH 7 for $O_2/H_2O$) (Fang et al. 2011). Similarly, nitrite could also be used as an effective electron acceptor (Ucar et al. 2017). In addition to this, cathodes can act as electron donors for denitrification in the form of hydrogen liberated from water hydrolysis near the cathode (Gregory et al. 2004). This process can take place in a single reactor when the anodic and cathodic chambers are separated by a proton exchange membrane. Indeed, the use of nitrate as an oxidant in MFC for coupled removal of organic (via oxidation in the anodic chamber) and nitrogen (via reduction of nitrate/nitrite in the cathodic chamber) has previously been reported (Clauwaert et al. 2007, Virdis et al. 2010, Chen et al. 2010, Fan et al. 2011). In this sense, MFCs represent a promising technology to treat organic matter and nitrogen with electricity generation as a byproduct of the process.

## 11.3.2 Bioelectrochemical Removal of Nitrogen: Operational Conditions and Microorganisms Involved

Briefly, the process of nitrogen removal from wastewater in MFC can be divided in two stages. Firstly, ammonium moves passively (as $NH_3$) or actively (as $NH_4^+$) from the anodic chamber to the cathodic chamber through the ion exchange membrane. Second, microorganisms oxidize ammonium to nitrate and denitrify nitrate to molecular nitrogen (Palanisamy et al. 2019). Therefore, bioelectrochemical removal of nitrogen in the cathodic chamber occurs when microorganisms use the electrons generated in the anodic chamber to perform nitrate/nitrite reduction (Sevda et al. 2018). Microorganisms facilitating the bioelectrochemical removal of nitrogen are thought to be common denitrifiers found in wastewater treatment systems, such as *Thauera* or *Azoarcus*, and ammonium oxidizers such as *Nitrosomonas* (Guo et al. 2019). However, it has been found that operational conditions greatly affect the diversity of these microorganisms, so a specific and indispensable microbial group for bioelectrochemical reduction of nitrate/nitrite has not been proposed to date.

Among these operational conditions, dissolved oxygen has proven to be particularly important. It has been reported that dominant denitrifiers in the cathodic chamber of MFCs change with increasing dissolved oxygen. In this sense, *Pirelulla* genus would dominate under hypoxic conditions but with decreasing abundance in favor of *Thermomonas*, *Azospira* and *Phaeodactylibacter/Povalibacter* as the oxygen concentration in the chamber increased (Guo et al. 2019). This was found to correlate with the performance of the MFC, which demonstrated higher nitrate removal and ammonium generation at the lower oxygen concentration conditions tested (Guo et al. 2019).

The nature of the carbon source entering the cathodic chamber is also an important parameter, capable of selecting for either autotrophic or heterotrophic denitrifiers. It has been hypothesized that autotrophic denitrification in MFCs allows for higher energy output but lower nitrate removal efficiency, while the opposite occurs for heterotrophic denitrification (Ding et al. 2018). This was confirmed in subsequent experiments where cathodic chambers were fed with 2 g $L^{-1}$ of either sodium bicarbonate (autotrophic) or sodium acetate (heterotrophic) as carbon sources. Voltage generation was approximately 3-fold higher for the autotrophic medium compared to the output with heterotrophic medium. However, nitrate degradation kinetics were significantly faster for the heterotrophic medium (from 350 mg $L^{-1}$ to negligible values in 5 h) when compared to that of the autotrophic medium (from 350 mg $L^{-1}$ to 325 in 5 h, with 50 mg $L^{-1}$ observed after 60 h of operation) (Vijay et al. 2019). Ecological analyses revealed the microbial communities in the autotrophic and heterotrophic cathodic chambers also differed. In the autotrophic system, the most important bacterial members related to bioelectrochemical metabolism were identified at genus level as *Thauera*, *Acholeplasma* and *Paracoccus*. In the heterotrophic system the dominant bioelectrochemical denitrifiers were affiliated at genus level with *Alcaligenes*, *Klebsiella*, *Bacteroides*, *Parabacteroides* and *Coprococcus*. It was noted that both systems presented an abundance of important genera, namely *Pseudomonas* and *Alkaliphilus* (Vijay et al. 2019). Interestingly, subsequent quantification of all denitrification associated genes suggested a higher abundance within the autotrophic medium related community.

The inoculum used for the start-up of the MFC has also been shown to be an influential factor in the denitrification capacity of these systems when comparing denitrifying, denitratation and anammox sludges (Ding et al. 2018). Under high COD/N ratio (~5) loadings, the denitrifying and denitratation sludges achieved higher performances in nitrogen removal (converting 600-700 mg $L^{-1}$ day$^{-1}$) when compared with the anammox sludge derived inoculum (converting 90 mg $L^{-1}$ day$^{-1}$). In contrast, when operating at a lower COD/N ratio (~1), the conversion rates were reduced for the denitrification (550 mg $L^{-1}$ day$^{-1}$) and denitratation (340 mg $L^{-1}$ day$^{-1}$) sludges but over 3 fold higher for the anammox sludge (335 mg $L^{-1}$ day$^{-1}$). When the COD/N ratio was 0, all systems displayed a lower conversion rate with 100 mg $L^{-1}$ day$^{-1}$ for denitrifying and anammox sludges and 47 mg $L^{-1}$ day$^{-1}$ for the denitratation sludge. At the end of the experiment, at COD/N of 0, the dominant genera in the three systems inoculated with different sludges were found to differ considerably (Ding et al. 2018).

Temperature represents another important parameter for the nitrogen removal performance and microbial community structure in MFC cathodes. Wang and colleagues operated an MFC at different temperatures (25, 30, 35, 40 and 45°C) to observe the effects on the bioprocess and reported an optimum operational temperature in terms of electricity generation, chemical oxygen demand and nitrogen removal of 35°C (Wang et al. 2018). In tandem with the impacts of operational temperature, the adaptation of inoculum to temperature is also an important factor. It has been shown that MFCs inoculated with cold-adapted (~8°C) sludge have higher electricity generation outputs when operating at low temperature (8°C) than at high temperature (25°C) (Gonzalez-Martinez et al. 2018b). Also, the archaeal and bacterial communities in the systems varied greatly between the two operational conditions.

As occurs in other bioreactor technologies, hydraulic retention time and pollutant loading have a very strong effect on bioreactor performance. In the case of denitrification in MFCs, experiments suggested that lower hydraulic retention (from 41.6 to 8.3 h) times resulted in lower nitrate removal performances (Kondaveeti et al. 2019). Also, increasing nitrate loading rates (from 0.02 to 0.08 Kg-NO$_3^-$-N m$^{-3}$ day$^{-1}$) decreased nitrate removal performance when operating under similar hydraulic retention times (Kondaveeti et al. 2019).

### 11.3.3 Future Perspectives for Bioelectrochemical Systems

The most promising feature of bioelectrochemical technology application to wastewater treatment is the recovery of energy during the process (Katuri et al. 2014). It has been proposed that the implementation of this technology to the treatment of mainstream urban wastewater treatment could, therefore, help reduce operational energy inputs (Katuri et al. 2019). Several different approaches for implementation at the pilot-scale have been accomplished, and the development of bioelectrochemical technologies for full-scale application is receiving considerable interest (Baeza et al. 2017, Cotterill et al. 2017).

The use of standard BESs as stand-alone modules for the treatment of urban wastewater to meet legislated discharge standards is unfeasible. Thus, different configurations have been employed to enable its effective incorporation into strategies for the treatment of urban wastewater by the means of BES. One of the most successful to date consisted of a flat-panel MFC (Park et al. 2017). In this configuration, operating at a short hydraulic retention time of 2 h, COD and total nitrogen were removed up to 85% and 94%, respectively, after eight months of operation (Park et al. 2017). New developments in MFC technology optimization are also emerging, such as designs facilitating removal of the ion exchange membrane separating the anodic and cathodic chambers. It has been shown that the presence of this membrane causes several problems for the MFC such as higher cost, increased biofouling, low ion exchange capacity and high exchange of oxygen and substrates between the anodic and cathodic chambers, which decrease the performance of the

MFC (Leong et al. 2013). New membraneless MFC systems have been developed and shown to provide high removal capacities for nitrogen (91% $NH_4^+$, 87% TN) and COD (97%) when treating urban wastewater (Zhou et al. 2018). The development of anaerobic electrochemical membrane bioreactors has also been investigated. The integration of a BES with a membrane bioreactor (MBR) was found to enhance the performance of the MBR with respect to nitrogen removal (Gajaraj and Hu 2014). A similar effect of enhanced nutrient removal capacity was observed when combining an MFC with an osmotic membrane system and aerated bioreactor (Feng et al. 2015, Hou et al. 2017). Hybrid MFC-MBRs introduce a membrane in the cathodic chamber and offer advantages over the conventional MBR, such as enhanced treatment efficiency, lower energy consumption and mitigation of ion exchange membrane fouling (Fakhry'l-Razi et al. 2010, Gao et al. 2018, Wang et al. 2019). Moreover, experiments using hollow fiber MFC-MBRs have reported that the electric current generated helps prevent biofouling of the hollow-fiber membrane in the system (Ailijiang et al. 2016). Similar results have been observed when operating an air-biocathode MFC-MBR (Ding et al. 2018) and for hybrid MEC-MBR (Shi et al. 2019). In addition to the hybrid MFB-MBR systems, there are also configurations in which both bioreactors operate separately with the MFR or MEC upstream the MBR (Wang et al. 2019), such as MBR/sludge MFC (Borea et al. 2017), osmotic MBR/MFC (Hou et al. 2016) or anaerobic fluidized bed MBR/MFC (Ren et al. 2014).

However, the most promising prospect for the BESs is their development in combination with anammox technologies. This combination derives from the wastewater treatment implementation known as the A-B process. In the A-B process, the treatment of wastewater is separated into two stages, referred to as A-stage and B-stage. The purpose of the A-stage is mainly the removal of organic matter, while the purpose of the B-stage is primarily the removal of inorganic nutrients (Katuri et al. 2019). It has been demonstrated that the A-B process can operate successfully when an anaerobic treatment by methanogenesis is used as A-stage and an anaerobic ammonium oxidation process is used as B-stage (Gu et al. 2018, Hoekstra et al. 2019). Following this model, a BES could be used as an A-stage platform, providing it is capable of delivering the COD removal required for the anammox system to function properly (Katuri et al. 2019). Improvement of COD removal in the BES process has been achieved by the implementation of granular activated carbon (Liu et al. 2014a, 2014b).

Other technological configurations have demonstrated the possibility of anammox removal of nitrogen via a nitrifyinNo g bioelectrochemical system (niBES) (Qu et al. 2014, Vilajeliu-Pons et al. 2018). In this configuration, a two-chamber MEC is used. In the anode chamber, an ammonium-rich buffer solution (ranging from 223 to 54 g-N $L^{-1}$ $day^{-1}$) is fed under anaerobic conditions. In the cathode chamber, the same buffer without ammonium is fed under anoxic conditions. The system successfully produced electric current with concomitant conversion of ammonium to nitrogen gas and low residual amounts of nitrate (0.05 mg-N $L^{-1}$) and nitrite (0.45 mg-N $L^{-1}$) (Vilajeliu-Pons et al. 2018). The dominant biological contributors to nitrogen removal were identified as *Nitrosomonas* and *Empedobacter* genera with *Thermomonas* genus also being implicated (Qu et al. 2014, Vilajeliu-Pons et al. 2018). The performance of this system and its lower energy costs offer certain advantages over traditional aerobic nitrification/denitrification systems and anammox systems alike, thus it could be a very promising technology for continued future development (Vilajeliu-Pons et al. 2018).

Another area of promising development is the combination of MEC and anaerobic digestion processes. Several experimental setups were found to enhance the removal of propionate when a MEC was integrated with the anaerobic digestion process (Hari et al. 2016a, 2016b), which would help solve operational issues commonly encountered in anaerobic digestion, such as short-chain fatty acids accumulation (Katuri et al. 2019). Another advantage is the enhanced generation of $CH_4$. It has been found that a combination of anaerobic digestion and BES increases the production of $CH_4$ in the range of 22% to 230% under optimum conditions (Bo et al. 2014, Feng et al. 2015). In addition to this, combined anaerobic digestion and MEC has been shown to reduce the time required for system start-up (up to 4-fold faster) (Park et al. 2018). Overall, the application of MEC to anaerobic digestion might offer a way forward for the development of more efficient anaerobic digestion processes.

There is also an interest in combining constructed wetlands/plant systems with BES for the treatment of urban wastewater (Guadarrama- Pérez et al. 2019) (Figure 11.5).

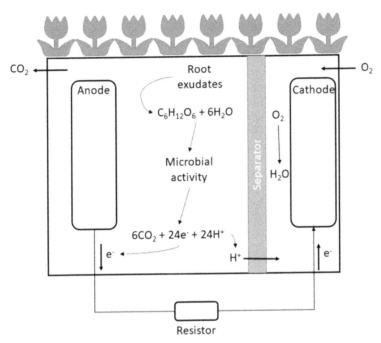

**FIGURE 11.5** Schematic of a constructed wetland-microbial fuel cell system (adapted from Guadarrama-Pérez et al. 2019).

The combined technology, referred to as constructed wetland-MFC (CW-MFC), is designed to offer a process in which sunlight is used to generate bioelectricity via the activity of microorganisms living in the rhizosphere of plants (Nitisoravut and Regmi 2017). The basic premise is that electrically active microorganisms use the root exudates of plants to generate electrons, which are captured by the anode and sent to the cathode, wherein the generation of electricity occurs in presence of an electron acceptor (Sophia and Sreeja 2017). The CW-MFC system also has the potential for the cost-effective removal of nitrogen. It has been shown that CW-MFC can remove ammonium (up to 77%) when treating swine manure (Liu et al. 2019). A CW-MFC was also found to efficiently remove nitrogen anaerobically (up to 85%) from synthetic wastewater while generating energy at the same time (Li et al. 2019). Moreover, it has been reported that the addition of an MFC to a CW generated several advantages for the treatment of wastewater including higher removal of nitrogen (Tao et al. 2020) and lower emissions of the greenhouse gases $CH_4$ and $N_2O$ (Wang et al. 2019). To date, published research on CW-MFC systems remains limited, but the process offers a wide array of opportunities regarding environmentally-friendly and cost-effective wastewater treatment coupled with the generation of electricity (Guadarrama-Pérez et al. 2019).

## 11.4   REFERENCES

Ailijiang, N., Chang, J., Liang, P., Li, P., Wu, Q., Zhang, X., et al. 2016. Electrical stimulation on biodegradation of phenol and responses of microbial communities in conductive carriers supported biofilms of the bioelectrochemical reactor. Bioresource Technology 201: 1-7. doi: 10.1016/j.biortech.2015.11.026.

Baeza, J.A., Martínez-Miró, À., Guerrero, J., Ruiz, Y. and Guisasola, A. 2017. Bioelectrochemical hydrogen production from urban wastewater on a pilot scale. Journal of Power Sources 356: 500-509. doi: 10.1016/j.jpowsour.2017.02.087.

Bhattacharjee, A.S., Motlagh, A.M., Jetten, M.S.M. and Goel, R. 2016. Methane dependent denitrification-from ecosystem to laboratory-scale enrichment for engineering applications. Water Research 99: 244-252. doi: 10.1016/j.watres.2016.04.070.

Bo, T., Zhu, X., Zhang, L., Tao, Y., He, X., Li, D., et al. 2014. A new upgraded biogas production process: coupling microbial electrolysis cell and anaerobic digestion in single-chamber, barrel-shape stainless steel reactor. Electrochemistry Communications 45: 67-70. doi: 10.1016/j.elecom.2014.05.026.

Borea, L., Puig, S., Monclús, H., Naddeo, V., Colprim, J. and Belgiorno, V. 2017. Microbial fuel cell technology as a downstream process of a membrane bioreactor for sludge reduction. Chemical Engineering Journal 326: 222-230. doi: 10.1016/j.cej.2017.05.137.

Cai, C., Hu, S., Chen, X., Ni, B.J., Pu, J. and Yuan, Z. 2018. Effect of methane partial pressure on the performance of a membrane biofilm reactor coupling methane-dependent denitrification and anammox. Science of the Total Environment 639: 278-285. doi: 10.1016/j.scitotenv.2018.05.164.

Chen, C., Shihu, H., Jianhua, G., Ying, S., Guo-Jun, X. and Zhiguo, Y. 2015. Nitrate reduction by denitrifying anaerobic methane oxidizing microorganisms can reach a practically useful rate. Water Research 7: 211-217. doi: 10.1016/j.watres.2015.09.026.

Chen, G.W., Choi, S.J., Cha, J.H., Lee, T.H. and Kim, C.W. 2010. Microbial community dynamics and electron transfer of a biocathode in microbial fuel cells. Korean Journal of Chemical Engineering 27: 1513-1520. doi: 10.1007/s11814-010-0231-6.

Chen, J., Dick, R., Lin, J.G. and Gu, J.D. 2016. Current advances in molecular methods for detection of nitrite-dependent anaerobic methane oxidizing bacteria in natural environments. Applied Microbiology and Biotechnology 100: 9845-9860. doi: 10.1007/s00253-016-7853-5.

Chen, J. and Gu, J.D. 2017. Faunal burrows alter the diversity, abundance, and structure of AOA, AOB, Anammox and n-damo communities in coastal mangrove sediments. Microbial Ecology 74: 140-156. doi: 10.1007/s00248-017-0939-5.

Chen, X., Guo, J., Xie, G.J., Liu, Y., Yuan, Z. and Ni, B.J. 2015. A new approach to simultaneous ammonium and dissolved methane removal from anaerobic digestion liquor: a model-based investigation of feasibility. Water Research 85: 295-303. doi: 10.1016/j.watres.2015.08.046.

Clauwaert, P., Rabaey, K., Aelterman, P., De Schamphelaire, L., Pham, T.H., Boeckx, P., et al. 2007. Biological denitrification in microbial fuel cells. Environmental Science and Technology 41: 3354-3360. doi: 10.1021/es062580r.

Cogert, K.I., Ziels, R.M. and Winkler, M.K.H. 2019. Reducing cost and environmental impact of wastewater treatment with denitrifying methanotrophs, Anammox, and mainstream anaerobic treatment. Environmental Science and Technology 53: 12935-12944. doi: 10.1021/acs.est.9b04764.

Cotterill, S.E., Dolfing, J., Jones, C., Curtis, T.P. and Heidrich, E.S. 2017. Low temperature domestic wastewater treatment in a microbial electrolysis cell with 1 m 2 anodes: towards system scale-up. Fuel Cells 17: 584-592. doi: 10.1002/fuce.201700034.

Ding, A., Fan, Q., Cheng, R., Sun, G., Zhang, M. and Wu, D. 2018. Impacts of applied voltage on microbial electrolysis cell-anaerobic membrane bioreactor (MEC-AnMBR) and its membrane fouling mitigation mechanism. Chemical Engineering Journal 333: 630-635. doi: 10.1016/j.cej.2017.09.190.

Du, Z., Li, H. and Gu, T. 2007. A state of the art review on microbial fuel cells: a promising technology for wastewater treatment and bioenergy. Biotechnology Advances 25: 464-482. doi: 10.1016/j.biotechadv.2007.05.004.

Ettwig, K.F., Butler, M.K., Le Paslier, D., Pelletier, E., Mangenot, S., Kuypers, M.M.M., et al. 2010. Nitrite-driven anaerobic methane oxidation by oxygenic bacteria. Nature 464: 543-548. doi: 10.1038/nature08883.

Fakhry'l-Razi, A., Pendashteh, A., Abidin, Z.Z., Abdullah, L.C., Biak, D.R.A. and Madaeni, S.S. 2010. Application of membrane-coupled sequencing batch reactor for oilfield produced water recycle and beneficial re-use. Bioresource Technology 101: 6942-6949. doi: 10.1016/j.biortech.2010.04.005.

Fang, C., Min, B. and Angelidaki, I. 2011. Nitrate as an oxidant in the cathode chamber of a microbial fuel cell for both power generation and nutrient removal purposes. Applied Biochemistry and Biotechnology 164: 464-474. doi: 10.1007/s12010-010-9148-0.

Feng, Y., Zhang, Y., Chen, S. and Quan, X. 2015. Enhanced production of methane from waste activated sludge by the combination of high-solid anaerobic digestion and microbial electrolysis cell with iron-graphite electrode. Chemical Engineering Journal 259: 787-794. doi: 10.1016/j.cej.2014.08.048.

Frutos, O.G., Quijano, G., Aizpuru, A. and Muñoz, R. 2018. A state-of-the-art review on nitrous oxide control from waste treatment and industrial sources. Biotechnology Advances 36(4): 1025-1037, ISSN 0734-9750. https://doi.org/10.1016/j.biotechadv.2018.03.004.

Fu, L., Ding, J., Lu, Y.Z., Ding, Z.W., Bai, Y.N. and Zeng, R.J. 2017. Hollow fiber membrane bioreactor affects microbial community and morphology of the DAMO and Anammox co-culture system. Bioresource Technology 232: 247-253. doi: 10.1016/j.biortech.2017.02.048.

Fu, L., Zhang, F., Bai, Y.N., Lu, Y.Z., Ding, J., Zhou, D., et al. 2019. Mass transfer affects reactor performance, microbial morphology, and community succession in the methane-dependent denitrification and anaerobic ammonium oxidation co-culture. Science of the Total Environment 651: 291-297. doi: 10.1016/j.scitotenv.2018.09.184.

Gajaraj, S. and Hu, Z. 2014. Integration of microbial fuel cell techniques into activated sludge wastewater treatment processes to improve nitrogen removal and reduce sludge production. Chemosphere 117: 151-157. doi: 10.1016/j.chemosphere.2014.06.013.

Gao, C., Liu, L. and Yang, F. 2018. Novel carbon fiber cathode membrane with Fe/Mn/C/F/O elements in bio-electrochemical system (BES) to enhance wastewater treatment. Journal of Power Sources 379: 123-133. doi: 10.1016/j.jpowsour.2018.01.037.

Gonzalez-Martinez, A., Muñoz-Palazon, B., Rodriguez-Sanchez, A. and Gonzalez-Lopez, J. 2018a. New concepts in anammox processes for wastewater nitrogen removal: recent advances and future prospects. FEMS Microbiology Letters 365: 1-10. doi: 10.1093/femsle/fny031.

Gonzalez-Martínez, A., Chengyuan, S., Rodriguez-Sanchez, A., Pozo, C., Gonzalez-Lopez, J. and Vahala, R. 2018b. Application of microbial fuel cell technology for wastewater treatment and electricity generation under Nordic countries climate conditions: study of performance and microbial communities. Bioresource Technology 270: 1-10. doi: 10.1016/j.biortech.2018.09.014.

Gregory, K.B., Bond, D.R. and Lovley, D.R. 2004. Graphite electrodes as electron donors for anaerobic respiration. Environmental Microbiology 6: 596-604. doi: 10.1111/j.1462-2920.2004.00593.x.

Gu, J., Yang, Q. and Liu, Y. 2018. Mainstream anammox in a novel A-2B process for energy-efficient municipal wastewater treatment with minimized sludge production. Water Research 138: 1-6. doi: 10.1016/j.watres.2018.02.051.

Guadarrama-Pérez, O., Gutiérrez-Macías, T., García-Sánchez, L., Guadarrama-Pérez, V.H. and Estrada-Arriaga, E.B. 2019. Recent advances in constructed wetland-microbial fuel cells for simultaneous bioelectricity production and wastewater treatment: a review. International Journal of Energy Research 43: 5106-5127. doi: 10.1002/er.4496.

Gude, V.G. 2016. Wastewater treatment in microbial fuel cells - an overview. Journal of Cleaner Production 122: 287-307. doi: 10.1016/j.jclepro.2016.02.022.

Guo, J., Cheng, J., Li, B., Wang, J. and Chu, P. 2019. Performance and microbial community in the biocathode of microbial fuel cells under different dissolved oxygen concentrations. Journal of Electroanalytical Chemistry 833: 433-440. doi: 10.1016/j.jelechem.2018.12.015.

Hari, A.R., Katuri, K.P., Gorron, E., Logan, B.E. and Saikaly, P.E. 2016a. Multiple paths of electron flow to current in microbial electrolysis cells fed with low and high concentrations of propionate. Applied Microbiology and Biotechnology 100: 5999-6011. doi: 10.1007/s00253-016-7402-2.

Hari, A.R., Katuri, K.P., Logan, B.E. and Saikaly, P.E. 2016b. Set anode potentials affect the electron fluxes and microbial community structure in propionate-fed microbial electrolysis cells. Scientific Reports 6: 1-11. doi: 10.1038/srep38690.

Haroon, M.F., Hu, S., Shi, Y., Imelfort, M., Keller, J., Hugenholtz, P., et al. 2013. Anaerobic oxidation of methane coupled to nitrate reduction in a novel archaeal lineage. Nature 500: 567-570. doi: 10.1038/nature12375.

He, Z., Geng, S., Shen, L., Lou, L., Zheng, P., Xu, X., et al. 2015a. The short- and long-term effects of environmental conditions on anaerobic methane oxidation coupled to nitrite reduction. Water Research 68: 554-562. doi: 10.1016/j.watres.2014.09.055.

He, Z., Wang, J., Zhang, X., Cai, C., Geng, S., Zheng, P., et al. 2015b. Nitrogen removal from wastewater by anaerobic methane-driven denitrification in a lab-scale reactor: heterotrophic denitrifiers associated with denitrifying methanotrophs. Applied Microbiology and Biotechnology 99: 10853-10860. doi: 10.1007/s00253-015-6939-9.

Hoekstra, M., Geilvoet, S.P., Hendrickx, T.L.G., van Erp Taalman Kip, C.S., Kleerebezem, R. and van Loosdrecht, M.C.M. 2019. Towards mainstream anammox: lessons learned from pilot-scale research at WWTP Dokhaven. Environmental Technology (United Kingdom) 40: 1721-1733. doi: 10.1080/09593330.2018.1470204.

Hou, D., Lu, L. and Ren, Z.J. 2016. Microbial fuel cells and osmotic membrane bioreactors have mutual benefits for wastewater treatment and energy production. Water Research 98: 183-189. doi: 10.1016/j. watres.2016.04.017.

Hou, D., Lu, L., Sun, D., Ge, Z., Huang, X., Cath, T.Y., et al. 2017. Microbial electrochemical nutrient recovery in anaerobic osmotic membrane bioreactors. Water Research 114: 181-188. doi: 10.1016/j. watres.2017.02.034.

Hu, Z., Speth, D.R., Francoijs, K.J., Quan, Z.X. and Jetten, M.S.M. 2012. Metagenome analysis of a complex community reveals the metabolic blueprint of anammox bacterium "Candidatus Jettenia asiatica." Frontiers in Microbiology 3: 1-9. doi: 10.3389/fmicb.2012.00366.

Hu, Z. and Ma, R. 2016. Distribution and characteristic of nitrite-dependent anaerobic methane oxidation bacteria by comparative analysis of wastewater treatment plants and agriculture fields in northern China. Peer J 4: e2766. doi: 10.7717/peerj.2766.

Hu, Z., Ru, D., Wang, Y., Zhang, J., Jiang, L., Xu, X., et al. 2019. Optimization of a nitrite-dependent anaerobic methane oxidation (n-damo) process by enhancing methane availability. Bioresource Technology 275: 101-108. doi: 10.1016/j.biortech.2018.12.035.

Huang, L., Regan, J.M. and Quan, X. 2011. Electron transfer mechanisms, new applications, and performance of biocathode microbial fuel cells. Bioresource Technology 102: 316-323. doi: 10.1016/j. biortech.2010.06.096.

Hui, C., Guo, X., Sun, P., Lin, H., Zhang, Q., Liang, Y., et al. 2017. Depth-specific distribution and diversity of nitrite-dependent anaerobic ammonium and methane-oxidizing bacteria in upland-cropping soil under different fertilizer treatments. Applied Soil Ecology 113: 117-126. doi: 10.1016/j.apsoil.2017.02.005.

Jain, A. and He, Z. 2018. Cathode-enhanced wastewater treatment in bioelectrochemical systems. npj Clean Water 1: 1-5. doi: 10.1038/s41545-018-0022-x.

Joss, A., Salzgeber, D., Eugster, J., König, R., Rottermann, K., Burger, S., et al. 2009. Full-scale nitrogen removal from digester liquid with partial nitritation and anammox in one SBR. Environmental Science and Technology 43(14): 5301-5306.

Kampman, C., Hendrickx, T.L.G., Luesken, F.A., van Alen, T.A., Op den Camp, H.J.M., Jetten, M.S.M., et al. 2012. Enrichment of denitrifying methanotrophic bacteria for application after direct low-temperature anaerobic sewage treatment. Journal of Hazardous Materials 227-228, 164-171. doi: 10.1016/j. jhazmat.2012.05.032.

Kampman, C., Temmink, H., Hendrickx, T.L.G., Zeeman, G. and Buisman, C.J.N. 2014. Enrichment of denitrifying methanotrophic bacteria from municipal wastewater sludge in a membrane bioreactor at 20°C. Journal of Hazardous Materials 274: 428-435. doi: 10.1016/j.jhazmat.2014.04.031.

Kampschreur, M.J., Temmink, H., Kleerebezem, R., Jetten, M.S.M. and van Loosdrecht, M.C.M. 2009. Nitrous oxide emission during wastewater treatment (2009). Water Research 43(17): 4093-4103.

Katuri, K.P., Werner, C.M., Jimenez-Sandoval, R.J., Chen, W., Jeon, S., Logan, B.E., et al. 2014. A novel anaerobic electrochemical membrane bioreactor (AnEMBR) with conductive hollow-fiber membrane for treatment of low-organic strength solutions. Environmental Science and Technology 48: 12833-12841. doi: 10.1021/es504392n.

Katuri, K.P., Ali, M. and Saikaly, P.E. 2019. The role of microbial electrolysis cell in urban wastewater treatment: integration options, challenges, and prospects. Current Opinion in Biotechnology 57: 101-110. doi: 10.1016/j.copbio.2019.03.007.

Kondaveeti, S., Kang, E., Liu, H. and Min, B. 2019. Continuous autotrophic denitrification process for treating ammonium-rich leachate wastewater in bioelectrochemical denitrification system (BEDS). Bioelectrochemistry 130: 107340. doi: 10.1016/j.bioelechem.2019.107340.

Lee, K.C. and Rittmann, B.E. 2002. Applying a novel autohydrogenotrophic hollow-fiber membrane biofilm reactor for denitrification of drinking water. Water Research 36(8): 2040-2052. ISSN 0043-1354, https:// doi.org/10.1016/S0043-1354(01)00425-0).

Leong, J.X., Daud, W.R.W., Ghasemi, M., Liew, K. and Ben Ismail, M. 2013. Ion exchange membranes as separators in microbial fuel cells for bioenergy conversion: a comprehensive review. Renewable Sustainable Energy Reviews 28: 575-587. doi: 10.1016/j.rser.2013.08.052.

Li, H., Qu, Y., Tian, Y. and Feng, Y. 2019. The plant-enhanced bio-cathode: root exudates and microbial community for nitrogen removal. Journal of Environmental Science (China) 77: 97-103. doi: 10.1016/j. jes.2018.06.018.

Li, W., Lu, P., Chai, F., Zhang, L., Han, X. and Zhang, D. 2018. Long-term nitrate removal through methane-dependent denitrification microorganisms in sequencing batch reactors fed with only nitrate and methane. AMB Express 8(1): 108. doi: 10.1186/s13568-018-0637-9.

Liu, F., Sun, L., Wan, J., Tang, A., Deng, M. and Wu, R. 2019. Organic matter and ammonia removal by a novel integrated process of constructed wetland and microbial fuel cells. RSC Advances 9: 5384-5393. doi: 10.1039/c8ra10625h.

Liu, J., Zhang, F., He, W., Yang, W., Feng, Y. and Logan, B.E. 2014a. A microbial fluidized electrode electrolysis cell (MFEEC) for enhanced hydrogen production. Journal of Power Sources 271: 530-533. doi: 10.1016/j.jpowsour.2014.08.042.

Liu, J., Zhang, F., He, W., Zhang, X., Feng, Y. and Logan, B.E. 2014b. Intermittent contact of fluidized anode particles containing exoelectrogenic biofilms for continuous power generation in microbial fuel cells. Journal of Power Sources 261: 278-284. doi: 10.1016/j.jpowsour.2014.03.071.

Liu, T., Hu, S., Yuan, Z. and Guo, J. 2019. High-level nitrogen removal by simultaneous partial nitritation, anammox and nitrite/nitrate-dependent anaerobic methane oxidation. Water Research 166: 115057. doi: 10.1016/j.watres.2019.115057.

Luesken, F.A., Sánchez, J., van Alen, T.A., Sanabria, J., Op den Camp, H.J.M., Jetten, M.S.M., et al. 2011a. Simultaneous nitrite-dependent anaerobic methane and ammonium oxidation processes. Applied Environmental Microbiology 77: 6802-6807. doi: 10.1128/AEM.05539-11.

Luesken, F.A., Van Alen, T.A., Van Der Biezen, E., Frijters, C., Toonen, G., Kampman, C., et al. 2011b. Diversity and enrichment of nitrite-dependent anaerobic methane oxidizing bacteria from wastewater sludge. Applied Microbiology and Biotechnology 92: 845-854. doi: 10.1007/s00253-011-3361-9.

Luesken, F.A., Wu, M.L., Op den Camp, H.J.M., Keltjens, J.T., Stunnenberg, H., Francoijs, K.J., et al. 2012. Effect of oxygen on the anaerobic methanotroph "Candidatus Methylomirabilis oxyfera": kinetic and transcriptional analysis. Environmental Mirobiology 14: 1024-1034. doi: 10.1111/j.1462-2920.2011.02682.x.

Ma, R., Hu, Z., Zhang, J., Ma, H., Jiang, L. and Ru, D. 2017. Reduction of greenhouse gases emissions during anoxic wastewater treatment by strengthening nitrite-dependent anaerobic methane oxidation process. Bioresource Technology 235: 211-218. doi: 10.1016/j.biortech.2017.03.094.

Meng, H., Wang, Y.F., Chan, H.W., Wu, R.N. and Gu, J.D. 2016. Co-occurrence of nitrite-dependent anaerobic ammonium and methane oxidation processes in subtropical acidic forest soils. Applied Microbiology and Biotechnology 100: 7727-7739. doi: 10.1007/s00253-016-7585-6.

Mook, W.T., Aroua, M.K.T., Chakrabarti, M.H., Noor, I.M., Irfan, M.F. and Low, C.T.J. 2013. A review on the effect of bio-electrodes on denitrification and organic matter removal processes in bio-electrochemical systems. Journal of Industrial Engineering and Chemistry 19: 1-13. doi: 10.1016/j.jiec.2012.07.004.

Muñoz-Palazon, B., Rodriguez-Sanchez, A., Castellano-Hinojosa, A., Gonzalez-Lopez, J., van Loosdrecht, M.C.M., Vahala, R., et al. 2018. Quantitative and qualitative studies of microorganisms involved in full-scale autotrophic nitrogen removal performance. AIChE Journal 64(2): 457-467. doi: 10.1002/aic.15925.

Nitisoravut, R. and Regmi, R. 2017. Plant microbial fuel cells: a promising biosystems engineering. Renewable Sustainable Energy Reviews 76: 81-89. doi: 10.1016/j.rser.2017.03.064.

Palanisamy, G., Jung, H.Y., Sadhasivam, T., Kurkuri, M.D., Kim, S.C. and Roh, S.H. 2019. A comprehensive review on microbial fuel cell technologies: processes, utilization, and advanced developments in electrodes and membranes. Journal of Cleaner Production 221: 598-621. doi: 10.1016/j.jclepro.2019.02.172.

Park, J., Lee, B., Tian, D. and Jun, H. 2018. Bioelectrochemical enhancement of methane production from highly concentrated food waste in a combined anaerobic digester and microbial electrolysis cell. Bioresource Technology 247: 226-233. doi: 10.1016/j.biortech.2017.09.021.

Park, Y., Park, S., Nguyen, V.K., Yu, J., Torres, C.I., Rittmann, B.E., et al. 2017. Complete nitrogen removal by simultaneous nitrification and denitrification in flat-panel air-cathode microbial fuel cells treating domestic wastewater. Chemical Engineering Journal 316: 673-679. doi: 10.1016/j.cej.2017.02.005.

Puig, S., Serra, M., Vilar-Sanz, A., Cabré, M., Bañeras, L., Colprim, J., et al. 2011. Autotrophic nitrite removal in the cathode of microbial fuel cells. Bioresource Technology 102: 4462-4467. doi: 10.1016/j.biortech.2010.12.100.

Qu, B., Fan, B., Zhu, S. and Zheng, Y. 2014. Anaerobic ammonium oxidation with an anode as the electron acceptor. Environmental Microbiology Reports 6: 100-105. doi: 10.1111/1758-2229.12113.

Raghoebarsing, A.A., Pol, A., Van De Pas-Schoonen, K.T., Smolders, A.J.P., Ettwig, K.F., Rijpstra, W.I.C., et al. 2006. A microbial consortium couples anaerobic methane oxidation to denitrification. Nature 440: 918-921. doi: 10.1038/nature04617.

Ren, L., Ahn, Y. and Logan, B.E. 2014. A two-stage microbial fuel cell and anaerobic fluidized bed membrane bioreactor (MFC-AFMBR) system for effective domestic wastewater treatment. Environmental Science and Technology 48: 4199-4206. doi: 10.1021/es500737m.

Ren, Y., Hao Ngo, H., Guo, W., Wang, D., Peng, L., Ni, B.J., et al. 2019. New perspectives on microbial communities and biological nitrogen removal processes in wastewater treatment systems. Bioresource Technology 297: 122491. doi: 10.1016/j.biortech.2019.122491.

Rosenbaum, M., Aulenta, F., Villano, M. and Angenent, L.T. 2011. Cathodes as electron donors for microbial metabolism: which extracellular electron transfer mechanisms are involved? Bioresource Technology 102: 324-333. doi: 10.1016/j.biortech.2010.07.008.

Sevda, S., Sreekishnan, T.R., Pous, N., Puig, S. and Pant, D. 2018. Bioelectroremediation of perchlorate and nitrate contaminated water: a review. Bioresource Technology 255: 331-339. doi: 10.1016/j.biortech.2018.02.005.

Shen, L.-D., Wu, H.-S. and Gao, Z.-Q. 2014. Distribution and environmental significance of nitrite-dependent anaerobic methane-oxidising bacteria in natural ecosystems. Applied Microbiology and Biotechnology 99: 133-142. doi: 10.1007/s00253-014-6200-y.

Shen, L.-D., He, Z.-F., Wu, H.-S. and Gao, Z.-Q. 2015. Nitrite-dependent anaerobic methane-oxidising bacteria: unique microorganisms with special properties. Current Microbiology 70: 562-570. doi: 10.1007/s00284-014-0762-x.

Shen, L., Ouyang, L., Zhu, Y. and Trimmer, M. 2019. Spatial separation of anaerobic ammonium oxidation and nitrite-dependent anaerobic methane oxidation in permeable riverbeds. Environmental Microbiology 21: 1185-1195. doi: 10.1111/1462-2920.14554.

Shen, L.D., He, Z.F., Zhu, Q., Chen, D.Q., Lou, L.P., Xu, X.Y., et al. 2012. Microbiology, ecology, and application of the nitrite-dependent anaerobic methane oxidation process. Frontiers in Microbiology 3: 1-5. doi: 10.3389/fmicb.2012.00269.

Shi, S., Xu, J., Zeng, Q., Liu, J., Hou, Y. and Jiang, B. 2019. Impacts of applied voltage on EMBR treating phenol wastewater: performance and membrane antifouling mechanism. Bioresource Technology 282: 56-62. doi: 10.1016/j.biortech.2019.02.113.

Sophia, A.C. and Sreeja, S. 2017. Green energy generation from plant microbial fuel cells (PMFC) using compost and a novel clay separator. Sustainable Energy Technology Assessments 21: 59-66. doi: 10.1016/j.seta.2017.05.001.

Stultiens, K., Cruz, S.G., van Kessel, M.A.H.J., Jetten, M.S.M., Kartal, B. and Op den Camp, H.J.M. 2019. Interactions between anaerobic ammonium- and methane-oxidizing microorganisms in a laboratory-scale sequencing batch reactor. Applied Microbiology and Biotechnology 103: 6783-6795. doi: 10.1007/s00253-019-09976-9.

Tao, M., Guan, L., Jing, Z., Tao, Z., Wang Yue Luo, H. and Wang, Yin. 2020. Enhanced denitrification and power generation of municipal wastewater treatment plants (WWTPs) effluents with biomass in microbial fuel cell coupled with constructed wetland. Science of the Total Environment 709: 136159. doi: 10.1016/j.scitotenv.2019.136159.

Ucar, D., Zhang, Y. and Angelidaki, I. 2017. An overview of electron acceptors in microbial fuel cells. Frontiers in Microbiology 8: 1-14. doi: 10.3389/fmicb.2017.00643.

van Kessel, M.A., Stultiens, K., Slegers, M.F., Guerrero Cruz, S., Jetten, M.S., Kartal, B., et al. 2018. Current perspectives on the application of N-damo and anammox in wastewater treatment. Current Opinion in Biotechnology 50: 222-227. doi: 10.1016/j.copbio.2018.01.031.

Vijay, A., Chhabra, M. and Vincent, T. 2019. Microbial community modulates electrochemical performance and denitrification rate in a biocathodic autotrophic and heterotrophic denitrifying microbial fuel cell. Bioresource Technology 272: 217-225. doi: 10.1016/j.biortech.2018.10.030.

Vilajeliu-Pons, A., Koch, C., Balaguer, M.D., Colprim, J., Harnisch, F. and Puig, S. 2018. Microbial electricity driven anoxic ammonium removal. Water Research 130: 168-175. doi: 10.1016/j.watres.2017.11.059.

Virdis, B., Rabaey, K., Rozendal, R.A., Yuan, Z. and Keller, J. 2010. Simultaneous nitrification, denitrification and carbon removal in microbial fuel cells. Water Research 44: 2970-2980. doi: 10.1016/j.watres.2010.02.022.

Wang, D., Wang, Y., Liu, Y., Ngo, H.H., Lian, Y., Zhao, J., et al. 2017a. Is denitrifying anaerobic methane oxidation-centered technologies a solution for the sustainable operation of wastewater treatment Plants? Bioresource Technology 234: 456-465. doi: 10.1016/j.biortech.2017.02.059.

Wang, J., Hua, M., Li, Y., Ma, F., Zheng, P. and Hu, B. 2019. Achieving high nitrogen removal efficiency by optimizing nitrite-dependent anaerobic methane oxidation process with growth factors. Water Research 161: 35-42. doi: 10.1016/j.watres.2019.05.101.

Wang, S., Zhao, J., Liu, S., Zhao, R. and Hu, B. 2018. Effect of temperature on nitrogen removal and electricity generation of a dual-chamber microbial fuel cell. Water, Air and Soil Pollution 229: 244. doi: 10.1007/s11270-018-3840-z.

Wang, X., Tian, Y., Liu, H., Zhao, X. and Peng, S. 2019. The influence of incorporating microbial fuel cells on greenhouse gas emissions from constructed wetlands. Science of the Total Environment 656: 270-279. doi: 10.1016/j.scitotenv.2018.11.328.

Wang, Y., Wang, D., Yang, Q., Zeng, G. and Li, X. 2017b. Wastewater opportunities for denitrifying anaerobic methane oxidation. Trends in Biotechnology 35: 799-802. doi: 10.1016/j.tibtech.2017.02.010.

Wang, Y.F., Dick, R.P., Lorenz, N. and Lee, N. 2019. Interactions and responses of n-damo archaea, n-damo bacteria and anammox bacteria to various electron acceptors in natural and constructed wetland sediments. International Biodeterioration and Biodegradation 144: 104749. doi: 10.1016/j.ibiod.2019.104749.

Watanabe, K. 2008. Recent developments in microbial fuel cell technologies for sustainable bioenergy. Journal of Bioscience and Bioengineering 106: 528-536. doi: 10.1263/jbb.106.528.

Wu, J. and Zhang, Y. 2017. Evaluation of the impact of organic material on the anaerobic methane and ammonium removal in a membrane aerated biofilm reactor (MABR) based on the multispecies biofilm modeling. Environmental Science and Pollution Research 24: 1677-1685. doi: 10.1007/s11356-016-7938-9.

Zhou, Y., Zhao, S., Yin, L., Zhang, J., Bao, Y. and Shi, H. 2018. Development of a novel membrane-less microbial fuel cell (ML-MFC) with a sandwiched nitrifying chamber for efficient wastewater treatment. Electroanalysis 30: 2145-2152. doi: 10.1002/elan.201800232.

Zhu, B., Sánchez, J., Van Alen, T.A., Sanabria, J., Jetten, M.S.M., Ettwig, K.F., et al. 2011. Combined anaerobic ammonium and methane oxidation for nitrogen and methane removal. Biochemical Society Transactions 39: 1822-1825. doi: 10.1042/BST20110704.

# 12

# Influence of Cover Crops on Nitrogen Cycling and the Soil Microbial Community

Antonio Castellano-Hinojosa, Clayton J. Nevins and
Sarah L. Strauss[*]

## 12.1 INTRODUCTION

Nitrogen (N) availability is a major limiting factor for plant growth and crop production (Ågren et al. 2012, Rütting et al. 2018). In addition, cultivation reduces soil organic carbon (SOC) and organic N (Ramesh et al. 2019). These limitations generally require significant N inputs by farmers to ensure adequate N availability for crops. Increased interest in improving the sustainability of agriculture and availability of N has led to renewed attention in cover cropping as an agricultural practice with benefits for growers and the environment (Couëdel et al. 2018, García-González et al. 2018, Groff 2015, Kaye and Quemada 2017, Ordóñez-Fernández et al. 2018, Roth et al. 2018, Vukicevich et al. 2016, White et al. 2016). According to the Soil Science Society of America, cover crops are defined as "a close-growing crop that provides soil protection, seeding protection, and soil improvement between periods of normal crop production, or between trees in orchards and vines in vineyards. When plowed under and incorporated into the soil, cover crops may be referred to as green manure crops" (Glossary of Soil Science Terms 2008).

Historically, cover crops have been used to provide specific production and environmental benefits beyond those involved in N-cycling, including control of weeds, decreased runoff and soil erosion, improved phosphorus retention, increased SOC, and maintenance of soil aggregate stability (Hubbard et al. 2013, White et al. 2016). The below-ground microbial communities associated with cover crops are also fundamental for maintaining and improving soil health, enhancing crop productivity, improving accessibility to low-abundance nutrients, and coping with a range of (a) biotic stressors that affect their associated cash crop (Mercado-Blanco et al. 2018, Vukicevich et al. 2016).

One of the primary benefits of cover crops is their impact on the N-cycle, including providing additional N, reducing nitrate ($NO_3^-$) leaching, and improving soil N retention. The influence of cover crops on soil N depends on biotic (i.e., cover crop species and management) and abiotic (i.e., irrigation management, temperature, light, soil N availability, and application of N fertilizers) factors. In cropping systems, the main routes of belowground N transfer can be characterized as (i) the exudation of soluble N compounds by plant roots and uptake by receivers (Thilakarathna et al. 2016), (ii) decomposition of above- and below-ground legume residues followed by the uptake

---

[*] University of Florida IFAS Southwest Florida Research and Education Center, 2685 State Rd 29N, Immokalee, FL.
[*] Corresponding author: strauss@ufl.edu

of released N by neighboring plants (Fustec et al. 2010), and (iii) transfer of N mediated by plant-associated mycorrhizae (He et al. 2009). The benefits of cover crops on soil nutrients in general (Blanco-Canqui et al. 2015, Cherr et al. 2006, Dabney et al. 2001, Kaspar and Singer 2011) and considerations for manipulating them in agroforestry systems (Munroe and Isaac 2014, Vukicevich et al. 2016) and herbaceous crops (Thilakarathna et al. 2016) have been previously reported.

However, the influence of cover crops on each of the key transformations of N-cycling and their associated soil microbial community has not been reviewed. With the development of fast and efficient DNA sequencing technologies (e.g., high-throughput sequencing of the bacterial 16S rRNA gene and fungal ITS region) and the application of molecular-based methods, such as the absolute and relative quantification of microbial gene abundance (e.g., quantitative PCR or qPCR), the process of unraveling how changes in the structure and composition of the soil microbial community in cropping systems, with emphasis on N-cycling (Mercado-Blanco et al. 2018), has begun. Thus, a summarization of the existing knowledge about potential cover crops' effects on N-cycling is needed to provide a broader understanding of cover crop impacts on soil and agricultural production and identify knowledge gaps that require further research.

## 12.2 COVER CROP IMPACTS ON SOIL NITROGEN CYCLING

### 12.2.1 Nitrogen Inputs

#### 12.2.1.1 *Fixation by Legumes*

There are approximately 20,000 species in the Leguminosae plant family (Peix et al. 2015), and a subset of plants in this family are routinely incorporated into cover crop rotations. Legumes can form symbiotic relationships with rhizobia (Shamseldin et al. 2017), gram-negative aerobic bacteria that are free-living in many soils and are critical to the contribution of legumes to the N-cycle. To ensure the availability of rhizobia to infect legume cover crops, inoculation of cover crop seeds with plant-specific rhizobia is recommended, especially when the legume being planted has not been grown in the field in the past two to three years (Shamseldin et al. 2017, Velázquez et al. 2017). Nevertheless, competition between resident rhizobia and adaptability of inoculated microbes have been reported, which may affect rhizobia migration and colonization dynamics (Checcucci et al. 2017, Miao et al. 2018).

Legumes excrete carbohydrate-containing plant proteins from their roots, including flavonoids and lectins such as phytoagglutinins, that are favored by and attract rhizobia (Downie 2010). In response, the rhizobia secrete Nod factors that recognize the plant roots and bind to the plant root hair cells (Geurts et al. 2005, Wang et al. 2018). This leads to a swelling of the root hairs, which encompasses the rhizobia. The root cortex cells divide, and nodules are formed which are then colonized by the rhizobia. The rhizobia multiply in the nodules using C sources from the plant for energy (Jones et al. 2007). In return, they complete biological dinitrogen gas ($N_2$) fixation, the conversion of atmospheric $N_2$ into plant-available ammonia ($NH_3$). The $NH_3$ is almost immediately converted into ammonium ($NH_4^+$) and is used by the plant to create proteins. When the cover crop is terminated and the cover crop biomass is incorporated into the soil, microorganisms begin the process of N mineralization by decomposing the organic N in the legume biomass. After mineralization, this N is available in the soils for future plant uptake (Figure 12.1).

Legumes are routinely grown as cover crops in agroecosystems after cash crop harvest and before planting the succeeding cash crop. This is practiced with the expectation that the legume will add N to the soil after the biomass undergoes N mineralization. However, there are many factors that impact the potential soil N addition from legume cover crops. For example, legume species do not equally provide N to soil (Stagnari et al. 2017, Vitousek et al. 2013). While it is currently not known why there are differences between species, numerous factors including climatic conditions,

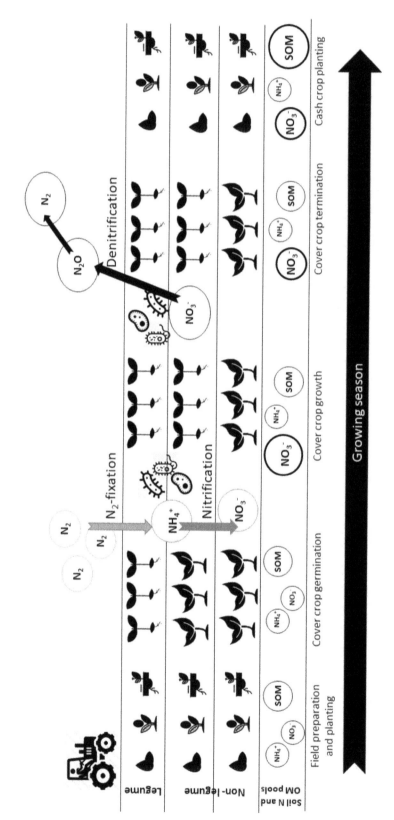

**FIGURE 12.1** The contribution of cover crops to the cycle of SOM formation. Bacteria utilize cover crop plant biomass as a C source. Over time, substrates become limiting, and the bacteria begin to starve and die. The cell fragments interact with soil minerals and form a substantial part of the SOM. Eventually, decomposition of SOM will contribute to the formation of ammonium and/or nitrate. Plant roots, both the cover crops and cash crops, release as exudates that also contribute to the SOM.

soil pH, and salinity may play a role. For example, summer annual legumes such as cowpeas (*Vigna unguiculata*), soybeans (*Glycine max*), and sunn hemp (*Crotalaria juncea*) typically have lower N concentrations than winter annual legumes including crimson clover (*Trifolium incarnatum*) and hairy vetch (*Vicia villosa*). However, summer legume cover crops can produce higher amounts of N-enriched biomass than winter legume cover crops because summer legumes can grow rapidly in late summer, resulting in a greater effect on soil N (Blanco-Canqui et al. 2015). Differences in biological $N_2$-fixation are also found between varieties of the same legume species. For example, Mokgehle et al. (2014) found significant differences in the N-fixed by 25 varieties of groundnut (*Arachis hypogaea*), which ranged between 76 and 188 kg ha$^{-1}$.

The amount of N added to the soil by a legume is also dependent on soil and growing conditions (Bordeleau and Prévost 1994). Before legumes establish a symbiotic relationship with rhizobia to receive N, the plant assimilates plant-available N ($NO_3^-$ and $NH_4^+$) from the soil (Flynn and Idowu 2015) (Figure 12.2). Most of this N is returned to the soil during legume decomposition. However, in low organic matter soil, N fertilizer application at the time of legume germination may be required to meet the N demand of the legume before $N_2$-fixation occurs (Flynn and Idowu 2015). Nevertheless, the N contribution from biological $N_2$-fixation to subsequent crop cycles is higher in situations of low N fertilization (Preissel et al. 2015). Indeed, high rates of N fertilization commonly suppress $N_2$-fixation since N-fixing bacteria are typically facultative (Reed et al. 2011).

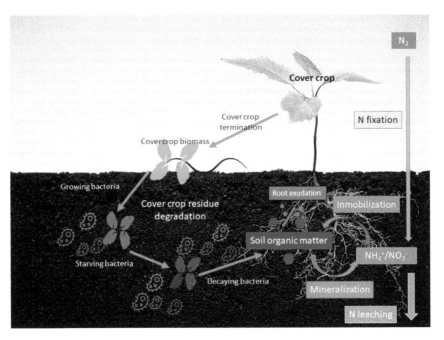

**FIGURE 12.2** Schematic representation of the influence of cover crops on N-cycling and the soil microbial community during a growing season. Legume cover crops increase $NH_4^+$ pool in soil trough their symbiotic association with $N_2$-fixing microorganisms which convert atmospheric N into plant-available $NH_4^+$. Nitrifying microorganisms convert $NH_4^+$ to $NO_3^-$, which is then partially reduced by denitrifying bacteria to gaseous N forms such as $N_2O$ and $N_2$. Thus, a substantial part of the $NO_3^-$ pool remains available for the cash crop planting. Changes in the SOM pool when planting cover crops are illustrated in Figure 1.

In natural grasslands, it has been observed that increasing the number of legume species from one to four in an area provides additional N benefits (Finn et al. 2013). Applying this idea to cover crop mixtures, it has been demonstrated that carefully planned mixtures of a few complementary species (one to three legume species) may provide the same or more cover crop services than a mixture with two-three legume species (Blesh 2018). However, it is challenging to pinpoint an optimal number of legume species in cover mixtures because, just as with cash crops, there is variability in cover crop establishment and growth by year due to weather and management practices. This leads to year-to-year differences in the contribution of each species to a cover crop mixture that can impact the desired cover crop service production (Blesh, 2018). Future research should investigate the contribution of different legume cover crop species to N-cycling when grown in legume mixtures.

### 12.2.1.2 Availability of N to Cropping Systems

In vegetable cropping systems, legume cover crops such as alfalfa (*Medicago sativa*) (Dhont et al. 2006), pea (*Pisum sativum*) (Ketterings et al. 2015), clover (*Trifolium* sp.) (Ketterings et al. 2015, Vyn et al. 2000), soybean, sunn hemp (Finney et al. 2016), and vetch (*Vicia villosa*) (Alonso-Ayuso et al. 2014, Clark et al. 1997, Gabriel et al. 2013, Poffenbarger et al. 2015a, Seo et al. 2000) can accumulate between 5-30 g N/m² during fall and winter seasons. One study found that after four cycles of cover and cash crop rotations, soil total N (TN) content increased by 258 and 279 kg ha⁻¹ where soybean and sunn hemp cover crops were planted, respectively (Blanco-Canqui et al. 2015). However, the amount of N that is available for the subsequent crop can depend on cover crop incorporation method after termination and termination timing. The amount of cover crop N that remains available for the subsequent crop can be roughly estimated by dividing the above-ground legume cover crop N by two if the cover crops are conventionally tilled or no-tilled in Southern climates or by four if they are no-tilled in Northern climates (Clark 2007). However, when cover crops are not incorporated into the soil, the available N might be only one-third of the cover crop N (Gabriel et al. 2013, Seo et al. 2006). The timing between termination and subsequent crop planting can also impact the amount of available N. For example, the fraction of cover crop N that becomes available to subsequent vegetable crops has been shown to increase from 27-41% by prolonging the time (from 6 to 20 days) between cover crop death and subsequent crop planting (White et al. 2016). This is due to the long decomposition period of cover crop residues between cover crop termination and physiological maturity of the subsequent cash crop, which allows for the complete turnover of cover crop residue C through the microbial biomass. Additional information on mineralization and immobilization are in the sections that follow.

In tree cropping systems, the introduction of cover crops such as *Stylosanthes guianensis* and *Paspalum natatu* (Cui et al. 2015), crown vetch (*Coronilla varia*) (Capó-Bauçà et al. 2019, Cui et al. 2015, Gao et al. 2019, Liu et al. 2019, Pereg et al. 2018, Gong et al. 2018, Zhao et al. 2018a), vetch (Pereg et al. 2018), *Avena* sp., *Cynodon dactylon*, *Medicago polymorpha* and *Hordeum murinum* (Capó-Bauçà et al. 2019), wheat (*Triticum aestivum*) (Gao et al. 2019) and *Flemingia macrophylla* (Liu et al. 2019) have been related to increased SOC and N content in apple (*Malus pumila*) (Gong et al. 2018), guava (*Psidium guajava*) (Cui et al. 2015), grapevine (*Vitis vinifera*) (Capó-Bauçà et al. 2019, Pereg et al. 2018) rubber (*Eucalyptus urophylla*) (Liu et al. 2019), and walnut (*Juglans regia*)-based (Gao et al. 2019) agroforestry systems. However, additional studies are needed to make generalizations regarding the impact of cover crops on soil N availability for perennial tree crops due to the complexity of agroforestry systems and the limited number of trees species and cover crops examined.

## 12.2.2 Nitrogen Immobilization and Mineralization

Cover crops are terminated prior to cash crop planting and the biomass is either left on the surface of the soil (no-tillage) or incorporated into the soil using a tillage application. After termination, the rate of decomposition of the cover crop biomass determines when the N from cover crops is added to the soil and can differ based on management practices, cover crop species, and tillage application (Melkonian et al. 2017, Poffenbarger et al. 2015b). The soil microbial community mediates the decomposition of the cover crop residue as microorganisms utilize the N to assemble proteins and nucleic acids necessary to break down the plant material. Bacteria and archaea have narrower C:N ratios (C:N 3-6) than fungi (C:N 5-17) and require more N per unit of biomass (Cleveland and Liptzin 2007, Strickland and Rousk 2010). However, cover crops do not always have enough N in their biomass to meet the N demands of the soil microbial community to facilitate decomposition (Cabrera et al. 2005). For example, grasses such as annual ryegrass and winter ryegrass have wide C:N ratios ranging from 30-60 parts C to 1 part N (Finney et al. 2016, Wagger. et al. 1998). This influx of C into the soil requires more N for microbial activity than is supplied by the cover crop. In this case, the soil microbial community must immobilize N from the soil solution to meet internal N demands to decompose the cover crop. This organic N in the microbial biomass is released to the soil after cover crop decomposition and microbial decay.

Nitrogen mineralization occurs when microbial biomass N demands are met by the cover crop residue, allowing rapid decomposition to occur without N being immobilized from the soil or cover crop. Legumes, which have narrower C:N ratios than non-legume cover crops, are mineralized more rapidly than non-legumes, such as grasses (Lynch et al. 2016, Sievers and Cook 2018). Compared to non-legume cover crops, legumes have a greater potential to provide resources similar to a fertilizer because their residues usually have C:N ratios lower than 20:1 (Abiven et al. 2005, Corbeels et al. 2003, Thilakarathna et al. 2016). A C:N ratio greater than 35:1 results in a higher N mineralization process followed by a slow N release rate, whereas a C:N ratio lower than this carries a faster N release rate (Trinsoutrot et al. 2000). In fact, one method to lower the C:N of the residues to promote higher N availability to the subsequent crop is to plant grass-legume cover crop mixtures (Blesh et al. 2019, Chapagain and Riseman 2014, 2015, Thilakarathna et al. 2016, White et al. 2017). For example, vetch-rye cropping rotation lowered the C:N ratio from about 60:1 for rye to approximately 17:1 for the mixture, while vetch alone had a C:N ratio of approximately 11:1 (Clark 2007). This reduction in C:N also occurred regardless of the termination method used, as cover crops (rye + hairy vetch) decreased the C:N ratio from 23:1 to 17:1 (Lawson et al. 2015). The decreased C:N ratio of this mixture can be attributed to the cover crops being planted together, as there was a slower decrease in the C:N ratio when cover crops were planted independently (Poffenbarger et al. 2015a). It should be noted that decreases in the C:N ratio of cover crops in mixture treatments is dependent on the establishment of the legume cover crop in the mixture, which can be outcompeted by the non-legume during establishment (Brainard et al. 2011, Hayden et al. 2014).

### 12.2.2.1 *Cover Crops and Nitrogen Scavenging*

Cover crops scavenge $NO_3^-$ from the soil and convert it to organic N compounds, preventing leaching and improving N supply to subsequent cash crops (Kaye et al. 2019, Schipanski et al. 2014). Both grass and brassica cover crop species are more effective than legumes at absorbing and immobilizing N (Alonso-Ayuso et al. 2014, Dabney et al. 2001, Gabriel et al. 2012, Kaye et al. 2019, Shelton et al. 2017, Thapa et al. 2018). Although bicultures have reported some success, studies that measure N supply in mixtures with more than two species are not as well reported (Finney et al. 2017, White et al. 2017). Recently, Kaye et al. (2019) showed that leaching rates were 80% lower in a three-species mixture of a pea, red clover, and rye cover crops than fallow plots, but differences in maize N uptake were not found. A summary of studies comparing the effect of cover crop treatments on grain yield with no cover crop (control) is presented in Table 12.1.

**TABLE 12.1** Summary of studies comparing the effect of cover crop application with no cover crops (controls) on nitrous oxide emission. $N_2O$ emissions were calculated as either cumulative emissions (g $N_2O$-N $ha^{-1}$ unless otherwise indicated) or fluxes (g $N_2O$-N $ha^{-1}$ $d^{-1}$ unless otherwise indicated)

| Cover Crop Species | Cash Crops | Soil Texture | $N_2O$ Calculation | $N_2O$ Sampling Period | Reference |
|---|---|---|---|---|---|
| Rye, peas, rye-grass, and barley | Potato and cereal | Loamy sand | Cumulative emissions | 19 and 53 days | Baggs et al. (2000) |
| Bean or rye | Maize | Silt loam | Fluxes and cumulative emissions | 65 days | Baggs et al. (2003) |
| Hairy vetch and rye | Maize | Coarse loamy | Cumulative emissions (mg N $core^{-1}$) | 55 days | Rosecrance and Teasdale (2000) |
| *Sesbania aculeata* | Rice-wheat | Sandy soil | Fluxes and cumulative emissions (kg $N_2O$-N $ha^{-1}$) | 6 months | Aulakh et al. (2001) |
| Vetch | Rice | Clay | Fluxes ($\mu$g $N_2O$-N $ha^{-1}$ $d^{-1}$) and cumulative emissions (kg $N_2O$-N $ha^{-1}$) | 6 months | Xiong et al. (2002) |
| Winter rye | Maize | Silt loam | Fluxes and cumulative emissions | 4 months | Sarkodie-Addo et al. (2003) |
| Sesbania aculeata | Maize | Silty clay loam | Fluxes (mg $N_2O$-N $ha^{-1}$ $d^{-1}$) and cumulative emissions (mg N-$m^2$) | 30 days | Millar and Baggs (2004) |
| Rye | Maize-soybean | Fine loamy | Fluxes and cumulative emissions (mg $N_2O$-N $ha^{-1}$ $d^{-1}$) | 2 years | Parkin and Kaspar (2006) |
| Trios and rye | Vineyard | Loam | Fluxes | 4 months | Kerri and Belina (2008) |
| Legumes and non-legumes | Maize-soybean | Loamy | Fluxes (kg $N_2O$-N $ha^{-1}$ $yr^{-1}$) over 4 months | 4 months | Farahbakhshazad et al. (2008) |
| Winter rye | Maize-soybean | Sandy loam | Fluxes | 2 years | Fronning et al. (2008) |
| Rye and hairy vetch | Rice | Clay loam | Fluxes | 2 years | Zhaorigetu et al. (2008) |
| Rye | Maize-soybean | Silt loam | Fluxes (nmol $m^{-2}$ $s^{-1}$) | 2 years | Bavin et al. (2009) |
| Rye | Maize-soybean | Clay loam | Fluxes (mg $N_2O$-N $m^{-2}$ $d^{-1}$) and cumulative emissions (kg $N_2O$-N $ha^{-1}$) | 30 days | Jarecki et al. (2009) |
| Hairy vetch | Silage corn | Silt loam | Cumulative emissions | 2 years | Alluvione et al. (2010) |
| Hairy vetch and winter pea | Tomato | Coarse loamy | Fluxes ($\mu$g $N_2O$-N $m^{-2}$ $h^{-1}$) | 1 year | Kallenbach et al. (2010) |

*(Contd.)*

**TABLE 12.1** Contd.

| Rye | Maize-soybean-wheat | Fine loamy | Fluxes | 1 year | McSwiney et al. (2010) |
|---|---|---|---|---|---|
| Legume mixture | Vineyard | Silty clay | Fluxes ($\mu$g $N_2$O-N ha$^{-1}$ d$^{-1}$) and cumulative emissions (kg $N_2$O-N ha$^{-1}$) over 6 months | 6 months | Garland et al. (2011) |
| Mustard | Tomato | Silt loam | Fluxes (mg $N_2$O-N m$^{-2}$ h$^{-1}$) | 6 months | Barrios-Masias et al. (2011) |
| Rye | Maize | Coarse loamy | Fluxes (ng $N_2$O-N cm$^{-2}$ h$^{-1}$) | 2 years | Dietzel et al. (2011) |
| Fodder radish | Spring barley | Loamy sand | Cumulative emissions mg $N_2$O-N kg$^{-1}$) | After 10-year treatment | Petersen et al. (2011) |
| Rye | Maize-soybean | Loam | Fluxes and cumulative emissions (kg $N_2$O-N ha$^{-1}$) | 6 months | Mitchell et al. (2013) |
| Mustard | Spring barley, sugar beet, and oil seed rape | Sandy loam | Fluxes and cumulative emissions (kg $N_2$O-N ha$^{-1}$) | 2 years | Abdalla et al. (2012), (2014) |
| Oats and mustard | Tomato | Silt loam | Fluxes ($\mu$g $N_2$O-N m$^{-2}$ h$^{-1}$) | 2 years | Smukler et al. (2012) |
| Barley, vetch, and rape | Maize | Silty clay | Fluxes and cumulative emissions | 5 months | Sanz-Cobena et al. (2014) |
| Barley and hairy vetch | Rice | Silty loam | Fluxes (mg $N_2$O-N ha$^{-1}$ d$^{-1}$) and cumulative emissions (kg $N_2$O-N ha$^{-1}$) | 120 days | Pramanik et al. (2014); Hwang et al. (2017) |
| Rye | Maize-soybean | Fine loamy | Fluxes (mg $N_2$O-N m$^{-2}$ h$^{-1}$) | 2 years | Negassa et al. (2015) |
| Barley and vetch | Maize | Silty clay | Fluxes (mg $N_2$O-N m$^{-2}$ d$^{-1}$) and cumulative emissions (kg $N_2$O-N ha$^{-1}$) | 240 days | Guardia et al. (2016) |

### 12.2.2.2  *Agricultural Practices*

Conservation agriculture, defined as "a minimal soil disturbance (no-till) and permanent soil cover combined with rotations" (Food and Agriculture Organization of the United Nations 2020), is a recent agricultural management system that is gaining popularity among farmers in many parts of the world (Hobbs et al. 2008). Legume cover crops are suitable for sustainable conservation agriculture, as they are functional during growth and as crop residue. In no-tillage systems, SOC accumulates over time with the greatest accumulation occurring in the first 0-15 cm layer (Guo et al. 2016, Jagadamma et al. 2019, H. Zheng et al. 2018). The rate of SOC sequestration with no-tillage has been estimated to increase from -0.07 to 0.48 Mg C ha$^{-1}$ y$^{-1}$ compared to conventional tillage and may increase an additional 0.12-0.22 Mg C ha$^{-1}$ y$^{-1}$ when cover crops are used (Franzluebbers and Follett 2005). Cover crop cultivation for one (Delgado et al. 2010; Nevins et al. 2020), two

(Sanchez et al. 2019), three (Büchi et al. 2018), five (Nivelle et al. 2016), and eighteen (Zanatta et al. 2019) years increased SOC, total N and C, and N-cycling enzyme activity is reduced and/or no-tillage treatments. Many countries already rely on conservation agriculture using cover crops [for reviews see Baker et al. (2007), Giller et al. (2015), Stagnari et al. (2017), Food and Agriculture Organization of the United Nations (2019)]. For example, Brazil has implemented conservation agriculture systems using soybeans, a legume crop. In North America, cover crop legumes such as lentil (*Lens culinaris*), chickpea (*Cicer arietinum*), and pea and bean (*Vicia faba*) play a major role in conservation agriculture. In Europe, common bean (*Phaseolus vulgaris*) is the most commonly cultivated legume cover crop in conservation agriculture followed by pea, soybean, and fava bean. Regardless of the legume cover crop, the positive effects of conservation agriculture on SOM and N availability are often limited to the first 20 cm depth (Baker et al. 2007, Frasier et al. 2016, García-González et al. 2018, VandenBygaart et al. 2003).

Cover crop termination strategies have been reported to differentially impact N-cycling in cropping systems. The most common termination strategies of cover crops are 1) chemically by application of herbicides (i.e., glyphosate and/or paraquat), 2) naturally by frost (i.e., winter-kill), and 3) mechanically by rolling the cover crops with a roller-crimper. Of these methods, mechanical termination of cover crops with a roller-crimper may provide additional benefits to N availability by enhancing N mineralization (Mirsky et al. 2017, Parr et al. 2014, Romdhane et al. 2019). The roller-crimper terminates a cover crop by crimping the stems while leaving the root system and soil undisturbed (Kornecki et al. 2009). Cover crop residues remain on the soil surface and act as a mulch that suppresses weeds. Nitrogen from the roll-crimped cover crop residues would be more available to the cash crop through microbial mineralization (Wallace et al. 2017).

### 12.2.3 Nitrogen Losses

#### 12.2.3.1 Nitrous Oxide Emissions

Agricultural soils are the largest source of nitrous oxide ($N_2O$), a greenhouse gas, to the atmosphere (Stocker et al. 2013). $N_2O$ fluxes from agricultural soils mainly result from nitrification and denitrification processes (Figure 12.2) (Butterbach-Bahl et al. 2013). Nitrification, the aerobic oxidation of $NH_4^+$ to $NO_3^-$, is carried out by ammonia-oxidizing archaea and bacteria harboring the *amoA* gene. During this process, $N_2O$ can be formed by two biochemical pathways. First, as a bio-product during the $NH_4^+$ oxidation, hydroxylamine is spontaneously decomposed to $N_2O$. This process is regarded as the main source of $N_2O$ from nitrification. Second, it can be formed by nitrifier denitrification (Wrage-Mönnig et al. 2018), where $N_2O$ is an intermediate of the reduction of $NO_2^-$ to $N_2$. In general, denitrification is the anaerobic process by which $NO_3^-$ is sequentially reduced to $N_2$ via the formation of nitrite ($NO_2^-$), nitric oxide (NO), and $N_2O$ by the nitrate-, nitrite-, nitric oxide-, and nitrous oxide-reductase enzymes, encoded by the *napA/narG*, *nirK/nirS*, *norB*, and *nosZ* structural genes, respectively.

Cover crop effects on $N_2O$ emissions depend on multiple factors, such as cover crop species (Rosecrance et al. 2000), tillage system (Petersen et al. 2011), and irrigation management (Kallenbach et al. 2010). However, there is no consensus in the literature regarding the effect of cover crops on $N_2O$ emissions (Muhammad et al. 2019). A summary of studies comparing the effect of cover crop treatments on nitrous oxide emission with no cover crop (control) is presented in Table 12.2. Nitrogen availability following legume cover crop termination can increase nitrification activity in soils, and subsequent denitrification rates could increase $N_2O$ emissions (Rosecrance et al. 2000). Indeed, higher $N_2O$ emissions from legume cover crops compared to non-legumes have been reported (Basche et al. 2014). Carbon inputs from cover crops may also enhance denitrification as the majority of denitrifying bacteria are facultative aerobic heterotrophs (Mitchell et al. 2013).

**TABLE 12.2** Summary of studies comparing the effect of cover crop treatments with no cover crop (control) on grain yield

| Cover Crop Species | Grain Yield (Mg ha$^{-1}$) | Cash Crops | Soil Texture | Tillage | Year(s) After Experiment Start | Cover Crop Planting Time | Reference |
|---|---|---|---|---|---|---|---|
| Control | 3.6 | Maize | Sand | No-till | 2 | Winter | Ewing et al. (1991) |
| Crimson clover | 2.7 | | | | | | |
| Control | 7.7 | Maize | Gravelly loam | No-till | 1 | Winter | Decker et al. (1994) |
| Hairy vetch | 8.7 | | | | | | |
| Winter pea | 8.9 | | | | | | |
| Crimson clover | 8.1 | | | | | | |
| wheat | 7.2 | | | | | | |
| Control | 4.83 | Maize | Silt loam | No-till | 3 | Winter | Reinbott et al. (2004) |
| Oat | 4.56 | | | | | | |
| Hairy vetch | 5.06 | | | | | | |
| Winter pea | 5.19 | | | | | | |
| Hairy vetch + oat | 4.88 | | | | | | |
| Winter pea + oat | 4.47 | | | | | | |
| Control | 6.59 | Sorghum | | | | | |
| Oat | 6.57 | | | | | | |
| Hairy vetch | 7.33 | | | | | | |
| Winter pea | 7.06 | | | | | | |
| Hairy vetch + oat | 6.94 | | | | | | |
| Winter pea + oat | 6.64 | | | | | | |
| Control | 10.55 | Maize | Clay loam | No-till | 1 | Winter | Andraski and Bundy (2005) |
| Oat | 9.91 | | | | | | |
| Triticale | 10.69 | | | | | | |
| Rye | 10.47 | | | | | | |
| Control | 6.3 | Maize | Loamy sand | Conventional | 3 | Winter | Balkcom and Reeves (2005) |
| Sunn hemp | 6.9 | | | | | | |

*(Contd.)*

**TABLE 12.2** Contd.

| | | | | | | | | |
|---|---|---|---|---|---|---|---|---|
| Control | 9.70 | Maize | Silt loam | No-till | 3 | | Winter | Duiker and Curran (2005) |
| rye | 10.02 | | | | | | | |
| Control | 10.8 | Maize | Silty clay loam | Conventional | 4 | | Winter | Maughan et al. (2009) |
| Spring oats and cereal rye | 11.5 | | | | | | | |
| Control | 11.1 | Maize | Silt loam | No-till | 3 | | Winter | Henry et al. (2010) |
| Red clover | 11.7 | | | | | | | |
| Control | 16.9 | Maize | Silt loam | Conventional | 2 | | Winter | Salmeron et al. (2010) |
| Barley | 13.9 | | | | | | | |
| Winter rapessed | 14.2 | | | | | | | |
| Vetch | 17.8 | | | | | | | |
| Control | 13.39 | Maize | Loam | No-till | 2 | | Fall | Reese et al. (2014) |
| Radish + winter canola + turnip + lentil + pea + millet | 13.02 | | | | | | | |

The incorporation method can also impact $N_2O$ losses. In irrigated systems, cover crops incorporated into the soil may have higher emissions than cover crops left on the surface since denitrification is favored under $O_2$-limiting conditions (Basche et al. 2014). In vegetable cropping systems, high C inputs from non-legume cover crops stimulated more $N_2O$ emission compared to legume cover crops (Guardia et al. 2016, Mitchell et al. 2013), especially when the cover crops were incorporated into the soil (Sanz-Cobena et al. 2014). This was probably due to the higher C:N ratio of the non-legumes residues that led to a greater proportion of $N_2O$ losses when compared to legume residues. In general, conventional tillage increased $N_2O$ emissions from crop residues compared to reduced tillage (Petersen et al. 2011) in part through increasing soil porosity and gas diffusivity, allowing greater $N_2O$ release to the atmosphere before further reduction to $N_2$. Tillage of crop residues also provides easily available N and degradable C for denitrifying bacteria in the soil, increasing the potential for denitrification. Overall, the risk of high $N_2O$ fluxes with cover crops increases when large amounts of plant biomass with readily available N and degradable C are incorporated into the soil or left on the surface. However, intercropping of cover crops in a maize-sunflower rotation mitigated $N_2O$ emissions by 77.4% (barley) and 91.9% (vetch) compared to the traditional fall-winter fallow treatment in a ten-year field experiment (Guardia et al. 2019).

The magnitude of $N_2O$ emissions from soils cultivated with cover crops differs greatly among cropping systems and cover crops species with values ranging from 1.4-50.2 kg N ha$^{-1}$ year$^{-1}$ (Basche et al. 2014). The high variability in the cover crop effects on $N_2O$ emissions (Table 12.2) is dependent on several environmental factors, including soil management (no-tillage versus conventional tillage), soil amendment (C and N inputs), soil characteristics (bulk density and pH), and weather conditions. The percentage of applied N lost as $N_2O$ emission varies from 1.2-7.2% in corn-soybean and soybean-winter rye cropping systems but can increase up to 28.3% in rice-wheat cropping rotations (Basche et al. 2014). In vegetable cropping systems, cover crops induced increases or decreases of $N_2O$ emissions have been estimated to be approximately 100 kg N ha$^{-1}$ year$^{-1}$ relative to fallow soils (Kaye and Quemada 2017).

### 12.2.3.2 Reduction of Nitrogen Leaching

Cover crops reduce the potential for $NO_3^-$ leaching in soils by providing protective cover to the soil and increasing the opportunity for water infiltration (Ruffatti et al. 2019, Thapa et al. 2018). Compared to a fallow soil, winter wheat (*Triticum aestivum*), mustard (*Brassica rapa*), oat (*Avena sativa*), and rye cover crops appear to have greater effectiveness (70%) than legumes (hairy vetch, purple vetch (*Vicia Americana*), clover) (23%) in reducing $NO_3^-$ leaching likely due to faster and deeper root growth along with improved cool-season growth (Blanco-Canqui et al. 2013, Dabney et al. 2001, Kaspar et al. 2012, 2001, Meisinger et al. 1991, Shipley et al. 1992, Wortman et al. 2012). In a review of 31 published studies, Tonitto et al. (2006) found $NO_3^-$ leaching was reduced by 70% and 40% in non-legume and legume cover-cropped systems relative to fallow soil, respectively. $NO_3^-$ leaching was 50% lower when barley and rapeseed were planted in typical maize-fallow rotation, whereas vetch cover crop alone did not change N leaching relative to fallow controls (Gabriel et al. 2013). A recent meta-analysis using 238 observations from 28 studies concluded that cover crop mixtures with both legumes and non-legumes reduced $NO_3^-$ leaching as effectively as non-legumes alone but significantly more than legumes alone (Thapa et al. 2018). Compared to monocultures, mixtures of non-legume and legume cover crops were more effective at reducing $NO_3^-$ leaching, though they did not necessarily enhance the N supply for the subsequent crop (Vogeler et al. 2019).

## 12.3 INFLUENCE OF COVER CROPS ON THE SOIL MICROBIAL COMMUNITY ACTIVITY AND DIVERSITY

As noted throughout this review, cover cropping plays a key role in the enhancement of N-cycling. Cover cropping has been shown to increase the diversity of microbial communities and shift soil microbial community structure in vegetable (Finney et al. 2017, Nevins et al. 2018, Romdhane et al. 2019, Schmidt et al. 2015) and tree cropping (Capó-Bauçà et al. 2019, Moreno et al. 2009, Sofo et al. 2014, Zheng et al. 2018a,b, 2019) agroecosystems. However, understanding how cover crop management affects N-cycling soil microbial communities and how these interactions might impact soil N-cycling is largely unknown.

In general, microbial enzymatic activity, as determined by urease activity, increased under wheat-spring pea (Bolton et al. 1985), crimson clover-corn (Kirchner et al. 1993), vetch-wheat rotations (Mullen et al. 1998), sunflower-soybean-corn (Wortman et al. 2012) oat/radish/vetch-soybean/corn (Chavarría et al. 2016), and rye-corn (Nevins et al. 2020) rotations. However, when examining the impact of cover crops on specific N-cycling gene abundances, results are less consistent. After ten years of treatment, the introduction of vetch increased the abundance of diazotrophic (*nifH* gene) communities in vineyard soils (Pereg et al. 2018). However, the combined application of non-legume cover crops and compost increased the abundance and diversity of *nirK*-type denitrifiers in an apple orchard (Jones et al. 2017). The cover crop crown vetch (*Securigera varia*) increased the abundance of nitrification (*amoA*) and denitrification (*nirB* and *nirD*) genes in an intercropping apple (*Malus pumila*) orchard compared to a control treatment (Zheng et al. 2019).

Cover crops can also positively affect the abundance of nitrification and nitrate reduction genes (Hai et al. 2009, Tu et al. 2017, Y. Wang et al. 2014). For example, genes related to the production of glutamate (*gltB* and *GDH2*) are more abundant in cover crop treatments (Wutzler et al. 2017, Zheng et al. 2019). Recently, Graf et al. (2019) showed that intercropping with a legume (*Medicago sativa*) and a grass (*Dactylis glomerata*) species decreases the $N_2O$ emission potential by changing the abundance of root-associated $N_2O$-reducing microbial communities. The legume increased N availability and exerted a stronger effect on the $N_2O$ reducing communities than the grass.

Although it has been demonstrated that cover crops can increase functional complementarity and redundancy of bacterial communities (Alahmad et al. 2019), the effect of cover crops on the overall structure and composition of soil N-related microbial communities is largely unknown. After intercropping of crown vetch for nine years in an apple orchard, correlations of keystone bacterial and fungal species with N were not found (Zhao et al. 2018b). Nevertheless, among the dominant taxa, Proteobacteria, Acidobacteria, and Actinobacteria were the major drivers of N-cycling in the soil. These taxa have been reported to dominate in soil microenvironments that contain decomposing plant residues and participate in N and P reactions (Schmidt et al. 2015, Tan et al. 2013, Tu et al. 2017).

Two-year wheat-pea rotation under different tillage treatments (no-tillage, conventional, conventional tillage with stubble incorporated, and no-tillage with stubble retained) showed no differences in the bacterial and fungal diversity in the rhizosphere and bulk soil among treatments (Essel et al. 2019). No-tillage had the greatest number of bacterial (Actinobacteria, Proteobacteria, Chloroflexi, Acidobacteria, and Planctomycetes) and fungal (Ascomycota and Basidiomycota) taxa which were linked to improved soil quality. On the other hand, cover crop management practices were more important drivers of the abundance of N-cycling microbial communities than cover crop mixtures (Romdhane et al. 2019). Bacterial genera such as *Kaistobacter* (Alphaproteobacteria), *Solirubrobacterales*, *Gaiellaceae*, *Rubrobacter* (Actinobacteria), *Bacillus* (Firmicutes), and *Xiphinematobacter* (Verrucomicrobia) were positively correlated to the total N content. Members of *Kaistobacter* genus may be involved in the mitigation of $NO_3^-$ leaching (Chen et al. 2018), and previous studies have found positive correlations between the relative abundance of *Solirubrobacterales* (Tu et al. 2017), *Gaiellaceae* (Hermans et al. 2017), and *Bacillus* (Meena et al. 2017) to the C:N ratio in soils. Finally, members of the *Rubrobacter* genus have been described to be able to fix $N_2$ from the atmosphere (Norman et al. 2017).

## 12.4 CONCLUSIONS

With increasing production costs and greater awareness of the importance of soil quality for agriculture, there is a renewed interest in the use of cover crops. In addition to increasing SOM, reducing weed pressure, and controlling erosion, cover crops can significantly increase N retention and availability. However, many questions remain regarding the optimization and application of cover crops and mixtures for different cash crops. For example, the amount of N provided to the soil and subsequent cash crop varies based on legume species and season, but the reason for these differences and the influence on $N_2$-fixation rates is unknown. The majority of existing cover crop studies have focused on cereal and/or grain crops, and while there are examples of significant impacts of cover crop use with tree crops, less is known about the varieties, timing, and contribution of cover crops to these perennial agroecosystems. Finally, although the impact on soil N-cycling by cover crops has been established, the specific influence on the microbes responsible for much of this N-cycling is only beginning to be elucidated. Additional research on how cover crops alter the soil microbial community composition and activity will allow for greater predictions about N availability and cycling for subsequent cash crops.

## 12.5 ACKNOWLEDGMENTS

Support for the authors of this chapter was provided in part by the USDA NIFA Hatch project #1011186.

## 12.6 REFERENCES

Abiven, S., Recous, S., Reyes, V. and Oliver, R. 2005. Mineralisation of C and N from root, stem and leaf residues in soil and role of their biochemical quality. Biology and Fertility of Soils 42: 119-128.

Ågren, G.I., Wetterstedt, J.Å.M. and Billberger, M.F.K. 2012. Nutrient limitation on terrestrial plant growth - modeling the interaction between nitrogen and phosphorus. New Phytologist 194: 953-960.

Alahmad, A., Decocq, G., Spicher, F., Kheirbeik, L., Kobaissi, A., Tetu, T., et al. 2019. Cover crops in arable lands increase functional complementarity and redundancy of bacterial communities. Journal of Applied Ecology 56: 651-664.

Alonso-Ayuso, M., Gabriel, J.L. and Quemada, M. 2014. The kill date as a management tool for cover cropping success. PLoS One 9: e109587.

Baker, J.M., Ochsner, T.E., Venterea, R.T. and Griffis, T.J. 2007. Tillage and soil carbon sequestration-what do we really know? Agriculture, Ecosystems and Environment 118: 1-5.

Basche, A.D., Miguez, F.E., Kaspar, T.C. and Castellano, M.J. 2014. Do cover crops increase or decrease nitrous oxide emissions? A meta-analysis. Journal of Soil and Water Conservation 69: 471-482.

Blanco-Canqui, H., Holman, J.D., Schlegel, A.J., Tatarko, J. and Shaver, T.M. 2013. Replacing fallow with cover crops in a semiarid soil: effects on soil properties. Soil Science Society of America Journal 77: 1026.

Blanco-Canqui, H., Shaver, T.M., Lindquist, J.L., Shapiro, C.A., Elmore, R.W., Francis, C.A., et al. 2015. Cover crops and ecosystem services: insights from studies in temperate soils. Agronomy Journal 107: 2449.

Blesh, J. 2018. Functional traits in cover crop mixtures: biological nitrogen fixation and multifunctionality. Journal of Applied Ecology 55: 38-48.

Blesh, J., VanDusen, B.M. and Brainard, D.C. 2019. Managing ecosystem services with cover crop mixtures on organic farms. Agronomy Journal 111: 826.

Bolton, H., Elliott, L.F. and Papendick, R. 1985. Soil microbial biomass and selected soil enzyme activities: effect of fertilization and cropping practices. Soil Biology and Biochemistry 17: 297-302.

Bordeleau, L.M. and Prévost, D. 1994. Nodulation and nitrogen fixation in extreme environments. Plant and Soil 161: 115-125.

Brainard, D.C., Bellinder, R.R. and Kumar, V. 2011. Grass–legume mixtures and soil fertility affect cover crop performance and weed seed production. Weed Technology 25: 473-479.

Büchi, L., Wendling, M., Amossé, C., Necpalova, M. and Charles, R. 2018. Importance of cover crops in alleviating negative effects of reduced soil tillage and promoting soil fertility in a winter wheat cropping system. Agriculture, Ecosystems and Environment 256: 92-104. https://doi.org/10.1016/j.agee.2018.01.005.

Butterbach-Bahl, K., Baggs, E.M., Dannenmann, M., Kiese, R. and Zechmeister-Boltenstern, S. 2013. Nitrous oxide emissions from soils: how well do we understand the processes and their controls? Philosophical Transactions of the Royal Society B. Biological Sciences 368: 20130122.

Cabrera, M.L., Kissel, D.E. and Vigil, M.F. 2005. Nitrogen mineralization from organic residues. Journal of Environment Quality 34: 75.

Capó-Bauçà, S., Marqués, A., Llopis-Vidal, N., Bota, J. and Baraza, E. 2019. Long-term establishment of natural green cover provides agroecosystem services by improving soil quality in a Mediterranean vineyard. Ecological Engineering 127: 285-291.

Chapagain, T. and Riseman, A. 2014. Barley–pea intercropping: effects on land productivity, carbon and nitrogen transformations. Field Crops Research 166: 18-25.

Chapagain, T. and Riseman, A. 2015. Nitrogen and carbon transformations, water use efficiency and ecosystem productivity in monocultures and wheat-bean intercropping systems. Nutrient Cycling in Agroecosystems 101: 107-121.

Chavarría, D.N., Verdenelli, R.A., Serri, D.L., Restovich, S.B., Andriulo, A.E., Meriles, J.M., et al. 2016. Effect of cover crops on microbial community structure and related enzyme activities and macronutrient availability. European Journal of Soil Biology 76: 74-82.

Checcucci, A., DiCenzo, G.C., Bazzicalupo, M. and Mengoni, A. 2017. Trade, diplomacy, and warfare: the Quest for elite rhizobia inoculant strains. Frontiers in Microbiology 8: 2207.

Chen, S., Wang, F., Zhang, Y., Qin, S., Wei, S., Wang, S., et al. 2018. Organic carbon availability limiting microbial denitrification in the deep vadose zone. Environmental Microbiology 20: 980-992.

Cherr, C.M., Scholberg, J.M.S. and McSorley, R. 2006. Green manure approaches to crop production. Agronomy Journal 98: 302.

Clark, A.J. 2007. Managing Cover Crops Profitably. (A.J. Clark, Ed.) (3rd ed.). Network, Beltsville.

Clark, Andrew, J., Decker, A.M., Meisinger, J.J. and McIntosh, M.S. 1997. Kill date of vetch, rye, and a vetch-rye mixture: I. cover crop and corn nitrogen. Agronomy Journal 89: 427.

Cleveland, C.C. and Liptzin, D. 2007. C.N.P stoichiometry in soil: is there a "Redfield ratio" for the microbial biomass? Biogeochemistry 85: 235-252.

Corbeels, M., O'Connell, A.M., Grove, T.S., Mendham, D.S. and Rance, S.J. 2003. Nitrogen release from eucalypt leaves and legume residues as influenced by their biochemical quality and degree of contact with soil. Plant and Soil 250: 15-28.

Couëdel, A., Alletto, L., Tribouillois, H. and Justes, É. 2018. Cover crop crucifer-legume mixtures provide effective nitrate catch crop and nitrogen green manure ecosystem services. Agriculture, Ecosystems and Environment 254: 50-59.

Cui, H., Zhou, Y., Gu, Z., Zhu, H., Fu, S. and Yao, Q. 2015. The combined effects of cover crops and symbiotic microbes on phosphatase gene and organic phosphorus hydrolysis in subtropical orchard soils. Soil Biology and Biochemistry 82: 119-126.

Dabney, S.M., Delgado, J.A. and Reeves, D.W. 2001. Using cover crops to improve soil and and water quality. Communications in Soil Science and Plant Analysis 32: 1221-1250.

Delgado, J.A., Del Grosso, S.J. and Ogle, S.M. 2010. 15N isotopic crop residue cycling studies and modeling suggest that IPCC methodologies to assess residue contributions to N2O-N emissions should be reevaluated. Nutrient Cycling in Agroecosystems 86: 383-390. https://doi.org/10.1007/s10705-009-9300-9.

Dhont, C., Castonguay, Y., Nadeau, P., Belanger, G., Drapeau, R., Laberge, S., et al. 2006. Nitrogen reserves, spring regrowth and winter survival of field-grown alfalfa (Medicago sativa) defoliated in the autumn. Annals of Botany 97: 109-120.

Downie, J.A. 2010. The roles of extracellular proteins, polysaccharides and signals in the interactions of rhizobia with legume roots. FEMS Microbiology Reviews 34: 150-170.

Essel, E., Xie, J., Deng, C., Peng, Z., Wang, J., Shen, J., et al. 2019. Bacterial and fungal diversity in rhizosphere and bulk soil under different long-term tillage and cereal/legume rotation. Soil and Tillage Research 194: 104302.

Finn, J.A., Kirwan, L., Connolly, J., Sebastià, M.T., Helgadottir, A., Baadshaug, O.H., et al. 2013. Ecosystem function enhanced by combining four functional types of plant species in intensively managed grassland mixtures: a 3-year continental-scale field experiment. Journal of Applied Ecology 50: 365-375.

Finney, D.M., White, C.M. and Kaye, J.P. 2016. Biomass production and carbon/nitrogen ratio influence ecosystem services from cover crop mixtures. Agronomy Journal 108: 39.

Finney, D.M., Murrell, E.G., White, C.M., Baraibar, B., Barbercheck, M.E., Bradley, B.A., et al. 2017. Ecosystem services and disservices are bundled in simple and diverse cover cropping systems. Agricultural and Environmental Letters 2: 1-5.

Flynn, R. and Idowu, J. 2015. Nitrogen Fixation by Legumes. New Mexico State University Cooperative Extension Services, Guide A-129.

Food and Agriculture Organization of the United Nations. 2019. Food and agriculture data. http://faostat3.fao.org/home/E.

Food and Agriculture Organization of the United Nations. 2020. Conservation Agriculture. http://www.fao.org/ag/ca/1a.html.

Franzluebbers, A. and Follett, R. 2005. Greenhouse gas contributions and mitigation potential in agricultural regions of North America: introduction. Soil and Tillage Research 83: 1-8.

Frasier, I., Noellemeyer, E., Figuerola, E., Erijman, L., Permingeat, H. and Quiroga, A. 2016. High quality residues from cover crops favor changes in microbial community and enhance C and N sequestration. Global Ecology and Conservation 6: 242-256.

Fustec, J., Lesuffleur, F., Mahieu, S. and Cliquet, J.-B. 2010. Nitrogen rhizodeposition of legumes. A review. Agronomy for Sustainable Development 30: 57-66.

Gabriel, J.L., Garrido, A. and Quemada, M. 2013. Cover crops effect on farm benefits and nitrate leaching: linking economic and environmental analysis. Agricultural Systems 121: 23-32.

Gao, P., Zheng, X., Wang, L., Liu, B. and Zhang, S. 2019. Changes in the soil bacterial community in a chronosequence of temperate walnut-based intercropping systems. Forests 10: 299.

García-González, I., Hontoria, C., Gabriel, J.L., Alonso-Ayuso, M. and Quemada, M. 2018. Cover crops to mitigate soil degradation and enhance soil functionality in irrigated land. Geoderma 322: 81-88.

Geurts, R., Fedorova, E. and Bisseling, T. 2005. Nod factor signaling genes and their function in the early stages of Rhizobium infection. Current Opinion in Plant Biology 8: 346-352.

Giller, K.E., Andersson, J.A., Corbeels, M., Kirkegaard, J., Mortensen, D., Erenstein, O., et al. 2015. Beyond conservation agriculture. Frontiers in Plant Science 6: 870.

Glossary of Soil Science Terms 2008. Soil Science Society of America | Digital Library.

Graf, D.R.H., Saghaï, A., Zhao, M., Carlsson, G., Jones, C.M. and Hallin, S. 2019. Lucerne (Medicago sativa) alters $N_2O$-reducing communities associated with cocksfoot (Dactylis glomerata) roots and promotes $N_2O$ production in intercropping in a greenhouse experiment. Soil Biology and Biochemistry 137: 107547.

Groff, S. 2015. The past, present, and future of the cover crop industry. Journal of Soil and Water Conservation 70: 130A-133A.

Guardia, G., Abalos, D., García-Marco, S., Quemada, M., Alonso-Ayuso, M., Cárdenas, L.M., et al. 2016. Effect of cover crops on greenhouse gas emissions in an irrigated field under integrated soil fertility management. Biogeosciences 13: 5245-5257.

Guardia, G., Aguilera, E., Vallejo, A., Sanz-Cobena, A., Alonso-Ayuso, M. and Quemada, M. 2019. Effective climate change mitigation through cover cropping and integrated fertilization: a global warming potential assessment from a 10-year field experiment. Journal of Cleaner Production 241: 118307.

Guo, L.-J., Lin, S., Liu, T.-Q., Cao, C.-G. and Li, C.-F. 2016. Effects of conservation tillage on topsoil microbial metabolic characteristics and organic carbon within aggregates under a rice (Oryza sativa L.) – wheat (Triticum aestivum L.) cropping system in central China. Plos One 11: e0146145.

Hai, B., Diallo, N.H., Sall, S., Haesler, F., Schauss, K., Bonzi, M., et al. 2009. Quantification of key genes steering the microbial nitrogen cycle in the rhizosphere of sorghum cultivars in tropical agroecosystems. Applied and Environmental Microbiology 75: 4993-5000.

Hayden, Z.D., Ngouajio, M. and Brainard, D.C. 2014. Rye-vetch mixture proportion tradeoffs: cover crop productivity, nitrogen accumulation, and weed suppression. Agronomy Journal 106: 904-914.

He, X., Xu, M., Qiu, G.Y. and Zhou, J. 2009. Use of 15N stable isotope to quantify nitrogen transfer between mycorrhizal plants. Journal of Plant Ecology 2: 107-118.

Hermans, S.M., Buckley, H.L., Case, B.S., Curran-Cournane, F., Taylor, M. and Lear, G. 2017. Bacteria as emerging indicators of soil condition. Applied and Environmental Microbiology 83: e02826-16.

Hobbs, P.R., Sayre, K. and Gupta, R. 2008. The role of conservation agriculture in sustainable agriculture. Philosophical Transactions of the Royal Society B. Biological Sciences 363: 1491.

Hubbard, R.K., Strickland, T.C. and Phatak, S. 2013. Effects of cover crop systems on soil physical properties and carbon/nitrogen relationships in the coastal plain of southeastern USA. Soil and Tillage Research 126: 276-283.

Jagadamma, S., Essington, M.E., Xu, S. and Yin, X. 2019. Total and active soil organic carbon from long-term agricultural management practices in West Tennessee. Agricultural and Environmental Letters 4: 180062.

Jones, J., Savin, M.C., Rom, C.R. and Gbur, E. 2017. Denitrifier community response to seven years of ground cover and nutrient management in an organic fruit tree orchard soil. Applied Soil Ecology 112: 60-70.

Jones, K.M., Kobayashi, H., Davies, B.W., Taga, M.E. and Walker, G.C. 2007. How rhizobial symbionts invade plants: the Sinorhizobium–Medicago model. Nature Reviews. Microbiology 5: 619.

Kallenbach, C.M., Rolston, D.E. and Horwath, W.R. 2010. Cover cropping affects soil $N_2O$ and $CO_2$ emissions differently depending on type of irrigation. Agriculture, Ecosystems and Environment 137: 251-260.

Kaspar, T.C., Radke, J.K. and Laflen, J.M. 2001. Small grain cover crops and wheel traffic effects on infiltration, runoff, and erosion. Journal of Soil and Water Conservation 56: 160-164.

Kaspar, T.C. and Singer, J.W. 2011. The Use of Cover Crops to Manage Soil. Madison, WI: American Society of Agronomy and Soil Science Society of America.

Kaspar, T.C., Jaynes, D.B., Parkin, T.B., Moorman, T.B. and Singer, J.W. 2012. Effectiveness of oat and rye cover crops in reducing nitrate losses in drainage water. Agricultural Water Management 110: 25-33.

Kaye, J., Finney, D., White, C., Bradley, B., Schipanski, M., Alonso-Ayuso, M., et al. 2019. Managing nitrogen through cover crop species selection in the U.S. mid-Atlantic. Plos One 14: e0215448.

Kaye, J.P. and Quemada, M. 2017. Using cover crops to mitigate and adapt to climate change. A review. Agronomy for Sustainable Development 37: 4.

Ketterings, Q.M., Swink, S.N., Duiker, S.W., Czymmek, K.J., Beegle, D.B. and Cox, W.J. 2015. Integrating cover crops for nitrogen management in corn systems on northeastern U.S. dairies. Agronomy Journal 107: 1365.

Kirchner, M.J., Wollum, A.G. and King, L.D. 1993. Soil microbial populations and activities in reduced chemical input agroecosystems. Soil Science Society of America Journal 57: 1289.

Kornecki, T.S., Price, A.J., Raper, R.L. and Bergtold, J.S. 2009. Effectiveness of different herbicide applicators mounted on a roller/crimper for accelerated rye cover crop Termination. Applied Engineering in Agriculture 25: 819.

Lawson, A., Cogger, C., Bary, A. and Fortuna, A.-M. 2015. Influence of seeding ratio, planting date, and termination date on rye-hairy vetch cover crop mixture performance under organic management. Plos One 10: e0129597.

Liu, C., Jin, Y., Hu, Y., Tang, J., Xiong, Q., Xu, M., et al. 2019. Drivers of soil bacterial community structure and diversity in tropical agroforestry systems. Agriculture, Ecosystems and Environment 278: 24-34.

Lynch, M.J., Mulvaney, M.J., Hodges, S.C., Thompson, T.L. and Thomason, W.E. 2016. Decomposition, nitrogen and carbon mineralization from food and cover crop residues in the central plateau of Haiti. Springer Plus 5: 973.

Meena, V.S., Maurya, B.R., Meena, S.K., Meena, R.K., Kumar, A., Verma, J.P., et al. 2017. Can bacillus species enhance nutrient availability in agricultural soils? pp. 367-395. In: Islam, M., Rahman, M., Pandey, P., Jha, C. and Aeron, A. (eds.). Bacilli and Agrobiotechnology. Springer International Publishing.

Meisinger, J.J., Hargrove, W.L., Mikkelsen, R.B., Williams, J.R. and Benson, V.W. 1991. In: Hargrove, W.L. (ed.). Effects of Cover Crops on Groundwater Quality. Ankeny, IA: Soil and Water Conservation Society.

Melkonian, J., Poffenbarger, H.J., Mirsky, S.B., Ryan, M.R. and Moebius-Clune, B.N. 2017. Estimating nitrogen mineralization from cover crop mixtures using the precision nitrogen management model. Agronomy Journal 109: 1944.

Mercado-Blanco, J., Abrantes, I., Barra Caracciolo, A., Bevivino, A., Ciancio, A., Grenni, P., et al. 2018. Belowground microbiota and the health of tree crops. Frontiers in Microbiology 9: 1006.

Miao, Yang-yang, Shi, Shang-li, Nie, Zhong-nan, Kang, Wen-juan and Frazier, Taylor P. 2018. Inoculation treatments affect the migration and colonisation of rhizobia in alfalfa (Medicago sativa L.) plants. Acta Agriculturae Scandinavica Section B. Soil and Plant Science 68: 199-212.

Mirsky, S.B., Ackroyd, V.J., Cordeau, S., Curran, W.S., Hashemi, M., Reberg-Horton, S.C., et al. 2017. Hairy vetch biomass across the eastern United States: effects of latitude, seeding rate and date, and termination timing. Agronomy Journal 109: 1510.

Mitchell, D.C., Castellano, M.J., Sawyer, J.E. and Pantoja, J. 2013. Cover crop effects on nitrous oxide emissions: role of mineralizable carbon. Soil Science Society of America Journal 77: 1765.

Mokgehle, S.N., Dakora, F.D. and Mathews, C. 2014. Variation in $N_2$ fixation and N contribution by 25 groundnut (Arachis hypogaea L.) varieties grown in different agro-ecologies, measured using 15N natural abundance. Agriculture, Ecosystems and Environment 195: 161-172.

Moreno, B., Garcia-Rodriguez, S., Cañizares, R., Castro, J. and Benítez, E. 2009. Rainfed olive farming in south-eastern Spain: Long-term effect of soil management on biological indicators of soil quality. Agriculture, Ecosystems and Environment 131: 333-339.

Muhammad, I., Sainju, U.M., Zhao, F., Khan, A., Ghimire, R., Fu, X., et al. 2019. Regulation of soil $CO_2$ and $N_2O$ emissions by cover crops: a meta-analysis. Soil and Tillage Research 192: 103-112.

Mullen, M.D., Melhorn, C.G., Tyler, D.D. and Duck, B.N. 1998. Biological and biochemical soil properties in no-till corn with different cover crops. Journal of Soil and Water Conservation 53: 219-224.

Munroe, J.W. and Isaac, M.E. 2014. $N_2$-fixing trees and the transfer of fixed-N for sustainable agroforestry: a review. Agronomy for Sustainable Development 34: 417-427.

Nevins, C.J., Nakatsu, C. and Armstrong, S. 2018. Characterization of microbial community response to cover crop residue decomposition. Soil Biology and Biochemistry 127: 39-49.

Nevins, C.J., Lacey, C. and Armstrong, S. 2020. The synchrony of cover crop decomposition, enzyme activity, and nitrogen availability in a corn agroecosystem in the Midwest United States. Soil and Tillage Research 197: 104518.

Nivelle, E., Verzeaux, J., Habbib, H., Kuzyakov, Y., Decocq, G., Roger, D., et al. 2016. Functional response of soil microbial communities to tillage, cover crops and nitrogen fertilization. Applied Soil Ecology 108: 147-155.

Norman, J.S., King, G.M. and Friesen, M.L. 2017. Rubrobacter spartanus sp. nov., a moderately thermophilic oligotrophic bacterium isolated from volcanic soil. International Journal of Systematic and Evolutionary Microbiology 67: 3597-3602.

Ordóñez-Fernández, R., Repullo-Ruibérriz de Torres, M.A., Márquez-García, J., Moreno-García, M. and Carbonell-Bojollo, R.M. 2018. Legumes used as cover crops to reduce fertilisation problems improving soil nitrate in an organic orchard. European Journal of Agronomy 95: 1-13.

Parr, M., Grossman, J.M., Reberg-Horton, S.C., Brinton, C. and Crozier, C. 2014. Roller-crimper termination for legume cover crops in North Carolina: impacts on nutrient availability to a succeeding corn crop. Communications in Soil Science and Plant Analysis 45: 1106-1119.

Peix, A., Ramírez-Bahena, M.H., Velázquez, E. and Bedmar, E.J. 2015. Bacterial associations with legumes. Critical Reviews in Plant Sciences 34: 17-42.

Pereg, L., Morugán-Coronado, A., McMillan, M. and García-Orenes, F. 2018. Restoration of nitrogen cycling community in grapevine soil by a decade of organic fertilization. Soil and Tillage Research 179: 11-19.

Petersen, S.O., Mutegi, J.K., Hansen, E.M. and Munkholm, L.J. 2011. Tillage effects on $N_2O$ emissions as influenced by a winter cover crop. Soil Biology and Biochemistry 43: 1509-1517.

Poffenbarger, H.J., Mirsky, S.B., Weil, R.R., Kramer, M., Spargo, J.T. and Cavigelli, M.A. 2015a. Legume proportion, poultry litter, and tillage effects on cover crop decomposition. Agronomy Journal 107: 2083.

Preissel, S., Reckling, M., Schläfke, N. and Zander, P. 2015. Magnitude and farm-economic value of grain legume pre-crop benefits in Europe: a review. Field Crops Research 175: 64-79.

Ramesh, T., Bolan, N.S., Kirkham, M.B., Wijesekara, H., Kanchikerimath, M., Srinivasa, C., et al. 2019. Soil organic carbon dynamics: Impact of land use changes and management practices: a review. *In*: Advances in Agronomy (Vol. 156, pp. 1-107). Academic Press Inc. Cambridge, Massachusetts, USA.

Reed, S.C., Cleveland, C.C. and Townsend, A.R. 2011. Functional ecology of free-living nitrogen fixation: a contemporary perspective. Annual Review of Ecology, Evolution, and Systematics 42: 489-512.

Romdhane, S., Spor, A., Busset, H., Falchetto, L., Martin, J., Bizouard, F., et al. 2019. Cover crop management practices rather than composition of cover crop mixtures affect bacterial communities in no-till agroecosystems. Frontiers in Microbiology 10: 1618.

Rosecrance, R.C., McCarty, G.W., Shelton, D.R. and Teasdale, J.R. 2000. Denitrification and N mineralization from hairy vetch (Vicia villosa Roth) and rye (Secale cereale L.) cover crop monocultures and bicultures. Plant and Soil 227: 283-290.

Roth, R.T., Ruffatti, M.D., O'Rourke, P.D. and Armstrong, S.D. 2018. A cost analysis approach to valuing cover crop environmental and nitrogen cycling benefits: a central Illinois on farm case study. Agricultural Systems 159: 69-77.

Ruffatti, M.D., Roth, R.T., Lacey, C.G. and Armstrong, S.D. 2019. Impacts of nitrogen application timing and cover crop inclusion on subsurface drainage water quality. Agricultural Water Management 211: 81-88.

Rütting, T., Aronsson, H. and Delin, S. 2018. Efficient use of nitrogen in agriculture. Nutrient Cycling in Agroecosystems 110: 1-5.

Sanchez, I.I., Fultz, L.M., Lofton, J. and Haggard, B. 2019. Soil biological response to integration of cover crops and nitrogen rates in a conservation tillage corn production system. Soil Science Society of America Journal 83: 1356-1367.

Sanz-Cobena, A., García-Marco, S., Quemada, M., Gabriel, J.L., Almendros, P. and Vallejo, A. 2014. Do cover crops enhance $N_2O$, $CO_2$ or $CH_4$ emissions from soil in Mediterranean arable systems? Science of The Total Environment 466-467: 164-174.

Schipanski, M.E., Barbercheck, M., Douglas, M.R., Finney, D.M., Haider, K., Kaye, J.P., et al. 2014. A framework for evaluating ecosystem services provided by cover crops in agroecosystems. Agricultural Systems 125: 12-22.

Schmidt, O., Horn, M.A., Kolb, S. and Drake, H.L. 2015. Temperature impacts differentially on the methanogenic food web of cellulose-supplemented peatland soil. Environmental Microbiology 17: 720-734.

Seo, J., Lee, H., Hur, I. and Kim, S. 2000. Use of hairy vetch green manure as nitrogen fertilizer for corn production. Korean Journal of Crop Science 45: 294-299.

Seo, J.-H., Meisinger, J.J. and Lee, H.-J. 2006. Recovery of nitrogen-15–labeled hairy vetch and fertilizer applied to corn. Agronomy Journal 98: 245.

Shamseldin, A., Abdelkhalek, A. and Sadowsky, M.J. 2017. Recent changes to the classification of symbiotic, nitrogen-fixing, legume-associating bacteria: a review. Symbiosis 71: 91-109.

Shelton, R.E., Jacobsen, K.L. and McCulley, R.L. 2017. Cover crops and fertilization alter nitrogen loss in organic and conventional conservation agriculture systems. Frontiers in Plant Science 8: 2260.

Shipley, P.R., Messinger, J.J. and Decker, A.M. 1992. Conserving residual corn fertilizer nitrogen with winter cover crops. Agronomy Journal 84: 869.

Sievers, T. and Cook, R.L. 2018. Aboveground and root decomposition of cereal rye and hairy vetch cover crops. Soil Science Society of America Journal 82: 147.

Sofo, A., Ciarfaglia, A., Scopa, A., Camele, I., Curci, M., Crecchio, C., et al. 2014. Soil microbial diversity and activity in a Mediterranean olive orchard using sustainable agricultural practices. Soil Use and Management 30: 160-167.

Stagnari, F., Maggio, A., Galieni, A. and Pisante, M. 2017. Multiple benefits of legumes for agriculture sustainability: an overview. Chemical and Biological Technologies in Agriculture 4: 2.

Stocker, T., Qin, D., Plattner, G.-K., Tignor, M., Allen, S., Boschung, J., et al. 2013. Climate Change 2013: The Physical Science Basis — IPCC. Cambridge, MA.

Strickland, M.S. and Rousk, J. 2010. Considering fungal: bacterial dominance in soils - methods, controls, and ecosystem implications. Soil Biology and Biochemistry 42: 1385-1395.

Tan, H., Barret, M., Mooij, M.J., Rice, O., Morrissey, J.P., Dobson, A., et al. 2013. Long-term phosphorus fertilisation increased the diversity of the total bacterial community and the phoD phosphorus mineraliser group in pasture soils. Biology and Fertility of Soils 49: 661-672.

Thapa, R., Mirsky, S.B. and Tully, K.L. 2018. Cover crops reduce nitrate leaching in agroecosystems: a global meta-analysis. Journal of Environment Quality 47: 1400.

Thilakarathna, M.S., McElroy, M.S., Chapagain, T., Papadopoulos, Y.A. and Raizada, M.N. 2016. Belowground nitrogen transfer from legumes to non-legumes under managed herbaceous cropping systems. A review. Agronomy for Sustainable Development 36: 58.

Tonitto, C., David, M.B. and Drinkwater, L.E. 2006. Replacing bare fallows with cover crops in fertilizer-intensive cropping systems: a meta-analysis of crop yield and N dynamics. Agriculture, Ecosystems and Environment 112: 58-72.

Trinsoutrot, I., Recous, S., Bentz, B., Line`res, M., Che`neby, D. and Nicolardot, B. 2000. Biochemical quality of crop residues and carbon and nitrogen mineralization kinetics under nonlimiting nitrogen conditions. Soil Science Society of America Journal 64: 918.

Tu, Q., He, Z., Wu, L., Xue, K., Xie, G., Chain, P., et al. 2017. Metagenomic reconstruction of nitrogen cycling pathways in a $CO_2$-enriched grassland ecosystem. Soil Biology and Biochemistry 106: 99-108.

VandenBygaart, A.J., Gregorich, E.G. and Angers, D.A. 2003. Influence of agricultural management on soil organic carbon: a compendium and assessment of Canadian studies. Canadian Journal of Soil Science 83: 363-380.

Velázquez, E., Carro, L., Flores-Félix, J.D., Martínez-Hidalgo, P., Menéndez, E., Ramírez-Bahena, M.-H., et al. 2017. The legume nodule microbiome: a source of plant growth-promoting bacteria. pp. 41-70. In: Kumar, V., Kumar, M., Sharma, S. and Prasad, R. (eds). Probiotics and Plant Health. Singapore: Springer Singapore.

Vitousek, P.M., Menge, D.N.L., Reed, S.C. and Cleveland, C.C. 2013. Biological nitrogen fixation: rates, patterns and ecological controls in terrestrial ecosystems. Philosophical Transactions of the Royal Society B. Biological Sciences 368: 1621.

Vogeler, I., Hansen, E.M., Thomsen, I.K. and Østergaard, H.S. 2019. Legumes in catch crop mixtures: effects on nitrogen retention and availability, and leaching losses. Journal of Environmental Management 239: 324-332.

Vukicevich, E., Lowery, T., Bowen, P., Úrbez-Torres, J.R. and Hart, M. 2016. Cover crops to increase soil microbial diversity and mitigate decline in perennial agriculture. A review. Agronomy for Sustainable Development 36: 48.

Vyn, T.J., Faber, J.G., Janovicek, K.J. and Beauchamp, E.G. 2000. Cover crop effects on nitrogen availability to corn following wheat. Agronomy Journal 92: 915.

Wagger, M.G., Cabrera, M.L. and Ranells, N.N. 1998. Nitrogen and carbon cycling in relation to cover crop residue quality. Journal of Soil and Water Conservation 53: 214-218.

Wallace, J., Williams, A., Liebert, J., Ackroyd, V., Vann, R., Curran, W., et al. 2017. Cover crop-based, organic rotational no-till corn and soybean production systems in the Mid-Atlantic United States. Agriculture 7: 34.

Wang, Q., Liu, J. and Zhu, H. 2018. Genetic and molecular mechanisms underlying symbiotic specificity in legume-rhizobium interactions. Frontiers in Plant Science 9: 313.

Wang, Y., Zhu, G., Song, L., Wang, S. and Yin, C. 2014. Manure fertilization alters the population of ammonia-oxidizing bacteria rather than ammonia-oxidizing archaea in a paddy soil. Journal of Basic Microbiology 54: 190-197.

White, C.M., Finney, D.M., Kemanian, A.R. and Kaye, J.P. 2016. A model–data fusion approach for predicting cover crop nitrogen supply to corn. Agronomy Journal 108: 2527.

White, C.M., DuPont, S.T., Hautau, M., Hartman, D., Finney, D.M., Bradley, B., et al. 2017. Managing the trade off between nitrogen supply and retention with cover crop mixtures. Agriculture, Ecosystems and Environment 237: 121-133.

Wortman, S., Francis, C., Bernards, M., Brijber, R. and Lindquist, J. 2012. Optimizing cover crop benefits with diverse mixtures and an alternative termination method. Agron. J. 104: 1425-1435.

Wrage-Mönnig, N., Horn, M.A., Well, R., Müller, C., Velthof, G. and Oenema, O. 2018. The role of nitrifier denitrification in the production of nitrous oxide revisited. Soil Biology and Biochemistry 123: A3-A16. https://doi.org/10.1016/j.soilbio.2018.03.020.

Wutzler, T., Zaehle, S., Schrumpf, M., Ahrens, B. and Reichstein, M. 2017. Adaptation of microbial resource allocation affects modelled long term soil organic matter and nutrient cycling. Soil Biology and Biochemistry 115: 322-336.

Zanatta, J.A., Vieira, F.C.B., Briedis, C., Dieckow, J. and Bayer, C. 2019. Carbon indices to assess quality of management systems in a Subtropical Acrisol. Scientia. Agricola. 76: 501-508.

Zheng, H., Liu, W., Zheng, J., Luo, Y., Li, R., Wang, H., et al. 2018. Effect of long-term tillage on soil aggregates and aggregate-associated carbon in black soil of Northeast China. Plos One 13: e0199523.

Zheng, W., Gong, Q., Zhao, Z., Liu, J., Zhai, B., Wang, Z., et al. 2018. Changes in the soil bacterial community structure and enzyme activities after intercrop mulch with cover crop for eight years in an orchard. European Journal of Soil Biology 86: 34-41.

Zheng, W., Zhao, Z., Gong, Q., Zhai, B. and Li, Z. 2018a. Effects of cover crop in an apple orchard on microbial community composition, networks, and potential genes involved with degradation of crop residues in soil. Biology and Fertility of Soils 54: 743-759.

Zheng, W., Zhao, Z., Gong, Q., Zhai, B. and Li, Z. 2018b. Responses of fungal–bacterial community and network to organic inputs vary among different spatial habitats in soil. Soil Biology and Biochemistry 125: 54-63.

Zheng, W., Zhao, Z., Lv, F., Wang, R., Gong, Q., Zhai, B., et al. 2019. Metagenomic exploration of the interactions between N and P cycling and SOM turnover in an apple orchard with a cover crop fertilized for 9 years. Biology and Fertility of Soils 55: 365-381.

# Index